METHODS IN MOLECULAR BIOLOGY

For further volumes:
http://www.springer.com/series/7651

For over 35 years, biological scientists have come to rely on the research protocols and methodologies in the critically acclaimed *Methods in Molecular Biology* series. The series was the first to introduce the step-by-step protocols approach that has become the standard in all biomedical protocol publishing. Each protocol is provided in readily-reproducible step-by-step fashion, opening with an introductory overview, a list of the materials and reagents needed to complete the experiment, and followed by a detailed procedure that is supported with a helpful notes section offering tips and tricks of the trade as well as troubleshooting advice. These hallmark features were introduced by series editor Dr. John Walker and constitute the key ingredient in each and every volume of the *Methods in Molecular Biology* series. Tested and trusted, comprehensive and reliable, all protocols from the series are indexed in PubMed.

The Plant Cell Wall

Methods and Protocols

Second Edition

Edited by

Zoë A. Popper

Botany and Plant Science, School of Natural Sciences, The National University of Ireland Galway, Galway, Ireland; Ryan Institute for Environmental, Marine and Energy Research, School of Natural Sciences, The National University of Ireland Galway, Galway, Ireland

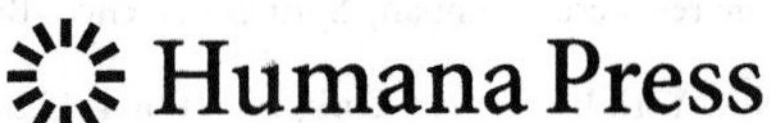

Editor
Zoë A. Popper
Botany and Plant Science
School of Natural Sciences
The National University of Ireland Galway
Galway, Ireland

Ryan Institute for Environmental
Marine and Energy Research
School of Natural Sciences
The National University of Ireland Galway
Galway, Ireland

ISSN 1064-3745 ISSN 1940-6029 (electronic)
Methods in Molecular Biology
ISBN 978-1-0716-0619-3 ISBN 978-1-0716-0621-6 (eBook)
https://doi.org/10.1007/978-1-0716-0621-6

Cover Caption: The image is a transverse section through the stipe of the macroalga, Laminaria digitata, which has been immunolabelled with a monoclonal antibody that recognizes the polysaccharide β-D-(1,3),(1,4)-glucan. The cell walls of the hyphal cells, used to transport compounds throughout the alga, are clearly labelled.

This Humana imprint is published by the registered company Springer Science+Business Media, LLC, part of Springer Nature.
The registered company address is: 1 New York Plaza, New York, NY 10004, U.S.A.

Preface

Where I live and work in the West of Ireland, as in many places in the world—both urban and rural—walls are a major, and very obvious, feature of the landscape (*see* Fig. 1). More fundamentally, but less obviously, we are also surrounded by the cell walls that envelope, with a few minor exceptions, the cells of plants and algae, beautifully rendered in [1], and which are pivotal to their function and survival. The cell wall, and its constituent polysaccharides and proteins, controls nearly all plant- and algal-based biological and biophysical processes including expansive cell growth, development, cell–cell communication, and interactions with and adaptations to the abiotic and biotic environment [2–4]. As a result, cell walls are highly complex, and their components may vary between different groups of plants and algae, and may even differ between different cells of a single organism, and/or at different stages of development [5–8].

Not only are cell walls essential to the living plant or alga, they, and their components have significant applications. Cellulose, a major cell wall polymer, is the most abundantly occurring natural biopolymer, with other cell wall components being among the next most abundant. They represent the most abundant source of renewable energy. Plants, in various forms, have therefore been used as a source of fuel for millennia, but their cell wall components are now specifically receiving interest as next-generation biofuels. Other wall components, such as pectins from plants, and alginates and carrageenans from algae, are employed by the food industry for their gelling and emulsifying properties. Understanding cell walls is therefore not only fundamental to the plant sciences, but it is also pertinent to

Fig. 1 A stone wall in the West of Ireland

aspects of human and animal nutrition, agriculture, food, and pharmaceutical industries. Furthermore, advanced cell wall analysis is key to developing novel, or improving current, uses of plants and algae.

Comprehensive cell wall analysis requires a multidisciplinary approach "from chemistry to biology" and employs many tools and techniques. This volume describes some methods that are currently used to investigate cell wall biochemistry, biomechanical properties, chemistry, and biology. Compared to the first edition of *The Plant Cell Wall: Methods and Protocols,* there are an additional 17 entirely new chapters, in several cases describing the latest technological advances in the field. The remaining 12 chapters within this volume have been revised and updated. Each chapter is written by leading experts in cell wall research and with the aim that the protocol(s) can be carried out by a novice to that method and/or to cell wall research. The techniques included in this volume range from those that can be used to characterize the cell wall composition and structure, to plant tissue culture techniques, protoplast isolation, and genetic manipulation, to the investigation of enzyme activities, and in situ localization of wall components, to bioinformatics. These methods have all provided powerful new insights into cell wall biochemistry and metabolism.

While this volume aims to describe a wide range of cell wall-directed protocols, the reader is also directed to other volumes that they will find extremely valuable to their understanding of cell walls [4], and [9] which provides user-friendly descriptions of many cell wall-directed methods that are not included in this volume and form an essential part of the repertoire towards understanding cell walls.

As in the preface to the first volume, I would like to wish the reader enjoyment, excitement, and success in their experiments.

Galway, Ireland ***Zoë A. Popper***

References

1. Brodersen CR, Roddy AB (2016) New frontiers in the three-dimensional visualisation of plant structure and function. Am J Bot 103:184–188
2. Hamann T (2012) Plant cell wall integrity maintenance as an essential component of biotic stress response mechanisms. Front Plant Sci 3:77. https://doi.org/10.3389/fpls.2012.00077
3. Houston K, Tucker MR, Chowdhury J, Shirley N, Little A (2016) The plant cell wall: a complex and dynamic structure as revealed by the responses of genes under stress conditions. Front Plant Sci 10. https://doi.org/10.3389/fpls.2016.00984
4. Albersheim P, Darvill A, Roberts K, Sederoff R, Staehelin A (2016) Plant cell walls: From chemistry to biology. Garland Science, Taylor-Francis Group LLC, New York
5. Popper ZA, Michel G, Hervé C, Domozych DS, Willats WGT, Tuohy MG, Kloareg B, Stengel DB (2011) Evolution and diversity of plant cell walls: from algae to flowering plants. Annu Rev Plant Biol 62:567–590
6. Fangel JU, Ulvskov P, Knox JP, Mikkelsen MD, Harholt J, Popper ZA, Willats WG (2012) Cell wall evolution and diversity. Front Plant Sci 3:152
7. Sorensen I, Domozych DS, Willats WGT (2010) How have plant cell walls evolved? Plant Physiol 153: 366–372
8. Popper ZA, Tuohy MG (2010) Beyond the green: understanding the evolutionary puzzle of plant and algal cell walls. Plant Physiol 153:373–383
9. Fry SC (2000) The growing plant cell wall: chemical and metabolic analysis. Reprint edition, Blackburn, Caldwell, NJ. ISBN 1-930665-08-3

Acknowledgements

There are so many people to whom I owe thanks that it is with some trepidation that I write acknowledgements least I forget anyone. Many who know me will understand that completing this volume has not been a smooth process; therefore, there are some people whom I would like to specifically highlight because they have been instrumental to this book. My first and foremost thanks go to the Series Editor, John M. Walker, the Executive Editor, Patrick Marton, and Assistant Editor, Anna Rakovsky, at Springer Protocols, and all of the contributors to this volume, for their patience and for trusting me to finish.

Thank you to my parents, Jan and Peter Popper, and my brother, Simon, for your encouragement, and to my colleagues and friends at NUI Galway, particularly the head of Botany and Plant Science, Dagmar Stengel, for your support, and Aline Horwath, for amiably, enthusiastically, and expertly taking on my teaching during my maternity leaves. Thank you to my husband Heinz-Peter Nasheuer for all your help, for your energy, and for being an amazing papa to our girls. Finally, thank you to my miracle daughters, Noa and Leonie, happy masters of the micro- and pico-schlaf, respectively, who turned our world upside down, and whose arrivals both delayed and facilitated completion of this volume.

Contents

Contributors

GORDON G. ALLISON • *Institute of Biological, Environmental and Rural Sciences, Aberystwyth University, Aberystwyth, UK*
RHIANNON BALAZIC • *Department of Biological Sciences, Northern Illinois University, DeKalb, IL, USA*
WILLIAM BARRETT • *Institute of Biological, Environmental and Rural Sciences, Aberystwyth University, Aberystwyth, UK*
SOPHIE BERNARD • *UNIROUEN, Laboratoire Glyco-MEV, Normandie Université, Rouen, France; UNIROUEN, PRIMACEN, Normandie Université, Rouen, France*
PETER BOCK • *Department of Nanobiotechnology, Institute of Biophysics, BOKU-University of Natural Resources and Life Sciences, Vienna, Austria*
TRACEY J. BELL • *The Ferrier Research Institute, Victoria University of Wellington, Gracefield Research Centre, Petone, New Zealand*
MICHAEL J. BUDZISZEK JR • *Department of Biological Sciences, Center for Biotechnology and Life Sciences, University of Rhode Island, Kingston, RI, USA*
INGO BURGERT • *Department of Civil, Environmental and Geomatic Engineering, Institute for Building Materials (IfB), Zürich, Switzerland*
VINCENT BURLAT • *Laboratoire de Recherche en Sciences Végétales, UMR 5546 Université Paul Sabatier-Toulouse III/CNRS, Auzeville Tolosane, France*
JOSÉ CARLI • *Institute of Biological, Environmental and Rural Sciences, Aberystwyth University, Aberystwyth, UK*
ROMAIN CASTILLEUX • *UNIROUEN, Laboratoire Glyco-MEV, Normandie Université, Rouen, France*
ARIELLE M. CHAVES • *Department of Biological Sciences, Center for Biotechnology and Life Sciences, University of Rhode Island, Kingston, RI, USA*
HÉLÈNE SAN CLEMENTE • *Laboratoire de Recherche en Sciences Végétales, UMR 5546 Université Paul Sabatier-Toulouse III/CNRS, Auzeville Tolosane, France*
SÍLVIA COIMBRA • *Departamento de Biologia, Faculdade de Ciências, Universidade do Porto, Porto, Portugal; Requimte-Associated Laboratory for Green Chemistry, Porto, Portugal*
DANIEL J. COSGROVE • *Department of Biology, Penn State University, University Park, PA, USA*
MÁRIO LUÍS DA COSTA • *Departamento de Biologia, Faculdade de Ciências, Universidade do Porto, Porto, Portugal; Requimte-Associated Laboratory for Green Chemistry, Porto, Portugal*
RICARDO M. F. DA COSTA • *Institute of Biological, Environmental and Rural Sciences, Aberystwyth University, Aberystwyth, UK; Faculty of Sciences and Technology, Department of Life Sciences, Centre for Functional Ecology, University of Coimbra, Coimbra, Portugal*
MARTA DERBA-MACELUCH • *Department of Forest Genetics and Plant Physiology, Swedish University of Agricultural Sciences, Umea, Sweden*
CHRISTOS S. DIMOS • *Department of Biological Sciences, Center for Biotechnology and Life Sciences, University of Rhode Island, Kingston, RI, USA*
DAVID S. DOMOZYCH • *Department of Biology and Skidmore Microscopy Imaging Center, Skidmore College, Dana Science Center, Saratoga Springs, NY, USA*

AZEDDINE DRIOUICH • *UNIROUEN, Laboratoire Glyco-MEV, Normandie Université, Rouen, France; UNIROUEN, PRIMACEN, Normandie Université, Rouen, France; Structure Fédérative de Recherche (Normandie-Végétale) FED 4277 – Université de Rouen Normandie, Mont Saint Aignan Cedex, France*

DELPHINE DUFFIEUX • *Station Biologique de Roscoff, Sorbonne Universités, CNRS, Integrative Biology of Marine Models (LBI2M), Roscoff, France*

PAUL DUPREE • *Department of Biochemistry, University of Cambridge, Cambridge, UK*

JONATAN U. FANGEL • *Department of Plant and Environmental Sciences, Faculty of Science, University of Copenhagen, Copenhagen, Denmark*

ELISABETH FITZEK • *Department of Biological Sciences, Northern Illinois University, DeKalb, IL, USA*

LENKA FRANKOVÁ • *The Edinburgh Cell Wall Group, Institute of Molecular Plant Sciences, The University of Edinburgh, Edinburgh, UK*

ALFRED D. FRENCH • *Southern Regional Research Center, U.S. Department of Agriculture, New Orleans, LA, USA*

STEPHEN C. FRY • *The Edinburgh Cell Wall Group, Institute of Molecular Plant Sciences, School of Biological Sciences, The University of Edinburgh, Edinburgh, UK*

HIROO FUKUDA • *Department of Biological Sciences, Graduate School of Science, University of Tokyo, Tokyo, Japan*

YU GAO • *Department of Viticulture and Oenology, Faculty of AgriSciences, Institute for Wine Biotechnology, Stellenbosch University, Matieland, South Africa; Department of Plant Science, Center for Viticulture and Enology, School of Agriculture and Biology, Shanghai Jiao Tong University, Shanghai, China*

MICHAEL J. GIDLEY • *ARC Centre of Excellence in Plant Cell Walls, Centre for Nutrition and Food Sciences, Queensland Alliance for Agriculture and Food Innovation, The University of Queensland, Brisbane, QLD, Australia*

NOTBURGA GIERLINGER • *Department of Nanobiotechnology, Institute of Biophysics, BOKU-University of Natural Resources and Life Sciences, Vienna, Austria*

CHESSA A. GOSS • *Department of Biological Sciences, Center for Biotechnology and Life Sciences, University of Rhode Island, Kingston, RI, USA*

FLORENCE GOUBET • *BASF Belgium Coordination Center Comm.V., Innovation Center Gent, Ghent, Belgium*

PHILIP J. HARRIS • *School of Biological Sciences, The University of Auckland, Auckland, New Zealand*

CÉCILE HERVÉ • *Station Biologique de Roscoff, Sorbonne Universités, CNRS, Integrative Biology of Marine Models (LBI2M), Roscoff, France*

STEFAN J. HILL • *Scion, Rotorua, New Zealand*

LAURENT HOFFMANN • *Laboratoire de Recherche en Sciences Végétales, UMR 5546 Université Paul Sabatier-Toulouse III/CNRS, Auzeville Tolosane, France*

KUNINORI IWAMOTO • *Department of Biological Sciences, Graduate School of Science, University of Tokyo, Tokyo, Japan*

ELISABETH JAMET • *Laboratoire de Recherche en Sciences Végétales, UMR 5546 Université Paul Sabatier-Toulouse III/CNRS, Auzeville Tolosane, France*

KATJA SALOMON JOHANSEN • *University of Copenhagen, Geosciences and Natural Resources Management, Frederiksberg, Denmark*

GLENN P. JOHNSON • *Southern Regional Research Center, U.S. Department of Agriculture, New Orleans, LA, USA*

Anna Kärkönen • *Natural Resources Institute Finland (Luke), Production Systems, Plant Genetics, Helsinki, Finland; Department of Agricultural Sciences, Viikki Plant Science Centre, University of Helsinki, Helsinki, Finland*

Tobias Keplinger • *Department of Civil, Environmental and Geomatic Engineering, Institute for Building Materials (IfB), Zürich, Switzerland*

J. Paul Knox • *Centre for Plant Sciences, Faculty of Biological Sciences, University of Leeds, Leeds, UK*

Kirsten Krause • *Department of Arctic and Marine Biology, Faculty of Biosciences, Fisheries and Economics, UiT The Arctic University of Norway, Tromsø, Norway*

Virginia Lai • *Department of Biological Sciences, Center for Biotechnology and Life Sciences, University of Rhode Island, Kingston, RI, USA*

Olivier Leroux • *Department of Biology, Ghent University, Ghent, Belgium*

Anna Lietz • *Department of Biology, Skidmore College, Saratoga Springs, NY, USA*

Xiao Liu • *Department of Environmental and Plant Biology, Molecular and Cellular Biology Program, Ohio University, Athens, OH, USA; Russ College of Engineering and Technology, Center for Intelligent, Distributed and Dependable Systems, Ohio University, Athens, OH, USA*

Patricia Lopez-Sanchez • *ARC Centre of Excellence in Plant Cell Walls, Centre for Nutrition and Food Sciences, Queensland Alliance for Agriculture and Food Innovation, The University of Queensland, Brisbane, QLD, Australia; Product Design, Agrifood, Bioeconomy and Health, RISE Research Institutes of Sweden, Gothenburg, Sweden*

Youssef Manasfi • *UNIROUEN, Laboratoire Glyco-MEV, Normandie Université, Rouen, France*

Susan E. Marcus • *Centre for Plant Sciences, Faculty of Biological Sciences, University of Leeds, Leeds, UK*

Batirtze Prats Mateu • *Department of Nanobiotechnology, Institute of Biophysics, BOKU-University of Natural Resources and Life Sciences, Vienna, Austria*

Savannah McKenna • *Department of Biological Sciences, Ohio University, Athens, OH, USA*

Ewa J. Mellerowicz • *Department of Forest Genetics and Plant Physiology, Swedish University of Agricultural Sciences, Umea, Sweden*

Laurence D. Melton • *School of Chemical Sciences, The University of Auckland, Auckland, New Zealand*

Deirdre Mikkelsen • *ARC Centre of Excellence in Plant Cell Walls, Centre for Nutrition and Food Sciences, Queensland Alliance for Agriculture and Food Innovation, The University of Queensland, Brisbane, QLD, Australia*

John P. Moore • *Department of Viticulture and Oenology, Faculty of AgriSciences, Institute for Wine Biotechnology, Stellenbosch University, Matieland, South Africa*

Andrew Mort • *Department of Biochemistry and Molecular Biology, Oklahoma State University, Stillwater, OK, USA*

Gang Ning • *Department of Biology, Penn State University, University Park, PA, USA; Huck Institutes of the Life Sciences, Penn State University, University Park, PA, USA*

Stian Olsen • *Department of Arctic and Marine Biology, Faculty of Biosciences, Fisheries and Economics, UiT The Arctic University of Norway, Tromsø, Norway*

Marisa S. Otegui • *Department of Botany, University of Wisconsin, Madison, WI, USA; Department of Genetics, University of Wisconsin-Madison, Madison, WI, USA; Laboratory of Molecular and Cellular Biology, University of Wisconsin-Madison, Madison, WI, USA*

FABIAN PFRENGLE • *Department of Chemistry, University of Natural Resources and Life Sciences, Vienna, Austria; Department of Biomolecular Systems, Max Planck Institute of Colloids and Interfaces, Potsdam, Germany*

ZOË A. POPPER • *Botany and Plant Science, School of Natural Sciences, The National University of Ireland Galway, Galway, Ireland; Ryan Institute for Environmental, Marine and Energy Research, School of Natural Sciences, The National University of Ireland Galway, Galway, Ireland*

SANDRA C. RAIMUNDO • *Department of Biology, Skidmore College, Saratoga Springs, NY, USA*

ELEANORE RITTER • *Department of Biology, Skidmore College, Saratoga Springs, NY, USA*

CHRISTOPHE RITZENTHALER • *Institut de Biologie Moléculaire des Plantes (IBMP), UPR2357 CNRS, Université de Strasbourg, Strasbourg, France*

ALISON W. ROBERTS • *Department of Biological Sciences, Center for Biotechnology and Life Sciences, University of Rhode Island, Kingston, RI, USA*

MARC ROPITAUX • *UNIROUEN, Laboratoire Glyco-MEV, Normandie Université, Rouen, France*

DAVID ROUJOL • *Laboratoire de Recherche en Sciences Végétales, UMR 5546 Université Paul Sabatier-Toulouse III/CNRS, Auzeville Tolosane, France*

ARJA SANTANEN • *Department of Agricultural Sciences, Viikki Plant Science Centre, University of Helsinki, Helsinki, Finland*

CORINNE SCHMITT-KEICHINGER • *INRA, SVQV UMR_A 1131, Université de Strasbourg, Colmar, France*

ALLAN M. SHOWALTER • *Department of Environmental and Plant Biology, Molecular and Cellular Biology Program, Ohio University, Athens, OH, USA*

MARÍA-TERESA SOLÍS • *Biological Research Center, CIB-CSIC, Madrid, Spain*

LI SUN • *Department of Biology and Skidmore Microscopy Imaging Center, Skidmore College, Saratoga Springs, NY, USA*

PILAR S. TESTILLANO • *Biological Research Center, CIB-CSIC, Madrid, Spain; Department of Genetics, Microbiology and Physiology, Complutense University of Madrid, Madrid, Spain*

MARY L. TIERNEY • *Department of Plant Biology, University of Vermont, Burlington, VT, USA*

BERKE TINAZ • *Department of Biology, Skidmore College, Saratoga Springs, NY, USA*

JOHAN TRYGG • *Computational Life Science Cluster (CLiC), Department of Chemistry, Umeå University, Umeå, Sweden*

REINA J. VEENHOF • *Botany and Plant Science, School of Natural Sciences, The National University of Ireland Galway, Galway, Ireland; Ryan Institute for Environmental, Marine and Energy Research, School of Natural Sciences, The National University of Ireland Galway, Galway, Ireland*

MAÏTÉ VICRÉ-GIBOUIN • *UNIROUEN, Laboratoire Glyco-MEV, Normandie Université, Rouen, France; Structure Fédérative de Recherche (Normandie-Végétale) FED 4277 – Université de Rouen Normandie, Mont Saint Aignan Cedex, France*

MELANÉ A. VIVIER • *Department of Viticulture and Oenology, Faculty of AgriSciences, Institute for Wine Biotechnology, Stellenbosch University, Matieland, South Africa*

DONGJIE WANG • *ARC Centre of Excellence in Plant Cell Walls, Centre for Nutrition and Food Sciences, Queensland Alliance for Agriculture and Food Innovation, The University of Queensland, Brisbane, QLD, Australia; College of Food Engineering and Biotechnology, Tianjin University of Science and Technology, TEDA, Tianjin, China*

LONNIE R. WELCH • *Russ College of Engineering and Technology, Center for Intelligent, Distributed and Dependable Systems, Ohio University, Athens, OH, USA*

WILLIAM G. T. WILLATS • *School of Agriculture, Food and Rural Development, Newcastle University, Newcastle-Upon-Tyne, UK*

XIANGMEI WU • *Department of Biochemistry and Molecular Biology, Oklahoma State University, Stillwater, OK, USA*

YANBIN YIN • *Department of Biological Sciences, Northern Illinois University, DeKalb, IL, USA; Department of Food Science and Technology, University of Nebraska-Lincoln, Lincoln, NE, USA*

YUNZHEN ZHENG • *Department of Biology, Penn State University, University Park, PA, USA*

ANSCHA J. J. ZIETSMAN • *Department of Viticulture and Oenology, Faculty of AgriSciences, Institute for Wine Biotechnology, Stellenbosch University, Matieland, South Africa*

Chapter 1

High-Voltage Paper Electrophoresis (HVPE)

Stephen C. Fry

Abstract

HVPE is an excellent and often overlooked method for obtaining objective and meaningful information about cell-wall "building blocks" and their metabolic precursors. It provides not only a means of analysis of known compounds but also an insight into the charge and/or mass of any unfamiliar compounds that may be encountered. It can be used preparatively or analytically. It can achieve either "class separations" (e.g., delivering all hexose monophosphates into a single pool) or the resolution of different compounds within a given class (e.g., ADP-Glc from UDP-Glc; or GlcA from GalA).

All information from HVPE about charge and mass can be obtained on minute traces of analytes, especially those that have been radiolabeled, for example by in-vivo feeding of a ^{3}H- or ^{14}C-labeled precursor. HVPE does not usually damage the substance under investigation (unless staining is used), so samples of interest can be eluted intact from the paper ready for further analysis. Although HVPE is a technique that has been available for several decades, recently it has tended to be sidelined, possible because the apparatus is not widely available. Interested scientists are invited to contact the author about the possibility of accessing the Edinburgh apparatus.

Key words Charge–mass ratio, Electrophoresis, Hydroxyproline oligoarabinosides, Inorganic ions, Ionization, Monosaccharides, Nucleotide sugars, Oligosaccharides, Radiolabeling, Sugar-phosphates, Uronic acids

1 Introduction

High-voltage paper electrophoresis (HVPE) is a seriously undervalued method for the analysis of small (<2500 Da), hydrophilic, charged compounds such as uronic acids, amino sugars, sugar phosphates, sugar sulfates, sugar nucleotides, ascorbate metabolites, Krebs-cycle intermediates, and inorganic ions. In addition, it can be useful for normally uncharged compounds, such as neutral sugars, that can be given a charge by complexing with ions such as borate or molybdate. HVPE is a very rapid separation method, typical run-times being 30–60 min with 10–20 samples typically being run simultaneously. This chapter reviews the use of HVPE to investigate the small-molecule building blocks (monosaccharides, amino acids, etc.) of which cell-wall polysaccharides and

Zoë A. Popper (ed.), *The Plant Cell Wall: Methods and Protocols*, Methods in Molecular Biology, vol. 2149,
https://doi.org/10.1007/978-1-0716-0621-6_1, © Springer Science+Business Media, LLC, part of Springer Nature 2020

glycoproteins are composed; the small, intraprotoplasmic precursors of wall polymers such as sugar nucleotides; and the small to medium-sized solutes present in the apoplast (the aqueous solution that permeates the cell wall) such as oligosaccharides, ascorbate degradation products, and inorganic ions.

The electrophoretic mobilities of compounds on HVPE are not simply random values that need to be determined empirically. On the contrary, a compound's electrophoretic mobility is governed by a definable property, related to its charge–mass ratio, in a highly predictable way, discussed below [1].

Ion-exchange chromatography is in some ways comparable to HVPE and is an alternative method for analysis of charged compounds. For example, an anion-exchange matrix will bind anions, which can subsequently be released from the chromatography column by a gradient of increasing ionic strength or changing pH. This provides some evidence that the compound thus obtained had a negative charge. However, there is the potential pitfall that a compound might adsorb to an anion-exchange resin by some means *other than* electrostatic bonding, such as hydrophobic interaction. In contrast, a compound can migrate toward the anode on HVPE (relative to a neutral marker) *only* if it possesses a negative charge.

HVPE can be used analytically (e.g., 20 small samples on thin paper) or preparatively (a single large sample "streak-loaded" on a sheet of thick paper). The separated compounds can be detected either by staining (which is usually destructive) or by fluorescence, autoradiography (for ^{14}C, ^{35}S, or ^{32}P) or fluorography (for ^{3}H), after which the sample can usually be recovered for further work. Alternatively the compounds can be eluted from a preparative paper ready for further analysis, for example by MS, NMR, or bioassay. Most of the recommended electrophoresis buffers are volatile, so the eluted compounds simply need to be dried, not specially desalted, prior to further analysis. Another advantage of HVPE is that—like paper chromatography—it provides a very convenient means of archiving samples that have been separated but not yet analyzed: what you are storing is a dried paper with zones of potentially interesting compounds on it; this is much more convenient than column chromatography separations in which liquid fractions need to be stored.

2 Materials

2.1 HVPE Apparatus: With Immiscible Coolants

The basic principle of HVPE, a form of zone electrophoresis, is that the sample is dried on a sheet of paper which is then wetted with an aqueous buffer and subjected to a voltage gradient. The compounds in the sample migrate as zones toward the anode or cathode according to their net charge. The higher the voltage applied, the

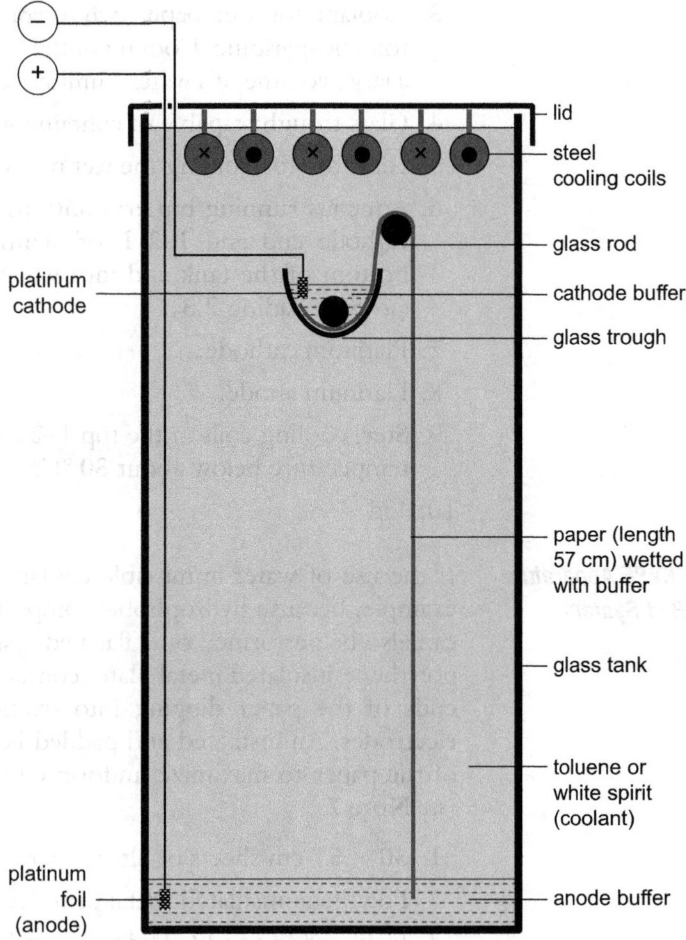

Fig. 1 Apparatus for high-voltage paper electrophoresis using a liquid coolant such as white spirit or toluene. This type of apparatus is available at the University of Edinburgh but not widely elsewhere; interested scientists are invited to contact the author about the possibility of accessing the Edinburgh apparatus. The tank is shown in cross-section. ● = glass rod, **+** = anode, − = cathode, ⊗⊙⊗⊙⊗⊙ = cooling coil (suspended from the lid). Cold tap-water flows through the coil to cool the white spirit or toluene. The anode and cathode are platinum wires encased in glass tubes and connected at the tips to platinum foil

faster the compounds migrate; fast migration minimizes diffusion of the spots. However, an excessively high voltage would overheat the wet paper and possibly degrade the compounds of interest. A cooling system must therefore also be used.

1. HVPE equipment: apparatus set up as shown in Fig. 1 (*see* **Note 1**).
2. 42 × 57-cm sheets of filter paper (*see* **Note 2**): Two types of paper are commonly used: Whatman No. 1 and the thicker Whatman 3CHR (formerly called 3MM) (*see* **Notes 3** and **4**).

3. Coolant for wet paper: white spirit, toluene, or 20:1 (v/v) toluene–pyridine. Cooling of the wet paper is best provided by a large volume of a water-immiscible liquid (*see* **Notes 5** and **6**).
4. Glass trough: capable of containing ~350 mL.
5. Glass bar for holding the wet paper in the glass trough.
6. Aqueous running buffer: ~350 mL in the glass trough at the cathode end and 1–2 L of identical running buffer at the bottom of the tank and into which the platinum anode dips (*see* Subheading 2.3).
7. Platinum cathode.
8. Platinum anode.
9. Steel cooling coils in the top 1–2 cm of the coolant to keep its temperature below about 30 °C.
10. Lid.

2.2 HVPE Apparatus: Flat-Bed System

If the use of water-immiscible coolant liquids is not feasible, for example, because hydrophobic compounds are of interest, HVPE can also be performed on a flat-bed system. The paper is laid on a polythene-insulated metal plate (containing cooling coils) with the ends of the paper dipping into troughs containing buffer and electrodes. An insulated and padded lid is tightly clamped on top of the paper to maximize uniform contact with the cooling plate (*see* **Note 7**).

1. 30 × 57-cm sheets of filter paper.
2. Polythene-insulated metal plate (containing cooling coils).
3. Insulated and padded lid to maximize uniform contact with the cooling plate (*see* **Note 6**).
4. Troughs containing buffer (*see* Subheading 2.3).

2.3 Recommended Buffers and Coolants

Whenever possible, volatile buffers are used so that after the run the electrophoretograms can be freed of buffer salts simply by drying. The three volatile buffers routinely used in our laboratory are listed in Table 1. The buffer concentrations recommended are a compromise between the need to provide adequate buffering capacity (the higher the better) and the need to avoid excessive heating during the run (a concentrated buffer has a higher conductivity and thus draws a higher current, giving excessive heating). Approximate pK_a values of the compounds employed in the buffers are H_2SO_4, 2.0 (second ionization); formic acid, 3.7; acetic acid, 4.7; pyridine, 5.3; borate, 9.2; molybdate, 6.0. White spirit (painters' turpentine substitute) is used as the coolant for HVPE at pH 2.0 or 3.5. A toluene–pyridine mixture is used for cooling the pH 6.5 buffer because pure toluene or white spirit would extract a high proportion of the pyridine (which is largely in its un-ionized form at

Table 1
Recommended HVPE buffer–coolant–marker combinations[a]

Buffer[b]	Composition	Coolant	Suitable colored negative markers	Suitable colored positive markers
Volatile buffers				
pH 2.0	Formic acid–acetic acid–H_2O (1:4:45 by vol.)	White spirit	Orange G [picrate would be lost into the white spirit]	N^{ε}-2,4-Dinitrophenyl-lysine, methyl green, methyl violet
pH 3.5	Acetic acid–pyridine–H_2O (10:1:189 by vol.)[c]	White spirit	Orange G, picrate	Methyl green
pH 6.5[d]	Acetic acid–pyridine–H_2O (10:1:189 by vol.)	Toluene–pyridine (20:1)	Orange G, picrate	Methyl green
Sugar-complexing buffers				
Borate, pH 9.4	1.9% w/v borax ($Na_2B_4O_7{\cdot}10H_2O$), pH adjusted with NaOH	White spirit	Orange G, picrate	[Not required]
Molybdate, pH 3–5	2.0% w/v $Na_2MoO_4{\cdot}2H_2O$, pH adjusted with formic acid or H_2SO_4	White spirit	Orange G, bromophenol blue, picrate	[Not required]

[a]In most buffer systems, a suitable neutral marker is glucose, which is revealed by staining. A better alternative is a nonionic fluorescent compound, such as feruloyl-arabinose [28], which can be used at trace concentrations as an internal marker and visualized by its autofluorescence without staining. On HVPE in borate or molybdate buffers, glucose is unsuitable as a neutral marker since it may bind oxyanions; nonbinding alternatives include 2,3,4,6-tetra-*O*-methylglucose

[b]Except for pH 3.5, the buffer specified is used both at the electrodes and for wetting the paper

[c]The composition of the electrode buffer is given; it should be diluted with an equal volume of water for wetting the paper

[d]For quantitatively accurate conclusions about charge–mass ratio to be drawn, the effective pH of the buffer in the paper during electrophoresis must be known. In the case of the pH 6.5 buffer, it sometimes happens that insufficient pyridine is added to the coolant (toluene), so that although the buffer used for wetting the paper had been accurately adjusted to pH 6.5 the pH of the buffer within the paper gradually falls during the run because some of the pyridine partitions from the paper into the toluene. A good test compound with which to show whether the pH in the paper during the run was correct is histidine, whose imido group has a pK_a of 6.0. At pH 6.5, histidine has a net charge of +0.24, increasing strongly at slightly lower pH values. On the other hand, lysine has a constant net charge of approximately +1.00 at all pH values between 4.5 and 7.5. At pH 6.5, His should move toward the cathode (relative to a neutral marker) at about 0.23× the rate of Lys. If His runs faster than this, more pyridine should be added to the toluene coolant

pH 6.5) from the wet paper into the coolant, thus decreasing the pH of the buffer during the run.

Sodium borate and sodium molybdate are not volatile, so compounds that have been purified by HVPE in these buffers and eluted from the paper with water require desalting, for example by re-electrophoresis in a volatile buffer such as at pH 2.0.

2.4 Buffers Used in Sample Preparation Prior to Electrophoresis

Since nonvolatile salts (including buffers) can interfere in electrophoresis, these should be avoided during sample preparation. Any buffer used, for example to control the pH of an enzyme, should preferably be *volatile*. For the pH range 2–6, this can be similar to one of the mixtures given in Table 1—if necessary diluted so that pyridine, which at high concentrations may inhibit enzymes, does not exceed ~1% v/v. For pH values in the 6–8 range, 1% v/v methylpyridines (adjusted to the desired pH with acetic acid) form useful volatile buffers, for example 2,6-dimethylpyridine (lutidine; $pK_a \approx 6.7$) or 2,4,6-trimethylpyridine (collidine; $pK_a \approx 7.4$).

2.5 Wetting the Paper

1. Tissue paper.
2. Pipette.
3. Two glass rods.
4. Neutral marker: examples include glucose, [^{14}C]glucose, or 5-*O*-feruloyl arabinose.
5. Mobile marker: typically orange G (Table 1).
6. Glass plate: this should be slightly larger than the paper to be loaded.
7. Wash bottle.
8. Buffer (*see* Subheading 2.4).

2.6 Detection of Analytes

1. Aniline hydrogen phthalate stain: 16 g phthalic acid in 490 mL acetone, 490 mL diethyl ether and 20 mL dH_2O is used to prepare a stock solution. 0.5 mL of aniline is added to 100 mL of stock solution immediately before use [2].
2. $AgNO_3$ stain [2] solution A: 0.8 g $AgNO_3$ is dissolved in 1.6 mL water, then diluted into in 104 mL acetone; if necessary, a little extra water added dropwise to redissolve the $AgNO_3$.
3. $AgNO_3$ stain solution B: 1 L 96% (v/v) ethanol + 12.5 mL 10 M NaOH.
4. $AgNO_3$ stain solution C: 10% w/v sodium thiosulfate.
5. NH_3 vapor.
6. Folin–Ciocalteu reagent.
7. Ninhydrin stain: 0.5% (w/v) ninhydrin in acetone.
8. Ninhydrin–isatin stain: dissolve 270 mg ninhydrin plus 130 mg isatin in 100 mL acetone. Immediately before using the stain, add 2.5 mL triethylamine [2].
9. −80 °C freezer.
10. Fluor solution: 7% (w/v) PPO in ether.
11. Autoradiography film.

12. Film for fluorography: autoradiography film preflashed with a photographic flash gun at such a distance that the background of the film will be slightly fogged when developed.
13. pH indicator: 1 L 96% (v/v) ethanol containing 0.4 g bromophenol blue and 0.4 mL collidine (stir the bromophenol blue overnight to thoroughly dissolve).
14. Solution for removing pyridine: acetic acid–toluene 1:20 (v/v).
15. Solution for removing acetic acid: acetone–methanol (4:1) (v/v).

3 Methods

3.1 Layout of Electrophoretogram

1. If it is known that all compounds of interest in the samples are neutral or negatively charged, the samples are loaded 12 cm from the cathode end of the paper. Conversely, if all compounds of interest are neutral or positively charged, they are loaded 9 cm from the anode end. If anions and cations are both of interest, the samples are loaded near the middle of the paper.
2. For analytical HVPE, the solutions are typically loaded at up to 12 and 30 μL per sample on Whatman No. 1 and 3CHR papers respectively. If necessary, multiple 12- or 30-μL portions are applied, with drying after each application.
3. A small amount of a visible mobile marker (e.g., 5 μg orange G; Table 1) is either premixed with the sample (i.e., used as an internal marker) or loaded as a series of spots alternating with the samples (i.e., as an external marker).
4. A neutral marker: preferably internal (e.g., glucose, [^{14}C]glucose, or 5-*O*-feruloyl-arabinose; revealed by staining, autoradiography or autofluorescence, respectively), should also be included so that the extent of electroendo-osmosis (*see* Subheading 3.3 below) can be monitored.

3.2 Wetting and Running the Electrophoretogram

1. The paper, with samples loaded, is laid on a large glass plate and wetted with electrophoresis buffer (Fig. 2 top). The majority of the paper area can be wetted quite quickly, for example by use of a wash bottle.
2. Any excess buffer is very lightly blotted from the electrophoretogram with dry tissue paper; care must be taken not to crease the wet paper during this process.
3. Wetting in the vicinity of the samples should be done last, and very carefully, with a pipette so that the samples do not diffuse excessively. This can be achieved if the area of the paper that includes the origin line is suspended between two glass rods.

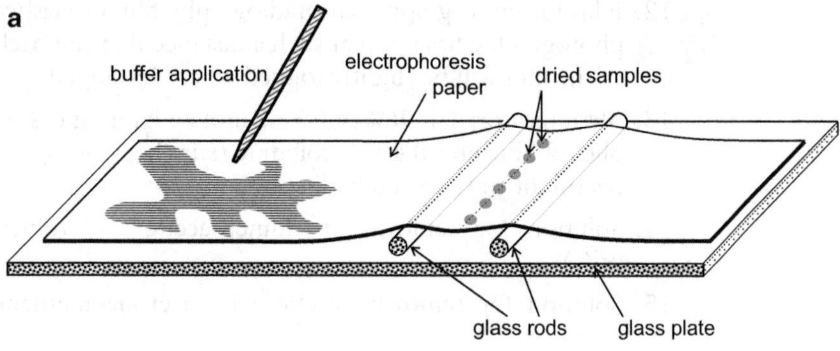

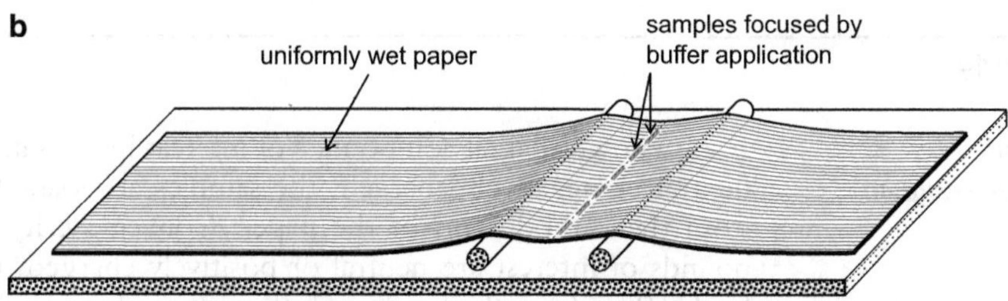

Fig. 2 Method for wetting the electrophoretogram with running buffer

With practice, the wetting of this part of the paper can be made to focus the sample spot into a narrow band, instead of a disc, along the origin line (Fig. 2 bottom).

4. With the apparatus shown in Fig. 1, and with the samples loaded near the end of the paper, we typically perform HVPE at 4.5 kV for ~30 min (*see* **Note 8**). With a full-size sheet of Whatman No. 1 paper (57 × 42 cm), appropriately wetted, 4.5 kV delivers a current of about 150 mA (pH 2.0 buffer), 80 mA (pH 3.5) or 130 mA (pH 6.5) (*see* **Note 9**).
5. It should be noted that the pH of the buffer has an effect on the charge of the analytes (*see* **Note 10**, Tables 2 and 3).

3.3 Reporting the Electrophoretic Mobility of an Analyte

The mobility (m) of a substance in HVPE is often quoted relative to an easily detectable mobile marker (e.g., orange G) and a neutral one (e.g., glucose). The neutral marker is important since neutral compounds often move slightly away from the origin, partly by electroendoosmosis and partly by gravity. If orange G is used, the electrophoretic mobility is quoted as m_{OG}, where

$$m_{OG} = \frac{(\text{distance moved by compound}) - (\text{distance moved by glucose})}{(\text{distance moved by orange G}) - (\text{distance moved by glucose})}$$

A compound that remains near the origin while the neutral marker moves toward the anode should therefore usually be recorded as having electrophoretic mobility toward the cathode, as exemplified by AMP on HVPE at pH 2.0 (Fig. 3a). This assumes

Table 2
Typical pK_a values of some functional groups involved in plant cell-wall and apoplast biochemistry

Functional group	Q[a]	pK_a value(s)	Examples of compounds in pK_a range stated[b]
–COOH	–	1.3	Oxalic acid (1st ionization)
		1.8–2.5	α-Carboxy group of amino acids, oxaloacetic (1st), pyruvic, diketogulonic
		3.0–5.2	Most typical carboxylic acids, e.g. acetic, gluconic, glucuronic, glucaric (both), tartaric (both), malic (both), ferulic, side chain of glutamic and aspartic, oxalic (2nd), citric (1st and 2nd)
		5.0–6.5	Tricarboxylic and a few dicarboxylic acids, e.g. citric (3rd), succinic (2nd)
$-SO_3H$	–	1.3	Cysteic acid
Φ–OH[c]	–	8.5–10.5	Phenolic –OH of tyrosine, isodityrosine, dityrosine (2nd), feruloyl esters, ferulate, coumarate
		6.7	Phenolic –OH of dityrosine (1st)
Amino	+	10–11	Most typical amino groups, e.g. of methylamine, lysine (side-chain), putrescine (2nd); also imino group of proline
		8.5–10	α-Amino group of amino acids; amino group of putrescine (1st); also imino group of hydroxyproline
		7–8	Amino group of amino-sugars
Phosphate ester	–	1–2	Phosphate group of Glc-6-P, dihydroxyacetone phosphate, etc. (1st)
	2–	6–6.5	Phosphate group of Glc-6-P, dihydroxyacetone phosphate etc. (2nd)
Imidazole	+	6.0	Imidazole ring of histidine
Inorganic cations	+	>13	K^+, Na^+, Ca^{2+}, Mg^{2+}, etc.
Inorganic phosphate	–, 2–, 3–	2.2, 7.2, 12.4	$H_2PO_4^-$
Inorganic sulfate	–, 2–	<0, 2.0	SO_4^{2-}

[a] Q = sign of charge when ionized
[b] 1st, 2nd, both, and so on refer to compounds possessing more than one of the functional group mentioned
[c] Φ = benzene ring

that a compound remaining at the origin has not been insolubilized there (as would happen for example with cellohexaose, which strongly hydrogen-bonds to the cellulose of paper)—an assumption that can easily be tested by rinsing in water if there is any doubt. m_{OG} values estimated from Fig. 3a, with fructose as the neutral marker, are: ATP, 0.57; ADP, 0.36; AMP, −0.05; and (by definition) fructose, 0.00; orange G, 1.00.

Table 3
Behaviour of the principal ionizable groups present in cell-wall components and precursors (amino, carboxy, and phosphate) on HVPE at pH 2.0 and 6.5

Observed net → charge ↓	Positive at pH 2.0	Little or no net charge at pH 2.0	Negative at pH 2.0
Positive at pH 6.5	$-NH_2$ group(s) present. Fewer or no acidic groups. *Lysine, putrescine, K^+, Mg^{2+}*	n/a[a]	n/a[a]
Little or no net charge at pH 6.5	$-NH_2$ group(s) present. Equal –COOH groups. *Serine, isodityrosine*	No –COOH, $-NH_2$, phosphate or sulfate groups. *Glucose, glucitol*	n/a[a]
Negative at pH 6.5	$-NH_2$ group(s) present. More –COOH groups. *Aspartate, glutamate*	Weak acid group(s) present such as –COOH, e.g. *galacturonate, malate, tartrate, citrate*, or enediol, e.g. *ascorbate*	Strong acid group(s) present, e.g. phosphate, sulfate, or a low-pK_a –COOH. Few or no $-NH_2$ groups. *Glucose 6-phosphate, oxalate, inorganic sulfate*

The sample is run by HVPE at pH 2.0 and 6.5, and the direction of migration (if any) at these two pH values leads to the conclusions entered in the Table. Specific examples of compounds in each category are given in italics
[a]Would be an implausible result

3.4 Calibrating with Electrophoretic Mobilities of "Knowns"

As mentioned above, the mobility of a compound during HVPE is dictated by its charge and mass. More precisely, mobility is proportional to the $Q{:}M_r^{2/3}$ ratio, where Q is the net charge and M_r to the power of 2/3 is a surrogate for the molecule's relative surface area [1]. It is useful to prepare a calibration curve of the relationship between m_{OG} and $Q{:}M_r^{2/3}$ ratio for a few "knowns," run as markers. The M_r of an authentic marker is usually given by the suppliers (but remember to subtract any contribution due to a counter-ion such as the Cl^- in glucosamine hydrochloride, or any water of crystallization).

If the pK_a of the "known" has been published (for example, see http://research.chem.psu.edu/brpgroup/pKa_compilation.pdf), then its Q can be calculated. The pK_a is the pH at which 50% of the molecules are ionized (at the group under consideration) and 50% are not. For example, glyceric acid has a pK_a of 3.5; therefore, if a dilute solution of this compound is present in pH 3.5 electrophoresis buffer, 50% of the molecules at any moment are glycerate anions ($CH_2OH.CHOH.COO^-$, represented as A^-), while the other 50% are un-ionized ($CH_2OH.CHOH.COOH$, represented as HA). In other words, at pH 3.5, the reaction

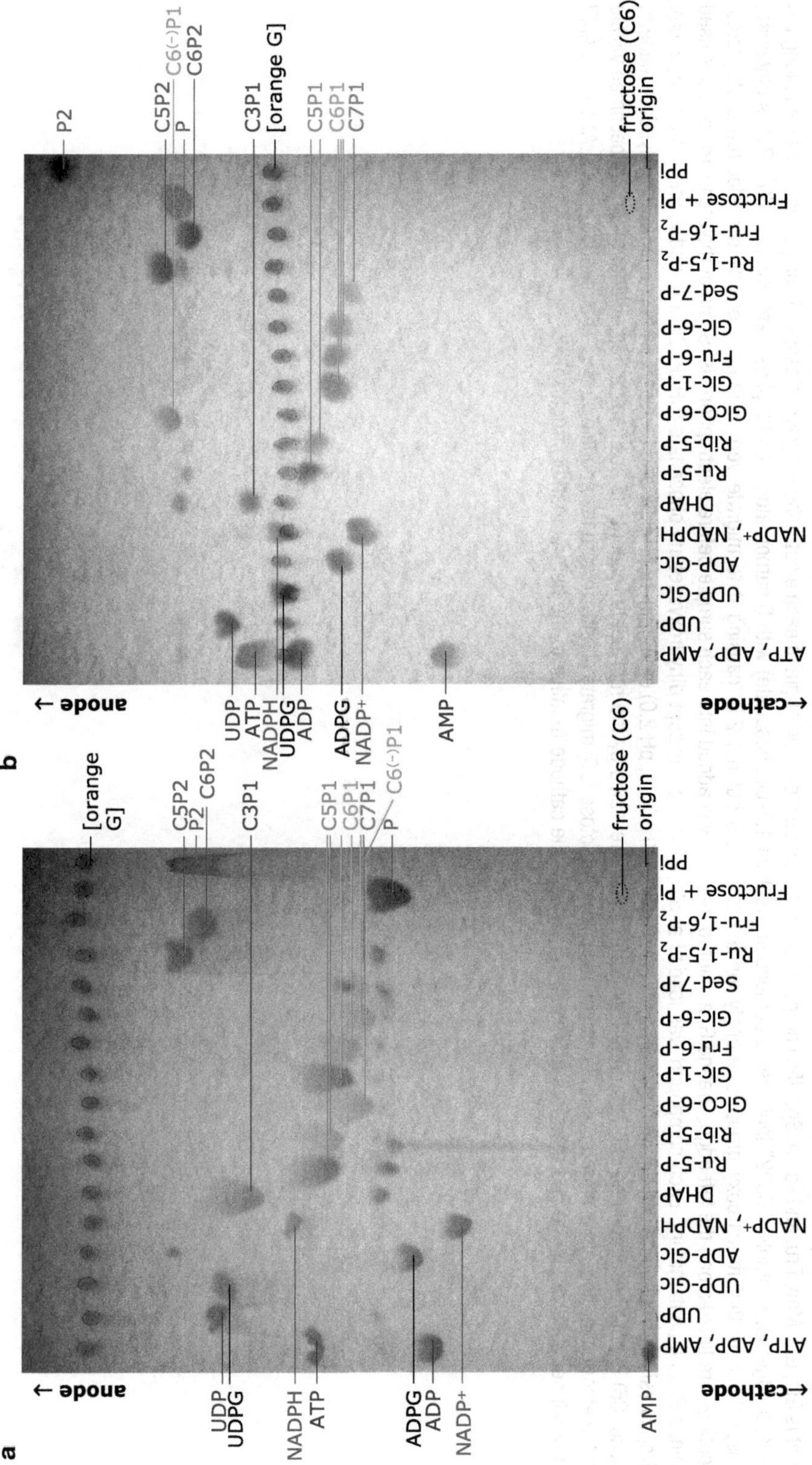

a
b
← anode
cathode→
ATP, ADP, AMP
UDP
UDP-Glc
ADP-Glc
NADP+, NADPH
DHAP
Ru-5-P
Rib-5-P
GlcO-6-P
Glc-1-P
Fru-6-P
Glc-6-P
Sed-7-P
Ru-1,5-P2
Fru-1,6-P2
Fructose + Pi
PPi
[orange G]
C5P2
P2
C6P2
C3P1
C5P1
C6P1
C7P1
P
C6(-)P1
fructose (C6)
origin
UDP
UDPG
NADPH
ATP
ADPG
ADP
NADP+
AMP

Fig. 3 HVPE at pH 2.0 and 6.5 used for separating classes of phosphorylated metabolites involved in cell-wall biosynthesis. Here we are aiming not to resolve all possible phosphorylated metabolites, but to place them into classes of related compounds (sharing approximately the same charge–mass ratio) prior to more detailed analysis. For example, the hexose monophosphate pool can later be eluted from the electrophoretogram and hydrolyzed to determine whether the sugar present is Glc, Gal, Man, Fru, and so on. Specific compounds run are listed along the origin of each electrophoretogram; classes of compounds are listed along the right-hand edge: for example, "C5P2" indicates a compound (ribulose 1,5-bisphosphate) with 5 carbon atoms and 2 phosphate groups. "$C6^{(-)}P1$" is gluconate 6-phosphate: note that its carboxy group is strongly ionized at pH 6.5 but not 2.0, causing it to migrate very differently on the two electrophoretograms. Nucleotides are listed along the left edge. Orange G, a colored marker, was added into each sample before electrophoresis. Some of the standard solutions used contained P_i as a contaminant. Each compound was loaded at 50 μg per spot except dihydroxyacetone phosphate (DHAP; 25 μg), and P_i and PP_i (15 μg each). Electrophoresis was conducted on Whatman no. 1 paper at 4.5 kV for 30 min (**a**, pH 2.0) or 35 min (**b**, pH 6.5), and the spots were stained with molybdate reagent. Fructose (C6) is not phosphorylated and is therefore not immediately revealed by the molybdate reagent; its position (dotted outline) gradually becomes visible when the electrophoretogram is stored. Neutral compounds such as fructose (C6) migrate slightly toward the anode owing to electroendoosmosis. At pH 2.0, AMP has a small net positive charge and therefore moves slightly toward the cathode (relative to the neutral marker, fructose)

$$H^+ + A^- \leftrightarrow HA$$

is precisely balanced with half the molecules on the left and half on the right. If a buffer of pH 2.0 had been used (i.e., with a 32-fold higher concentration of H^+ than at pH 3.5), then the equilibrium would have been pushed toward the right and there would have been much more HA and much less A^-. Conversely, if a buffer of pH 6.5 had been used, then the vast majority of the molecules would have been in the form A^-. The ratio of $[A^-]$ to [HA] at any given pH is given by the equation

$$\log\{[A^-]/[HA]\} = pH - pK_a.$$

Thus, with glucuronic acid ($pK_a = 3.2$) in pH 3.5 buffer, log $\{[A^-]/[HA]\} = 0.3$, so $[A^-]/[HA] \approx 2.0$, and therefore 67% is present as the glucuronate anion and 33% as uncharged glucuronic acid. This can conveniently be described as GlcA having a net charge $Q = -0.67$ at pH 3.5. This and similar values can be read off Fig. 4.

With weak *bases* such as glucosamine (Glc–NH_2), similar rules apply, but raising the pH *decreases* the % ionization, for example

$$Glc - NH_3{}^+ \leftrightarrow H^+ + Glc - NH_2$$

The equation for weak bases is

$$\log\{[B]/[BH^+]\} = pH - pK_a$$

where, in this example, B is Glc–NH_2 and BH^+ is Glc–$NH_3{}^+$.

In compounds with two or more ionizable groups, each group must be treated separately, and the net charge for the whole molecule is then calculated by addition of the individual charges. For example, with fumaric acid [which has two carboxy (–COOH) groups, with pK_a values of 3.0 and 4.4 respectively] in an electrophoresis buffer of pH 3.5, it is calculated that the "first" and "second" carboxy groups contribute partial charges of -0.76 and -0.11 respectively, and thus the compound has net charge $Q = -0.87$ at this pH. As a second example, leucine has a carboxy group with $pK_a = 2.3$ and an amino (–NH_2) group with $pK_a = 9.7$; thus in a buffer at pH 2.0 these two ionizable groups contribute partial charges of -0.33 and very nearly $+1.00$ respectively, giving leucine a net charge $Q = +0.67$.

3.5 Interpreting Electrophoretic Mobilities of "Unknowns"

Armed with a graph plotting m_{OG} against $Q{:}M_r^{2/3}$ for several "knowns," we can interpret the m_{OG} values of unknown compounds, run under the same conditions, in terms of their $Q{:}M_r^{2/3}$ ratios. If either of the parameters, Q or M_r, can be assumed, then we can estimate the other. For example, if the charge is known to be -1 at the pH of the electrophoresis buffer, then compound's M_r can be estimated. Likewise, if the M_r is known (e.g., because we

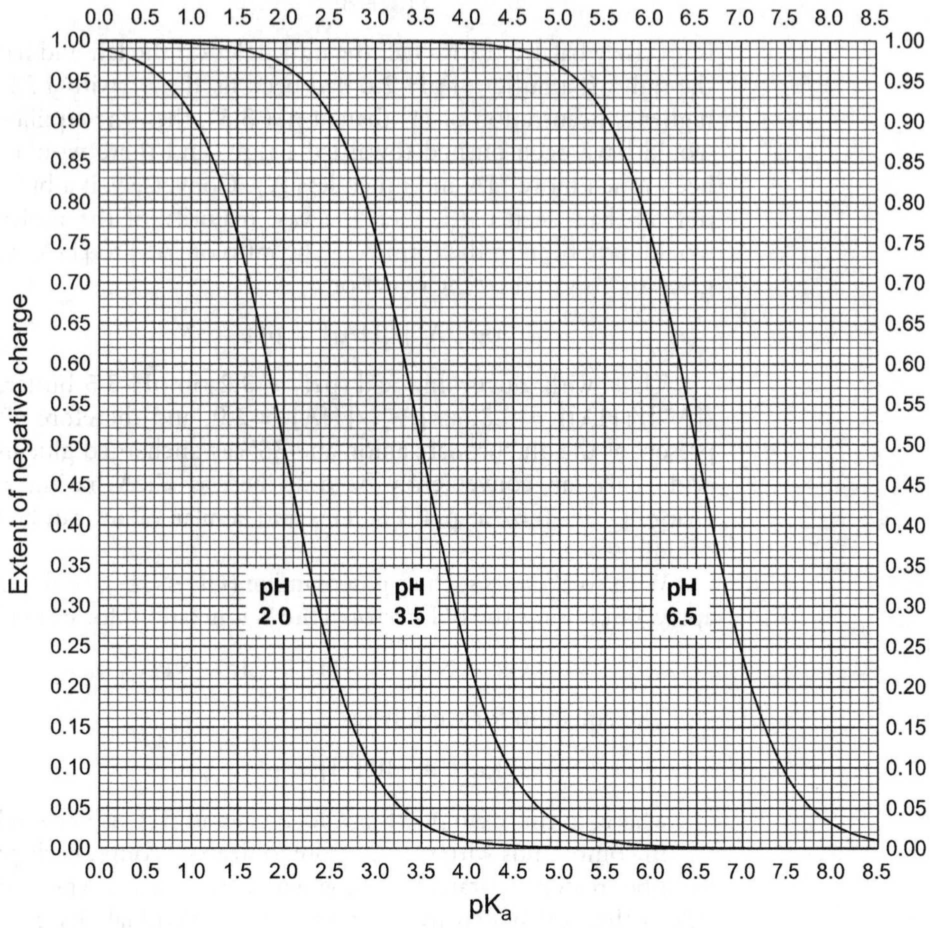

Fig. 4 Relationship between pH and pK_a. The degree of ionization of any anionic group with a pK_a between 0.0 and 8.5 at each of the three recommended pH values for routine HVPE. For example, the graph shows that a carboxy group with pK_a = 6.0 has a charge of −0.76 in pH 6.5 buffer. The graph can also be used for cationic groups, but an amino group with pK_a = 6.0 has a charge of 1.00–0.76 = +0.24 in pH 6.5 buffer

know it is a hexuronic acid), then we can estimate the pK_a, which may identify which specific uronic acid it is.

On electrophoresis at pH 2.0, amino groups are fully positively charged. Therefore, a compound with a single amino group, and no groups that possess an appreciable negative charge at that pH, can be estimated for size by HVPE if we have a calibration curve. An example of a calibration curve is given in Fig. 5. The compounds tested here were reductively aminated sugars [oligosaccharidyl–1-amino-1-deoxyalditols (OADs), prepared from glucose and various authentic oligosaccharides of DP 2–9], which can be assumed to have $Q = +1.00$ at pH 2.0. The reference compounds were glucose and glucosamine, so the y-axis on this occasion reports m_{GlcN} rather than m_{OG}. The graph exhibits a good approximation to a straight

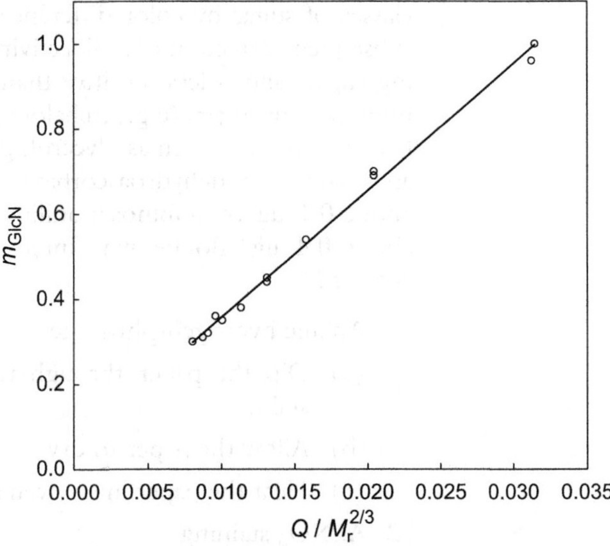

Fig. 5 A calibration curve plotting electrophoretic mobility against the $Q{:}M_r^{2/3}$ ratio for several "knowns" (reductively aminated sugars). 1-Amino-1-deoxyalditols were obtained by reaction of glucose or an oligosaccharide in the presence of $NaCNBH_3$ + NH_4HCO_3. The *x*-axis shows the $Q{:}M_r^{2/3}$ ratio calculated on the basis that $Q = +1.00$; the *y*-axis plots the observed electrophoretic mobility (as m_{GlcN}) at pH 2.0. The top data-point is for glucosamine itself ($m_{GlcN} = 1.00$, by definition). Other data points refer to (in order of decreasing m_{GlcN}) reductively aminated glucose, maltose, cellobiose, maltotriose, maltotetraose, isomaltotetraose, maltopentaose, maltohexaose, XXXG, maltoheptaose, XXLG, and XLLG [29]

line, so this calibration curve can be used to estimate the size of other OADs, prepared from unknown oligosaccharides.

At pH 6.5, most carboxy groups are almost fully negatively ionized, so each carboxy group can be assumed to have $Q \approx -1$. Again, then, if the approximate M_r is known, a suitable calibration curve allows us to "count" the carboxy groups. Thus, for example, the C_4 compound tartaric acid (with two COOH groups) is readily distinguished from threonic acid (also a C_4 compound but with one COOH group). Likewise, if the number of carboxy groups is known, then the M_r can be roughly estimated. These approaches have proved useful in characterizing novel apoplastic metabolites of ascorbate by HVPE at pH 6.5 [3].

3.6 Detection of Analytes: Staining

Staining can be used after HVPE to detect many cell-wall components with reasonable sensitivity, though usually destructively. Several methods are described in detail by Fry [2].

3.6.1 Staining for Sugars

Sugars are readily detected by staining with aniline hydrogen-phthalate or $AgNO_3$. The former is quicker, distinguishes different classes of sugar by color differences, and is compatible with the subsequent detection of radioactivity, but fails to detect nonreducing sugars and is less sensitive than $AgNO_3$. $AgNO_3$ also detects nonreducing sugars (e.g., trehalose and sucrose) as well as "sugar-like" compounds such as glycerol, galactonate, threonate, tartrate, ascorbate, and dehydroascorbate. $AgNO_3$ can detect down to about 0.1 μg of arabinose; aniline hydrogen-phthalate down to about 0.4 μg. Borate may interfere with $AgNO_3$ staining (*see* **Note 11**).

1. Aniline hydrogen-phthalate
   (a) Dip the paper through the aniline hydrogen-phthalate stain.
   (b) Allow the paper to dry.
   (c) Heat the paper in an oven at 105 °C for 5 min.
2. $AgNO_3$ staining
   (a) Work in subdued light throughout. Dip the electrophoretogram through solution A (*see* Subheading 2.6, **item 2**).
   (b) Dry for ~15 min.
   (c) Dip the electrophoretogram with a smooth continuous motion through solution B (*see* Subheading 2.6, **item 2**).
   (d) Dry ~15 min.
   (e) Repeat **steps c** and **d** until the spots can be seen clearly.
   (f) Dip the paper through solution C (*see* Subheading 2.6, **item 2**) and then wash in tap water for 1–2 h.
   (g) Dry the paper.

3.6.2 Staining for Phosphates

Phosphates, including sugar phosphates and NDP sugars, are detected by molybdate reagent (Fig. 3), although this is relatively insensitive (down to 2 μg of Glc-6-P) and is mainly used for localization of markers run alongside (or mixed with) radioactive samples of interest.

3.6.3 Staining for Phenolics

Phenolics can often be visualized by their autofluorescence, sometimes intensified by exposure of the paper to NH_3 vapor, or less sensitively by spraying with commercially available Folin–Ciocalteu phenol reagent followed by exposure of the paper to NH_3 vapor until the yellow background is decolorized.

3.6.4 Staining for Amino Acids

Amino acids can be detected very sensitively with ninhydrin (dip the paper through the ninhydrin solution, dry for ~30 min, and record the α-amino acid and imino acid spots; then heat at 105 °C for 5–20 and record any additional (non-α-)amino compounds, e.g., γ-aminobutyric acid and putrescine). The staining is rendered more specific for hydroxyproline by inclusion of isatin [2]. For analysis of hydroxyproline oligoarabinosides, the electrophoretogram should be acid-pretreated in situ to produce free hydroxyproline:

1. Dip the paper quickly through butan-1-ol–acetone–water–6 M HCl (45:45:5:2.5 by vol.).
2. Allow to drip in a fume hood for ~ 1 min.
3. Heat in a well-ventilated oven for 60 min at about 80 °C.
4. **Steps 1–3** may be repeated for maximal yield of free hydroxyproline; if the compound of interest is already free hydroxyproline, start at **step 5**.
5. Dip the paper through freshly prepared isatin–ninhydrin–triethylamine solution (care—the paper is brittle after an acid pretreatment).
6. Dry the paper 20 min in a fume hood.
7. Heat in a well-ventilated oven for 20 min at about 60 °C.
8. Scan the paper.
9. Wash paper in running tap-water for 1 h to remove most amino acids (aspartate remains).
10. Dry and scan again to record the imino acids (proline, orange; hydroxyproline, pink) and aspartate (blue).

3.6.5 Staining Relatively Inert Anions and Cations

Anions such as oxalate, citrate, or inorganic sulfate, that are not phosphates or carbohydrate-related, and therefore cannot be stained with molybdate or $AgNO_3$, can be detected if the paper is dipped through an ethanolic solution of a pH indicator (0.04% bromophenol blue with 0.04% collidine in ethanol). The same indicator solution can also be used to detect nonreactive cations such as K^+, Na^+, Ca^{2+}, and Mg^{2+}. Anions and cations show up as yellow and blue spots, respectively, on a greenish background. If the background appears too yellow or too blue, the indicator solution can be modified by slightly increasing or decreasing, respectively, the collidine concentration. Before use of a pH indicator, the electrophoretogram should be freed of any traces of pyridine, acetic acid, and formic acid (by dipping through acetone/methanol (4:1) and redrying for 1–16 h in an oven at about 70 °C).

3.7 Detection of Radioactive Analytes

Radioactivity is often detected on paper electrophoretograms by autoradiography (for ^{14}C, ^{32}P, ^{33}P and ^{35}S) or fluorography (^{3}H).

3.7.1 Autoradiography

1. Expose the electrophoretogram to film.
2. After incubating in the dark develop the film in a darkroom.

3.7.2 Fluorography

1. The paper is then dipped through a fluor (7% w/v PPO in ether) and dried.
2. The film is preflashed (*see* Subheading 2.6, **item 8**).
3. The electrophoretogram is exposed to the film at a low temperature (e.g., −80 °C) (*see* **Note 12**).

3.8 Detection of Analytes: Bioassay

Substances separated on the electrophoretogram can be eluted for further analysis (*see* Subheading 3.9)—including bioassays (e.g., for oligosaccharins), in which case the paper should be carefully freed of all traces of the liquids used during electrophoresis (acetic acid, pyridine, trace contaminants from the white spirit, etc.). Anionic analytes on the paper (e.g., oligogalacturonides) will be present as their pyridinium salts; thus, after electrophoresis, the dried paper is dipped in toluene–acetic acid (20:1 v/v) and redried to remove the pyridine, then dipped in acetone–methanol (4:1) and redried to remove the acetic acid. If necessary, the paper can be very thoroughly washed in pure toluene (e.g., by descending paper chromatography with the toluene as the solvent) to remove contaminants picked up from the coolant (white spirit or practical-grade toluene).

3.9 Elution Method

Substances separated on the electrophoretogram can be eluted, usually in water, for further analysis. Elution is achieved in a minimal volume of eluent by the syringe-barrel method [4].

1. The relevant zone of the electrophoretogram is cut out with scissors and tightly packed into a plastic syringe barrel, which is then suspended in a plastic centrifuge tube (Fig. 6).
2. The paper is wetted with just enough water to moisten it.
3. The assembly is bench-centrifuged (e.g., 4000 × *g*), causing the eluate to collect in the bottom of the centrifuge tube.
4. Moistening and centrifugation are repeated, typically 4–5 times, until the analyte of interest has been eluted. Simultaneously eluting an orange G marker spot can indicate when elution is complete.

3.10 Specific Examples

3.10.1 Monomeric Sugar Acids and Related Metabolites

Monomeric uronic acids, aldonic acids, aldaric acids, Krebs-cycle intermediates and many ascorbate metabolites possess 1–3 ionizable carboxy groups. At pH 6.5, most carboxy groups are almost fully ionized: any carboxy group with a pK_a less than 5.2 will be >95% fully ionized (Fig. 4). Thus, the advantage of HVPE at pH 6.5 is that it can be used to "count" the compound's carboxy groups if its approximate M_r is known, or to estimate its M_r if the number of carboxy groups is known.

At pH 6.5 there is little resolution between different members of a given class (e.g., hexuronic acids, which all have the same M_r and a single carboxy group and are thus grouped together into a tight zone on the electrophoretogram). On the other hand, pH 3.5

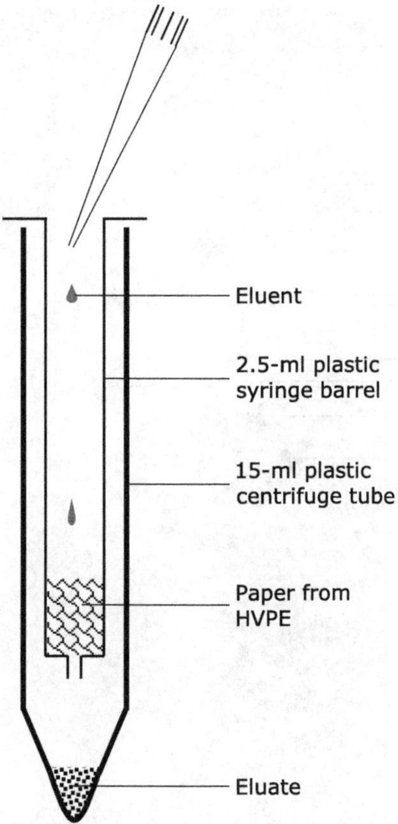

Fig. 6 Method for the elution of analytes from electrophoresis paper

is reasonably close to the pK_a of many carboxy groups. Therefore, small differences in pK_a, such as that exist between isomeric uronic acids (e.g., GalA, GlcA, ManA, and GulA), enable their separation from each other (Fig. 7) [5].

3.10.2 Ascorbate Metabolites

Many of the ascorbate metabolites occurring in the apoplast, although similar in size monomeric uronic acids, and likewise carboxylic acids, are much stronger acids—that is, they have a low pK_a. Therefore, these metabolites (but not dehydroascorbic acid or ascorbate itself) migrate on HVPE at pH 2.0 far faster than the great majority of carboxylic acids. Examples are oxalate, oxalyl threonates, cyclic oxalyl threonates, 2,3-diketogulonate, and 2-carboxy-L-xylonate and its monolactones [3, 6]. HVPE at pH 3.5 resolves certain isomers of these (e.g., 3-*O*-oxalyl threonate from 3-*O*-oxalyl threonate) [7]. Differences in electrophoretic mobility between pH 2.0 and 6.5 (where the metabolites are partially and almost fully ionized respectively) are a very powerful tool for characterizing ascorbate catabolites.

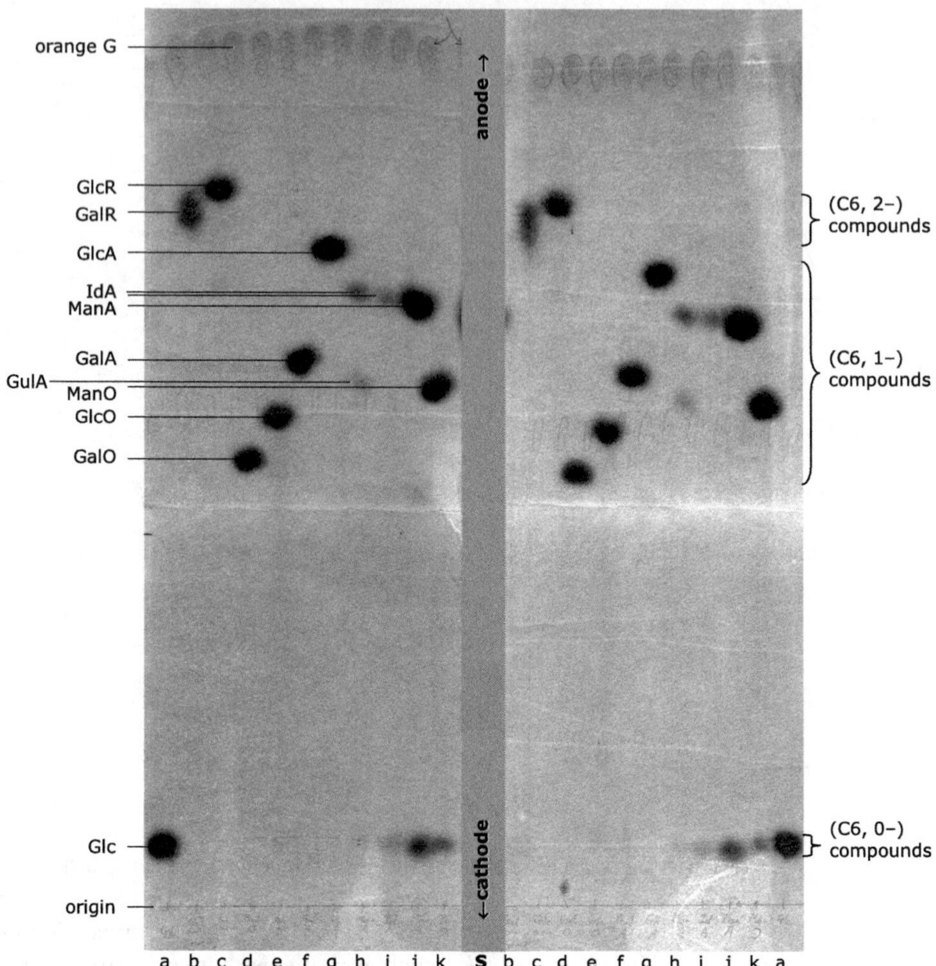

Fig. 7 HVPE at pH 3.5 used for separating specific sugar-acids, differing subtly in their pK_a values. The specific compounds are listed along the left edge; classes of compounds are listed on the right: for example, "(C6, 1−)" indicates a compound (e.g., galacturonic acid) with 6 carbon atoms and 1 negatively ionized group. Orange G, a colored marker, was loaded between each sample. Each compound was loaded at 10 μg per spot except h and i, which were crude preparations of unknown concentration; h is a mixture of ManA and GulA. Electrophoresis was conducted on Whatman 3CHR paper at 4.5 kV for 50 min, and the spots were stained with $AgNO_3$. Several of the sugar-acids are accompanied by neutral lactones, comigrating with glucose. The suffixes -A, -O, and -R indicates -uronic acids, -onic acids, and -aric acids: for example, GalA = galacturonic acid, GalO = galactonic acid, GalR = galactaric acid. The left and right halves are from a single, wide electrophoretogram, illustrating the good reproducibility of m_{OG} values; the compounds shown acted as external markers for a radioactive experimental sample (central band "S"; not shown in full)

3.10.3 Acidic Oligosaccharides

HVPE at pH 6.5, where uronic acid carboxy groups are almost fully ionized, is valuable for determining whether an acidic oligosaccharide of known size (degree of polymerization, as estimated for example by gel-permeation or paper chromatography) consists only of acidic residues (e.g., galacturonobiose) or of a mixture of neutral and acidic residues (e.g., α-D-glucuronosyl-(1→3)-L-galactose) [8]. Excellent separation was obtained of three sugars possessing one glucuronate residue and 0, 1, or 2 neutral sugar residues [9]. As above, the charge–mass ratio is obtained. For example, if the approximate M_r is known, the charge can be estimated, as illustrated by the use of HVPE to determine whether an oxidized radioactive xyloglucan oligosaccharide (XGO) was XGO–gluconate (1−) or XGO–glucarate (2−) [10].

Although acidic oligosaccharides are usually run at pH 6.5, giving the carboxylic acids an almost full negative charge, HVPE at lower pH values, even 2.0, can be useful, separating acidic disaccharides of similar size but different pK_a values; for example, lepidimoic acid, a pectic disaccharide possessing an unsaturated residue (generated by lyase action) has a lower pK_a (is faster-migrating at pH 2.0) than a similarly sized pectic disaccharide, xylosyl-galacturonate produced by hydrolase action [11].

3.10.4 Acidic Polysaccharides

Paper is not generally recommended for electrophoresis of polysaccharides because they often have some affinity for cellulose and therefore tend to streak. However, rhamnogalacturonan-II (RG-II) is small enough (effectively, a large oligosaccharide) to produce a well-defined, tight spot on HVPE at pH 2.0 [12]. In addition, larger polysaccharides can be electrophoresed on noncellulosic supports such as Whatman GF/A glass fibre [13].

3.10.5 Sugar Phosphates and Sugar Sulfates

Since phosphate and sulfate are strongly acidic groups, they remain appreciably ionized during HVPE at pH 2.0 (Fig. 3a), unlike most of the carboxylic acid groups present in carbohydrates. Thus, it is possible to obtain an excellent class-separation of sugar phosphates from nonphosphorylated sugars at this pH [14]. Here we are aiming not to resolve all possible phosphorylated metabolites from each other, but to place them into classes of related compounds (sharing approximately the same charge–mass ratio) prior to more detailed analysis. For example, the hexose monophosphate pool can later be eluted from the electrophoretogram and hydrolyzed; this will reveal whether the sugar present is Glc, Gal, Man, Fru, and so on. At pH 2.0, all the sugar phosphates are tightly clustered, though, as expected, the tendency is for some resolution to occur on the basis of size: in order of decreasing mobility toward the anode, we have triose-phosphates, tetrose-phosphates, pentose-phosphates, hexose-phosphates, and heptose-phosphates (Fig. 3a). Gluconic acid 6-phosphate comigrates with glucose 6-phosphate because the carboxy group of the former is not appreciably ionized at pH 2.0. Sugar bisphosphates, such as ribulose 1,5-bisphosphate and fructose 1,6-bisphosphate, migrate considerably faster than the corresponding monophosphates because both phosphate groups ionize.

At pH 6.5 (Fig. 3b), each phosphate group carries a stronger negative charge than at pH 2.0 and there is a more reliable correlation between size and mobility (mobility of triose- > tetrose- > pentose- > hexose- > heptose-phosphates). In this case, however, gluconic acid 6-phosphate runs much faster than glucose 6-phosphate because its carboxy group is almost fully ionized at pH 6.5.

Since sugar nucleotides and sugar phosphates are generally used in vitro, and present in vivo, at low concentrations, essentially the same techniques can be used for analytical and "preparative" electrophoresis. Often, the compounds under investigation will be radiolabeled (*see* **Note 13**).

Often the isolation of a given class of compounds by HVPE at pH 6.5, for example, the hexose monophosphates, provides the metabolic information required for the project in hand, since Fru-6-P, Glc-6-P, Glc-1-P, and Gal-1-P are all readily interconverted in vivo. However, if necessary, a further class separation is easily achieved by graded hydrolysis: the aldose 1-phosphates (Glc-1-P, Gal-1-P, Man-1-P, etc.) are completely hydrolyzed to the free monosaccharides under mildly acidic conditions (0.1 M trifluoroacetic acid at 100 °C for 25 min), whereas aldose 6-phosphates (e.g., Glc-6-P, Man-6-P), ketose 6-phosphates (e.g., Fru-6-P), and ketose 1-phosphates (e.g., Fru-1-P, although this probably does not occur in plants) are not. Thus, the sugars from the former class are obtained as free monosaccharides after mild acid hydrolysis; thereafter the monosaccharides can be obtained from the remaining unhydrolyzed hexose monophosphates belonging to the latter class by digestion with a commercial phosphatase preparation [14].

The major hexose bisphosphate (Fru-1,6-P_2) is not appreciably hydrolyzed by mild acid—unlike the UDP-sugars, which release the monosaccharide almost quantitatively. Mild acid removes one phosphate group (the one attached to the anomeric carbon) from Glc-1,6-P_2, Man-1,6-P_2, and Fru-2,6-P_2, leaving a hexose 6-phosphate.

Similar advice applies to phosphorylated sugars other than hexoses, but in the case of pentoses, for example, for "6" read "5."

3.10.6 Nucleotides, Including NDP-Sugars and CoA-Thioesters

As with sugar monophosphates versus bisphosphates at pH 2.0, there is excellent resolution within a given series of nucleotides: in order of decreasing mobility toward the anode, we have ATP, ADP, and AMP (Fig. 3a). Sugar-nucleotides tend to migrate fairly close to the corresponding nucleoside diphosphates (e.g., UDP-Glc near UDP, and ADP-Glc near ADP). Some of the nucleoside moieties (adenosine, uridine, etc.) differ in charge at pH 2.0 depending on the presence of amino groups. Adenosine has such a significant positive charge at pH 2.0 that (despite its negatively charged phosphate group) AMP has a small net positive charge and therefore moves slightly toward the cathode (relative to the neutral marker). Thus, ADP-hexoses are very well resolved from UDP-hexoses (Fig. 3a). GDP-hexoses migrate only slightly faster than ADP-hexoses at pH 2.0 (data not shown).

At pH 6.5, UDP-glucose resolves well from UDP-glucuronate [15] because the GlcA residue carries a full negative charge. However, at pH 2.0, UDP-glucose runs only slightly slower than UDP-glucuronic acid because the GlcA residue is scarcely ionized at such a low pH. Thus, at pH 2.0, it is possible to obtain very useful class separations of UDP-sugars (including UDP-Glc and UDP-GlcA) in one zone and ADP-sugars + GDP-sugars in a second zone. These two zones can be eluted from the paper for further analysis, for example by mild acid hydrolysis (0.1 M trifluoroacetic acid at 100 °C for 25 min) followed by paper chromatography or TLC (or HVPE at pH 3.5 for the uronic acids) to resolve the diverse monosaccharides—for example from the UDP-sugar pool, containing UDP-Glc, UDP-D-Gal, UDP-Ara, UDP-Xyl, UDP-Rha, UDP-GlcA, and UDP-GalA. Note that the ribose moiety of NDP-sugars is not released by mild acid hydrolysis.

At both pH values, 2.0 and 6.5, NADPH migrates faster toward the anode than $NADP^+$ because the former lacks the extra positive charge (note that both NADPH and so-called "$NADP^+$" both possess a net negative charge; the "+" is only relative) (Fig. 3). At pH 3.5 and 6.5, in contrast to pH 2.0, NDP-sugars consistently migrate slower than the corresponding NDPs [16].

HVPE at pH 3.5 is useful for class-separating ADP-hexoses ($m_{OG} \approx 0.58$) from GDP-hexoses ($m_{OG} \approx 0.75$). These two classes can subsequently be further analyzed, for example by acid hydrolysis of the GDP-sugar zone to yield the monosaccharides (typically D-mannose, L-fucose, and L-galactose).

Thioesters of coenzyme-A (acetyl-CoA, feruloyl-CoA, etc.) are conveniently analyzed by HVPE at pH 2.0 [17], in which buffer they all possess a substantial net negative charge. Commercially available dodecanoyl-CoA is a useful marker for feruloyl-CoA, which has about the same charge–mass ratio. [^{14}C]Cinnamate metabolites with an appreciable net negative charge at pH 2.0 are very likely to be CoA conjugates since nothing else would give them such a charge. They also migrate very rapidly at pH 3.5 (m_{OG} values: CoA, 1.0; feruloyl-CoA, 0.61). They can be further characterized by mild alkaline hydrolysis, yielding the former acyl residue in free form (ferulate, acetate, etc.).

3.10.7 Amino-Sugars

The amino groups of ManN, GalN, and GlcN have pK_a values of about 7.3, 7.7, and 7.8 respectively. These compounds thus carry almost a full positive charge at all pH values commonly used for HVPE and so are not resolved except slightly at pH 6.5.

Nevertheless, HVPE at pH 2.0, 3.5, or 6.5 permits the quick and easy class separation of sugar amines from acidic and neutral sugars, including *N*-acetyl amino-sugars (e.g., GlcNAc). Similarly, HVPE at pH 2.0 allows excellent class separation of amino-sugar derivatives with crystal violet (CV) as a cationic marker and orange G (OG) as an anionic marker [18]:

$m_{CV} = 1.26$: GlcN, GalN;

$m_{CV} = 0.20$: GlcN-1-P, GlcN-6-P, GalN-1-P;

$m_{CV} = m_{OG} = 0$: GlcNAc, GalNAc;

$m_{OG} = 0.57$: GlcNAc-1-P, GlcNAc-6-P, GalNAc-1-P;

$m_{OG} = 0.76$: UDP-GlcNAc, UDP-GalNAc.

Amino-sugars and their derivatives can be stained with aniline hydrogen-phthalate (unless carrying a phosphate group at position 1) or ninhydrin (unless acetylated on the N).

3.10.8 Amino Acids and Polyamines

α-Amino acids (i.e., compounds with an amino group and a carboxy group attached to the same carbon, the simplest being glycine) have at least one amino group (contributing at least one full positive charge at pH 2.0, 3.5, and 6.5), and an α-carboxy group which bears a full or partial negative charge depending on the pH. If the amino group is free (not involved in a peptide bond), the α-carboxy group has a very low pK_a (in the range 1.8–2.5) and is thus 61–24% fully anionic at pH 2.0, unlike most other carboxy groups which are almost neutral at that pH. The side-chain carboxy groups of Asp and Glu have much higher pK_a values and are only ~1% charged at pH 2.0 but highly charged at pH 3.5 and 6.5. Thus, on HVPE at pH 2.0, all the common α-amino acids have a net positive charge; Arg, Lys and His have a particularly large one owing to the presence of a second positively ionizing group.

Cysteic acid (an oxidation product of Cys) is the only commonly encountered α-amino acid with a net negative charge at pH 2.0: this is due to its negatively ionizing $-SO_3H$ group (pK_a = 1.3). Other than cysteic acid, hydroxyproline stands out on HVPE at pH 2.0 as the slowest-migrating major "amino acid" (strictly an imino acid) because its carboxy group is unusually acidic (pK_a = 1.8). Hydroxyproline mono-, di-, tri-, and tetra-arabinosides, obtained from some cell-wall glycoproteins such as extensins, migrate progressively slower still [19].

HVPE at pH 2.0 followed by staining with Folin and Ciocalteu's phenol reagent is useful for the detection of tyrosine and its oxidative coupling products (isodityrosine, pulcherosine, diisodityrosine), since these are among the very few cationic phenols. Others that do exist include tyramine and *N*-feruloyl-putrescine.

4-Aminobutyrate (GABA) is a significant stress metabolite in plants, sometimes found in the apoplast and thus relevant to wall metabolism. Its carboxy group is not attached to the same C atom as its amino group and thus has a more typical pK_a value (4.0, as opposed to 2.3 for 2-aminobutyrate). Therefore, at pH 2.0, GABA has a net charge of about +1.0 and migrates toward the cathode about 90% as rapidly as lysine, which has two amino groups.

Polyamines possess hydrocarbon chains, usually with two primary amino groups; in addition there may be one or more secondary amino groups. Major examples in plants are putrescine (a diamine), spermidine (a triamine), and spermine and its isomer thermospermine (tetraamines) [20]. Their presence in solution in the apoplast and covalently bonded to wall polymers has been reported. Polyamines migrate particularly rapidly toward the cathode during HVPE, not only at pH 2.0 but also at pH 6.5 since the amino groups are essentially fully ionized, there are no negatively ionizing groups, and they have low M_r. HVPE thus enables a very convenient class separation, resolving polyamines from all other common phytochemicals. All the common polyamines have $m_{lysine} \approx 1.6$ at pH 2.0, and stain slowly with ninhydrin. Resolution of the various polyamines from each other by HVPE requires a higher pH buffer, closer to the pK_a of the amino groups, for example 0.2 M ammonium carbonate (pH 8.7); however, better separation of spermine, spermidine and putrescine is obtained by paper chromatography in butan-1-ol–acetic acid–pyridine–water (4:1:1:2).

It has been suggested that polyamines such as putrescine may be linked via amide (isopeptide) bonds to pectic GalA residues [21]. Model compounds with which to search for the natural occurrence of such linkages were synthesized chemically [e.g., *N*-D-galacturonoyl-putrescinamide (GalA–Put) and *N*,*N*′-di-D-galacturonoyl-putrescinamide (GalA–Put–GalA)]. In addition, promising diagnostic "fragments" were isolated by Driselase digestion of artificially putrescine-conjugated homogalacturonan [yielding products such as Put–$GalA_3$ and $GalA_3$–Put–$GalA_3$]. These various novel glycoconjugates were characterized largely by HVPE on the basis of the following rules:

- Carboxy groups: almost fully ionized negatively at pH 6.5, partially at pH 3.5, almost un-ionized at pH 2.0.
- Amino groups: fully ionized positively in all three buffers.
- Amide (–CONH–) groups: un-ionized in all three buffers.

Similar techniques were used in the preparation and characterization of N^{ϵ}-D-galacturonoyl–L-lysine and related conjugates, which are useful model compounds with which to test for the possible occurrence of pectin–extensin amide linkages [22].

Glutathione (GSH, a tripeptide) and its disulfide-bridged oxidation-product (GSSG), both of which have been reported in the apoplast, can also be resolved quickly and cleanly from many co-occurring compounds by HVPE. Prelabeling with [^{35}S]sulfate is particularly helpful as plants possess few extracellular sulfur compounds. GSH and GSSG have a net positive charge at pH 2.0 and net negative charge at pH 6.5. Selected m_{OG} values at pH 6.5 are: GSH, 0.74; GSSG, 0.88; cysteic acid, 1.29; inorganic sulfate, 2.8; cysteine, 0.00; cystine, 0.00; methionine, 0.00 [23].

3.10.9 Electrophoresis of Neutral Sugars

Neutral sugars and alditols do not migrate on HVPE in ordinary buffers, for example at pH 2.0, 3.5, and 6.5. However, such sugars can reversibly be given a negative charge by complexing with oxyanions such as borate, molybdate, tungstate, stannate, or aluminate [2, 24]. Complexing is a very rapid and simple procedure: the sugar sample is loaded in the normal way (Fig. 2), and the paper is wetted with an aqueous solution of the oxyanion, which is used as the electrophoresis buffer (Table 1).

Borate binds to suitably orientated pairs of –OH groups, which occur in most sugars and alditols, conferring a negative charge. Binding is optimal at high pH (e.g., 9.4) and progressively weaker at lower pH values. The borate–sugar dissociation constant (and thus the average number of borate ions bound per unit mass of sugar at equilibrium) determines the electrophoretic mobility. The method is useful for distinguishing oligosaccharides that differ in bond position [(1 → 2), (1 → 3), (1 → 4), etc.] [2, 25, 26]. There may also be good separation of pairs of oligosaccharides that differ only in anomeric configuration (e.g., maltose versus cellobiose). Wall-derived oligosaccharides can often be freed of contaminating maltooligosaccharides on the basis of the slow migration of the latter in borate buffer [27].

Molybdate reversibly binds to alditols and related compounds possessing certain patterns of –OH groups; binding is optimal at pH 2 and progressively weaker at higher pH values. HVPE in molybdate is particularly effective at resolving reducing oligosaccharides (e.g., the xyloglucan-derived nonasaccharide, XXFG), most of which cannot bind molybdate, from the corresponding reduced oligosaccharides (e.g., XXFGol), which can bind it specifically at the glucitol moiety. Thus, XXFGol is mobile in molybdate buffer, whereas XXFG is not. The m_{OG} value of a reduced oligosaccharide is strongly influenced by the position(s) at which the rest of the oligosaccharide is attached to the alditol moiety, and can therefore give valuable information on sugar–sugar linkages in novel radiolabeled oligosaccharides [2, 28].

Neutral sugars can also be given an artificial positive charge by covalent derivatization, for example by addition of an *iso*-propyl 2-aminoacridone (pAMAC) group, which also renders the sugars fluorescent [16], or by reductive amination with inorganic ammonium ions, which does not [29]. These derivatives then migrate at a speed inversely related to DP (Fig. 5).

pAMAC labeling is also valuable for fluorescently tagging oligosaccharides that had been subjected in vivo to oxidation by reactive oxygen species (ROS) [30], especially the hydroxyl radical (•OH), and thus contained a glycosulose residue (mid-chain sugar unit possessing a free keto group). The pAMAC group has little effect on electrophoretic mobility at pH 6.5 (thus, pAMAC-labeled acidic oligosaccharides still migrate as anions), but always confer a net positive charge at pH 2.0.

3.10.10 Inorganic Ions

Many inorganic anions and cations migrate extremely rapidly on HVPE owing to their high charge–mass ratios, and so short run-times are recommended. Buffers of pH 2.0 and 6.5 are both valuable, often giving different electrophoretic mobilities (Fig. 8). Predicting the mobilities is complicated by the fact that inorganic cations have a hydration shell; for example, Na^+ (atomic wt 23) has a much larger hydration shell than K^+ (atomic wt 39), and therefore K^+ has the faster mobility toward the cathode despite being "heavier" according to the periodic table and having exactly the same charge (+1). Bromophenol blue is a valuable and very general indicator for inorganics, which may not be detected by the more specific stains discussed.

4 Notes

1. The applied voltage is typically 4.5 kV.
2. The effective path-length is a little less than 57 cm, the two ends of the paper being submerged in buffer.
3. HVPE can conveniently handle up to 20 samples per sheet if the samples are spot-loaded on No. 1 or 3CHR paper at, respectively, <0.3 or <1.0 μmol of total ions (including any nonvolatile salts) per spot. Nonionic compounds (e.g., glucose or glycerol) can be present in higher amounts as these do not interfere in the ionization and electrophoresis processes.
4. For preparative work, a single sample containing up to about 25 μmol total ions (equivalent to ~5 mg of galacturonic acid) can be streak-loaded per sheet of 3CHR.
5. If the use of water-immiscible coolant liquids is not feasible, for example, because hydrophobic compounds are of interest, HVPE can also be performed on a flat-bed system.
6. One potential disadvantage of this method is that nonpolar compounds (e.g., ferulic acid) might partition into the coolant and be lost off the paper. However, the great majority of metabolites of interest in cell-wall research are hydrophilic

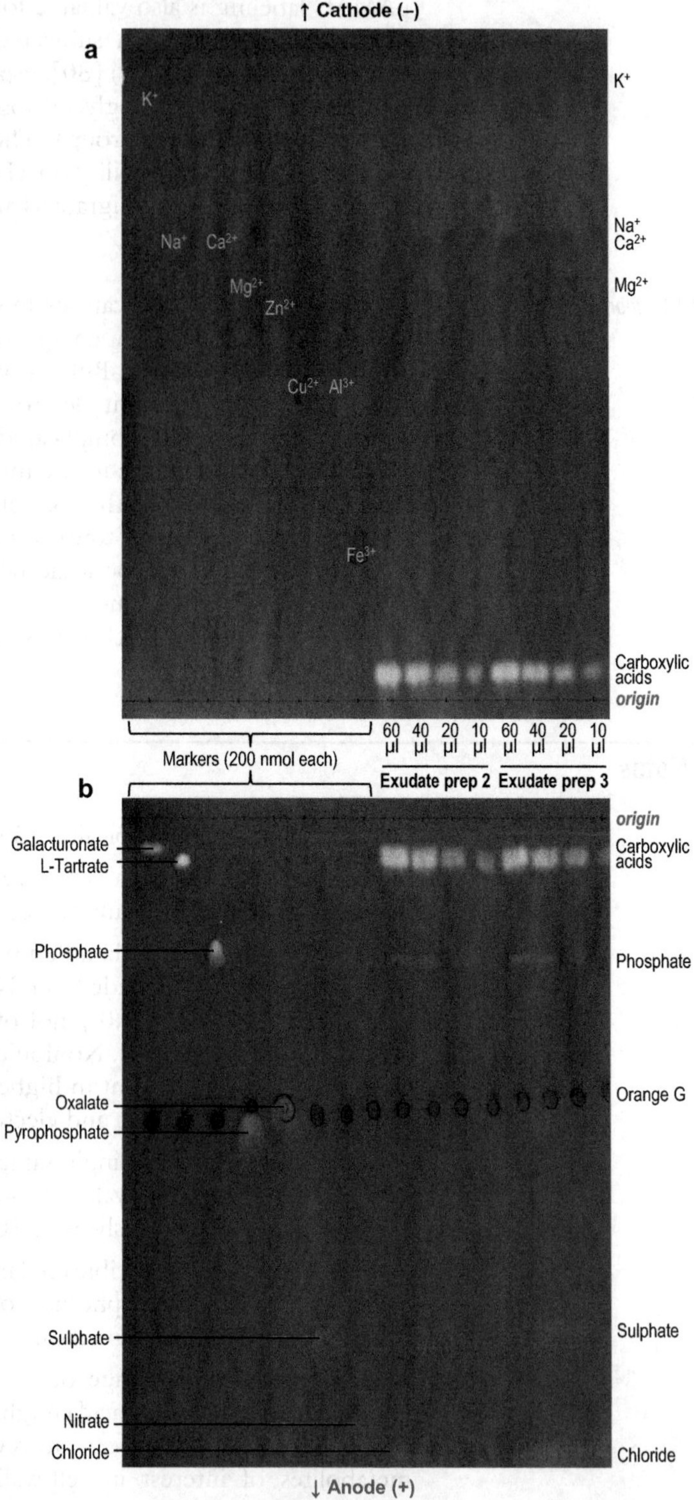

Fig. 8 HVPE of general anions and cations, especially inorganics, present in the apoplast. (**a**) Electrophoresis toward the cathode at pH 2.0 and 3.0 kV for 25 min,

enough to be retained in the aqueous buffer in the paper; even isoleucine (the most "hydrophobic" of the 20 common amino acids) and feruloylated sugars are retained.

7. In this laboratory's experience, this method provides less uniform cooling and therefore greater irregularity in the migration of replicate samples across the width of the paper.
8. The high voltage and short run-time minimize diffusion of the analytes on the wet paper.
9. This entails a considerable heating effect (e.g., at pH 2.0, power = 4500 V × 0.15 A = 675 W). The thicker 3CHR paper draws an even greater current, and the voltage must be correspondingly reduced (and the run time increased) so that the power remains at <750 W.
10. The charges borne by compounds during electrophoresis depend on the pK_a values of the ionizable groups and the pH of the buffer. Approximate pK_a values are listed in Table 2. Deductions that can be drawn from pH-dependent shifts in electrophoretic mobility, are summarized in Table 3.
11. Borate may interfere with $AgNO_3$ staining; to avoid this on electrophoretograms that have been run in borate buffer, the alkaline solution normally used for $AgNO_3$ staining should be replaced by 80% (v/v) ethanol containing 2% (w/v) NaOH and 4% (w/v) pentaerythritol.
12. Flashing and cooling are not beneficial for autoradiography.
13. Labeling with stable isotopes (e.g., ^{2}H or ^{13}C) can also be very useful to provide internal markers for quantification of phosphorylated metabolites by LC-MS/MS [31].

Acknowledgments

I thank numerous past and present colleagues and students for electrophoretic data and experience presented in this chapter and the UK BBSRC for financial support.

Fig. 8 (continued) (**b**) toward the anode at pH 2.0 and 3.0 kV for 30 min. Each sample contained a trace of Orange G as an internal marker. After dipping in bromophenol blue, cations and anions show up blue and yellow, respectively, against a blue–green background. Left half, authentic markers; right half, ions exuded by cress seeds into the apoplast during imbibition

References

1. Offord RE (1966) Electrophoretic mobilities of peptides on paper and their use in the determination of amide groups. Nature 211:591–593
2. Fry SC (2000) The growing plant cell wall: chemical and metabolic analysis. Reprint Edition. The Blackburn Press, Caldwell, NJ, p xviii + 333. ISBN 1-930665-08-3
3. Green MA, Fry SC (2005) Vitamin C degradation in plant cells via enzymatic hydrolysis of 4-*O*-oxalyl-L-threonate. Nature 433:83–88
4. Eshdat Y, Mirelman D (1972) An improved method for the recovery of compounds from paper chromatograms. J Chromatogr 65:458–459
5. Wright K, Northcote DH (1975) An acidic oligosaccharide from maize slime. Phytochemistry 14:1793–1798
6. Parsons HT, Yasmin T, Fry SC (2011) Alternative pathways of dehydroascorbic acid degradation *in vitro* and in plant cell cultures: novel insights into vitamin C catabolism. Biochem J 440:375–383
7. Parsons HT, Fry SC (2012) Oxidation of dehydroascorbic acid and 2,3-diketogulonate under plant apoplastic conditions. Phytochemistry 75:41–49
8. Popper ZA, Sadler IH, Fry SC (2003) α-D-Glucuronosyl-(1→3)-L-galactose, an unusual disaccharide from polysaccharides of the hornwort *Anthoceros caucasicus*. Phytochemistry 64:325–335
9. Smith CM, Fry SC, Gough KC, Patel AJF, Glenn S, Hawes WS, Goldrick M, Roberts IS, Whitelam GC, Andrew PW (2014) Recombinant plants provide a new approach to the production of bacterial polysaccharide for vaccines. PLoS One 9:e88144
10. Takeda T, Miller JG, Fry SC (2008) Anionic derivatives of xyloglucan function as acceptor but not donor substrates for xyloglucan endotransglucosylase activity. Planta 227:893–905
11. Iqbal A, Miller JG, Murray L, Sadler IH, Fry SC (2016) The pectic disaccharides lepidimoic acid and β-D-xylopyranosyl-(1→3)-D-galacturonic acid occur in cress-seed exudate but lack allelochemical activity. Ann Bot 117:607–623
12. Voxeur A, Fry SC (2014) Glycosylinositol phosphorylceramides (GIPCs) from *Rosa* cell cultures are boron-bridged in the plasma membrane and form complexes with rhamnogalacturonan II. Plant J 79:139–149
13. Thompson JE, Fry SC (2000) Evidence for covalent linkage between xyloglucan and acidic pectins in suspension-cultured rose cells. Planta 211:275–286
14. Sharples SC, Fry SC (2007) Radio-isotope ratios discriminate between competing pathways of cell wall polysaccharide and RNA biosynthesis in living plant cells. Plant J 52:252–262
15. Kärkönen A, Fry SC (2006) Novel characteristics of UDP-glucose dehydrogenase activities in maize: non-involvement of alcohol dehydrogenases in cell wall polysaccharide biosynthesis. Planta 223:858–870
16. Vreeburg RAM, Airianah OB, Fry SC (2014) Fingerprinting of hydroxyl radical-attacked polysaccharides by *N*-isopropyl 2-aminoacridone labelling. Biochem J 463:225–237
17. Fry SC, Willis SC, Paterson AEJ (2000) Intraprotoplasmic and wall-localised formation of arabinoxylan-bound diferulates and larger ferulate coupling-products in maize cell-suspension cultures. Planta 211:679–692
18. Piro G, Perotto S, Bonfante-Fasolo P, Dalessandro G (1988) Metabolism of D-[U-^{14}C]glucosamine in seedlings of *Calluna vulgaris* (L) Hull. J Plant Physiol 132:695–701
19. Lamport DTA (1967) Hydroxyproline-*O*-glycosidic linkage of the plant cell wall glycoprotein extensin. Nature 216:1322–1324
20. Takahashi T, Kakehi J-I (2010) Polyamines: ubiquitous polycations with unique roles in growth and stress responses. Ann Bot 105:1–6
21. Lenucci M, Piro G, Miller JG, Dalessandro G, Fry SC (2005) Do polyamines contribute to plant cell wall assembly by forming amide bonds with pectins? Phytochemistry 66:2581–2594
22. Perrone P, Hewage C, Sadler IH, Fry SC (1998) N^{α}- and N^{ε}-D-galacturonoyl-L-lysine amides: properties and possible occurrence in plant cell walls. Phytochemistry 49:1879–1890
23. Kärkönen A, Warinowski T, Teeri TH, Simola LK, Fry SC (2009) On the mechanism of apoplastic H_2O_2 production during lignin formation and elicitation in cultured spruce cells; peroxidases after elicitation. Planta 230:553–567
24. Weigel H (1963) Paper electrophoresis of carbohydrates. Adv Carbohydr Chem 18:61–96
25. Dumville JC, Fry SC (2003) Gentiobiose: a novel oligosaccharin in ripening tomato fruit. Planta 216:484–495
26. Narasimham S, Harpaz N, Longmore G, Carver JP, Grey AA, Schachter H (1980) Control of glycoprotein synthesis: the purification

by preparative paper electrophoresis in borate of glycopeptides containing high mannose and complex oligosaccharide chains linked to asparagine. J Biol Chem 255:4876–4884

27. O'Looney N, Fry SC (2005) Oxaziclomefone, a new herbicide, inhibits wall expansion in maize cell-cultures without affecting polysaccharide biosynthesis, xyloglucan transglycosylation, peroxidase action or apoplastic ascorbate oxidation. Ann Bot 96:1097–1107
28. Wende G, Fry SC (1996) 2-*O*-β-D-Xylopyranosyl-(5-*O*-feruloyl)-L-arabinose, a widespread component of grass cell walls. Phytochemistry 44:1019–1030
29. Miller JG, Farkaš V, Sharples SC, Fry SC (2007) *O*-Oligosaccharidyl-1-amino-1-deoxyalditols as intermediates for fluorescent labelling of oligosaccharides. Carbohydr Res 342:44–54
30. Airianah OB, Vreeburg RAM, Fry SC (2016) Pectic polysaccharides are attacked by hydroxyl radicals in ripening fruit: evidence from a fluorescent fingerprinting method. Ann Bot 117:441–455
31. Arrivault S, Guenther M, Fry SC, Fuenfgeld MFF, Veyel D, Mettler-Altmann T, Stitt M, Lunn JE (2015) Synthesis and use of stable isotope labelled internal standards for quantification of phosphorylated metabolites by LC–MS/MS. Anal Chem 87:6896–6904

Chapter 2

Carbohydrate Gel Electrophoresis

Florence Goubet, Paul Dupree, and Katja Salomon Johansen

Abstract

Polysaccharide analysis using carbohydrate gel electrophoresis (PACE) relies on derivatization of reducing ends of sugars with a fluorophore, followed by electrophoresis under optimized conditions in polyacrylamide gels. PACE is a sensitive and simple tool for studying polysaccharide structure or quantity and also has applications in the investigation of enzyme specificity.

Key words Oligosaccharide, Monosaccharide, Polysaccharide, Pectin, Hemicellulose, Enzyme

1 Introduction

The PACE method is both simple and robust. It involves two main steps:

1. Conjugation of a fluorophore onto the reducing end of sugar using different types of fluorophores depending of the sugars under study. In the case of highly negatively charged sugar, the fluorophore used is uncharged (e.g., *2-Aminoacridone*; AMAC) whereas in the case of partially charged or neutral sugars, the fluorophore is charged (e.g., *8-aminonaphthalene-1,3,6-trisulfonic acid*; ANTS).
2. Electrophoresis at high voltage in thin polyacrylamide gels.

PACE method can provide information on both polysaccharide structure and substrate specificity of carbohydrate active enzymes.

The main advantages of the method are

- The equipment required is not expensive and easy to use.
- The sample to be analyzed does not need any type of clean-up prior to the derivatization. Each gel will be used once so the purity of the samples is not an issue for preserving the equipment.

Zoë A. Popper (ed.), *The Plant Cell Wall: Methods and Protocols*, Methods in Molecular Biology, vol. 2149,
https://doi.org/10.1007/978-1-0716-0621-6_2,

- There is a linear relationship between molar amounts of any sugar and the intensity of the signal. This provides data for long oligomers that are superior to the data obtained with the HPLC coupled with amperimetric detection.
- PACE can be used as a separation step prior to analysis using a mass spectrometry (MS).

2 Materials

2.1 Sample Preparation

1. 0.1 M ANTS in acetic acid–water (3/17, v/v).
2. 0.2 or 1 M $NaCNBH_3$ in DMSO.
3. 50 mM AMAC in acetic acid–DMSO (1.5/18.5, v/v).
4. 0.5 M $NaCNBH_3$ in water.
5. 0.1 M $NaCNBH_3$ in water.
6. 6 M urea in water.
7. Acetic acid–water–DMSO 3/17/20 v/v/v.
8. Monosaccharide or oligosaccharide standards: 1 mM sugars (e.g., of a range of monosaccharides and oligosaccharides) 5 μL is added to a 0.1–2 mL tube (depending of the reaction sample) and dried before derivatization.
9. Polysaccharides: 0.5 mg/mL of a range of polysaccharides in buffer.
10. Enzymes (*see* **Note 1**).
11. Incubator.
12. Centrifugal vacuum evaporator.

Buffers: For Enzyme Reaction

13. 0.1 M ammonium acetate adjusted to pH 4.5–6 with glacial acetic acid.
14. 10 mM Tris–HCl, pH 7–9.
15. 0.1 mM Tris–HCl, 1 mM $CaCl_2$, pH 8.

2.2 Electrophoresis

Buffers

1. 0.1 M Tris adjusted to pH 8.2 with boric acid.
2. 0.1 M Tris–HCl, pH 8.2.
3. 0.15 M Tris adjusted to pH 8.5 with 0.15 M glycine.

Gels

4. 29:1 (w/v) polyacrylamide–acrylamide: *N*,*N*-9-methylenebisacrylamide from Bio-Rad (Hertfordshire, UK).
5. Gel for analysis of neutral oligosaccharides: 20% (w/v) polyacrylamide gel containing 0.5% (w/v) *N*,*N*-9-methylene

bisacrylamide, 0.1 M Tris–borate pH 8.2. Stacking gel for analysis of neutral oligosaccharides: 8% (w/v) polyacrylamide, 0.2% *N*,*N*-9-methylenebisacrylamide, 0.1 M Tris–borate pH 8.2.

6. Gel for analysis of acidic oligosaccharides: 25% (w/v) polyacrylamide gel containing 0.8% (w/v) *N*,*N*-9-methylenebisacrylamide, 0.1 M Tris–borate pH 8.2. Stacking gel for analysis of acidic oligosaccharides: 10% (w/v) polyacrylamide, 0.4% *N*,*N*-9-methylenebisacrylamide, 0.1 M Tris–borate pH 8.2.
7. Hoefer SE 660 vertical slab gel electrophoresis apparatus (Amersham, Buckinghamshire, UK) with 24-cm plates, 0.75-mm spacer, and well of width 0.25 cm (other equipment can also be used).
8. Microsyringes.
9. Standard glass or low-fluorescent Pyrex plates.

2.3 Gel Imaging (One of the Following Systems Can Be Used)

1. MasterImager CCD camera system (Amersham).
2. G:BOX Chemi HR16 (Syngene, Cambridge, UK).
3. Standard UV transilluminator.

2.4 Gel Analysis

1. GeneTools software (Syngene, Cambridge UK).

2.5 Extraction from a Gel

1. Nanosep system.
2. MilliQ water.
3. 1% acetic acid.
4. Dialysis tubing.

3 Methods

3.1 Sample Preparation Hydrolysis of Pure Polysaccharides Using Enzyme Preparations

Polysaccharides (0.5 mg/mL; 100 mL) are suspended with enzymes (*see* **Note 1**) in a suitable buffer at a total volume of 250 mL. The suspension is incubated at room temperature (*see* **Note 2**) for 1 min to overnight (hydrolysis could be longer but special care must be taken to prevent contamination with fungi or bacteria). Different buffers can be used with a preference of ammonium acetate, which is volatile and thus leaves no salts behind. Buffers containing amino group (e.g., Tris buffer) can increase the background since they can react with the fluorophore.

For pH 4.5–6, the buffer used is 0.1 M ammonium acetate adjusted with glacial acetic acid; for pH 7–9, the buffer is 10 mM Tris–HCl. In order of study lyase activity, buffer of pH 8 in presence of 1 mM $CaCl_2$ is added in the reaction. Controls without substrates or enzymes are performed under the same conditions to identify any unspecific compounds present in the enzymes, polysaccharides and/or labeling reagents. The reactions are stopped by boiling for 30 min (*see* **Note 3**). The samples are dried using a centrifugal vacuum evaporator.

3.2 Sample Preparation: Hydrolysis of Plant Cell Wall Materials

To study the polysaccharides architecture of plant cell walls, highly purified and well defined enzymes with known activity are used to cleave the polysaccharides into smaller fragments. The resulting oligomers can only arise from polysaccharides containing the particular type of bonds which the enzyme can recognize and cleave. The amount of released products can be quantified or the pattern of bands can be used as an indicator for the presence or absence of a particular saccharide in the cell wall sample.

The plant cell wall contains a mixture of polysaccharides in different ratios (*see* **Note 4**). To study them, different protocols could be used. For the study of polygalacturonan, which are very abundant in the cell wall, only small amounts of cell wall material is needed (50 μg; [1]) whereas for the analysis of mannan, which is a minor compound of the cell wall, considerably more cell wall material is required (0.5 mg; [2]).

The specific compounds may in some cases be inaccessible to the enzyme. For example, most of polygalacturonases are not able to cleave the glycosidic bonds of highly esterified pectin. In this case, the accessibility of the enzyme can be improved by the removal of the methyl groups by either pretreatment with pectin methyl esterase or by incubation of the cell wall material in an alkaline solution [1]. Another example is xylan, which can be esterified and/or closely bound to the cell wall. To eliminate the esterification and partially solubilize it, a highly concentrated NaOH solution can be used [3].

The samples are dried using a centrifugal vacuum evaporator.

3.3 Analysis of Neutral Oligosaccharides Derivatized with ANTS

1. Derivatization is carried out in the tubes containing dried polysaccharides, oligosaccharides, or monosaccharides. ANTS are prepared in acetic acid–water (3/17, v/v) at a final concentration 0.1 M (made freshly or stored at −20 °C for at least 6 months). $NaCNBH_3$ (0.1 M, made freshly and used immediately, toxic) is solubilized in DMSO (in a fumehood) for ANTS derivatization.
2. To each dry sample, 5 μL of ANTS solution and 5 μL of the appropriate $NaCNBH_3$ solution are added. The volume can be slightly adjusted if large quantities of cell wall material is used and a very low amount of oligosaccharides produced (e.g., to detect mannan in Arabidopsis, 0.5 mg is needed and the oligosaccharide production is very low [2]). In this case, more solvent are added with a similar ratio (acetic acid–water–DMSO; 3/17/20 v/v/v) to keep the compounds in suspension.

3. The reagents are briefly mixed (using a vortex), centrifuged and incubated at 60 °C overnight. We previously used 1 M of $NaCNBH_3$ and incubated at 37 °C; this condition is optimal to study most of oligosaccharides. However, there is a poor recovery of oligosaccharides containing glucosamine using this condition. To increase their labelling, a decrease of $NaCNBH_3$ concentration, to 0.2 M, and incubation at 60 °C is required. Using this condition, all types of oligosaccharides were derivatized similarly.
4. The solution is dried in a centrifugal vacuum evaporator for 2 h at 40 °C (avoid using high temperatures that could increase the background). The derivatized sugars are resuspended in 100 μL of 6 M urea and stored before use at −20 °C, and are stable for at least 6 months (the background signal can increase afterward).
5. Samples (0.5–4 mL depending of the sugar concentration) are loaded to the gel using microsyringes (*see* **Note 4**). In all cases, an Hoefer SE 660 vertical slab gel electrophoresis apparatus (Amersham, Buckinghamshire, UK) is used with 24-cm plates, 0.75-mm spacer, and well of width 0.25 cm. Standard glass or low-fluorescence Pyrex plates is used. Electrophoresis is performed at 10 °C in all cases to avoid any heating. The 20% (w/v) polyacrylamide gel contained 0.5% (w/v) *N*,*N*-9-methylenebisacrylamide with a stacking gel (2 cm) of 8% (w/v) polyacrylamide and 0.2% (w/v) *N*,*N*-9-methylenebisacrylamide (*see* **Note 5**); both gels are made in 0.1 M Tris–borate pH 8.2 (*see* **Note 6**). The gels are cast and cooled at least 1 day before they are to be used in order to allow complete polymerization of the acrylamide. The gel is then stored overnight at 4 °C so that it will be cold and ready to use. Smaller electrophoresis equipment can also be used but some oligosaccharides may be less well separated [4].
6. The electrophoresis buffer system (cooled at 10 °C) is 0.1 M Tris adjusted to pH 8.2 with boric acid (Tris–borate; *see* **Note 6**). The samples are electrophoresed first at 200 V for 20 min and then at 1000 V for 90 min. The buffer can be used several times (*see* **Note 7**).

3.4 Analysis of Acidic Oligosaccharides Derivatized with AMAC

1. AMAC is prepared in acetic acid–DMSO (1.5/18.5, v/v) at 50 mM final concentration (made freshly to reduce the background). $NaCNBH_3$ (0.5 M; made freshly and used immediately; toxic) is solubilized in water. To each dry sample, 5 μL of AMAC solution and 5 μL of the appropriate $NaCNBH_3$ solution were added. The reagents are mixed, centrifuged, and incubated at 37 °C overnight. The solution is dried in a centrifugal vacuum evaporator for 2 h at 40 °C. The derivatized sugars are resuspended in 100 μL of 6 M urea.

2. To have a good derivatization and a better gel visualization and when the polysaccharide studied is contained in the cell wall at a low level, more derivatization solvent (to recover the material produced) and a lower concentration of urea are used. The samples can be stored for at least 1 month at −20 °C. Long-term storage will create background in the samples.
3. Samples (0.5–4 μL depending on the sugar concentration) are loaded to the gel using microsyringes (*see* **Note 4**). In all cases, an Hoefer SE 660 vertical slab gel electrophoresis apparatus (Amersham, Buckinghamshire, UK) is used with 24-cm plates, 0.75-mm spacer, and well of width 0.25 cm. Standard glass or low-fluorescence Pyrex plates is used. The 25% (w/v) polyacrylamide gel contained 0.8% (w/v) *N*,*N*-9-methylenebisacrylamide with a stacking gel (2 cm) of 10% (w/v) polyacrylamide and 0.4% (w/v) *N*,*N*-9-methylenebisacrylamide (*see* **Note 8**); both gels are made in 0.1 M Tris adjusted to pH 8.2 with HCl (Tris–Cl). The gels are cast and cooled at least 1 day before they are used in order to allow complete polymerization of the acrylamide. The gels is then stored overnight at 4 °C ready to be used. The acrylamide percent can adjusted to study all oligosaccharides as described in Goubet et al. [5] (*see* **Note 8**).
4. The electrophoresis buffer system (cooled at 10 °C) is 0.1 M Tris–Cl as the anode reservoir buffer and 0.15 M Tris adjusted to pH 8.5 with 0.15 M glycine as the cathode reservoir buffer (*see* **Note 6**). The samples are electrophoresed first at 200 V for 20 min and then at 1000 V for 2 h. The buffer can be used different times (*see* **Note 7**).

3.5 Gel Imaging

Different systems can be used to image the gels:

(a) Gels are scanned using a *MasterImager CCD camera system* (Amersham) with an excitation filter at 400 nm and a detection filter at 530 nm. Exposure time is optimized to increase sensitivity without saturating the intense bands. An image of the gel (resolution 100 μm) can be obtained and exported in a 16-bit file to be quantified.

(b) A *G:BOX Chemi HR16* (Syngene, Cambridge, UK) can also be used. The stacking gel is removed and a small amount of water is added onto each gel prior to imaging to flatten them out and reduce the wrinkling. The gels are then transferred to a G:BOX Chemi HR16 for imaging. Since the emission peak for ANTS is 356 nm and for AMAC is 420 nm the gels are imaged using short wavelength UV, with UV and short pass emission filters and without neutral fielding (http://www.syngene.com/PACE_Intl_Labmate_Article.pdf).

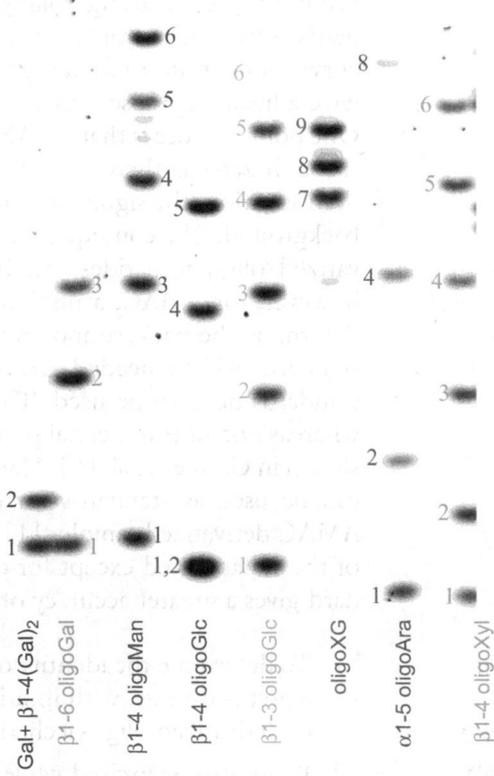

Fig. 1 Different oligosaccharides were derivatized with ANTS and separated in polyacrylamide gel. There is oligosaccharide separation by the composition, size and also by the glycosidic bond. Xyloglucan oligosaccharides (XXXG (DP 7), XXLG/XLXG (DP8) and XLLG (DP 9)), where G is an unsubstituted glucose residue, X is a xylose-substituted glucose residue, and L is a galactosylxylose-substituted glucose residue [8]. The number close to each band is the DP for each of the oligosaccharides—color coded by type of oligosaccharide

(c) The gel could also be visualized using a *standard UV transilluminator* (wavelength 360 nm), but this has been found to be less sensitive than the use of the MasterImager, particularly in the case of ANTS derivatization [6, 7].

3.6 Oligosaccharide Separation

All oligosaccharides appear as one clear band (Fig. 1) except for glucosamine and oligoglucosamines (oligochitosans). For these last compounds, each sugar is represented by two bands (data not shown). Compounds that cannot be separated are cellobiose and glucose as shown in Fig. 1. However, using different polyacrylamide conditions, those two compounds can be separated [9].

3.7 Gel Analysis

Quantification is performed using GeneTools software (Syngene, Cambridge UK), using rolling ball background detection. Standards (single or multiple) are run in each gel to obtain a standard curve for quantification of sugars in the samples. Derivatized sugars have a linear response between the concentration and signal level. One point to note is that for ANTS derivatization, the signal can go through zero as there is low background; however, with AMAC derivatization, the signal cannot go through zero due to higher gel background. The consequence is that quantification of ANTS derivatized oligosaccharides can be done using only one standard; however, for AMAC, a minimum of two standards is required to determine the background level. For accuracy in both cases, more standards will be needed. To obtain accurate quantification, pure standards need to be used. The monosaccharides are highly pure whereas not all commercial preparations of oligosaccharides are, as shown in Goubet et al. [6]. Man and (Man)3 or GalU and (GalU)3 can be used as standards for quantification for either ANTS or AMAC derivatized samples [1]. The band intensity is independent of the sugar tested except for oligochitosan. More than one standard gives a greater accuracy of quantification.

3.8 Extraction of Derivatized Oligosaccharides and MS Analysis

1. To determine the identity of oligosaccharides in specific bands, a preparative gel with multiple adjacent lanes loaded with 6 μL of derivatized oligosaccharides is prepared.
2. Bands can be excised while briefly viewing the gel under a UV transilluminator (wavelength 360 nm), and suspended in 1 mL of milliQ water.
3. To extract the oligosaccharides, acrylamide gel slices are partially crushed and subjected to three cycles of freeze-thawing. In the case of purification of esterified oligosaccharides, the pH needs to be slightly acid to reduce any demethylesterification process [10], so acetic acid solution (1%) replaces water to extract the compounds.
4. To eliminate any polyacrylamide fragments, the samples are filtered using the Nanosep system (MWCO of 10,000 Da; Pall, East Hills, NY) and then dialyzed against water using dialysis tubing (MWCO of 500 Da—note that derivatized monosaccharides will be lost).
5. The solution is dried and the pellet is suspended in 10–20 μL of water. Aliquots of the specific derivatized Me-OGA samples are loaded onto a gel to estimate their quantity.
6. These samples are then analyzed by different MS configurations as described by Goubet et al. [10].

4 Results

PACE can resolve many carbohydrates from each other based on their size and composition and function of the glycosidic bond (Fig. 1). Already in monosaccharide form, some isomers can be separated from each other (e.g., glucose (Glc) from galactose (Gal)). The oligosaccharides can also be separated from each other depending on their composition. For example, a dimer of arabinose (Ara) is clearly separated from the dimer of xylose (Xyl); and in this case both monosaccharides are pentose. Another example is that dimer of Gal which is separated from the dimer of mannose (Man) and in this case, the monosaccharide unit for both dimers is a hexose.

The glycosidic bond can also play a role in their separation. In the example given in Fig. 1 a β-1-3 oligoglucan can be separated from a β-1-4 oligoglucan with the same degree of polymerization. Similar observations are made for large oligosaccharides but to achieve better separation, different compositions of polyacrylamide gels need to be used. The same oligosaccharides with substitutions in different places on the backbone can also clearly be separated from each other as shown in Goubet et al. [10].

Similar observations were made studying charged oligosaccharides as described in Goubet et al. [5] and shown in Fig. 2. Carrageenans are polysaccharides of repeating disaccharide units of 3-linked β-D-galactopyranose and 4-linked α-D-galactopyranose. Three main structural groups are defined based on the presence or absence of α-D-anhydrogalactose in place of α-D-galactopyranose and on the position of sulphate groups as described in Liners et al. [11]. Oligosaccharides of iota- and kappa-carrageenans (kindly provided by F. Liners and P. Van Cutsem, University of Namur, Belgium) were analyzed by PACE. As previously shown, charged oligosaccharides have a size-dependent "turning point" [5]. As a consequence of this turning point, some large oligosaccharides could migrate to similar positions as smaller ones. In Fig. 2, two kappa-carrageenans with degree of polymerization (DP) of 4 and 6 respectively comigrated under the conditions used. To separate them, different polyacrylamide conditions can be used [5]. The two types of carrageenan are separated from each other so that PACE can be very useful to study carrageenan structure and also some slight changes of structures as already been shown for the methylation of polygalacturonan [10].

Recently, separation of saturated oligosaccharides versus unsaturated ones has also been shown possible using PACE technology [12]. This technique could be used for rapid screening of the mode of action of hydrolase and lyase activities.

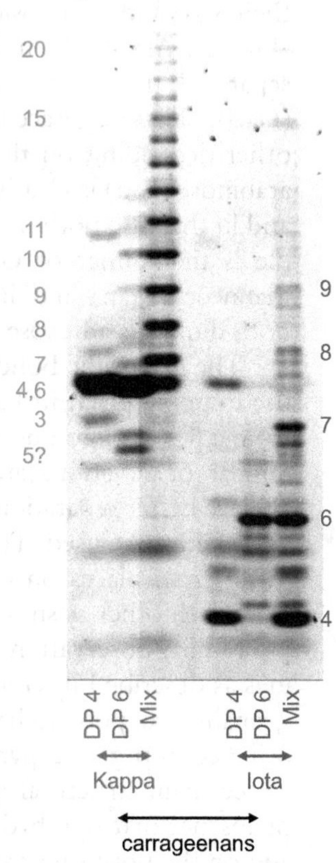

Fig. 2 Analysis of kappa- and iota-oligocarrageenans by PACE—derivatization was carried out using the AMAC fluorophore. The oligosaccharides separated by the size and the composition. The number close to each band is the DP of each oligosaccharide. "Mix" is a partial hydrolysis of the corresponding polysaccharides. Please note that the standards contain additional compounds than the ones indicated. For example, the kappa-carrageenan DP4 contains also two minor compounds that one has been identified as kappa-carrageenan DP3. The other one is unknown and could be a DP2. The number close to each band is the DP for each oligosaccharide—colour coded by type of oligosaccharide. The position of the kappa-penta-carrageenan (DP5) has been indicated as a possible but it has not confirmed

5 Conclusion

PACE is a versatile method for the separation and detection of any kind of carbohydrate with a reducing end. The power to separate each carbohydrate coupled with its high sensitivity makes PACE an attractive alternative to HPLC, TLC, and MS-based analysis.

Using this method, polysaccharide structure and quantity and enzyme characteristics can be obtained using simple equipment. Furthermore, PACE has successfully been combined with mass spectrometry of oligosaccharides [13, 14].

6 Notes

1. If the enzymes are contained in a mixture, pure polysaccharides are required to study the enzyme characteristics as described in Phalip et al. [15]. Many polysaccharides and oligosaccharides are available from Megazyme (http://www.megazyme.com/) and Sigma Aldrich (http://www.sigmaaldrich.com/sigma-aldrich/home.html) to study the enzyme specificity.
2. Higher temperatures may be appropriate depending on the properties of enzymes used (e.g., their thermostability [16]).
3. To avoid any product modification (e.g., degradation by some enzymes) during derivatization, the reaction is stopped. This will induce protein precipitation which is not an issue for PACE analysis except during loading. To avoid any problems with the microsyringe, suspend the compounds fully before loading.
4. Cell wall mass is accurately measured by using a cell wall suspension (i.e., 0.5 mg/mL; homogenized using a glass potter) and an aliquot is taken for the analysis. Before taking an aliquot, the suspension is well mixed since the cell wall easily sediments.
5. For acrylamide polymerization, TEMED and APS solutions are used. TEMED solution can be bought ready-to-use and can be stored at room temperature for at least 1 year. A solution of APS can be stored for a couple of months at 4 °C, but if polymerization starts to take a longer time, a fresh solution should be made.
6. A 10× stock solution of the following buffers can be made and diluted when needed. The Tris–glycine stock solution can be stored at 4 °C for at least 6 months. The Tris–HCl and Tris–borate solutions can be stored at room temperature for at least 6 months. Borate can precipitate if stored for too long or if the storage temperature is too cold.
7. The running solutions can be used several times. Borate salt can precipitate which leads to deteriorating electrophoresis materials. A sign of this is that the longer the electrophoresis takes to run, the less efficient the separation becomes and the higher the background. Prepare a fresh solution if this situation occurs.
8. To study charged oligosaccharides, 1–31% (w/v) polyacrylamide gel contained 0.5–1.1% (w/v) *N*,*N*-9-methylenebisacrylamide can be used (higher acrylamide percent can also be used) [5].

References

1. Barton CJ, Tailford L, Welchman H, Zhang Z, Gilbert HJ, Dupree P, Goubet F (2006) Enzymatic fingerprinting of Arabidopsis pectic polysaccharides using PACE-polysaccharide analysis by carbohydrate gel electrophoresis. Planta 224:163–174
2. Handford MG, Baldwin TC, Goubet F, Prime TA, Miles J, Yu X, Dupree P (2003) Localisation and characterisation of mannan cell wall polysaccharides in *Arabidopsis thaliana*. Planta 218:27–36
3. Brown DM, Goubet F, Wong VW, Goodacre R, Stephens E, Dupree P, Turner SR (2007) Comparison of five xylan synthesis mutants reveals new insight into the mechanisms of xylan synthesis. Plant J 52:1154–1168
4. Karousou E, Porta G, De Luca G, Passi A (2004) Analysis of fluorophore-labelled hyaluronan and chondroitin sulfate disaccharides in biological samples. J Pharm Biomed Anal 34:791–795
5. Goubet F, Morriswood B, Dupree P (2003) Analysis of methylated and unmethylated polygalacturonic acid structure by PACE: polysaccharide analysis using carbohydrate gel electrophoresis. Anal Biochem 321:174–182
6. Goubet F, Jackson P, Deery M, Dupree P (2002) Polysaccharide analysis using carbohydrate gel electrophoresis (PACE): a method to study plant cell wall polysaccharides and polysaccharide hydrolases. Anal Biochem 300:53–68
7. Fliegmann J, Mithöfer A, Wanner G, Ebel J (2004) An ancient enzyme domain hidden in the putative β-glucan elicitor receptor of soybean may play an active part in the perception of pathogen-associated molecular patterns during broad host resistance. J Biol Chem 279:1132–1114
8. Fanutti C, Gidley MJ, Reid GJS (1996) Substrate subsite recognition of the xyloglucan endo-transglycosylase or xyloglucanspecific endo-(1→4)-β-D-glucanase from the cotyledons of germinated nasturtium (*Tropaeolum majus* L.) seeds. Planta 200:221–228
9. Jackson P (1994) The analysis of fluorophore labeled glycans by high-resolution polyacrylamide gel electrophoresis. Anal Biochem 216:243–252
10. Goubet F, Ström A, Quéméner B, Stephens E, Williams MAK, Dupree P (2006) Resolution of the structural isomers of partially methylesterified oligogalacturonides by polysaccharide analysis using carbohydrate gel electrophoresis. Glycobiology 16:29–35
11. Liners F, Helbert W, Van Cutsem P (2005) Production and characterization of a phage-display recombinant antibody against carrageenans: evidence for the recognition of a secondary structure of carrageenan chains present in red algae tissues. Glycobiology 15:849–860
12. Phalip V, Goubet F, Carapito R, Jeltsch J-M (2009) Plant cell wall degradation with a powerful *Fusarium graminearum* enzymatic arsenal. J Microbiol Biotechnol 19:573–581
13. Tryfona T, Liang HC, Kotake T, Tsumuraya Y, Stephens E, Dupree P (2012) Structural characterization of Arabidopsis leaf arabinogalactan polysaccharides. Plant Physiol 160 (2):653–666
14. Bromley JR, Busse-Wicher M, Tryfona T, Mortimer JC, Zhang Z, Brown DM, Dupree P (2013) GUX1 and GUX2 glucuronyltransferases decorate distinct domains of glucuronoxylan with different substitution patterns. Plant J 74(3):423–434
15. Phalip V, Delalande F, Carapito C, Goubet F, Hatsch D, Leize-Wagner E, Dupree P, Dorsselaer AV, Jeltsch JM (2005) Diversity of the exoproteome of *Fusarium graminearum* grown on plant cell wall. Curr Genet 48:366–379
16. Palackal N, Brennan Y, Callen NC, Dupree P, Frey G, Goubet F, Hazlewood GP, Healey S, Kang YE, Kretz KA, Lee E, Tan X, Tomlinson GL, Verruto J, Wong VWK, Mathur EJ, Short JM, Robertson DE, Steer BA (2004) An evolutionary route to xylanase process fitness. Protein Sci 13:494–503

Chapter 3

Capillary Electrophoresis with Detection by Laser-Induced Fluorescence

Andrew Mort and Xiangmei Wu

Abstract

Capillary zone electrophoresis (CZE) has been rising in its importance in structural analysis of plant cell walls, characterization of enzymes that degrade polysaccharides, and profiling of oligosaccharides to characterize cell wall mutants. CZE with laser-induced fluorescence detection provides high separation efficiencies, high-speed analysis, with extremely small sample requirements. Here we describe the instrumentation we use and methods for attaching fluorescent labels to oligosaccharides so that they can be detected.

Key words Capillary zone electrophoresis (CZE), Laser induced fluorescence detection, Oligosaccharide, Separation, Profiling

1 Introduction

Plant cell walls contain polysaccharides with complex structures such as xyloglucans and pectins which vary somewhat from species to species, from cell type to cell type, and even at different locations within the wall around a single cell [1] With the fairly recent availability of pure enzymes [2, 3], one can hydrolyze these polymers into oligomers which can then be separated and characterized.

The oligomers can be labeled at their reducing ends by reacting them with a fluorescent amine at the aldehyde group to form an imine which can be selectively reduced with sodium cyanoborohydride to a stable secondary amine. We use two different aromatic trisulfonated amines; 8-aminonaphthalene-1,3,6-trisulfonate (ANTS), and 8-aminopyrene-1,3,6-trisulfonate (APTS). ANTS is quite inexpensive and in our hands gives few interfering peaks on the CZE trace. However, ANTS absorbs in the UV range and so, requires a UV laser for excitation. APTS is about 100 times as expensive and gives more interfering signals just from the reagent. It absorbs blue light and fluoresces green so there is little chance of interfering signals from plant compounds. The sensitivity of

Zoë A. Popper (ed.), *The Plant Cell Wall: Methods and Protocols*, Methods in Molecular Biology, vol. 2149,
https://doi.org/10.1007/978-1-0716-0621-6_3,

detection for APTS labeled oligomers excited with an Argon-Ion laser at 448 nm is about 50 times greater than that of an ANTS-labeled oligomer excited by a 325 nm Helium Cadmium laser [4]. The pyrene ring seems to make the APTS "sticky." Thus, the capillary needs cleaning frequently to avoid traily peaks.

Both ANTS and APTS have three negative charges on them which ensure that all labeled oligomers have a charge of −3. We use uncoated capillaries for the electrophoresis with a fairly high ionic strength buffer at a pH of 2.5. This low pH ensures that silicic acid groups on the surface of the capillary are protonated so there are no fixed charges and hence no electro-osmotic flow. A pH of 2.5 also causes almost complete protonation of uronic acids, thus both neutral and acidic oligosaccharides should only be charged because of the three sulfonic acid groups on the fluorescent amine label. Since all oligomers will have the same net charge, the driving force of the electric field should be the same for all oligomers. The frictional force on the oligomers will be the product of the velocity of the oligomer times its frictional coefficient. According to Stokes law the frictional coefficient $f = 6\pi\eta r$ where η is the viscosity of the medium and r is the Stokes radius of the oligomer. Thus the larger the hydrodynamic radius of the oligomer the slower it will move in the electric field. Electrophoresis of the labeled oligomers therefore allows one to distinguish them according to their size.

Using these methods one can profile the products of an enzyme digest of cell wall polymers to follow the progress of the digestion, or can compare the profile of products between wild type and potential cell wall mutants. Li et al. [5] have used an ABI 3730xl DNA sequencer with its 96 capillary array for profiling differences in structure between wild type and various mutants in xylan biosynthesis.

If one labels a pure oligosaccharide it can then be used to characterize the mode of action of an enzyme, or to detect the presence of enzymes that act on the oligomer, even in the presence of a very complex medium. For example we labeled an oligomer from a partial acid hydrolysate of rhamnogalacturonan and used this to detect and characterize the mode of action of rhamnogalacturonan lyase in intact cotton plants [6].

2 Materials

2.1 Preparation of Aminopyrene Trisulfonate (APTS)-Labeled Oligomers for Enzyme Characterization

1. 1 mg oligosaccharide in 20 μL dH_2O: purified oligomers such as xylohexaose (Megazyme International Ireland Ltd., Bray Business Park, Bray, Co. Wicklow, Ireland) should be used (*see* **Notes 1** and **2**).
2. 100 mM 8-aminopyrene-1,3,6-trisulfonate (APTS) (Molecular Probes, Carbosynth, or Sigma) in 25% acid (*see* **Note 3**).

3. 1 M sodium cyanoborohydride in dimethylsulfoxide (*see* **Note 4**).
4. 50 mM ammonium acetate buffer, pH 5.2: HPLC-grade acetonitrile 3:1 v/v, briefly degassed with a water aspirator.
5. Chromatography materials: HW-40S gel filtration material column (Toyopearl, Supelco) packed in a 100 × 10 mm stainless steel column.
6. Heating block or other reliable heat source set at 80 °C with holes suitable for 500 μL microfuge tubes.
7. Vortex.
8. Microfuge.
9. Speed vac concentrator.
10. Spectrophotometer.

2.2 Labeling of Digested Polysaccharides with Aminonaphthalene Trisulfonate (ANTS)

1. Polysaccharide solution: 10 mg/mL.
2. Internal standard (optional): 1 μg/μL cellobiose or maltose solution.
3. Appropriate buffer for the enzyme.
4. 4 mU of the appropriate enzyme.
5. Heating block.
6. Pipette.

Labeling Reagents

7. 23 mM 8-aminonaphthalene-1,3,6-trisulfonate (ANTS) (Molecular Probes) in 3% w/v acetic acid.
8. 1 M solution of sodium cyanoborohydride in dimethylsulfoxide.
9. 500 μL microfuge tubes.
10. Vortex.
11. Microfuge.

2.3 Determination of Enzyme Activity In Vitro

1. 1:100 dilution of stock APTS-labeled substrate prepared as below.
2. 1 μU of an appropriate hydrolytic enzyme.
3. Spectrophotometer.
4. Heating block.
5. Standard mixture of oligomers prepared either by specific enzyme digestion or partial TFA hydrolysis of an appropriate polysaccharide; for example, for determination of xylanase activity the oligomers should be generated from xylan.

2.4 In Situ Detection of Enzyme Activity Using APTS

1. 6 pmol/μL APTS-labeled substrate, prepared as described below.
2. Gas-tight syringe: 10 μL (1701 RNFS Hamilton Co., Reno, NV, USA) fitted with a needle made of a 15-cm section of 0.17-mm o.d. fused silica capillary (Alltech Associates, Inc., Deerfield, IL, USA).
3. Plant of interest.
4. Temperature-controlled plant growth chamber.
5. 125 mL Erlenmeyer flask.
6. Extracting solvent: 25 mM sodium acetate buffer, pH 5.2, ice-cold.
7. Water aspirator.
8. Paper towels.
9. Tissue.
10. 2-mL Reacti-Vial (Supelco, Inc., Bellefonte, PA, USA).
11. Centrifuge.

2.5 Capillary Electrophoresis Columns and Buffers for APTS and ANTS-Labeled Oligomer Separation

Capillary Columns

1. Column for APTS-labeled oligomer separation: A fused-silica capillary (TSP050375, Polymicro Technologies, http://www.polymicro.com) of internal diameter 50 μm and length 31 cm with a 3–4 mm window burned into the plastic coating 5.7 cm from the anode end. The window is formed by resting the capillary on a glass window etching device (pEZ-Window, J & W Scientific, Folsom, CA, USA) on a heating plate and covering the capillary at that point with a drop of concentrated sulfuric acid. The capillary is carefully inserted into the plastic tube for circulation of the coolant and assembled into the cartridge as instructed by the manufacturer (*see* **Note 5**).
2. Column for ANTS-labeled oligomer separation: A fused-silica capillary (TSP050375, Polymicro Technologies, http://www.polymicro.com) of internal diameter 50 μm and length 50 cm with a 3–4 mm window burned into the plastic coating at 30 cm from the anode end as described above.
3. Running buffer: 0.1 M sodium phosphate, pH 2.5. This buffer is made by slowly adding 0.1 M sodium monophosphate to 0.1 M phosphoric acid until the pH just reaches 2.5 (*see* **Note 6**).
4. Rinsing buffer: 1 M NaOH.

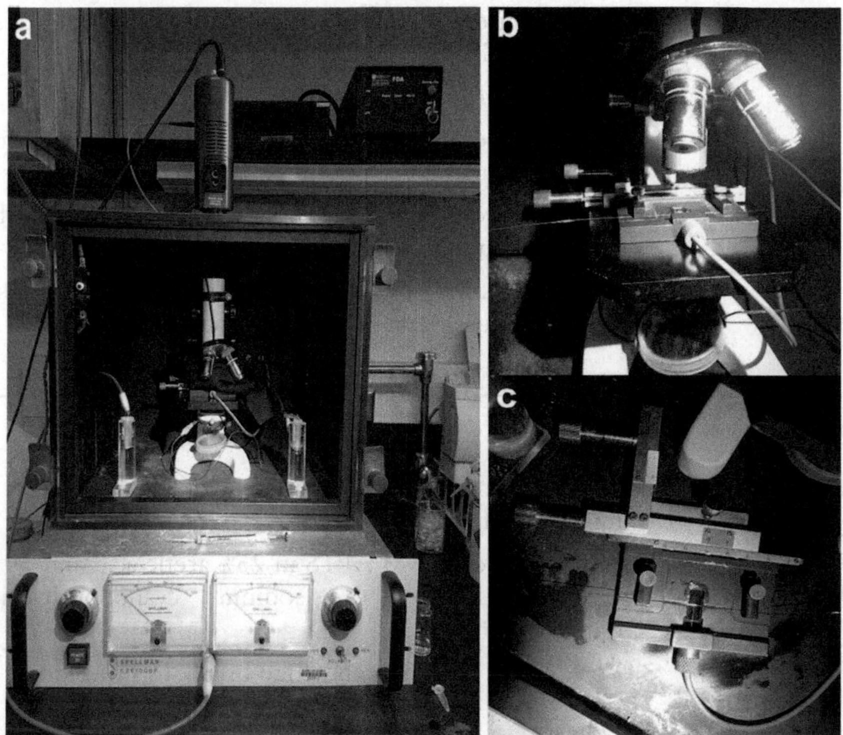

Fig. 1 Pictures showing the major components of the custom built CZE apparatus. (**a**) The complete setup excluding the computer and controller boxes for the attenuator and camera. Upper right, diode pumped laser. Upper centre intensified CCD camera mounted on top of an inexpensive microscope. Middle left, microswitch to ensure high voltage is disabled if the door to the safety and light excluding enclosure is not in place. Center, microscope. Lower center left and right, plastic holders for the 1.5 mL microfuge vial electrode chambers. Bottom, high-voltage power supply. (**b**) Close up view of the holder and X-Y positioner for alignment of the capillary under the microscope objective. (**c**) Close up of the capillary and the fibre optic held in a stainless steel cannula attached to an eccentric brass nut for movement of the fibre up and down so that it can be aimed directly at the center of the capillary

2.6 Capillary Electrophoresis System for ANTS-Labeled Oligosaccharides

A custom built CZE system consisting of a high voltage power supply, helium cadmium laser, optical system built on an inexpensive microscope, an intensified CCD camera for light detection, and a computer-controlled variable-light-attenuator. A complete description of the system is available [7], (Fig. 1). (Our helium cadmium laser has ceased to function and the manufacturer, Omnichrome, taken over by Melles Griot, no longer supports the laser. Thus, we have switched to a 5 mW Diode-Pumped Solid-State 355 nm laser from Laserglow Technologies, Toronto, Canada).

2.7 Capillary Electrophoresis System for APTS Labeled Oligosaccharides

1. Instrument: Biofocus-2000 (Bio-Rad laboratories, http://www.bio-rad.com) Capillary zone electrophoresis apparatus with laser-induced fluorescence detection. Unfortunately this instrument is no longer available. The P/ACE MQD system (Beckman Coulter, now part of Sciex) can be used in almost the same manner as described here.

2. Gas pressure: for injection 4.5 lb/in^2 of helium pressure for 0.22 s and for rinsing 80 lb/in^2.
3. Excitation: 488 nm light from a 5 mW argon ion laser.
4. Fluorescence emission collection: emission collected through a 520 nm narrow band pass filter.
5. Electrophoresis conditions: 15 kV/70–100 μA with the cathode at the inlet; controlled temperature of 20 °C.
6. Data processing: BioFocus 2000 System operating software and BioFocus system integration software.

3 Methods

3.1 Preparation of Aminopyrene Trisulfonate (APTS)-Labeled Substrate

1. Dissolve 1 mg of oligosaccharide in 20 μL of water in a 500-μL microcentrifuge tube.
2. Add 50 μL of 0.1 M APTS in 25% acetic acid and 50 μL of a 1 M solution of sodium cyanoborohydride in dimethylsulfoxide.
3. Vortex-mix well and centrifuge briefly to bring down the liquid, cap the vial tightly.
4. Heat for 60 min at 80 °C.
5. After the mixture has cooled to room temperature, dilute to about 200 μL with water and pass through a Toyopearl HW-40S (100 × 10 mm) gel filtration column eluted with 25% acetonitrile and 75% 50 mM ammonium acetate buffer v/v, pH 5.2 at 1 mL/min (*see* **Note 7**). The labeled substrate elutes at around 5–6 min and the salts and excess labeling reagents around 10 min. The fractions containing the labeled substrate and free label can be detected using a fluorescence or visible detector. However, at this scale one can observe by eye where each fraction is.
6. Pool the labeled substrate fractions and evaporate them to dryness in a speed vac concentrator.
7. Redissolve the labeled oligomer in 100 μL of water and store frozen.
8. The concentration of APTS labeled substrate can be determined based on the extinction coefficient of 17,160 M/cm at 455 nm [4].

3.2 Determination of the Mode of Action of an Enzyme

1. Prepare an appropriate labeled substrate as described above (*see* **Note 8**).
2. Estimate the amount of enzyme needed to degrade the substrate over a period of several hours.

3. Incubate an amount of substrate, which will give a total fluorescence intensity of at least 10–50 RFU with the enzyme in an appropriate buffer, taking aliquots after various lengths of time and stopping the reaction by heating at 80 °C for 10 min. For xylohexaose 1 μL of a 1:100 dilution of the stock solution of labeled substrate in a 25-μL incubation with around 1 μU of enzyme is about right.
4. Generate a standard mixture of labeled oligomers for comparison with the enzyme-produced products; for example, for xylanase take a xylan and digest it for a short time with a xylanase, or hydrolyze it for a short time in trifluoroacetic acid and then label the products as described above but using 1 μL 0.1 M APTS, 10 μL sample, and 10 μL cyanoborohydride solution (*see* **Note 9**).

3.3 Detection of Enzyme Activity in a Complex Medium Such as an Intact Plant Using APTS

Since APTS absorbs blue light and fluoresces green light, the CZE detection system is oblivious to all imaginable plant components. Thus if one adds an APTS labeled substrate to a complex system and then analyses the result using CZE with laser induced fluorescence detection only the undigested substrate and any degradation products which still contain the label will be visible.

1. Prepare an appropriate labeled substrate and dilute it so that it has a concentration of about 30 pmol per 5 μL.
2. Inject 5 μL of the solution into the intercellular spaces of an intact cotton cotyledon using a gas-tight 10 μL-syringe fitted with a needle made of a 15-cm section of 0.17-mm o.d. fused silica capillary.
3. Return the plant to the growth chamber for the desired incubation time.
4. Cut the cotyledon from the plant and place it in an Erlenmeyer flask containing about 30 mL of ice-cold extracting solvent (25 mM sodium acetate buffer, pH 5.2).
5. Apply vacuum for 2 min from a water aspirator.
6. Release the vacuum to cause infiltration of the cotyledon's intercellular spaces by the extracting solvent.
7. Repeat the evacuation and vacuum release two to three times.
8. Transfer the cotyledon to paper towels and blot with tissue.
9. Roll the cotyledon in a conical shape and put into a 2-mL Reacti-Vial reaching only half way to the bottom to avoid contact of the cotyledon with the intercellular wash fluid during centrifugation.
10. Centrifuge at 1500 × *g* for 15 min. About 0.3 mL of intercellular wash fluid per cotyledon will be collected.
11. Analyze the products by CZE and identify them by comparison to standard labeled oligomers.

3.4 Separation of APTs-Labeled Compounds on Capillary Zone Electrophoresis (CZE)

1. Load the samples, running buffer, rinsing buffer, NaOH, and water in the inlet carousel according to the "configuration" stored in the computer memory. Load a vial for waste and running buffer in the outlet carousel according to the "configuration" to be used.
2. Select a "method." For following degradation of xylohexaose we use:

   Inlet buffer: 0.1 M phosphate, pH 2.5.

   Outlet buffer: 0.1 M phosphate, pH 2.5.

   Cartridge temperature 20 °C.

   Inverse polarity.

   Voltage 15 kV.

   Current limit 100 μA.

   Run time 8 min.

   Prerun steps: high pressure rinse with NaOH for 30 s.

   High pressure rinse with wash buffer for 60 s.

   Inject 2 psi/s.

   Water dip to prevent carryover between samples.

   Run.

3.5 Profiling and Time Course Analysis of Oligosaccharides Produced by Enzyme Degradation of Polysaccharides

The range of products produced by digestion of a polysaccharide substrate with enzymes can be followed over time by taking small aliquots of the digestion mixture at suitable time points and derivatizing them with ANTS for subsequent analysis by CZE. To make the time course relatively quantitative one can add a constant amount of a commercially available disaccharide such as maltose or cellobiose as an internal standard for comparison of peak heights or areas.

3.5.1 Labeling with Aminonaphthalene Trisulfonate (ANTS)

1. Incubate 25 μL of a 10 mg/mL solution of the polysaccharide in an appropriate buffer with 4 mU of enzyme at the temperature optimum for the enzyme.
2. Take duplicate 1 μL aliquots at 0 time, 15 min, 30 min, 1 h, 2 h, and 4 h.
3. Put the aliquots in 500 μL microfuge tubes along with 1 μL internal standard, 1 μL cyanoborohydride solution, and 10 μL ANTS solution. Mix well and then centrifuge briefly to collect the liquid in the tip of the tube.
4. Heat for 1 h at 80 °C.

3.5.2 Separation of ANTS-Labeled Compounds by Capillary Zone Electrophoresis (CZE)

1. Turn on power to the laser, the camera, the attenuator controller, and the serial to parallel interface box.
2. Start the CZE data collection program on the computer.

3. Set the attenuation to 1 and view the image from the camera.
4. Draw a rectangle around the area that should be used for detection of the fluorescence.
5. Move the rectangle to the top right hand corner of the image and click on the background button.
6. Move the rectangle back over the image of the lumen of the capillary.
7. Designate a file name and path for data storage.
8. Rinse the capillary with running buffer, or with NaOH and then running buffer to remove contaminants from the walls of the capillary. We use a 250 μL gas tight syringe with a replaceable blunt ended needle adapted to press fit onto the capillary with a 1 cm piece of tubing fitting over the needle and a piece of Teflon tubing with ID of 360 nm pushed inside it.
9. Inject the sample by dipping the inlet end of the capillary into the microfuge tube and elevating it 15 cm above the outlet end of the capillary.
10. After 6 s lower the inlet to the same level as the outlet and quickly transfer the inlet end of the capillary to the cathode well (a 1.5 mL microfuge tube with two holes poked in its lid with a push pin, one for the inlet platinum wire electrode and the other for the capillary).
11. Turn on the high voltage to 18 kV with the negative electrode at the inlet. The current should be about 60 μA.
12. Start the run on the computer.
13. After the desired run time turn of the high voltage and click stop on the computer. The data will automatically be saved.

Figure 2 shows an example of the time course of reaction of endopolygalacturonase with pectic acid.

4 Notes

1. Megazyme has a selection of purified oligomers from plant polysaccharides as do Carbosynth (8 & 9 Old Station Business Park, Compton, Berkshire, RG20 6NE, UK) and Dextra Laboratories Ltd. (Science and Technology Centre, Earley Gate, Whiteknights Road, Reading, RG6 6BZ, UK).

   Many potential substrate oligosaccharides are not commercially available, so must be generated and purified by the individual investigator.

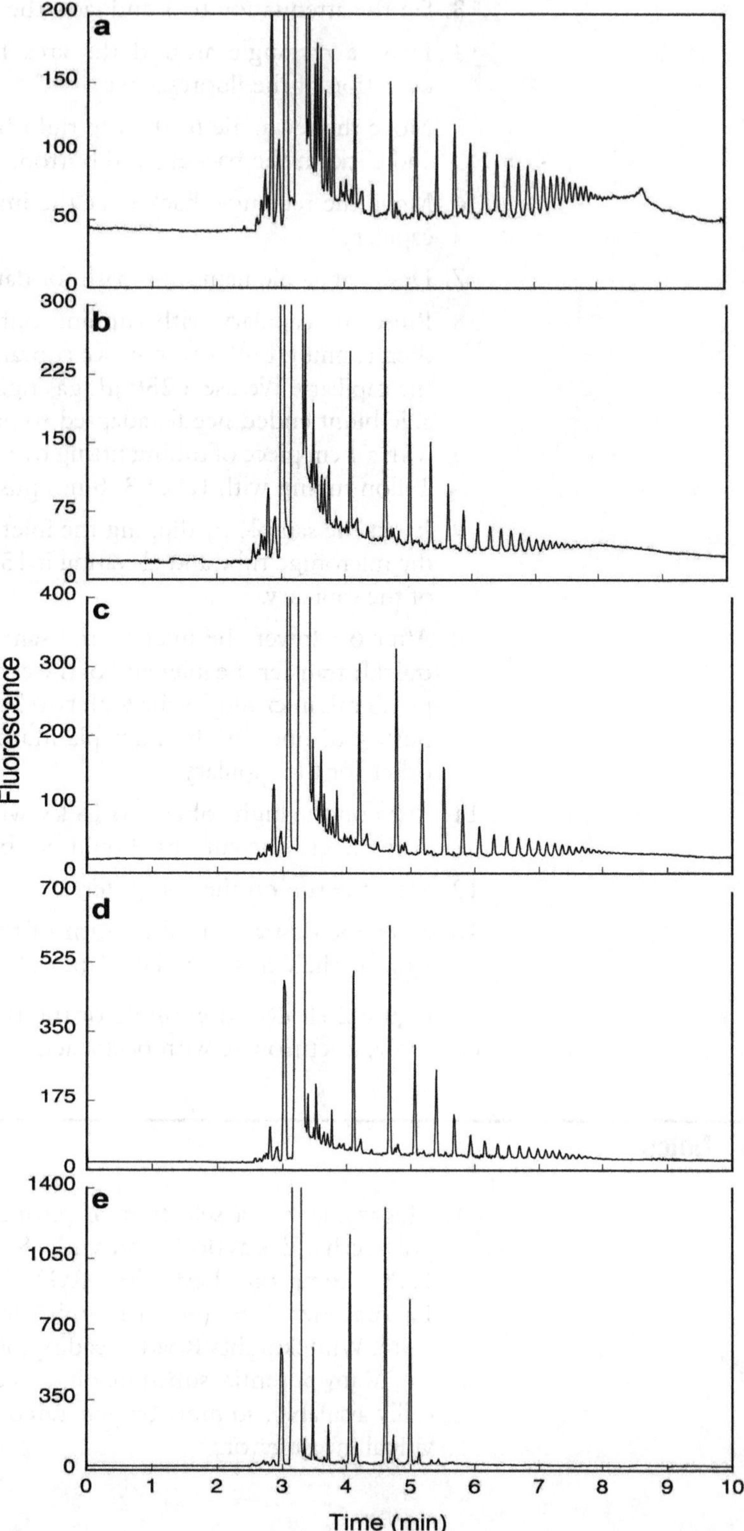

Fig. 2 A time course of degradation of pectic acid with endopolygalacturonase obtained from a *Pichia* clone expressing open reading frame AN 8327 [2]. (**a**) No

2. An example of purification of rhamnogalacturonan oligomers for investigation of rhamnogalacturonase and rhamnogalacturonan lyase is given in [6].
3. We add 200 μL of 25% acid to the 10 mg vial we receive the reagent from the manufacturer in to avoid the inevitable losses that would occur during transfer of the powder.
4. Use caution when preparing and handling this solution. The cyanoborohydride breaks down slowly to produce HCN and dimethylsulfoxide can carry dissolved substances through skin.
5. We have found the manufacturer's cartridges to be flimsy. They break easily when you tighten the seal at the inlet end to prevent arcing between the inlet electrode and the coolant compartment. A replacement cartridge incorporating the optical bench section of the original cartridge can be made by a competent machine shop and is much sturdier.
6. Too much sodium monophosphate makes the current through the capillary at the suggested operating voltage too high.
7. The acetonitrile is necessary to keep the APTS from adsorbing to the gel filtration material.
8. An increasing variety of labeled oligosaccharides is available from EnzymatiX http://enzymatix.us.com.
9. Submicrogram amounts of oligosaccharide mixtures can be derivatized in smaller volumes. We have constructed microliter sized reaction vessels. A description of their use is given in Carbohydrate Analysis by Modern Chromatography and Electrophoresis [8].

Fig. 2 (continued) enzyme, just substrate. The peak at 3.2 min is from the great excess of free ANTS which had not bound to sugar. Note the low levels of oligomers stretching out until around 9 min where the peaks all fuse together. Note also the increase in the relative fluorescence scale as the incubation times increase. (**b**) After 15 min of incubation a series of oligomers is observed with somewhat higher fluorescent intensity and the material moving very slowly is no longer apparent. Remember only the end of the oligomers are labeled so large oligomers only give a low fluorescence yield per mg. (**c**) After 30 min shorter oligomers predominate. (**d**) After 1 h of incubation the longer oligomers have all been degraded, and (**e**) after 24 h only the monomer, dimer, trimer and tetramer of GalA remain. Each GalA oligomer can give rise to two peaks because of the tendency of the labeled GalA residue to form a lactone between the carboxyl group and C-3. Thus part of the oligomer is in the lactone form and part as the free acid. This is especially apparent in panel (**e**). The pattern of degradation we see reflects a random attack of the polymer by the enzyme

References

1. Willats WGT, Knox JP, Mikkelsen JD (2006) Pectin: new insights into an old polymer are starting to gel. Trends Food Sci Technol 17:97–104
2. Bauer S, Vasu P, Persson S, Mort AJ, Somerville CR (2006) Development and application of a suite of polysaccharide-degrading enzymes for analyzing plant cell walls. Proc Nat Acad Sci U S A 103:11417–11422
3. De Vries RP, Visser J (2001) *Aspergillus* enzymes involved in degradation of plant cell wall polysaccharides. Microbiol Mol Biol Rev 65:497–522
4. Evangelista RA, Liu M, Chen F (1995) Characterization of 9-aminopyrene-1,4,6-trisulfonate-derivatized sugars by capillary electropherosis with laser-induced fluorescence detection. Anal Chem 67:2239–2245
5. Li X, Jackson P, Rubtsov DV et al (2013) Development and application of a high throughput carbohydrate profiling technique for analyzing plant cell wall polysaccharides and carbohydrate active enzymes. Biotechnol Biofuels 6:94
6. Naran R, Pierce ML, Mort AJ (2007) Detection and identification of rhamnogalacturonan lyase activity in intercellular spaces of expanding cotton cotyledons. Plant J 50:95–107
7. Merz JM, Mort AJ (1998) A computer-controlled variable light attenuator for protection and autoranging of a laser-induced fluorescence detector for capillary zone electrophoresis. Electrophoresis 19:2239–2242
8. Mort AJ, Pierce ML (2002) Preparation of carbohydrates for analysis by modern chromatography and electrophoresis. In: El Rassi Z (ed) Carbohydrate analysis by modern chromatography and electrophoresis. Elsevier, Amsterdam

Chapter 4

Theory and Practice in Measuring In-Vitro Extensibility of Growing Plant Cell Walls

Daniel J. Cosgrove

Abstract

This chapter summarizes four extensometer techniques for measuring cell wall extensibility in vitro and discusses how the results of these methods relate to the concept and ideal measurement of cell wall extensibility in the context of plant cell growth. These in-vitro techniques are particularly useful for studies of the molecular basis of cell wall extension. Measurements of breaking strength, elastic compliance and plastic compliance may be informative about changes in cell wall structure, whereas measurements of wall stress relaxation and creep are sensitive to both changes in wall structure and wall-loosening processes, such as those mediated by expansins and some lytic enzymes. A combination of methods is needed to obtain a broader view of cell wall behavior and properties connected with the concept of cell wall extensibility.

Key words Extensibility, Cell wall, Stress relaxation, Creep, Expansin

1 Introduction

Acting like a secure corset, the cell wall restrains the expansive tendency of the protoplast, giving plant cells their specific shape and size. In growing cells this mechanical restraint is rather more subtle than a corset, as the cell wall not only resists turgor pressure elastically but at the same time stretches slowly and irreversibly and often anisotropically (directionally). This is a controlled process in which the load-bearing network of cellulose microfibrils and wall matrix polymers yields irreversibly to the turgor-generated tensile forces in the cell wall [1–3]. As used here, the term “wall extensibility” refers to this ability to extend irreversibly, but a close reading of the literature reveals “extensibility” to be a multifaceted term, with various technical definitions (when defined at all) and a variety of methods for assessing its value. These concepts and methods were critically reviewed in [4, 5].

One key point to keep in mind is that elastic extensibility, measured by imposing rapid, reversible deformations or “strains” on the wall, differs from the extensibility that determines

Zoë A. Popper (ed.), *The Plant Cell Wall: Methods and Protocols*, Methods in Molecular Biology, vol. 2149,
https://doi.org/10.1007/978-1-0716-0621-6_4,

irreversible wall extension [6]. The confusion about this has grown recently with the use of new methods to measure elastic moduli by indentation procedures facilitated by modern atomic force microscopes (AFMs). A second point is that the way a mechanical force is applied to the cell wall makes a big difference in the resulting cell wall deformation. Cell walls will deform to lateral compressive forces, for example, when pushing an AFM probe against the outside of a cell wall, very differently from tensile forces, for example, stretching in the plane of the wall. This is so because cell walls are inherently anisotropic: cellulose is deposited as long reinforcing microfibrils in the plane of the cell wall, often in a cross polylamellate pattern [7]. Even the so-called isotropic cell walls are isotropic only in the plane of the cell wall; they are anisotropic when the third dimension (across the wall thickness) is considered. Forces applied normal to the plane of the cell wall (compressing cell wall thickness) cause a different mechanical deformation from those applied directly in the plane of the wall. With sophisticated models and experimental procedures, it is possible to estimate in-plane wall elastic moduli from compression measurements [8–10], but the values so obtained are estimates of elastic extensibility, not growth extensibility. A third point is that the time scale of the measurement matters a great deal for cell wall mechanics. Elastic deformations can be measured at time scales of nanoseconds, for example, with acoustic and Brillouin microscopy [11, 12]; microseconds to milliseconds, for example, by atomic force microscopy [10, 13–15]; or seconds, for example, by mechanical testers [16–18]. The kinds of polymer motions that are relevant for growth are in the range of 0.1–100 s or greater.

The current article presents an update focused on the utility and concepts underlying various measurements of wall properties in vitro, specifically by use of extensometers in which a cell wall specimen is clamped and extended in various modes to measure cell wall behavior that one might relate to cell wall extensibility in the context of cell growth. Alternative techniques designed for single cell and even subcellular measurements with living tissues have recently become common [13, 19, 20]; they use an AFM or similar indentation device to deform the surface of a plant cell or organ and measure the corresponding force–deflection curve. These measurements present special challenges in both technique and interpretation [5, 8, 13, 14, 21] which are beyond the scope of this article. For nanometer probes that penetrate into the cell wall, the force–deflection curve may be dominated by the pectins that control cell wall thickness [6, 22] rather than the cellulose scaffold that determines lateral shear during cell wall growth [23]. If the probe is large (~1 μm), it is likely that it both compresses the wall and deforms it, so the resulting force–deflection curve may reflect a complex pattern of wall deformation.

As a hydrated composite of complex polysaccharides, the growing plant cell wall has elastic, viscoelastic, and plastic properties which have been measured in the past half century by numerous static and dynamic loading methods (e.g., [24–27]). In these mechanical techniques, a physical force is applied to the cell wall and the resulting deformation is measured. They assess a mechanical stress/strain property that differs from wall extensibility in the context of cell growth [2, 5]. Underlying the concept of wall extensibility is the assumption that cell growth is limited by the ability of the cell wall to expand irreversibly and that this ability, in turn, is a function of cell wall stress, normally originating from cell turgor pressure. In this conceptual framework, wall extensibility is defined as the local slope of the curve relating growth rate to cell turgor pressure [4, 28].

The gap between this concept of wall extensibility and mechanical measurements stems from the fact that wall expansive growth is not a simple matter of inert polymer mechanics; it depends on dynamic wall-loosening processes that modify the matrix–cellulose network in a time-dependent way, promoting yielding of the polymeric network to wall stresses. In vivo, this results in cell wall creep (slow, time-dependent, irreversible extension), in which the cellulose microfibrils shift or separate from one another. We might call this "*chemorheological extensibility*" [6, 29] to emphasize the fact that the wall's flow (rheology) depends not only on cell wall structure and tensile stress, but also on dynamic processes of polymer creep catalyzed by cell wall-loosening agents such as expansins and lytic enzymes whose activity may be rapidly modulated by changes in wall pH, redox potential, hydration, ion concentrations, the supply of cell wall materials, and other ephemeral conditions within the cell wall space [30].

This turn of phrase, chemorheological extensibility, may more clearly denote the dynamic nature of wall extensibility, but it still leaves us with the problem of how to measure this subtle property. This issue is specifically relevant to situations where walls become more, or less, extensible during cell development and in response to hormones, light, gravity, dehydration and a variety of other environmental and biological stresses that influence plant growth. Measuring wall extensibility and understanding the molecular nature of its modulation are key aspects of understanding how plants control their cell growth by modifying their primary cell walls.

In-vivo methods for measuring wall extensibility, summarized in [4], have the advantage of fitting nicely into the biophysical theory of plant growth in quantitatively meaningful ways and they are sensitive to the cell's dynamic modification of the cell wall environment. Methods of measuring wall stress relaxation In-vivo, such as the pressure-block method [31] or the pressure probe method [32, 33], have some advantages over other in-vivo methods, but they require specialized equipment and moreover they have limited utility for investigating cell wall loosening by exogenous enzymes or for other studies where the molecular nature of wall extension is investigated. For such studies, isolated cell walls—that is, where the living cells of the tissue have been removed or otherwise made inactive—have particular merit because the results are not compromised by complicated responses of the living cells during the measurement and because a wider variety of treatments and methods may be applied to the isolated cell walls without concern about complex responses (even death!) of the living cells. These in-vitro methods are particularly well suited to studies of the biochemical underpinnings of cell wall enlargement and are the subject of this article.

One notable caveat with such in-vitro studies, however, is that a negative result (no measurable change in isolated wall properties) does not necessarily mean that in-vivo wall extensibility does not change during a particular growth response. For instance, cells may rapidly modulate cell wall pH and thereby affect expansin activity and increase wall extensibility, but such action would not be detected in the methods outlined below because the cell walls are removed from their living condition. Other examples where cell wall extensibility changes but is not detected by stress/strain methods are given in [5]. Thus, the results of in-vivo and in-vitro methods may disagree, likely pointing to an ephemeral change in wall-loosening activity rather than a structural change in the cell wall.

Four methods for assessing extensibility of isolated cell walls by use of extensometers are summarized below. Note that they neglect, by their very nature, the dynamic aspects of cellular modulation of cell wall yielding, but in the creep and stress relaxation methods (Subheadings 3.2.4 and 3.2.3) there is a chemorheological aspect because expansins and other wall modifying enzymes may remain active in the isolated walls (assuming no protein-denaturing treatments).

2 Materials

1. Plant material of interest: stems, hypocotyls, or plant organs of similar geometry, from the growing region of the plant. Slices of tissues may also be used. For stretching delicate materials such as etiolated *Arabidopsis* hypocotyls [34], it may be helpful to clamp two or three hypocotyl walls in parallel to measure an aggregate response. The plant material must be pretreated to disrupt the protoplasts (*see* **Notes 1** and **2**).
2. Extensometer: may be a commercial unit such as an Instron Universal tester, which is often found in many polymer laboratories, or custom-made [16, 35, 36] (*see* **Note 3**).

3. Carborundum powder (320 grit, Fisher Scientific, C192-500), thoroughly pre-washed to remove contaminants.
4. Microscope slides or other suitable glass plates.
5. Weight (50–400 g, depending on plant material).
6. Liquid nitrogen or −80 °C freezer.
7. Adhesive tape or cyanoacrylate adhesive ("Superglue").
8. Buffer: 20 mM sodium acetate, pH 4.5.

3 Methods

3.1 Preparation of Plant Tissues to Disrupt Protoplasts

To disrupt the protoplasts and remove the bulk water and cellular fluids within the tissue, a simple procedure is as follows (*see* **Note 4**):

1. Freeze the sample in liquid nitrogen and store in a −80 °C freezer.
2. Remove from freezer and permeabilize the cuticle by rubbing the sample between the thumb and forefinger with a slurry of well-washed carborundum. It is best to start this while the sample is still frozen (*see* **Note 5**).
3. Cut the wall specimen to size suitable for the extensometer clamps; for example, a length of 12 mm is appropriate for a sample in which 5 mm is to be held between the clamps. This allows 3 mm on either side to be gripped by each of the clamps. It is important that the growing region be stationed between the clamps.
4. Optional, depending on the intent of the experiments: treat the plant tissues to inactivate wall-bound enzyme activity, for example, by boiling (*see* **Note 6**).
5. Press the sample between two glass plates (i.e., microscope slides) for ~5 min under a weight to flatten the tissue and express the cell sap (*see* **Note 7**).

3.2 Measuring Cell Wall Mechanics

Four methods are detailed below for measuring (1) breaking strength, (2) elastic and plastic compliances, (3) stress relaxation, and (4) cell wall creep. These techniques measure different aspects of the cell wall mechanics, with the creep method coming perhaps closest to the measurement desired for in-vitro wall extensibility. All of the methods are essentially described as follows:

1. Clamp the prepared cell walls in an extensometer.
2. Apply a tensile force to the walls.
3. Either extend the walls at a defined rate or measure their extension.

A visual demonstration of the cell wall creep procedure may be found in [35].

3.2.1 Breaking Strength

The principle here is simple, if not simple minded: measure the force needed to break the cell wall. There are various geometrical means for causing breakage or mechanical failure of plant organs, with pulling (axial tension) or buckling (compression or bending) being the most straightforward for engineering analysis. For measurements of axial breaking strength, a suitable sample is clamped in an extensometer that can extend the distance between the clamps at a constant rate while measuring the force needed for the extension. It is important to clamp the sample in such a way that the clamp does not bite into and tear the sample and does not allow the sample to slip within the clamp. Close-up video of the sample during extension is helpful for troubleshooting such technical problems. Sometimes adhesive tape or cyanoacrylate adhesive ("Superglue") is used to help fix the walls to the clamping device [17, 37].

As the sample is extended the force increases and then levels off as the sample begins to fail and then drops quickly as the breakage is completed (Fig. 1a). Typically the force maximum is taken as the breaking strength, but one can also measure the percentage extension before failure and the area under the curve, which is the energy input for breakage [38]. The breaking force of plant wall samples is often found to be a function of the extension rate, so this parameter needs to be the same for valid comparisons. We typically use 3 mm/min. Another important detail: be sure that the sample stays wet during the measurement, as wall dehydration greatly increases cell wall strength.

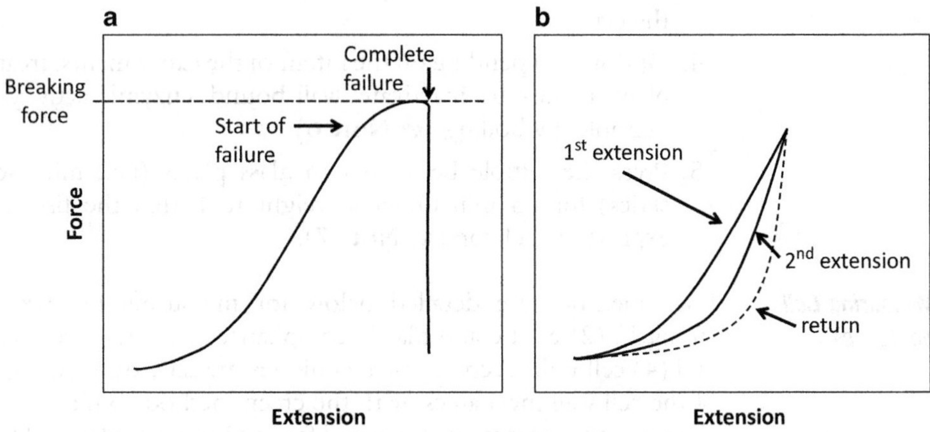

Fig. 1 Schematic diagrams of force–extension curves for cell walls measured in an extensometer for (**a**) tensile strength (breakage force) and (**b**) elastic and plastic moduli. For assessment of tensile strength the wall is extended until it fails, and the maximum force, the total extension and the area under the curve may be useful metrics of changes in cell wall structure. For assessment of cell wall moduli, the wall is extended, then returned to its original size and extended a second time. The slope of the second extension, near the end of the extension, is taken as the elastic modulus; its reciprocal is the elastic compliance. The slope of the first extension is the total modulus and the total compliance is its reciprocal. The plastic compliance is the difference between the total compliance and the elastic compliance. Note that extension is not linear and the slope varies with extension (strain)

The breaking strength of hypocotyls or inflorescence stems has been used to characterize *Arabidopsis* mutants with defects in cell wall composition or in wall assembly [37, 39–41], thereby drawing inferences about the structural role of a particular wall polymer. What such measurements mean for wall extensibility in the context of growth is more difficult to say, as there is at best only an indirect connection between the two concepts. Organ breakage occurs at the weakest point in the sample, which may be the middle lamella, that is, the adhesive layer of matrix polysaccharides and structural proteins that cements adjacent cell walls together. Changes in organ anatomy could also affect breaking strength. In contrast, wall extensibility depends on rearrangements within the matrix–cellulose network. Thus, while breaking strength may be informative about aspects of cell wall structure or the glue that holds cells together, it is not a reliable metric of wall extensibility. As a case it point, *Arabidopsis* hypocotyls from a xyloglucan-deficient mutant are weaker than wild type when assessed in mechanical tests of stiffness and breaking strength [40], yet they are less extensible in assays of cell wall creep [42]. Thus, tensile breaking strength may be informative about structural changes in cell walls or tissue architecture, but is in general a poor measure of wall extensibility.

3.2.2 Elastic and Plastic Compliances: Axial Extension

The principle of this method is similar to that described above for breaking strength, except that the sample is not extended to the breaking point, but is extended a small amount in two cycles. In the first cycle the sample is extended until a predetermined force is reached (well before the breakage point), then returned to the original length before a second extension is made (Fig. 1b). This second force–extension curve differs from the first one, but subsequent extensions cycles are reversible, at least to a first approximation, and so extension #2 is taken as an elastic extension. The difference between the first curve (total extension) and the second curve (elastic extension) gives the plastic, or irreversible, extension. The slope of the lines is estimated near the end of each extension cycle, to give Δforce/Δextension; this ratio is stiffness (*see* **Note 8**). When expressed per unit cross-sectional area, it becomes an engineering *modulus*. The higher the modulus, the stiffer the material. *See* **Note 9** for comparison of this macroscopic method with AFM-based indentation methods.

The modulus depends on many characteristics of the cell wall, but we do not have a comprehensive theory of this yet. Among the most important wall characteristics are: the number and bundling of cellulose microfibrils in wall cross-sectional area, the orientation of the microfibrils relative to the direction of extension, and the density, hydration and cross-linking of the matrix and its connection to the cellulose microfibrils. Recent studies have used finite element models to simulate the elastic behavior of a virtual cellulose–hemicellulose network constructed to mimic aspects of real cell walls [43, 44]. A very different type of model, based on the thermodynamics of hydrogen-bonded networks, was used to predict the plastic behavior of similar idealized cell walls [45]. These and other theoretical models make significant steps toward gaining molecular-scale insights into cell wall mechanics, but they are still limited and need further refinement and validation against the behavior of real cell walls.

The raw units for wall stiffness might be N/mm or g-force/mm. If the Δforce is divided by the cross-sectional area of the sample and the Δextension is calculated as fractional change in length, then the units can be readily converted to standard units of MPa (that is, stress divided by strain, or force per unit area divided by the fractional increase in length). These are the standard engineering units for modulus. Estimates of cross-sectional area, however, can be problematic (you cannot count cell lumens, the sample gets thinner as it extends, etc.), introducing errors into the absolute values of the moduli. Hence, such values reported in the literature should be examined with a critical eye. As long as the cross sections are similar for all samples, then a comparison of values is valid for reaching conclusions about relative changes among groups. On the other hand, if different groups have different cross-sectional areas, then the interpretation of the values can be more challenging, as differences in organ anatomy can influence the results. Similarly, if wall thicknesses differ among comparison groups, then the interpretation of the values needs careful thought. For instance, at equal force a thinner sample will have a larger stress than a thicker sample. The mechanical properties of cell walls are strongly nonlinear, so comparisons should be made at similar values of wall stress. This may mean one needs to apply a larger tensile force to a thicker tissue sample in order to equalize stress in the two measurements.

Because wall cross-sectional area in a sample is difficult to measure, a practical substitute is to use the cell wall mass per unit length of the sample: Cut the sample to 1 cm length, freeze and thaw to disrupt the cells, then wash and press the sample to remove cell contents. What remains are mostly cell walls, which can be weighed after drying. One may convert this to a volume by dividing by density of the cell wall polysaccharide (for cellulose it is 1.5 g/cm^3). An effective cross-sectional area for the mechanical measurement can be estimated by dividing this volume by the sample length.

The reciprocal of the force/extension value is known as a *compliance*, which corresponds to a type of extensibility, with units of strain/stress (or Δextension/Δforce for practical measurements). The elastic or plastic compliances that one measures with this technique are sometimes called extensibilities, but keep in mind that these are purely mechanical extensibilities that depend primarily on cell wall structure and sample thickness and do not include the chemorheological aspects of cell wall extensibility. As an example, α-expansin does not affect elastic or plastic compliances of cucumber hypocotyl cell walls [46]. These compliances are good reporters for changes in cell wall structure. For instance, treatment of cucumber hypocotyl cell walls with a family-12 endoglucanase caused large increases in both the elastic and plastic compliances of the cell walls [42].

3.2.3 Stress Relaxation

A crucial biophysical difference between growing and nongrowing cell walls is that the former undergo continuous stress relaxation (the physical face of cell wall loosening), which lowers cell turgor pressure and creates the water potential gradient necessary for sustained water uptake by the growing cell. Water uptake physically enlarges the cell and counter balances wall stress relaxation, so that turgor pressure stabilizes. This theory of wall relaxation was first enunciated qualitatively by Ray et al. [47], elaborated in specific quantitative terms by Cosgrove [48], and demonstrated experimentally in a series of studies in which water uptake into the growing cells was prevented, thereby allowing for stress relaxation to proceed unabated by water uptake, resulting in a decay in turgor pressure to the yield threshold [32, 33].

These techniques for measuring in-vivo wall stress relaxation have a counterpart for isolated cell walls [49–51], in which the isolated wall is clamped in an extensometer as above, rapidly extended until a predetermined force is attained, and then held to constant dimension while the holding force is monitored. The practical time scale for these measurements is from about 50 ms to 500 s. Longer times are possible if cell wall dehydration can be prevented. Shorter times are limited by the mechanics of the extensometer: it takes ~50 ms to extend the wall sample and to allow time for the induced vibrations (mechanical "ringing") in the sample to dampen out.

During the extension process, the cellulose–matrix network is elastically stretched; some of the wall polymers subsequently relax to lower energy states, resulting in wall stress relaxation. As a result, the holding force decays with time. The resulting decay in force, or stress relaxation, may be converted into a form known as a stress relaxation spectrum; a mathematically simpler operation, which approximates the relaxation spectrum, is to convert relaxation to log time scale and to plot the rate as a function of log time (Fig. 2), that is, $-\Delta\text{force}/\Delta\log$ time versus log time [36]. Much of this relaxation is the result of the viscoelastic nature of the cell wall material, that is, a passive physical response that depends on cell wall structure. This is the case for both growing and nongrowing

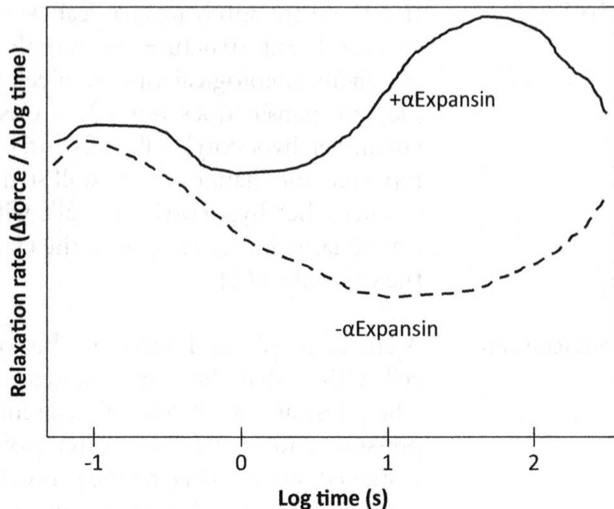

Fig. 2 Diagram illustrating stress relaxation spectra of heat-inactivated cucumber hypocotyl cell walls treated with buffer alone (−αExpansin) or with buffer containing α-expansins (+αExpansin) [46, 54]

walls, although the differences in wall structure in the two cases may result in different stress relaxation behaviors.

Additionally, any wall-loosening processes that are still active in isolated cell walls may result in additional stress relaxation not found in the nongrowing cell wall. Expansin-induced stress relaxation is readily demonstrated in these assays [52]; in acidic buffer the cell walls exhibit faster stress relaxation than in neutral buffer. This difference is eliminated by brief heat treatment, showing that the difference is not simply due to a pH-dependence of pectin rigidity in the cell wall. It is restored by addition of expansins, showing them to be the major mediators of wall stress relaxation in such isolated cell walls. On the other hand, a family-12 endoglucanase that increased elastic and plastic compliances (mentioned above) had no effect on the stress relaxation spectrum at times >1 s; its action on wall plasticity could be seen as increased stress relaxation at times <0.2 s [46]. Thus, stress relaxation assays are sensitive to changes in cell wall structure, expansin activity and potentially lytic enzymes that promote cell wall extension.

3.2.4 Creep

The fourth extensometer method measures cell wall creep, which is the time-dependent, irreversible extension of wall samples held at constant force. Because wall creep is a slow process, the measurement period is typically in the range of 30–150 min and the wall samples are clamped in a buffer-filled cuvette to prevent dehydration. Wall samples are typically clamped at a constant force in a

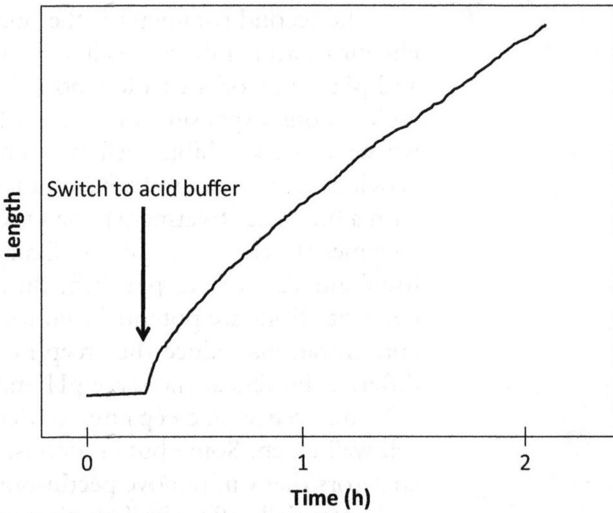

Fig. 3 Diagram illustrating the creep behavior of cucumber cell walls when clamped in a constant force extensometer at neutral pH and then switched to acidic buffer [36]

neutral buffer and after 10–15 min the buffer is switched an acid one, initiating rapid extension with gradually slows over 30–60 min to a constant or near constant rate (Fig. 3). A variant of this method is to clamp the walls in acidic buffer at low force which is insufficient to cause creep, then raise the force to a value high enough to cause wall creep.

Of the four in vitro methods described in this chapter, the creep method mimics the in-vivo wall extension process to the closest degree and it readily distinguishes between growing and nongrowing cell walls for many plant tissues. The difference between growing and nongrowing walls when measured with the creep method is much larger than measured by the three other extensometer techniques described above. Moreover, creep assays are sensitive to the activity of both expansins and the family-12 endoglucanase, making the creep assay more encompassing for detecting wall-loosening activities with different mechanisms of action and temporal signatures.

There are two critical parameters for cell wall creep measurements. First, one must establish a suitable force, which should be large enough to cause cell wall creep, but not so large that breakage becomes a problem. This is a matter of trial and error. In our hands, single *Arabidopsis* hypocotyls (~0.1 mm diameter) give good creep curves with 0.5 g-force; we have obtained good results clamping three *Arabidopsis* hypocotyls with 2.5 g-force; cucumber hypocotyls (~2 mm diameter) creep well with 20 g-force. These axial forces are smaller than the calculated axial forces that are generated internally by cell turgor pressure in the living tissues.

The second parameter is the buffer: its pH, ionic strength and chemical nature influence cell wall creep. Sodium acetate, at 20 mM and pH 4.5, works well for most tissues in our hands. At this pH endogenous expansins are active whereas pectin-methyl esterase, whose activity inhibits cell wall creep, is inactive. If enzymatic activities in the cell walls have not been inactivated (for instance with a brief heat treatment), then the activity of esterases and other enzymes attached to the cell wall may cause changes in the cell wall itself and also lead to pH drift. Buffer exchanges or higher buffer concentrations are potential solutions to this drift, but high buffer concentrations reduce the creep rate. We have tried a number of different buffers at the same pH and concentration and we found >2× differences in creep rates, evidently due to the anion effects on cell wall creep. Some buffer anions, such as citrate, act as divalent chelators that can remove pectin-bound calcium from the cell wall and potentially affect wall physical properties.

Takahashi et al. [53] constructed an elegant "programmable creep meter" in which the axial force starts at low values and gradually increases. In this way they were able to characterize creep rate as a function of applied force and to make estimates of the yield threshold for wall creep.

3.3 Conclusions

For reasons discussed in the text, it is unlikely that cell wall extensibility can be fully measured by in-vitro methods, as wall extensibility is based not only on wall mechanics but depends on wall-loosening processes that are sensitive to ephemeral conditions in the cell wall space. Nevertheless, the methods outlined here can provide positive evidence for changes in wall structure (indicated by elastic and plastic compliances) and changes in wall-bound wall-loosening activities such as expansins and lytic enzymes. These methods are particularly useful for investigations of the molecular basis of cell wall extension and its dependence on cell wall structure.

4 Notes

1. For these methods to be interpreted in terms of cell wall properties, it is important that the protoplast be disrupted so that turgor pressure and mechanically driven water flows within the tissue do not complicate the measurement. Turgor pressure generates wall tension, making the cell wall stiffer, much as a pressurized tire is stiffer than a flaccid one. As a result, the tensile properties of living tissues depend on both turgor pressure and cell wall structure. If protoplasts are intact, deformations can be partially limited by water flows from the protoplast, making for a more complicated interpretation of the results. For these reasons, these assays are simpler to interpret when the cell protoplasts are disrupted. On the other

hand, there may be circumstances where these complications are not deemed important, for example *see* ref. 17.

2. Sample variability typically requires 8–15 replicates to obtain adequate confidence limits of the measurements. The source of this variability is unknown but likely arises from differences in the plant materials. Therefore care in growing, harvesting, selecting, and preparing the plant materials is important for limiting variability.
3. Trace levels of metal ions such as copper, iron, and aluminum are potent inhibitors of cell wall creep and stress relaxation. Therefore, metal clamps and other metallic parts of the extensometer should be coated with epoxy, plastic, latex or other material to avoid interactions with the cell walls. This is most important in methods entailing long measurement periods, for example the creep method and the stress relaxation method.
4. In some studies researchers have boiled the cell walls in methanol, then rehydrated in buffer prior to measurement. This procedure disrupts the protoplast and inactivates most enzymes bound to the cell wall sample. However, the procedure does not inactivate expansins [36, 54] and also precipitates polysaccharides, potentially causing irreversible changes in wall mechanics. In most cases this type of pretreatment is best avoided.
5. Cuticle abrasion is only needed when buffers or other materials must be diffused into the wall samples or when the walls must be thoroughly washed. It is commonly omitted when measuring breaking strength and elastic and plastic compliances. Moreover, some samples such as *Arabidopsis* hypocotyls or wheat coleoptiles are so delicate that this step is omitted to avoid damage to the cell walls.
6. If wall-bound enzymes are not inactivated by heat or other treatment, care must be taken to limit enzymatic wall modification, most importantly pectin demethylation and polysaccharide hydrolysis, as such processes may modify the wall behaviors measured in these assays. The boiling treatment should be kept to as brief as possible, (5–15 s), to prevent weakening the middle lamella, which increases sample breakage.
7. We omit this step for very thin or delicate materials. On the other hand, for highly hydrated samples that tend to tear or split when clamped (think of trying to nail Jell-O to a wall), removal of sample fluids at this point can be very helpful. For new samples, the weight and time is determined by trial and error. For our experiments with cucumber hypocotyls, we arrange eight hypocotyl samples between a pair of slides and add 300–400 g of weight for 5 min.

8. The value of the target force is determined empirically for each type of wall material, based on a force–extension curve of the kind shown in Fig. 1a. Note that the slope of the curve is not constant, but initially increases, reaches a point of inflection, then decreases. We typically chose a target force close to or slightly above the point of inflection. The region used to estimate slope is also critical to the stiffness values. We usually pick the last 10% of the data in the curves to estimate slope.

9. A microscopic variant of the stress-strain method has been used for evaluating the mechanical properties of cell walls in single cells or parts of cells, such as at different parts of the apical dome of pollen tube [55] or the shoot apical meristem [56, 57]. The method, as typically deployed, does not separate elastic and plastic components, and the stiffness values obtained are useful for relative comparisons, but not for obtaining absolute values of wall modulus or compliance. The principal of the method is simple: a small probe is used to deform a local region of the cell wall and to measure the force on the wall. The resistance to such deformation is a complex function of cell wall stiffness, cell geometry and turgor pressure [8, 58]. Wall stiffness depends on the thickness of the wall and its modulus, which in turn depends on the arrangement of its structural elements (cellulose microfibril density, orientation, and interconnection by matrix polymers). The major advantage of the method is that it may be used at the single cell level and even to probe different parts of the cell, but this comes at the cost of some uncertainty about the physical interpretation of the measured values.

Acknowledgments

This chapter is based upon work supported as part of The Center for Lignocellulose Structure and Formation, an Energy Frontier Research Center funded by the U.S. Department of Energy, Office of Science, Office of Basic Energy Sciences under Award Number DE-SC0001090. The research on expansins and cell wall creep was supported by Award number DE-FG02-84ER13179 from the Department of Energy Office of Basic Energy Sciences.

References

1. Cosgrove DJ (2005) Growth of the plant cell wall. Nat Rev Mol Cell Biol 6:850–861
2. Cosgrove DJ (2018) Diffuse growth of plant cell walls. Plant Physiol 176:16–27
3. Chebli Y, Geitmann A (2017) Cellular growth in plants requires regulation of cell wall biochemistry. Curr Opin Cell Biol 44:28–35
4. Cosgrove DJ (1993) Wall extensibility: its nature, measurement and relationship to plant cell growth. New Phytol 124:1–23

5. Cosgrove DJ (2016) Plant cell wall extensibility: connecting plant cell growth with cell wall structure, mechanics, and the action of wall-modifying enzymes. J Exp Bot 67:463–476
6. Zhang T et al (2019) Disentangling loosening from softening: insights into primary cell wall structure. Plant J 100:1101–1117
7. Zhang T, Zheng Y, Cosgrove DJ (2016) Spatial organization of cellulose microfibrils and matrix polysaccharides in primary plant cell walls as imaged by multichannel atomic force microscopy. Plant J 85:179–192
8. Weber A et al (2015) Measuring the mechanical properties of plant cells by combining micro-indentation with osmotic treatments. J Exp Bot 66:3229–3241
9. Routier-Kierzkowska AL, Smith RS (2013) Measuring the mechanics of morphogenesis. Curr Opin Plant Biol 16:25–32
10. Beauzamy L, Derr J, Boudaoud A (2015) Quantifying hydrostatic pressure in plant cells by using indentation with an atomic force microscope. Biophys J 108:2448–2456
11. Elsayad K et al (2016) Mapping the subcellular mechanical properties of live cells in tissues with fluorescence emission–Brillouin imaging. Sci Signal 9:rs5
12. Scarcelli G et al (2015) Noncontact three-dimensional mapping of intracellular hydromechanical properties by Brillouin microscopy. Nat Methods 12:1132–1134
13. Yakubov GE et al (2016) Mapping nano-scale mechanical heterogeneity of primary plant cell walls. J Exp Bot 67:2799–2816
14. Braybrook SA (2015) Measuring the elasticity of plant cells with atomic force microscopy. Methods Cell Biol 125:237–254
15. Routier-Kierzkowska AL, Smith RS (2014) Mechanical measurements on living plant cells by micro-indentation with cellular force microscopy. Methods Mol Biol 1080:135–146
16. Saxe F, Burgert I, Eder M (2015) Structural and mechanical characterization of growing Arabidopsis plant cell walls. Methods Mol Biol 1242:211–227
17. Abasolo W et al (2009) Pectin may hinder the unfolding of xyloglucan chains during cell deformation: implications of the mechanical performance of Arabidopsis hypocotyls with pectin alterations. Mol Plant 2:990–999
18. Cleland RE (1984) The instron technique as a measure of immediate-past wall extensibility. Planta 160:514–520
19. Peaucelle A, Wightman R, Hofte H (2015) The control of growth symmetry breaking in the Arabidopsis hypocotyl. Curr Biol 25:1746–1752
20. Sampathkumar A et al (2014) Subcellular and supracellular mechanical stress prescribes cytoskeleton behavior in Arabidopsis cotyledon pavement cells. eLife 3:e01967
21. Bonilla MR et al (2015) Interpreting atomic force microscopy nanoindentation of hierarchical biological materials using multi-regime analysis. Soft Matter 11:1281–1292
22. Jarvis MC (1992) Control of thickness of collenchyma cell walls by pectins. Planta 187:218–220
23. Cosgrove DJ (2014) Re-constructing our models of cellulose and primary cell wall assembly. Curr Opin Plant Biol 22C:122–131
24. Hansen SL et al (2011) Mechanical properties of plant cell walls probed by relaxation spectra. Plant Physiol 155:246–258
25. Haughton PM, Sellen DB, Preston RD (1968) Dynamic mechanical properties of cell wall of *Nitella opaca*. J Exp Bot 19:1–12
26. Cleland RE, Haughton PM (1971) The effect of auxin on stress relaxation in isolated *Avena* coleoptiles. Plant Physiol 47:812–815
27. Probine MC, Barber NF (1966) The structure and plastic properties of the cell wall of Nitella in relation to extension growth. Aust J Biol Sci 19:439–457
28. Geitmann A, Ortega JKE (2009) Mechanics and modeling of plant cell growth. Trends Plant Sci 14:467–478
29. Ray PM (1987) Principles of plant cell expansion. In: Cosgrove DJ, Knievel DJ (eds) Physiology of cell expansion during plant growth (Symposium in Plant Physiology, Penn State Univ). American Society of Plant Physiologists, Rockville, pp 1–17
30. Cosgrove DJ (2016) Catalysts of plant cell wall loosening. F1000Res 5. https://doi.org/10.12688/f1000research.7180.1
31. Cosgrove DJ (1995) Measurements of wall stress relaxation in growing plant cells. Methods Cell Biol 49:231–243
32. Cosgrove DJ (1987) Wall relaxation in growing stems: comparison of four species and assessment of measurement techniques. Planta 171:266–278
33. Cosgrove DJ, Van Volkenburgh E, Cleland RE (1984) Stress relaxation of cell walls and the yield threshold for growth: demonstration and measurement by micro-pressure probe and psychrometer techniques. Planta 162:46–52
34. Park YB, Cosgrove DJ (2012) A revised architecture of primary cell walls based on biomechanical changes induced by substrate-specific endoglucanases. Plant Physiol 158:1933–1943

35. Durachko DM, Cosgrove DJ (2009) Measuring plant cell wall extension (creep) induced by acidic pH and by alpha-expansin. J Vis Exp (25):1263
36. Cosgrove DJ (1989) Characterization of long-term extension of isolated cell walls from growing cucumber hypocotyls. Planta 177:121–130
37. Reiter WD, Chapple CC, Somerville CR (1993) Altered growth and cell walls in a fucose-deficient mutant of Arabidopsis. Science 261:1032–1035
38. Wainwright SA et al (1976) Mechanical design in organisms. Edward Arnold, London, p 423
39. Ryden P et al (2003) Tensile properties of Arabidopsis cell walls depend on both a xyloglucan cross-linked microfibrillar network and rhamnogalacturonan II-borate complexes. Plant Physiol 132:1033–1040
40. Cavalier DM et al (2008) Disrupting two *Arabidopsis thaliana* xylosyltransferase genes results in plants deficient in xyloglucan, a major primary cell wall component. Plant Cell 20:1519–1537
41. Zhong RQ et al (2005) Arabidopsis fragile fiber8, which encodes a putative glucuronyltransferase, is essential for normal secondary wall synthesis. Plant Cell 17:3390–3408
42. Park YB, Cosgrove DJ (2012) Changes in cell wall biomechanical properties in the xyloglucan-deficient *xxt1/xxt2* mutant of Arabidopsis. Plant Physiol 158:465–475
43. Kha H et al (2010) WallGen, software to construct layered cellulose-hemicellulose networks and predict their small deformation mechanics. Plant Physiol 152:774–786
44. Yi H, Puri VM (2012) Architecture-based multiscale computational modeling of plant cell wall mechanics to examine the hydrogen-bonding hypothesis of the cell wall network structure model. Plant Physiol 160:1281–1292
45. Veytsman BA, Cosgrove DJ (1998) A model of cell wall expansion based on thermodynamics of polymer networks. Biophys J 75:2240–2250
46. Yuan S, Wu Y, Cosgrove DJ (2001) A fungal endoglucanase with plant cell wall extension activity. Plant Physiol 127:324–333
47. Ray PM, Green PB, Cleland RE (1972) Role of turgor in plant cell growth. Nature 239:163–164
48. Cosgrove DJ (1985) Cell wall yield properties of growing tissues. Evaluation by *in vivo* stress relaxation. Plant Physiol 78:347–356
49. Yamamoto R, Shinozak K, Masuda Y (1970) Stress-relaxation properties of plant cell walls with special reference to auxin action. Plant Cell Physiol 11:947–956
50. Yamamoto R, Kawamura H, Masuda Y (1974) Stress relaxation properties of the cell wall of growing intact plants. Plant Cell Physiol 15:1073–1082
51. Fujihara S, Yamamoto R, Masuda Y (1978) Viscolelastic properties of plant cell walls II. Effect of pre-extension rate of stress relaxation. Biorheology 15:77–85
52. McQueen-Mason SJ, Cosgrove DJ (1995) Expansin mode of action on cell walls. Analysis of wall hydrolysis, stress relaxation, and binding. Plant Physiol 107:87–100
53. Takahashi K et al (2006) Wall-yielding properties of cell walls from elongating cucumber hypocotyls in relation to the action of expansin. Plant Cell Physiol 47:1520–1529
54. McQueen-Mason S, Durachko DM, Cosgrove DJ (1992) Two endogenous proteins that induce cell wall extension in plants. Plant Cell 4:1425–1433
55. Zerzour R, Kroeger J, Geitmann A (2009) Polar growth in pollen tubes is associated with spatially confined dynamic changes in cell mechanical properties. Dev Biol 334:437–446
56. Milani P et al (2011) *In vivo* analysis of local wall stiffness at the shoot apical meristem in Arabidopsis using atomic force microscopy. Plant J 67:1116–1123
57. Kierzkowski D et al (2012) Elastic domains regulate growth and organogenesis in the plant shoot apical meristem. Science 335:1096–1099
58. Bolduc JE et al (2006) Finite-element analysis of geometrical factors in micro-indentation of pollen tubes. Biomech Model Mechanobiol 5:227–236

Chapter 5

Formation of Cellulose-Based Composites with Hemicelluloses and Pectins Using *Komagataeibacter* Fermentation

Deirdre Mikkelsen, Patricia Lopez-Sanchez, Dongjie Wang, and Michael J. Gidley

Abstract

Komagataeibacter xylinus synthesizes cellulose in an analogous fashion to plants. Through fermentation of *K. xylinus* in media containing cell wall polysaccharides from the hemicellulose and/or pectin families, composites with cellulose can be produced. These serve as general models for the assembly, structure, and properties of plant cell walls. By studying structure/property relationships of cellulose composites, the effects of defined hemicellulose and/or pectin polysaccharide structures can be investigated. The macroscopic nature of the composites also allows composite mechanical properties to be characterized.

The method for producing cellulose-based composites involves reviving and then culturing *K. xylinus* in the presence of desired hemicelluloses and/or pectins. Different conditions are required for construction of hemicellulose- and pectin-containing composites. Fermentation results in a floating mat or pellicle of cellulose-based composite that can be recovered, washed, and then studied under hydrated conditions without any need for intermediate drying.

Key words Arabinoxylan, β-glucan, Composites, *Gluconacetobacter xylinus*, Hemicellulose, *Komagataeibacter xylinus*, Pectin, Plant cell wall, Cellulose, Xyloglucan

1 Introduction

The cell walls of plants are typically complex in terms of their measured average composition, with variation being exhibited not only between different plant types but also between local tissue types and even within a single cell wall. While some information on the relationships between composition and properties of cell walls can be deduced through studies of, for example, plant mutants lacking defined compositional features, the isolation of plant cell wall material for the study of structure/property relationships has major limitations due to microheterogeneity within the plant and the need for harsh extraction conditions.

Zoë A. Popper (ed.), *The Plant Cell Wall: Methods and Protocols*, Methods in Molecular Biology, vol. 2149, https://doi.org/10.1007/978-1-0716-0621-6_5, © Springer Science+Business Media, LLC, part of Springer Nature 2020

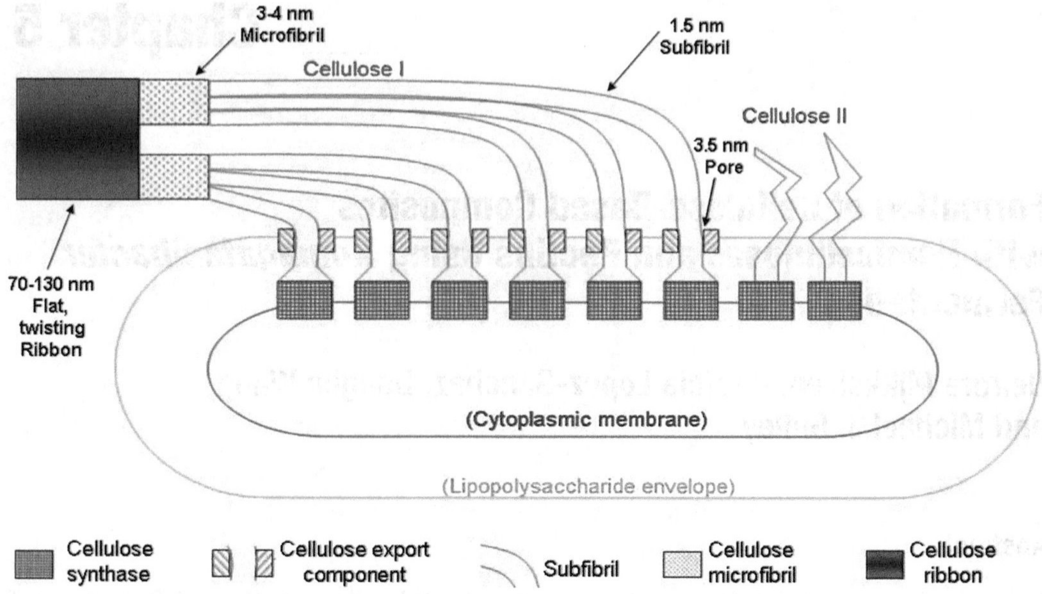

Fig. 1 Schematic illustration of cellulose biosynthesis by *Komagataeibacter xylinus* (adapted from Brown [1])

Komagataeibacter xylinus (previously known as *Gluconacetobacter xylinus* or *Acetobacter xylinus* or *Acetobacter xylinum*) has been used as a model for cellulose biosynthesis because it has the same general features of cellulose deposition as plants (Fig. 1). A transmembrane assembly of catalytic (cellulose synthase) and structural proteins produces strands of (1 → 4)-β-D-glucan that associate, first into small sub-fibrils, then into microfibrils which subsequently aggregate into a characteristic fibre ribbon (Fig. 1). Although plant primary wall cellulose usually forms thinner fibers and may have different microfibril geometries [2], all stages of cellulose synthesis are shared with *K. xylinus.* Cellulose secreted by bacteria into an aqueous fermentation medium is used as a model for plant cellulose secreted extracellularly into a nascent cell wall. In plants, cellulose is deposited into an environment containing a complex mixture of other cell wall polymers. In the bacterial model system, the extracellular environment is controlled and can be used to investigate the potential of each type of cell wall polymer to form composite structures with cellulose and to investigate the molecular, microscopic and macroscopic properties of the resulting materials. Validation of the system as a model comes from (a) microscopic observation of similar architectures in cellulose–xyloglucan composites [3] as in depectinated cell walls of, for example, onion [4], (b) responsiveness of hemicellulose–cellulose composites to application of expansin proteins [5] and xyloglucan endotransglycosylase [6] similar to that found or expected in plant tissues, (c) conversion of highly crystalline *K. xylinus* cellulose into a less crystalline form (characteristic of plant cell walls) in the presence of hemicelluloses [3, 7] and (d) understanding the influence of heavy metal toxicity on stress–strain relationships of cell walls (e.g., aluminum-induced inhibition of root elongation) [8]. While there are limitations to this model system, it is experimentally straightforward and provides insights into the effects of polymer components of defined chemistry that are not possible from cell walls isolated from plants [9–12]. In addition, this constructive approach has been adopted to investigate the interactions between plant cell wall components and polyphenols [13–16]. Also, in this age of minimally processed fresh produce consumerism, the influence certain plant cell wall polymers have on the attachment of food-borne pathogens have also been investigated using this model system [17, 18]. Furthermore, as cellulosic materials can be produced in multi-cm pieces, mechanical measurements can be made that are highly informative in defining the principles of cell wall materials properties [19–23, 35].

In the absence of polymers in the fermentation medium, a cellulose "mat" or pellicle is produced that is highly hydrated but mechanically tough. In outline, the formation of cellulose composites with hemicellulose or pectin is accomplished by fermenting *K. xylinus* in liquid fermentation containing the hemicellulose and/or pectin of interest. The cellulose produced by bacteria (Fig. 1) thus comes into contact with the added polymer(s) as soon as it is secreted from the bacteria. For hemicelluloses that can associate (bind) molecularly to cellulose, the fact that the crystallinity of cellulose is affected greatly [3, 7] is interpreted to mean that added polymers can access cellulose microfibrils prior to their aggregation into the final ribbon assembly (Fig. 1). The negative charge on pectins means that there is no or limited direct binding between the two backbones (although arabinan or galactan side chains may bind [24]), and composite formation typically requires the presence of a preformed pectin network [25]. Thus pectin–cellulose composites can be formed directly if the fermentation medium contains a weak gel which can be varied by choice of pectin type (particularly degree of methyl esterification) and calcium level.

Following fermentation, composites are recovered from the medium, and washed to remove bacteria and any polymers that are held nonspecifically. The isolated solid composite material can then be analyzed by chemical, spectroscopic, microscopic, or mechanical methods. Alternatively, pectin networks can be added after cellulose has been produced in HS medium by soaking the preformed cellulose mat in a pectin solution, followed by diffusion of calcium through the pectin–cellulose composite.

Isotopically labeled (e.g., ^{2}H or ^{13}C) cellulose material may also be produced and investigated by, for example, small angle neutron scattering techniques to study cellulose in its hydrated state [2], thus providing a platform to expand knowledge of the interaction mechanisms between cellulose and other plant cell wall components.

2 Materials

1. Laminar flow cabinet and appropriate facilities for observing good microbiological practice.
2. Bacterial strain: *Komagataeibacter xylinus* (formerly *Gluconacetobacter xylinus, Acetobacter xylinus* and *Acetobacter xylinum*) strain ATCC 53524 frozen (−80 °C) stock (*see* **Note 1**).
3. Growth medium: Hestrin and Schramm (HS) medium [26] containing (per L) 5 g peptone, 5 g yeast extract, 3.38 g $Na_2HPO_4.H_2O$, 1.15 g citric acid, and 20 g glucose (*see* **Note 2**). HS agar medium (containing 15 g/L agar) is used for maintenance and long-term storage of the bacterial strain. HS liquid medium is used for composite preparation. Preferably, the medium is made fresh as required and the starting pH is 5.0. Alternatively, sterile medium stored at 4 °C is stable for up to one week.
4. Long term storage of *K. xylinus*: Microbank™ preservation cryovials (Pro-Lab Diagnostics, Ontario, Canada).
5. Hemicellulose solutions: 1% (w/v) (*see* **Note 3**).
6. Pectin solutions: 1% (w/v) (*see* **Note 4**).
7. 0.125 M calcium chloride ($CaCl_2$) solution, sterilized by autoclaving at 121 °C for 15 min.
8. 2.5 mM, 12.5 mM and 125.0 mM $CaCl_2$ solutions, sterilized by autoclaving at 121 °C for 15 min.
9. Sterile culture containers (*see* **Note 5**) (*see* Fig. 2).
10. 0.02% (w/v) sodium azide solution made with deionized water (*see* **Note 6**).
11. 0.02% (w/v) sodium azide solution made with sterile 12.5 mM $CaCl_2$ solution (*see* **Note 6**).
12. Deionized water, sterilized by autoclaving at 121 °C for 15 min.

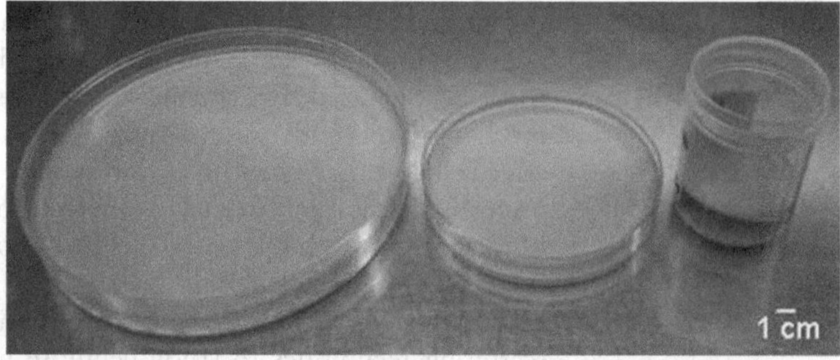

Fig. 2 Examples of sterile containers used for composite preparation. These include (from left to right) sterile 150 × 20 mm and 92 × 16 mm Petri dishes, as well as sterile screw lid containers with 70 mL capacity and 40 mm diameter

3 Methods

3.1 Revival and Maintenance of *K. xylinus*

1. Frozen stocks: Under aseptic conditions, open the cryovial and using a sterile needle, remove one bead to inoculate each HS agar plate (inoculate two plates). Use the bead to directly streak onto the solid medium to get isolated pure colonies [27]. Incubate at 30 °C for 3–4 days.
2. Maintenance of culture: Two plates of the culture are inoculated as one plate is reserved for starting a new working culture when required (*see* **Note 7**), and the other plate is used for routine transfers as required. The plates are wrapped with Parafilm® and stored at 4 °C. The cultures remain viable for 1 month.

3.2 Long-Term Storage of *K. xylinus* (See Note 8)

1. Under aseptic conditions open the screw cap of the Microbank™ cryovial.
2. Inoculate the cryopreservative fluid with a loopful of young colony growth (72 h) picked from a pure culture.
3. Close vial tightly and invert four to five times to emulsify organism. A vortex mixer must not be used.
4. The excess cryopreservative must be well aspirated leaving the inoculated beads as free of liquid as possible.
5. Inoculated cryovials are closed finger tight and stored at −80 °C (shelf-life 3–4 years).

3.3 Composite Preparation with Hemicelluloses

Bacterial cellulose–hemicellulose composites are constructed as follows.

1. Hemicellulose Solution: Prepare a 1% (w/v) solution by accurately weighing out 5 g of powdered substrate with a sterilized spatula into a sterile 1 L dry Pyrex beaker in a laminar flow cabinet. Add a sterile magnetic stirrer bar, followed by 400 mL of hot (90 °C) sterile deionized water (*see* **Note 9**). Immediately place the beaker containing the mixture on a magnetic hotplate-stirrer and heat at a setting of 100 °C with vigorous stirring. Loosely cover the beaker with aluminum foil, stirring and boiling the contents until the hemicellulose dissolves completely (*see* **Note 10**). Adjust the volume of the solution to 500 mL with sterile deionized water. Make solution fresh as required (*see* **Note 11**).
2. Modified HS broth medium: The medium is prepared as a concentrated (2×) solution by adding 5 g peptone, 5 g yeast extract, 3.38 g $Na_2HPO_4.H_2O$ and 1.15 g citric acid to 400 mL deionized water. The pH is adjusted to 5.0 with 10 M HCl, and the volume of the solution is adjusted to 460 mL with sterile deionized water. The medium is sterilized by autoclaving at 121 °C for 15 min. After cooling the sterile

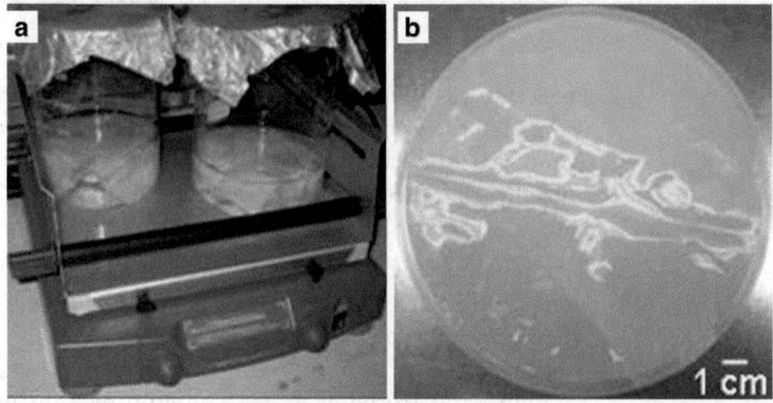

Fig. 3 Purification of hemicellulose composites by washing with gentle agitation (50 r.p.m.) at room temperature, with excess ice-cold sterile deionized water (**a**). Having removed the excess medium and bacterial cells, the purified pellicle is *white* (**b**)

medium to 55 °C, 40 mL of 0.22 μm filter-sterilized glucose solution (50% w/v), and 500 mL of the 1% (w/v) polysaccharide solution are added aseptically, giving final glucose and polysaccharide concentrations in the medium of 2.0% (w/v) and 0.5% (w/v) respectively.

3. For primary inoculum preparation: Inoculate 20 mL modified HS broth medium with a few colonies of bacteria from the HS agar plate used to maintain the strain. Incubate under static conditions at 30 °C for 72 h (*see* **Note 12**).
4. For scale-up preparation: Inoculate 18 mL modified HS broth medium with 2 mL primary inoculum and incubate without agitation at 30 °C for 48 h (arabinoxylan or β-glucan composite) or 72 h (for xyloglucan composite).
5. Harvesting of hemicellulose composites: After incubation, the composite pellicle is removed with forceps and washed at room temperature by gentle agitation (50 r.p.m.), in a sterile 3 L glass beaker containing excess ice-cold sterile deionized water (Fig. 3) (*see* **Note 13**).
6. Short-term preservation: Composites are stored in the hydrated state in 0.02% (w/v) sodium azide solution at 4 °C (*see* **Note 14**).

3.4 Composite Preparation with Pectins

3.4.1 Production of Cellulose in the Presence of a Pectin–Calcium Gel

The level of pectin incorporation within the cellulose network is dependent on the interaction between pectin and Ca^{2+} ions present in the medium. A preformed gel of the desired strength, which will allow for sufficient gelling as well as cellulose microfibril penetration, must be achieved in order to form a composite. The following detailed method describes the preparation of bacterial cellulose–pectin composites using pectin of degree of methyl esterification (DM) of 30%. Previous research [25] has determined that the highest pectin incorporation in cellulose composites occurs when using pectin of ~DM 30. When attempting to construct bacterial cellulose–pectin (0.5%) composites with other DM values, different $CaCl_2$ concentrations may be added to the modified HS medium as appropriate.

1. Pectin Solution: Prepare a 1% (w/v) solution by accurately weighing out in a laminar flow cabinet 7 g of powdered pectin with a sterilized spatula into a sterile 1 L Schott bottle containing 489.5 mL of sterile deionized water. Add a sterile magnetic stirrer bar, place the sealed bottle containing the mixture on a magnetic stirrer and stir vigorously overnight (*see* **Note 15**). Once the pectin is completely dissolved, the volume of the solution can be adjusted to 500 mL with sterile deionized water. Make solution fresh as required.
2. Modified HS broth medium: The medium is prepared as per **step 2** in Subheading 3.3, with the exception of 1% (w/v) pectin solution being added to the sterile medium instead of the hemicellulose solution.
3. For primary inoculum preparation: Inoculate 18 mL modified HS broth medium with two to three colonies of bacteria from the HS agar plate used to maintain the strain. Place the inoculated container on a platform shaker and under vigorous shaking (~250 r.p.m.), add 2 mL of 125.0 mM $CaCl_2$ solution, the final pectin and $CaCl_2$ in the medium being 0.5% (w/v) and 12.5 mM respectively. Shake the inoculated container for a further 5 min (*see* **Note 16**) (*see* Fig. 4). Incubate under static conditions at 30 °C for 72 h.

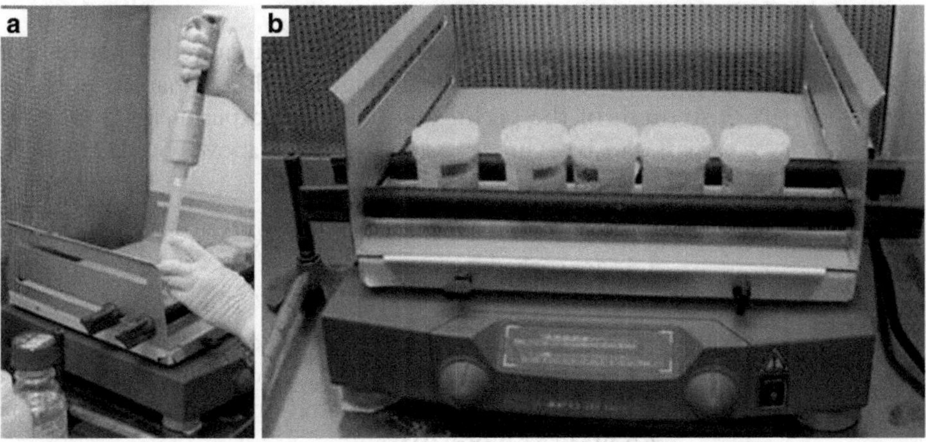

Fig. 4 Preparation of bacterial cellulose–pectin composites. The inoculated container is placed on a platform shaker and $CaCl_2$ solution (to give a final concentration of typically 12.5 mM) is added under vigorous shaking (**a**), followed by a further 5 min of shaking (**b**), to allow the HS-pectin medium to gel in a uniform manner. A uniform preformed gel results in a relatively homogeneous composite

4. For scale-up preparation: Inoculate 16 mL modified HS broth medium with 2 mL primary inoculum. As above, place the inoculated container on a platform shaker and under vigorous shaking (~250 r.p.m.), add 2 mL of 125.0 mM $CaCl_2$ solution, thus maintaining the final concentrations of pectin and $CaCl_2$ in the medium at the levels used in the primary inoculum. Shake the inoculated container for a further 5 min and incubate under static conditions at 30 °C for 72 h.
5. Harvesting of pectin composites: After incubation, remove the composite pellicle by carefully picking the pellicle up with the help of a spatula (*see* **Note 17**). Place the pectin composite in a sterile glass beaker containing excess ice-cold sterile $CaCl_2$ solution of the same concentration as the incubation medium (e.g., 12.5 mM), and wash at room temperature by gentle agitation (50 r.p.m.) (*see* **Note 18**).
6. Short-term preservation: Pectin composites are stored in the hydrated state in 0.02% (w/v) sodium azide solution containing the appropriate concentration of $CaCl_2$, at 4 °C (*see* **Note 19**).

The strength of the preformed pectin network might make it difficult for the bacteria to produce a homogeneous cellulose mat. In that case, the cellulose can be produced in the presence of a lower $CaCl_2$ concentration (e.g., 2.5 mM) and, after harvesting, further enriched in calcium by soaking the composites in, for example, 12.5 mM $CaCl_2$ solution. It should be noted that when using this method a thin layer of pectin–calcium gel is formed on the surface of the cellulose. This layer can be regarded as a middle lamella analogue, or gently removed with the help of a scalpel and tweezers.

3.4.2 Addition of Pectin Gel After Cellulose Production

For certain investigations, it might be required to have a constant cellulose network while changing other variables, such as the strength of the pectin gel. For these experiments, pure cellulose mats can be synthesized as described above (Subheading 3.4.1), and following the steps detailed below.

1. Purified bacterial cellulose pellicles (as described in Subheading 3.4) are soaked in excess volume of 0.5% (w/v) pectin solution in a beaker at room temperature overnight.
2. Samples are collected with the help of a spatula and each sample is placed into individual 40 mL containers and their surface sprayed with a low concentration $CaCl_2$ solution, that is, 2.5 mM (*see* **Note 20**).
3. A thin gel layer is formed immediately after the calcium is poured over the cellulose mat, whose surface has been in contact with the pectin solution (*see* **Note 20**). After 30 min

10 mL of $CaCl_2$ 12.5 mM is added on top of the composite and left in contact with the sample for 2 days at room temperature. Calcium diffuses slowly first through the pectin gel layer and then inside the composites. Calcium analysis confirms that calcium was inside of the composites after 2 days. After that the samples are collected with the help of a spatula and stored at 4 °C in excess 12.5 mM $CaCl_2$ with 0.02% NaN_3.

3.4.3 Composites Containing Xyloglucan and Pectin: Three Components Composites

Production of homogeneous three components composites is challenging due to the need for balancing pectin network strength with homogeneous dispersion of a soluble hemicellulose. However, homogeneous three component composites can be produced by using a low $CaCl_2$ solution in the medium containing pectin and xyloglucan and, after harvesting, further enriching the composites with calcium using a higher concentration of $CaCl_2$.

1. The three components composite is prepared as per Subheading 3.3 in combination with Subheading 3.4.1. In brief the cellulose is produced in HS medium containing the hemicellulose, pectin and $CaCl_2$ (2.5 mM) solutions.
2. Composites are carefully harvested with the help of a spatula and further enriched in calcium by soaking the composites in 12.5 mM $CaCl_2$ solution.
3. After that the samples are collected, washed with $CaCl_2$ 12.5 mM (as per Subheading 3.4.1). They are stored at 4 °C in excess 12.5 mM $CaCl_2$ with 0.02% NaN_3.

3.5 Isotopically Labeled Cellulose

Isotopically labeled cellulose material are produced as follows, where due to the high cost of the isotopically labeled carbon source, cellulose production is scaled down. Typically 12-well flat bottom cell culture plates can be used to produce pellicles. Here deuterated (carbon-bound ^2H) cellulose is used as an example.

1. The deuterated HS (d-HS) growth medium is prepared as follows. The medium is prepared as a concentrated (2×) solution by adding 0.55 g peptone, 0.55 g yeast extract, 0.372 g $Na_2HPO_4.H_2O$, and 0.127 g citric acid to 30 mL deionized water. The pH is adjusted to 5.0 with 10 M HCl, and the volume of the solution is adjusted to 50.6 mL with sterile deionized water. The medium is sterilized by autoclaving at 121 °C for 15 min. After cooling the sterile medium to 55 °C, 4.4 mL of 0.22 μm filter-sterilized deuterated glucose solution (50% w/v), and 55 mL of the 1% (w/v) polysaccharide solution are added aseptically, with the final glucose and hemicellulose concentrations in the medium being 2.0% (w/v) and 0.5% (w/v) respectively.

2. For primary inoculum preparation: Inoculate 20 mL d-HS broth medium with a few colonies of bacteria from the HS agar plate used to maintain the strain. Incubate under static conditions at 30 °C for 48 h (*see* **Note 12**).
3. For scale-up preparation: Inoculate 90 mL d-HS broth medium with 15 mL primary inoculum and statically incubate at 30 °C for 48 h. Once mixed by gently swirling by hand, aliquot 3 mL of the mixture into each of the wells in a 12 well plate to generate the desired number of pellicles.
4. Harvesting and storage of deuterated cellulose is carried out as described above in Subheading 3.3, **steps 5** and **6**.

4 Notes

1. *K. xylinus* ATCC 53524 is the strain of choice in our laboratory, as it produces cellulose that is chemically pure and highly crystalline [28]. In particular, this strain does not produce the water-soluble polysaccharide acetan, a heteropolymer containing glucose, mannose, glucuronic acid, and rhamnose in a molar ratio of 4:1:1:1 that is characteristic of some other *Komagataeibacter* strains [29, 30].
2. The medium detailed in this chapter is typically used in our laboratory. Alternatively, glucose can be substituted with other carbon sources such as mannitol or glycerol [28], as well as deuterated carbon sources including D-glucose [2] and D-glycerol (unpublished work).
3. The hemicelluloses routinely used in our laboratory include arabinoxylan (wheat, medium viscosity ~25 cSt), β-glucan (barley, medium viscosity ~28 cSt), and xyloglucan (tamarind seed; high viscosity ~6.5 dL/gram) (Megazyme International Ireland Ltd., County Wicklow, Ireland). They are stored at room temperature in a desiccator.
4. The pectins routinely used in our laboratory include commercially produced citrus extracted pectins of degree of methyl esterification (DM) 30–35 and 60–65. All pectins are stored at −20 °C as the DM is stable for up to a year. However, whether the pectin is stored at room temperature or at −20 °C, it is recommended that when preparing the pectin solution, the DM is routinely determined. This involves using titrimetry [31, 32]. Briefly, aliquots (1–5 mL) of a 1% pectin solution are titrated to between pH 7 and 8 with 0.02 M NaOH and the titer is recorded. Thereafter, 1 mL 0.5 M HCl solution is added and the solution is titrated again to pH 7–8. Blank values, obtained by substituting pectin with water, are

subtracted from the titer of pectin. The DM can be calculated directly from the titers using the following equation:

$$\mathrm{DM} - 100 \times V_s/V_t$$

where V_s is the hydrolyzed (or saponification) titer in milliliters, and V_t is the total titer (sum of initial titer plus hydrolyzed titer) in milliliters [31, 32].

5. The size of culture container depends on the desired size of the composite pellicle. Sterile yellow lid specimen containers (70 mL capacity; 40 mm diameter), standard size (92 × 16 mm) sterile Petri dishes and large size (150 × 20 mm) sterile Petri dishes are commonly used in our laboratory. If yellow lid specimen containers are used, care must be taken to ensure that the lid is screwed on only very loosely. As this organism is an obligate aerobe [30], this allows for adequate aeration of the culture medium and subsequently pellicle formation is not negatively impacted.
6. Great care must be taken when preparing 0.02% (w/v) sodium azide solution. Appropriate personal protective equipment (specified in the Material Safety Data Sheet) must be worn when weighing out the powder. Sterilized deionized water may be used for the preparation of the solution to be used for the hemicellulose composites. For pectin composites, sterile $CaCl_2$ solution of the appropriate concentration (e.g., 12.5 mM for DM 30 pectin composites) should be used when preparing the 0.02% (w/v) sodium azide solution. Once sodium azide is added, under no circumstances should these solutions be sterilized by autoclaving.
7. The culture from the revival plate is only subcultured once, and cultures over one month old are discarded. This approach is adopted to not only avoid spontaneous mutation of the bacterium, but also to consistently use healthy, viable cultures.
8. Microbank™ is a cryopreservation system commercially available from Pro-Lab Diagnostics (Ontario, Canada). Each vial contains a cryopreservative solution and approximately 25 porous beads which serve as carriers to support microorganisms. Using this long term storage solution, we have successfully stored our stock cultures for over 4 years and still have 100% success with revival as well as no contamination.
9. The hemicellulose solution cannot be sterilized by autoclaving or microfiltration as this may depolymerize or remove some of the polymer respectively. Thus, in order to prevent contamination, it is imperative to use sterilized equipment and diluent when making up the solution. Although it is common practice to add the polymer slowly to solvent, when working in a laminar flow cabinet, the air flow makes it difficult to handle fine powders.

Table 1
The effect of ethanol on the growth of *Komagataeibacter xylinus* ATCC 53524

HS-AX medium	Increase in cell growth ($\log_{10}$ CFU/mL)[a]	Final pH[b]	Cellulose yield at 48 h (g)
With EtOH	4.58	3.23	0.0353
Without EtOH	2.85	5.01	0.0356

[a]Values are the difference between cell growth at $t = 0$ h and $t = 48$ h. All values are presented as mean colony counts
[b]Initial pH = 5.0. Results are presented as means of triplicates

10. Wetting the plant cell wall polysaccharide with 40 mL of 95% (v/v) ethanol is recommended by the manufacturer. However, this is not appropriate as *K. xylinus* is able to utilize ethanol as a carbon source [33]. Subsequently, this results in a significant ($P < 0.05$) increase in bacterial cell numbers, but not in the rate of cellulose production (Table 1). By omitting substrate wetting with alcohol, the approximate solubilization times for arabinoxylan, β-glucan and xyloglucan are 30–60 min.
11. Arabinoxylan solutions typically have a slight off-white opalescent appearance, while xyloglucan and β-glucan solutions are sometimes very slightly turbid. This may be due to the presence of trace amounts of protein (Megazyme product datasheet).
12. Primary inoculum volumes depend on the desired scale up volume. Always prepare sufficient primary inoculum to ensure that exactly 10% (v/v) primary inoculum is used during scale up.
13. Washing of composite pellicle after harvesting is carried out to remove excess medium, polysaccharides nonspecifically trapped within the cellulose mat, and bacterial cells. This process is carried out until the pellicle changes from off-white to white color. This is typically achieved by carrying out at least six washes (3 × 30 min, followed by 3 × 10 min washes).
14. It is our experience that hemicellulose–cellulose composites are stable for up to approximately 2 months when stored at 4 °C. After this period, degradation is observed with the edges of the otherwise opaque pellicle becoming translucent—this observation is prominent in pellicles that have been stored at 4 °C for six months. Deuterated cellulose material appears to be stable at 4 °C for up to 6 months (the longest we have ever stored them prior to various analyses).
15. We choose to stir the pectin solution overnight as this ensures complete dissolution. Due to this overnight process, and the fact that the pectin solution cannot be sterilized by autoclaving or means of filtration, contamination is minimized/prevented by using sterilized equipment and diluent when making up the solution. Pectins are heat labile, so no heating is used during the dissolution process.

16. The inoculated container is placed on a platform shaker and $CaCl_2$ solution is added under vigorous shaking, followed by a further 5 min of shaking, as this allows the HS-pectin medium to subsequently gel under quiescent conditions in a uniform manner. Uniformity of the preformed gel is important for the formation of a relatively homogeneous composite.
17. Avoid using forceps when harvesting the bacterial cellulose–pectin composites, as the pellicles can be fragile and tear easily. Moreover, due to their highly extensible or "stretchy" nature, the use of forceps can impact on their shape. This is important to note when handing the samples for tensile stress/strain measurements, and also applies to other "weak" composites such as those produced after limited incubation times (e.g., 24 h).
18. Note that the purification of pectin composites by washing with $CaCl_2$ solution causes the pellicle to shrink slightly (*see* Fig. 5).
19. In order to ensure good storage conditions for pectin composites, it is critical to use 0.02% (w/v) sodium azide solution containing the appropriate concentration of $CaCl_2$. If sodium azide solution made up with deionized water is used, composites are stable for up to approximately 2 months when stored at 4 °C. After this period, pellicle degradation is observed, with the sodium azide solution becoming viscous and the pellicle losing some of its characteristic "lumpy" texture (due to precipitation of pectin from the composite)—this observation is prominent in pellicles that have been stored at 4 °C for 6 months.

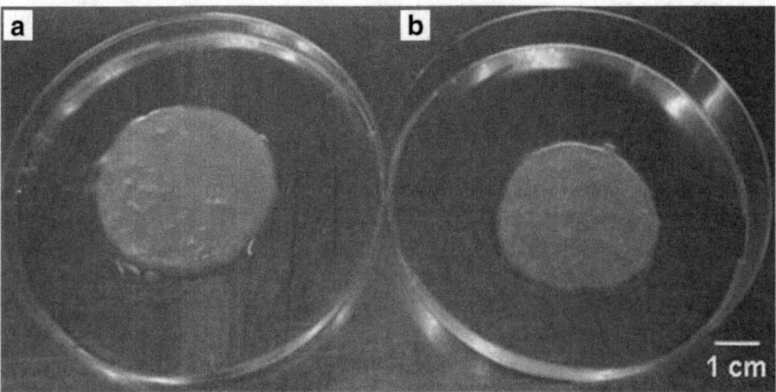

Fig. 5 Bacterial cellulose–pectin composites before (**a**) and after (**b**) washing with $CaCl_2$ solution (typically 12.5 mM), demonstrating typical shrinkage from 4.2 cm in width to 3.8 cm after washing the pellicle

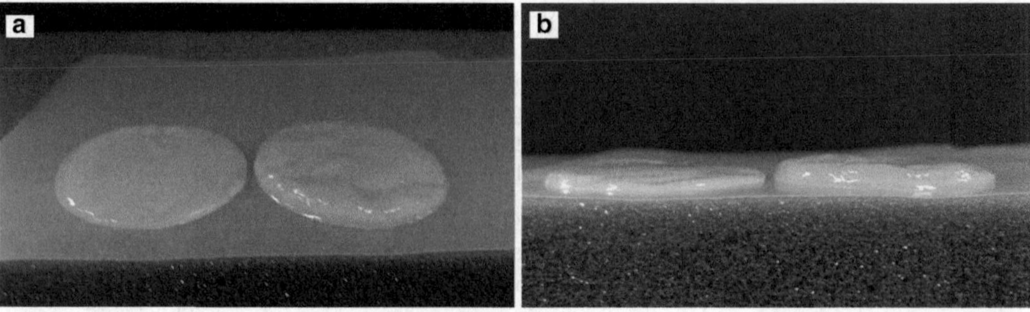

Fig. 6 Bacterial cellulose–pectin composites in which the pectin gel was introduced after soaking cellulose pellicles overnight (**a**) in an excess volume of pectin solution, and thereafter the pectin was gelled by addition of $CaCl_2$, resulting in an increase of pellicle thickness (**b**)

20. The method adopted here is modified from that of Schuster et al. [34]. A thin layer of pectin gel is formed on the surface of the bacterial cellulose pellicle (Fig. 6). The pectin gel layer is a weak gel; hence, it is difficult to measure its thickness. Also, the composites thickness varies depending on which pectin and what pectin concentration is used, and bacterial cellulose thickness varies between batches. However, pellicles produced in this manner are always 2–3 mm thick.

References

1. Brown RM (1989) Bacterial cellulose. In: Phillips GO, Kennedy JF, Williams PA (eds) Cellulose: structural and functional aspects. Ellis Horwood Ltd, New York, pp 145–151
2. Martínez-Sanz M, Mikkelsen D, Flanagan B, Gidley MJ, Gilbert E (2016) Multi-scale model for the hierarchical architecture of native cellulose hydrogels. Carbohydr Polym 147:542–555
3. Whitney SEC, Brigham JE, Darke A, Reid JSG, Gidley MJ (1995) *In vitro* assembly of cellulose/xyloglucan networks: ultrastructural and molecular aspects. Plant J 8:491–504
4. McCann MC, Wells B, Roberts K (1990) Direct visualization of cross-links in the primary plant-cell wall. J Cell Sci 96:323–334
5. Whitney SEC, Gidley MJ, McQueen-Mason SJ (2000) Probing expansin action using cellulose/hemicellulose composites. Plant J 22:327–334
6. Chanliaud E, DeSilva J, Strongitharm B, Jeronimidis G, Gidley MJ (2004) Mechanical effects of plant cell wall enzymes on cellulose/xyloglucan composites. Plant J 38:27–37
7. Whitney SEC, Brigham JE, Darke A, Reid JSG, Gidley MJ (1998) Structural aspects of the interaction of mannan-based polysaccharides with bacterial cellulose. Carbohydr Res 307:299–309
8. McKenna BA, Wehr JB, Mikkelsen D, Blamey FPC, Menzies NW (2016) Aluminium effects on mechanical properties of cell wall analogues. Physiol Plant 158:382–388
9. Gu J, Catchmark JM (2013) The impact of cellulose structure on binding interactions with hemicellulose and pectin. Cellulose 20:1613–1627
10. Gu J, Catchmark JM (2014) Roles of xyloglucan and pectin on the mechanical properties of bacterial cellulose composite films. Cellulose 21:275–289
11. Park YB, Lee CM, Kafle K, Park S, Cosgrove DJ, Kim SH (2014) Effects of plant cell wall matrix polysaccharides on bacterial cellulose structure studied with vibrational sum frequency generation spectroscopy and X-ray diffraction. Biomacromolecules 15:2718–2724
12. Mikkelsen D, Flanagan BM, Wilson SM, Bacic A, Gidley MJ (2015) Interactions of arabinoxylan and (1,3) (1,4)-beta-glucan with cellulose networks. Biomacromolecules 16:1232–1239
13. Padayachee A, Netzel G, Netzel M, Day L, Zabaras D, Mikkelsen D, Gidley M (2012)

Binding of polyphenols to plant cell wall analogues—part 1: anthocyanins. Food Chem 134:155–161

14. Padayachee A, Netzel G, Netzel M, Day L, Zabaras D, Mikkelsen D, Gidley M (2012) Binding of polyphenols to plant cell wall analogues—part 2: phenolic acids. Food Chem 135:2287–2292
15. Padayachee A, Netzel G, Netzel M, Day L, Mikkelsen D, Gidley M (2013) Lack of release of bound anthocyanins and phenolic acids from carrot plant cell walls and model composites during simulated gastric and small intestinal digestion. Food Funct 4:906–916
16. Phan ADT, Netzel G, Wang D, Flanagan B, D'Arcy B, Gidley MJ (2015) Binding of dietary polyphenols to cellulose: structural and nutritional aspects. Food Chem 171:388–396
17. Tan MSF, Rahman S, Dykes GA (2016) Pectin and xyloglucan influence the attachment of *Salmonella enterica* and Listeria monocytogenes to bacterial-cellulose-derived plant cell wall models. Appl Environ Microbiol 82:680–688
18. Tan MSF, Wang Y, Dykes GA (2013) Attachment of bacterial pathogens to a bacterial cellulose-derived plant cell wall model: a proof of concept. Foodborne Pathog Dis 10:992–994
19. Chanliaud E, Burrows KM, Jeronimidis G, Gidley MJ (2002) Mechanical properties of primary plant cell wall analogues. Planta 215:989–996
20. McKenna BA, Mikkelsen D, Wehr JB, Gidley MJ, Menzies NW (2009) Mechanical and structural properties of native and alkali-treated bacterial cellulose produced by *Gluconacetobacter xylinus* strain ATCC 53524. Cellulose 16:1047–1055
21. Lopez-Sanchez P, Cersosimo J, Wang D, Flanagan B, Stokes JR, Gidley MJ (2015) Poroelastic mechanical effects of hemicelluloses on cellulosic hydrogels under compression. PLoS One 10:e0122132
22. Lopez-Sanchez P, Rincon M, Wang D, Brulhart S, Stokes JR, Gidley MJ (2014) Micromechanics and poroelasticity of hydrated cellulose networks. Biomacromolecules 15:2274–2284
23. Whitney SEC, Gothard MGE, Mitchell JT, Gidley MJ (1999) Roles of cellulose and xyloglucan in determining the mechanical properties of plant cell walls. Plant Physiol 121:657–663
24. Zykwinska AW, Ralet MCJ, Garnier CD, Thibault JFJ (2005) Evidence for *in vitro* binding of pectin side chains to cellulose. Plant Physiol 139:397–407
25. Chanliaud E, Gidley MJ (1999) *In vitro* synthesis and properties of pectin/*Acetobacter xylinus* cellulose composites. Plant J 20:25–35
26. Hestrin S, Schramm M (1954) Synthesis of cellulose by *Acetobacter xylinum*. 2. Preparation of freeze-dried cells capable of polymerizing glucose to cellulose. Biochem J 58:345–352
27. Willey JM, Sherwood LM, Woolverton CJ (2008) Prescott, Harley, and Klein's microbiology. McGraw-Hill, New York, p 115
28. Mikkelsen D, Flanagan BM, Dykes GA, Gidley MJ (2009) Influence of different carbon sources on bacterial cellulose production by *Gluconacetobacter xylinus* strain ATCC 53524. J Appl Microbiol 107:576–583
29. Couso RO, Ielpi L, Dankert MA (1987) A xanthan-gum-like polysaccharide from *Acetobacter xylinum*. J Gen Microbiol 133:2123–2135
30. Kersters K, Lisdiyanti P, Komagata K, Swings J (2006) The Family Acetobaceraceae: the Genera Acetobacter, Acidomonas, Asaia, Gluconacetobacter, Gluconobacter and Kozakia. In: Dwokin M (ed) The Prokaryotes: an evolving electronic resource for the microbiological community. Springer Verlag, New York, pp 163–200
31. Marga F, Morvan C, Morvan H (1995) Pectins in normal and vitreous apple microplants cultured in liquid-medium. Plant Physiol Biochem 33:81–86
32. Wehr JB, Menzies NW, Blamey FPC (2004) Alkali hydroxide-induced gelation of pectin. Food Hydrocolloid 18:375–378
33. Ross P, Mayer R, Benziman M (1991) Cellulose biosynthesis and function in bacteria. Microbiol Rev 55:35–58
34. Schuster E, Eckardt J, Hermansson AM, Larsson A, Loren N, Altskar A, Strom A (2014) Microstructural, mechanical and mass transport properties of isotropic and capillary alginate gels. Soft Matter 10:357–366
35. Patricia Lopez-Sanchez, Marta Martinez-Sanz, Mauricio R. Bonilla, Dongjie Wang, Elliot P. Gilbert, Jason. R. Stokes, Michael. J. Gidley, (2017) Cellulose-pectin composite hydrogels: Intermolecular interactions and material properties depend on order of assembly. Carbohydrate Polymers 162:71–81

Chapter 6

Plant Tissue Cultures

Anna Kärkönen, Arja Santanen, Kuninori Iwamoto, and Hiroo Fukuda

Abstract

Plant tissue cultures are an efficient system to study cell wall biosynthesis in living cells in vivo. Tissue cultures also provide cells and culture medium from which enzymes and cell wall polymers can easily be separated for further studies. Tissue cultures with tracheary element differentiation or extracellular lignin formation have provided useful information related to several aspects of xylem and lignin formation. In this chapter, methods for nutrient medium preparation and callus culture initiation and its maintenance as well as those for protoplast isolation and viability observation are described. As a case study, we describe the establishment of a xylogenic culture of *Zinnia elegans* mesophyll cells.

Key words Callus culture, Initiation, Maintenance, Nutrient medium, Protoplast, Tracheary element

1 Introduction

Plant cells and organs can be cultivated in vitro in aseptic conditions [1]. Plant tissue cultures are an efficient system to study cell wall biosynthesis in living cells in vivo. Tissue cultures also provide cells and culture medium where enzymes and cell wall polymers can easily be separated for further studies. In vitro cultures allow for investigations to be conducted in controlled conditions independent of seasons. Factors related to cell wall formation can be studied, for example, by adding the compound of interest into the culture medium, and after incubation, the cells and the medium are collected for further analysis. The culture medium can be considered as a continuum of the apoplastic fluid that permeates the plant cell walls, since it contains the proteins and the cell wall polymers that are sloughed off from the cell wall. In a callus culture, cells grow mainly as a mass of undifferentiated cells, but there exist also some differentiated cells making the callus as an inhomogeneous mixture of cells. With certain growth regulators, organogenesis (shoots or roots) or somatic embryo formation may be induced. In the latter case, callus is embryogenic. Tissue cultures with tracheary element (TE) differentiation [2] and secondary cell

Zoë A. Popper (ed.), *The Plant Cell Wall: Methods and Protocols*, Methods in Molecular Biology, vol. 2149, https://doi.org/10.1007/978-1-0716-0621-6_6, © Springer Science+Business Media, LLC, part of Springer Nature 2020

wall or extracellular lignin formation [3–6] have provided useful information related to several aspects of xylem and lignin formation. One of the most famous xylogenic cultures is that of *Zinnia elegans* [2]. *Zinnia* system is useful for studying the sequence of events during xylem differentiation largely because the differentiation is highly frequent and synchronous, and all processes can be followed in single cells. Systematic gene expression analysis and molecular markers have revealed that many processes are common between in vitro and in situ TE formation [7, 8]. Studies using this system have clarified numbers of physiological, biochemical, cell biological, and molecular biological events underlying TE differentiation [9–11].

This paper describes the procedures for surface sterilization, callus culture initiation and maintenance, as well as protoplast preparation and viability observation. Finally, we describe a basic method for the establishment of a xylogenic culture of isolated *Zinnia* mesophyll cells.

2 Materials

2.1 Nutrient Medium

A nutrient medium is a source of nutrients that plants normally obtain from the soil. The medium contains also a carbon source (often 1–4% w/v sucrose) and growth regulators that plant cells need for cell divisions and growth in vitro. A gelling agent is usually added to make the medium solid [1]. As various species (even genotypes) have different nutritional requirements for the optimum growth, a wide variety of nutrient media have been developed for in vitro cultured plants. In order to select a medium for the species of interest, it is useful to make a literature search. In Table 1 we show some widely used media that can be the choices to start with (*see* **Note 1**). Additionally, explant types with successful callus initiation are listed, since the developmental stage of the plant has a great effect on the success of culture initiation.

Table 2 shows the nutrient salt compositions of the various media. Many macro- and microelement mixtures can be purchased commercially, or the stock solutions can be prepared from the nutrient salts. A mixture of macroelements can be prepared as 10 times concentrated (10×) stock solution, whereas those of microelements, vitamins and growth regulators can be made as 100–1000× stock solutions (*see* **Notes 2–5**). After combining all components of the medium except the gelling agent, adjust the pH, adjust to the final volume, add the gelling agent (e.g., agar) and autoclave the medium at 121 °C for 20 min. Let the medium cool to ca. 60 °C. In a laminar air-flow cabinet, filter-sterilize the heat-labile compounds (if any, *see* **Note 6**), and pour the medium into Petri dishes (ca. 25 mL medium/Petri dish with a diameter of 9 cm). *See* **Note 7**.

Table 1
Types of explants, some widely used nutrient media, and growth regulator concentrations used for successful callus culture initiation

Plant group	Explant	Medium	Growth regulators
Monocotyledonous plants	Immature and mature embryos Leaf and root segments of aseptically germinated seedlings	MS [14], N6 [24]	Auxin (2,4-D) 1.0–18 μM
Dicotyledonous plants	Young leaves, roots, stem segments Leaves, roots, and stem segments of aseptically germinated seedlings	MS, WPM [25]	Cytokinin (BA, 2iP, kinetin, zeatin) 0.1–40 μM + Auxin (NAA, 2,4-D) 0.5–10 μM
Gymnosperms	Cambial/xylem strips	MS	Cytokinin (BA, kinetin) 2–5 μM +
	Shoot tips	mod. Brown and Lawrence [3]	Auxin (2,4-D, NAA) 9–16 μM
	Zygotic embryos	mod. N6 [12]	2.4-D alone: 11 μM [3]

2.2 Surface Sterilization

1. 70% (v/v) ethanol.
2. Diluted Na-hypochlorite: NaClO, 1–2% (v/v) active chlorine, supplemented with a couple of drops of Tween 20.
3. Sterile distilled water.
4. 96% (v/v) ethanol for flaming.
5. Forceps.
6. Scalpels.
7. Sterile Petri dishes.
8. Parafilm®.

2.3 Maintenance

1. Fresh nutrient medium with a gelling agent or, alternatively, without the gelling agent.
2. 96% (v/v) ethanol for flaming.
3. Forceps.
4. Parafilm®.
5. Sterile 5-mL pipette tips with cut tips.
6. Sterile measuring cylinders (e.g., 25 mL in volume).
7. Orbital shaker (in the case of liquid cultures).
8. Temperature- and light-adjusted growth chamber.

Table 2
Nutrient medium constituents for plant tissue culture basal media [26]

	B5 [15]		mod. Brown and Lawrence [3]		MS [14]		N6 [24]		WPM [25]	
Macronutrients	mg/L	mM	mg/L	mM	mg/L	mM	mg/L	mM	mg/L	mM
NH_4NO_3			1650	20.6	1650	20.6			400	5
$(NH_4)_2SO_4$	134	1.0					463	3.5		
$Ca(NO_3)_2{\cdot}4H_2O$									556	2.4
KNO_3	2528	25	1900	18.8	1900	18.8	2830	28		
$MgSO_4{\cdot}7H_2O$	246	1.0	1900	7.7	370	1.5	185	0.75	370	1.5
KH_2PO_4			340	2.5	170	1.25	400	2.94	170	1.25
$NaH_2PO_4{\cdot}H_2O$	150	1.1								
$CaCl_2{\cdot}2H_2O$	150	1.0	22	0.15	440	3.0	166	1.1	96	0.65
K_2SO_4									990	5.7
Micronutrients		μM		μM		μM		μM		μM
H_3BO_3	3.0	49	30.9	500	6.2	100	1.6	26	6.2	100
KI	0.75	4.5	4.15	25	0.83	5.0	0.8	4.8		
$MnSO_4{\cdot}4H_2O$	13.2	59.2	31.2	140	22.3	100	4.4	19.7	22.3	100
$ZnSO_4{\cdot}7H_2O$	2.0	7.0	43.1	150	8.6	30	1.5	5.2	8.6	30
$CuSO_4{\cdot}5H_2O$	0.025	0.1	1.0	4.0[a]	0.025	0.1			0.25	1.0
$Na_2MoO_4{\cdot}2H_2O$	0.25	1.0	1.2	5.0	0.25	1.0			0.25	1.0
$CoCl_2{\cdot}6H_2O$	0.025	0.1	0.13	0.55	0.025	0.1				
$FeSO_4{\cdot}7H_2O$	27.8	100	27.8	100	27.8	100	27.8	100	27.8	100
$Na_2EDTA{\cdot}2H_2O$	37.2	100	37.2	100	37.2	100	37.2	100	37.2	100
Organic constituents:										
Myoinositol	100	560	20	111	100	560	100	560	100	560
Nicotinic acid	1.0	8.1	0.5	4.1	0.5	4.1	0.5	4.1	0.5	4.1
Pyridoxine–HCl	1.0	4.9	0.1	0.49	0.5	2.4	0.5	2.4	0.5	2.4
Thiamine–HCl	10	30	0.1	0.3	0.1	0.3	1.0	3.0	1.0	3.0
Glycine					2.0	26.6	2.0	26.6	2.0	26.6
	g/L	mM	g/L	mM	g/L	mM	g/L	mM	g/L	mM
Sucrose	20	58.4	30	87.6	30	87.6	20	58.4	20	58.4
pH		5.7		5.5		5.7		5.7		5.7

[a]L.B. Davin, personal communication

Table 3
Solutions for protoplast preparation and culturing

Preplasmolysis solution:	
B5/MS Macroelements	
B5/MS Microelements	
Sucrose	60 mM
Mannitol/sorbitol	0.3–0.5 M
pH 5.7 (*see* **Note 33**)	
Autoclave	
Enzyme solution (make fresh each time):	
0.5% (w/v) Cellulase and 0.2% (w/v) Macerase *or*	
0.1–4% (w/v) Cellulase, 0.05–2% (w/v) Pectolyase/Macerase, and 0.1–2% (w/v) Hemicellulase in the preplasmolysis solution	
Mix gently for 15–30 min to dissolve, filter-sterilize through syringe filters (0.2 μm pore size).	
Nutrient medium for protoplast culturing:	
B5/MS macroelements	
B5/MS microelements	
NaFe-EDTA	100 μM
B5/MS vitamins	
Sucrose	60 mM
Mannitol/sorbitol	0.3–0.5 M
Myoinositol	100 mg/L
Plant growth regulators:	
Auxin (2,4-D/NAA/IAA)	1–10 μM
Cytokinin (BA/2iP/kinetin/zeatin)	0.5–2.5 μM
pH 5.7 (*see* **Note 33**)	
Autoclave	

See Table 2 for B5/MS medium constituents

2.4 Protoplasts

1. Preplasmolysis solution, enzyme solution and nutrient medium for protoplast cultivation according to Table 3.
2. Syringes.
3. Syringe filters (0.2 μm pore size).
4. Forceps.
5. Scalpels.

6. 96% (v/v) ethanol for flaming.
7. Sterile nylon or steel sieves (70–100 μm pore size), screw cap centrifuge tubes.
8. 20% (w/v) sucrose solution (autoclaved).
9. Fuchs-Rosenthal modified hemocytometer.
10. Microscopic slides.
11. Cover glasses.
12. Agars with low melting point (m.p.) specifically designed for protoplast culturing (e.g., A8678 Agar washed, m.p. 25–27 °C; A7921 Agar purified, m.p. 30–35 °C, Sigma-Aldrich).
13. Sterile pipette tips.
14. Petri dishes.
15. Parafilm®.

Viability stains:

16. 5–10 mg/mL fluorescein diacetate in acetone (stock solution). This is diluted immediately prior to use by adding 20 μL of the stock solution to 1 mL of 0.65 M mannitol.
17. 0.025–0.25% (w/v) Evans blue in 0.65 M mannitol.
18. 0.025–0.25% (w/v) Methanol blue in 0.65 M mannitol.
19. 0.1% (w/v) Phenosafranine in 0.65 M mannitol.
20. 0.01–0.1% (w/v) Tinopal CBS-X (disodium 4,4′-bis[2-sulfostyryl]biphenyl) in 0.65 M mannitol.

2.5 *Zinnia* Cultures

2.5.1 Germination of *Zinnia* Seeds

1. 0.25% Na-hypochlorite.
2. Mesh strainer.
3. Vermiculite.
4. Plastic trays.
5. Growth chamber.
6. Liquid fertilizer (e.g., HYPONeX; N:P:K = 6:10:5 [HYPONeX Japan, Osaka]). Dilute 1:100 before use.

2.5.2 Isolation and Culture of Mesophyll Cells

1. Table 4 shows the composition of the nutrient medium (*see* **Note 8**). Frequency of TE differentiation is optimal when the nutrient medium is supplemented with 0.89 μM 6-benzyladenine (BA) and 0.54 μM 1-naphtaleneacetic acid (NAA). Medium without BA and/or NAA can be used for control cultures in which TE differentiation does not occur.
2. 0.1% Na-hypochlorite with 0.001% (w/v) Triton X-100.
3. Sterile distilled water.

Table 4
Nutrient medium for xylogenic culture of *Zinnia* mesophyll cells

Constituents	Concentration (mg/L)	Molarity
Macroelements		mM
KNO_3	2020	20
$MgSO_4 \cdot 7H_2O$	247	1
$CaCl_2 \cdot 2H_2O$	147	1
KH_2PO_4	68	0.5
NH_4Cl	54	1
Microelements I		μM
$MnSO_4 \cdot 4H_2O$	25	110
H_3BO_3	10	160
$ZnSO_4 \cdot 7H_2O$	10	35
$Na_2MoO_4 \cdot 2H_2O$	0.25	1
$CuSO_4 \cdot 5H_2O$	0.025	0.1
Microelements II		μM
$Na_2EDTA \cdot 2H_2O$	37	100
$FeSO_4 \cdot 7H_2O$	28	100
Organic growth factors I		μM
Myoinositol	100	560
Nicotinic acid	5	41
Glycine	2	27
Pyridoxine–HCl	0.5	2.4
Thiamine–HCl	0.5	1.5
Biotin	0.05	0.2
Organic growth factors II		μM
Folic acid	0.5	1.1
Growth regulators		μM
NAA	0.1	0.54
BA	0.2	0.89
	g/L	mM
Sucrose	10	29.2
D-mannitol	36.4	200
pH	5.5	

4. Sterile labware: Waring-type blender, stainless steel cups, nylon mesh (50–80 μm pore size), screw-cap centrifuge tubes, pipette tips, culture tubes (30 mm internal diameter (i.d.). × 200 mm, 18 mm i.d. × 180 mm or 12 mm i.d. × 105 mm) capped with aluminum foil.
5. Revolving drum.
6. Growth chamber.

2.5.3 Observations of Zinnia Cells

1. Glutaraldehyde.
2. 0.2 mg/mL 4′,6-diamidino-2-phenylindole (DAPI), 1 mM SYTO16 in DMSO (Molecular Probes).
3. Microscopic slides.
4. Cover glasses.
5. Hemocytometer.

3 Methods

3.1 Surface Sterilization

The idea of surface sterilization is to selectively kill microorganisms on the plant material without killing the plant tissue. For culture initiation, it is important to select a healthy plant tissue as an explant. If necessary, wash the plant organ with tap water and cut it to ca. 1 cm pieces. Seeds are surface-sterilized intact. Carry out the procedures aseptically in the laminar air-flow cabinet.

1. Pretreat the explants for 1 min in 70% (v/v) ethanol.
2. Transfer the pieces into a diluted 1–2% (v/v) Na hypochlorite solution supplemented with a couple of drops of Tween 20. Incubate for 5–30 minutes (*see* **Note 9**) with occasional shaking.
3. Rinse the explants carefully with sterile distilled water (three times with at least 1 min incubation in each rinse to wash all surface sterilant away). Use alcohol-flamed forceps to transfer the pieces from one solution into the other (*see* **Note 10**).
4. As the cut surfaces of the plant material are injured by the contact with the surface-sterilizing agent, cut the surfaces fresh with a sterile scalpel by using a half of a sterile Petri dish as a cutting board.
5. Aseptically dissect the tissue of interest (e.g., embryo, cambial strips) out of the seed/plant organ and place it onto the surface of the initiation medium. Use stereomicroscope in the laminar air-flow cabinet if needed. Seal the dish by using a strip of Parafilm®.

3.2 Growth Conditions

The temperature and light requirements depend on the plant species. If no information exists in literature in relation to in vitro growth conditions of the species (or related species) of your interest, it might be useful to choose the light and temperature conditions in which the plant grows in vivo. Usually, a constant temperature (e.g., +25 °C) is used; alternatively the temperature is reduced for night (+25 °C during day, +20 °C during night). The quality of light is obtained by selection of lamps. Some species, like Norway spruce, prefer fluorescent warm white lamps [5, 12].

The intensity of light and its rhythm are very important. It is basically by trial and error you can optimize these values unless some information is available in literature. In general, light intensities of 20–200 μmol/(m^2s) are used. Some in vitro cultures, however, are cultivated in the dark.

3.3 Maintenance

After 2–6 weeks in culture, callus growth becomes visible at the edges of the explant (Fig. 1). You have to subculture callus in order to supplement the cells with fresh nutrients and growth regulators. Subculture cells in 1- to 4-week intervals depending on the growth of the callus. You may need to modify the nutrient medium at this stage in relation to concentrations and types of nutrients and growth regulators (*see* **Notes 1** and **11**).

1. Subculture callus by transferring the freshest cells (usually at the edges of callus) onto the fresh medium with flamed forceps (*see* **Note 10**). The size of the inoculum should be kept in a constant size (ca. 0.9 × 0.9 × 0.5 cm; *see* **Note 12**). Do not transfer inoculums which are too small in size because it takes longer for cells to start dividing if they are too little.

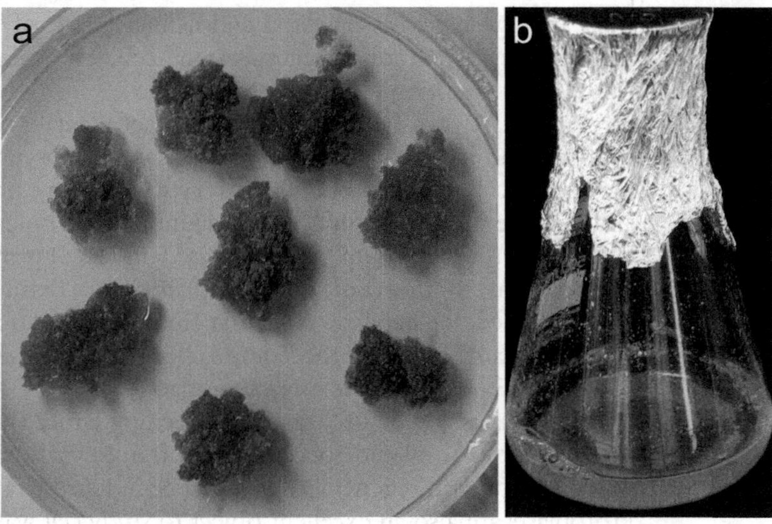

Fig. 1 (**a**) Callus culture of Norway spruce (*Picea abies*). (**b**) Cell suspension culture of Norway spruce composed of single cells and small cell aggregates

Alternatively, if you sub-culture inoculums which are too big the cells multiply very fast and enter the stationary phase early (they also fill the growth container). This means more frequent sub-culturing.

2. You can also transfer callus into liquid cultures (Fig. 1b). For this, make the nutrient medium without the gelling agent, aliquot it in 25 mL aliquots in 100 mL flasks (*see* **Note 13**), close the flasks with a double layer of aluminum foil and autoclave. Inoculate the most friable callus cells into the liquid medium (ca. 0.5 g of cells into 25 mL medium). It depends on the type of callus whether you get a fine cell suspension with single cells and small cell aggregates, or whether the callus grows in big clumps with no cell detachment.
3. For aeration, keep the cultures on an orbital shaker (100 rpm) in the same growth conditions as the cultures on the solid medium.
4. Subculture cells at regular intervals into fresh medium (see above) by letting cells to settle down to the bottom of the flask. Decant some culture medium off. Transfer ca. 5 mL cells into 20 mL of fresh medium, for example with the help of a 5-mL cut, autoclaved pipette tip, or with a sterile measuring cylinder.

3.4 Protoplasts

Protoplasts are plant cells that have their cell wall removed by digestion with plant cell wall-degrading enzymes pectinases, hemicellulases, and cellulases (Table 5). Protoplasts can be isolated enzymatically by two different ways. In a two-step method, cells are first separated to cell suspension with pectinases that digest the pectinous middle lamella between the cells. Then the remaining cell walls are digested with cellulases and hemicellulases. The one-step method uses a mixture of pectinases and cellulases simultaneously for cell wall digestion [13].

Protoplasts can be prepared from intact plant organs such as root tips and leaves, or from callus and suspension-cultured cells. Having no cell wall, protoplasts are very sensitive to osmotic stress, and must be handled in an isotonic/slightly hypertonic solution to prevent rupture. Nutrient medium requirements of protoplasts are quite similar to those of cultured plant cells. Extra calcium is supplemented to stabilize plasma membranes, and for different species, optimization of the composition of the often-used MS [14] and B5 [15] media is often essential.

Protoplasts can be used in plant breeding either through protoplast fusion of related species, or via transformation. After cell wall development, regenerable cells can be induced to form plants. Protoplasts are also an excellent object to study cell wall synthesis, or transport through cell membranes. Cell wall develops in

Table 5
Some commercially available cell wall–digesting enzymes utilized in protoplast isolation

Enzyme	Source	Supplier
Pectin-digesting enzymes:		
Driselase	*Basidiomycetes* sp.	Sigma-Aldrich
Macerase	*Rhizopus* sp.	Merck-Millipore
Macerozyme R-10	*Rhizopus* sp.	Sigma-Aldrich, Yakult Pharmaceutical Industry
Pectinase	*Aspergillus niger*	Sigma-Aldrich
Pectolyase Y-23	*A. japonicus*	Duchefa Biochemie, Sigma-Aldrich
Hemicellulose-digesting enzymes:		
Driselase	*Basidiomycetes* sp.	Sigma-Aldrich
Hemicellulase	*A. niger*	Sigma-Aldrich
Macerozyme R-10	*Rhizopus* sp.	Duchefa Biochemie, Yakult Pharmaceutical Industry
Viscozyme	*Aspergillus* sp.	Novozymes, Sigma-Aldrich
Cellulose-digesting enzymes:		
Cellulysin	*Trichoderma viride*	Merck-Millipore
Driselase	*Basidiomycetes* sp.	Sigma-Aldrich
Onozuka R-10	*T. viride*	Duchefa Biochemie, Serva, Yakult Pharmaceutical Industry
Onozuka RS	*Trichoderma* sp.	Duchefa Biochemie, Sigma-Aldrich

protoplasts normally during 24–36 h incubation, after which the cells are capable of division. Protoplasts lose their characteristic spherical shape once the cell wall formation is complete (Fig. 2).

3.4.1 Protoplast Isolation

Prepare the preplasmolysis and enzyme solutions according to Table 3. Then continue as described below.

Leaves:

1. Surface-sterilize young, fully expanded leaves as described in Subheading 3.1.
2. Cut the leaf into narrow sections with a sharp scalpel in a Petri dish that contains a small volume (10 mL) of the preplasmolysis solution. Peeling of the abaxial epidermis accelerates cell wall digestion by the enzymes as they enter intercellular spaces more easily (*see* **Note 14**).

Suspension-cultured cells:

3. Centrifuge actively growing cell suspension culture (10 mL) in the early logarithmic or exponential stage of growth for 5–10 min at 50–100 × *g* to separate the cells from the culture medium.

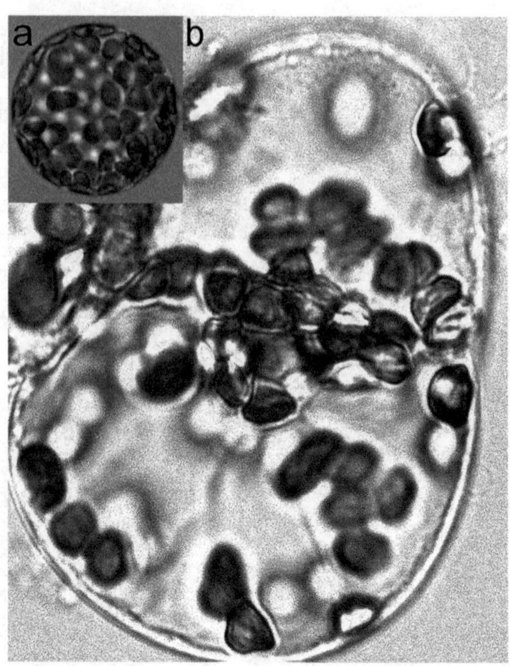

Fig. 2 (**a**) A protoplast made of *Nicotiana tabacum* leaves. (**b**) After a couple of days in culture, the cell wall has regenerated and the cell has divided. (Photograph courtesy of Enni Väisänen, Univ. of Helsinki)

4. After centrifugation, decant the medium and transfer the cells into a Petri dish with the preplasmolysis solution (see above).

 Callus culture:

5. Transfer actively growing callus cells from the edges of callus pieces into a Petri dish containing the preplasmolysis solution.
6. Incubate plant material in the preplasmolysis solution for 30 min; then replace the solution with the enzyme solution. Incubate for 0.5–20 h (*see* **Note 15**) in the dark at room temperature.
7. After incubation, shake the Petri dish gently to see that the material is digested; if not, incubate for 1–2 more hours.
8. Remove cell debris by pipetting protoplasts through a nylon or a steel sieve (70–100 μm pore size) into a sterile screw-cap centrifuge tube. Centrifuge for 5–10 min at 50–100 × *g*.
9. Resuspend the protoplast pellet in the preplasmolysis solution. Alternatively, fractionate protoplasts from the cell debris by pipetting the protoplast suspension on top of a 20% (w/v) sucrose solution.
10. Centrifuge for 5–10 min at 50–100 × *g*. The cell debris sediments to the bottom of the tube, and the protoplasts float at the interface of the sugar layer and the enzyme solution.

Transfer protoplasts on top of a fresh sucrose solution by pipetting, repeat washing for three times (*see* **Note 16**). Resuspend protoplasts in the nutrient medium at an appropriate density (*see* **Notes 17** and **18**).

3.4.2 Protoplast Viability Tests

Protoplast viability can be determined using different dyes that indicate viable or nonviable cells. Appropriate osmoticum has to be added to the staining solution to avoid protoplast bursting. Evans blue (EVB) is excluded from living cells and only dead cells are stained blue. Methanol blue (MB) enters both living and dead cells, but in living cells the dye is reduced to a colorless compound. Phenosafranine (PS) enters to dead protoplasts staining them red.

Fluorescent dyes: Fluorescein diacetate (FDA) accumulates inside protoplasts. In viable cells, FDA is cleaved to fluorescent fluorescein by an esterase. Tinopal CBS-X is capable of permeating only dead cells [16, 17].

1. Select the dye you will use in the viability staining. Prepare it as described in Subheading 2.4.
2. On a microscopic slide, mix equal volumes of the staining solution and the protoplast suspension, and overlay the sample with a cover glass.
3. Observe EVB, MB or PS in a light microscope and count the number of dead protoplasts per all protoplasts in some fields (*see* **Note 19**).
4. Observe FDA in a fluorescent microscope with the excitation and emission of 440–490 nm and 510 nm, respectively (FITC, fluorescein isothiocyanate filter combination). Living protoplasts have bright fluorescence (*see* **Note 20**).
5. Use excitation and emission of 334–385 nm and 420 nm, respectively, for Tinopal CBS-X. Viable protoplasts have blue fluorescence (*see* **Note 21**).

3.4.3 Culturing of Protoplasts

Protoplasts are usually cultured on semisolid, agar-containing medium or in liquid medium. The salts of MS [14] or B5 [15] medium supplemented with an extra osmoticum, such as sorbitol, mannitol, sucrose, or glucose, are usually suitable (Table 3). *See* **Note 17**.

1. Mix double density protoplast suspension with molten agar (*see* **Note 22**) at a double concentration as required in the final culture. Make sure that the agar is not too warm since this kills the protoplasts (the agar should be just above its melting point, ca. $\geq$30 °C).
2. Pipette the mixture quickly as small droplets (100–200 μL) or plate evenly onto a petri dish.
3. Seal the plates with Parafilm® and incubate in diffuse light (5–10 μmol/(m^2s)) at room temperature.

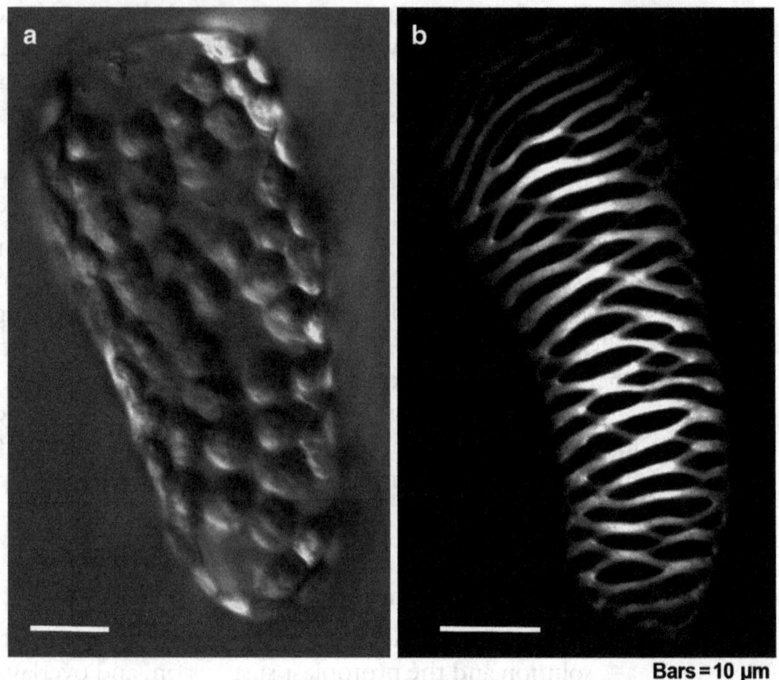

Fig. 3 A mesophyll cell and a TE formed in in vitro *Zinnia* xylogenic culture. (**a**) A single mesophyll cell just after isolation. (**b**) A TE with a thickened secondary cell wall

3.5 Special Case: Zinnia Cultures

Fukuda and Komamine established an in vitro experimental system in which single mesophyll cells of *Zinnia elegans* redifferentiate directly into TEs independently of cell divisions (Fig. 3) [2]. During TE formation, the structure of the cell wall undergoes dynamic changes, such as localized thickenings and lignification of secondary cell walls, partial degradation of primary cell walls, and perforation at the longitudinal end(s). In *Zinnia* xylogenic culture, concurrently with the secondary cell wall formation in developing TEs, active cell wall degradation takes place; and pectin is one of the most actively degraded substances [18]. Thus, taking advantage of the in vitro xylogenic culture system, also mechanisms concerning the structural changes of cell walls can be studied.

3.5.1 Germination of Zinnia Seeds

The first true leaves of 14-day-old seedlings of *Z. elegans* are used for the *in vitro* xylogenic culture. Mesophyll cells should be prepared from healthy leaves carefully grown under optimal conditions.

1. Surface-sterilize seeds of *Z. elegans* cv. Canary bird or Envy in 0.25% Na-hypochlorite solution for 10 min with occasional shaking.
2. Wash the seeds with running water for 10 min in a mesh strainer (*see* **Note 23**).

3. Sow seeds in moistened vermiculite (0.1 g of seeds/100 cm^2) in plastic trays (*see* **Note 24**).
4. Grow seedlings at 25 °C for 14 days under a cycle of 14 h light (approx. 100 μmol/(m^2s), white light from fluorescent lamps) and 10 h dark. Humidity in the growth chamber should be kept under 45%. Water when the surface of vermiculite is dry (*see* **Note 25**). Feed 100-fold diluted liquid fertilizer (HYPONeX; N:P:K = 6:10:5; HYPONeX Japan, Osaka) once on the fourth day after sowing.

3.5.2 Isolation and Culture of Mesophyll Cells

Because of weak attachment between mesophyll cells of *Z. elegans*, single mesophyll cells can be isolated by mechanical maceration using a Waring-type blender. Epidermal and vascular cells are removed by filtration of the cell homogenate through a nylon mesh because of their strong adhesion to each other. Steps of isolation and culture of mesophyll cells are described below.

1. Harvest first true leaves (80–120 leaves) that are 3–4 cm in length (*see* **Note 26**).
2. Surface-sterilize leaves for 10 min in 0.1% Na-hypochlorite solution supplemented with 0.001% (w/v) Triton X-100 with occasional stirring (*see* **Note 27**).
3. Rinse the leaves with autoclaved water for three times (*see* **Notes 23** and **28**).
4. Transfer the leaves into a 100-mL stainless steel cup containing 60 mL of the nutrient medium.
5. Macerate the leaves at 10,000 rpm for 40 s using a Waring-type blender (Fig. 4a, b, *see* **Note 29**).
6. Filter the homogenate through a nylon mesh (Fig. 4c, pore size 50–80 μm) by pipetting using a large-bore pipette. Wash the homogenate that remains on the nylon mesh with 40 mL of additional nutrient medium (*see* **Note 30**).
7. Centrifuge the filtrate at 200 × *g* for 1 min.
8. Remove and discard the supernatant with a pipette or by decantation. Suspend the pelleted cells in 80 mL of the nutrient medium by gentle shaking.
9. Centrifuge again at 200 × *g* for 1 min.
10. Resuspend the pelleted cells in the nutrient medium at a cell density of ca. 8 × 10^4 cells/mL. *See* **Note 18**.
11. Distribute the cell suspension into culture tubes (20 mL for a tube of 30 mm internal diameter (i.d.) × 200 mm, 3 mL for a tube of 18 mm i.d. × 180 mm and 1 mL for a tube of 12 mm i. d. × 105 mm) capped with aluminum foil.
12. Incubate cultures in darkness at 25–27 °C on a revolving drum at 10 rpm at an angle of elevation of 8° (Fig. 4d, *see* **Note 31**).

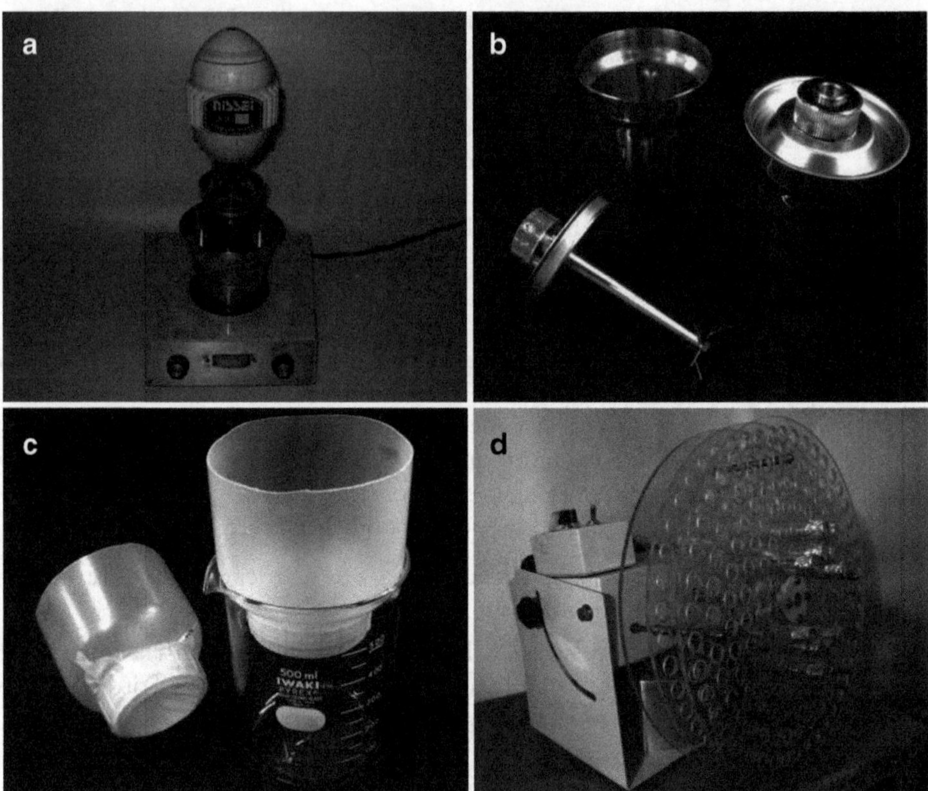

Fig. 4 Experimental apparatus used for the culture of *Zinnia* mesophyll cells. (**a**) A Waring-type blender (**b**) Two sets of the stainless steel cup and a blade used with the blender (**c**) A nylon mesh is attached to a cylinder and set on a glass beaker for use (**d**) A revolving drum, which is placed in a temperature-controlled incubator or a room

3.5.3 Determination of the Frequencies of TE Differentiation and Cell Division

At 72 h of culture, 30–50% of cells synchronously differentiate into TEs. These can be easily identified by characteristic patterns of secondary cell walls, which are observed even under a light microscope (*see* **Note 32**). Therefore, the number of TEs formed can be counted using hemocytometer without any pretreatment. The frequency of TE formation is determined as the number of TEs per number of living cells plus TEs. The frequency of cell divisions can be estimated from the number of septa, since initially all mesophyll cells are single.

3.5.4 Observation of Zinnia Cells

TEs are distinguishable from other cells by their peculiar cell wall thickenings seen under a light microscope as described above. TEs can also be detected by staining the lignified secondary cell walls with phloroglucinol hydrochloride [19] or with fluorochrome-conjugated wheat germ agglutinin [20]. Isolated cells of *Z. elegans* are suitable for observation under a fluorescence microscope and a confocal laser-scanning microscope as well.

Upon maturation of TEs, intracellular components including nuclei are lysed autonomously. This stage of differentiation can be monitored by staining of nuclei with a DNA-specific fluorochrome, 4′,6-diamidino-2-phenylindole (DAPI).

1. Fix the cells by adding glutaraldehyde to a final concentration of 2% (v/v).
2. Add 1/100 volume of 0.2 mg/mL DAPI and incubate briefly in the dark. Observe the nuclei under ultraviolet light using a fluorescence microscope.
3. To visualize the nuclei in living TEs, add 1/1000 volume of 1 mM SYTO16 and incubate for 10 min. Detect in fluorescent microscope. The dye is excited using a 488-nm line and the fluorescence is detected at 515–545 nm [21].

4 Notes

1. Sometimes it is useful to include an undefined mixture of organic substances (e.g., casein hydrolysate (0.1–1 g/L), coconut milk (3–10%, v/v)) to the nutrient medium for culture initiation. During the maintenance growth this substance is gradually omitted, if possible, as the exact composition of the mixture is not known and varies according to the lot.
2. Cytokinins (e.g., 6-benzyladenine (BA), 6-(γ,-γ-dimethylallylamino)purine (2iP), kinetin, zeatin) are usually dissolved in a few drops of alkali (1 M NaOH), then filled with water to the final volume. Auxins (e.g., 2,4-dichlorophenoxyacetic acid (2,4-D), 3-indoleacetic acid (IAA), 1-naphthaleneacetic acid (NAA)) are dissolved in a few drops of absolute ethanol. Boiling water (warmed in a water bath) is poured over to let the ethanol evaporate. After the solution has cooled to room temperature, adjust to the final volume. Store at +4 °C.
3. Store stock solutions of macroelements and microelements at +4 °C. Stock solutions of many macroelements (10×) can be autoclaved to increase their storage time.
4. Iron can be supplied as NaFe(III)EDTA chelate. Make a separate stock solution (100×) out of this. Store at +4 °C.
5. Mixtures of organic compounds, like vitamins, are prepared as 1000× stock solutions. Aliquot the stock solution (e.g., 1 mL aliquots) and store at −20 °C.
6. Heat-labile compounds (e.g., certain growth regulators, some amino acids) are added to the autoclaved medium (when cooled to ca. 60 °C) by filter sterilization through syringe filters (0.2 μm pore size).

7. Depending on the nutrient medium, you may be able to store the ready-made dishes for some time. After plating the agar-containing medium onto Petri dishes, let the medium solidify. Pack the dishes into clean plastic bags (the ones that contained the empty dishes) in a laminar air-flow cabinet, close the bags with tape. Store at room temperature, or at +4 °C, in the dark, agar-side down.
8. A mixture of macroelements can be stored as a 50× stock solution. Microelements I, microelements II, organic growth factors I, and organic growth factors II can be stored separately as 400× stock solutions. Microelements II should be autoclaved to form a chelate. Folic acid is dissolved by adding a small volume of NaOH.
9. For leaf material, 5–10 min in Na hypochlorite may be enough. For stem fragments and seeds, 20–30 min will be necessary.
10. After flaming, let the forceps cool down before touching the explants. Cooling can be done, for example, by dipping into sterile water or by pressing into the agar.
11. Sometimes the growth of callus ceases on the nutrient medium where the callus has previously grown well. It might help if you transfer cells onto a medium where either cytokinin or auxin is depleted. If the growth continues, cells have started to produce the corresponding plant hormone by themselves. This phenomenon is called habituation [22, 23].
12. At the beginning, especially if there is only little callus growth, it is good to transfer the whole explant with the new growth onto the fresh nutrient medium. Only when you have enough callus, separate it from the explant for subculturing.
13. You can use different volumes of liquid cultures but make sure that the medium to flask ratio is similar to 25 mL culture per 100 mL flask. This is to ensure that enough air space exists in the culture flask for gas exchange.
14. Epidermis can be easily removed from surface-sterilized leaves with the help of forceps and a scalpel. It is also possible to make protoplasts separately from the epidermal cell layer and mesophyll cells.
15. Duration of the incubation has to be determined for each plant material.
16. Alternatively, protoplasts can be fractionated with Ficoll (10% in 0.6 M mannitol) or with Percoll (20% with 0.25 M mannitol and 0.1 M $CaCl_2$).
17. Some plant species prefer glucose (100–200 mM) to sucrose.

18. Protoplast/cell density can be determined by Fuchs-Rosenthal modified hemocytometer. For many species, protoplast density of 1×10^3–1×10^5 is suitable.
19. Notice that most dyes are quite toxic, and prolonged incubation of protoplasts in the staining solution may kill them.
20. With FDA, you have to count the number of all protoplasts in the bright field since only the viable ones are detectable in the dark field.
21. Tinopal CBS-X-stained protoplasts can be counted by simultaneous illumination with UV and visible light.
22. Specifically designed agars for protoplast culture remain fluid down to their melting point, which is ca. $\geq$30 °C. Plating into agar facilitates further observation of protoplasts as they become stationary. Agar concentration should be low enough (1%, w/v) to give a soft medium. Agars with low melting point are recommended for temperature-sensitive protoplasts not successful in liquid culture.
23. Na-hypochlorite should be thoroughly removed after surface sterilization of seeds and leaves.
24. The size of the plastic trays we routinely use is $40 \times 30 \times 6.5$ cm, and the depth of vermiculite is about 4 cm. Seeds should be covered with a thin layer of vermiculite after sowing. Seeds and liquid fertilizer should be equally distributed in the plastic tray so that seedlings grow uniformly.
25. Since too much watering of seedlings often causes serious diseases, water only when the surface of the soil is dry. During watering, leaves should be kept dry. Use of leaves with splashes of water often leads to bacterial contamination in the subsequent cell culture.
26. During harvesting, healthy leaves without withered area and rough surface should be cut off by a pair of scissors sterilized with 70% (v/v) ethanol. Leaves harvested should be kept in water in a plastic container until the next step.
27. Leaves are damaged when they are soaked in the Na-hypochlorite solution for longer than 10 min.
28. All steps between the rinse of surface-sterilized leaves and the distribution of cell suspension to tubes must be done under aseptic conditions.
29. The optimal condition for maceration of leaves depends on the plant materials and the type of the blender used. Speed and time of maceration should be adjusted to keep the number of dead cells low and the number of living cells collected high. Single mesophyll cells can also be isolated using mortar and pestle with gentle maceration in the nutrient medium.

30. Application of the filtrated cell suspension onto the remaining homogenates on the nylon mesh increases the number of cells collected (optional step).
31. Alternatively, 25 mL of cell suspension can be incubated in a 100-mL flask using a rotary shaker at 40 rpm.
32. *Zinnia* cells fixed with 0.25% (v/v) glutaraldehyde can be stored at 4 °C for at least a few months without significant visual change.
33. pH can be adjusted to 5.7 with KOH. Alternatively, pH can be buffered to 5.7 with 0.5% (w/v) MES (2-[N-morpholino] ethanesulfonic acid)–KOH.

Acknowledgments

We thank University of Helsinki and the Academy of Finland (grant 251390 to A.K.) for the financial support of the work (A.K., A.S.). Teresa Laitinen (Univ. of Helsinki) is thanked for useful comments. Enni Väisänen (Univ. of Helsinki) is thanked for her kind permission for allowing us to include the protoplast figures.

References

1. Pierik RLM (1997) In vitro culture of higher plants, 4th edn. Kluwer Academic Publishers, Dordrecht, 348
2. Fukuda H, Komamine A (1980) Establishment of an experimental system for the tracheary element differentiation from single cells isolated from the mesophyll of *Zinnia elegans*. Plant Physiol 65:57–60
3. Eberhardt TL, Bernards MA, He L, Davin LB, Wooten JB, Lewis NG (1993) Lignification in cell suspension cultures of *Pinus taeda*. In situ characterization of a gymnosperm lignin. J Biol Chem 268:21088–21096
4. Brunow G, Ämmälahti E, Niemi T, Sipilä J, Simola LK, Kilpeläinen I (1998) Labelling of a lignin from suspension cultures of *Picea abies*. Phytochemistry 47:1495–1500
5. Kärkönen A, Koutaniemi S, Mustonen M, Syrjänen K, Brunow G, Kilpeläinen I, Teeri TH, Simola LK (2002) Lignification related enzymes in *Picea abies* suspension cultures. Physiol Plant 114:343–353
6. Kärkönen A, Koutaniemi S (2010) Lignin biosynthesis studies in plant tissue cultures. J Integr Plant Biol 52:176–185
7. Fukuda H (1997) Tracheary element differentiation. Plant Cell 9:1147–1156
8. Demura T, Tashiro G, Horiguchi G, Kishimoto N, Kubo M, Matsuoka N, Minami A, Nagata-Hiwatashi M, Nakamura K, Okamura Y, Sassa N, Suzuki S, Yazaki J, Kikuchi S, Fukuda H (2002) Visualization by comprehensive microarray analysis of gene expression programs during transdifferentition of mesophyll cells into xylem cells. Proc Natl Acad Sci U S A 99:15794–15799
9. Fukuda H (2004) Signals that govern plant vascular cell differentiation. Nat Rev Mol Cell Biol 5:379–391
10. Motose H, Sugiyama M, Fukuda H (2004) A proteoglycan mediates inductive interaction during plant vascular development. Nature 429:873–878
11. Ito Y, Nakanomyo I, Motose H, Iwamoto K, Sawa S, Dohmae N, Fukuda H (2006) Dodeca-CLE peptides as suppressors of plant stem cell. Science 313:842–845
12. Simola LK, Santanen A (1990) Improvement of nutrient medium for growth and embryogenesis of megagametophyte and embryo callus lines of *Picea abies*. Physiol Plant 80:27–35
13. Bajaj YPS (1996) Plant protoplasts and genetic engineering VII. Springer-Verlag, New York, p 317. ISBN 3-540-60876-1

14. Murashige T, Skoog F (1962) A revised medium for rapid growth and bio assays with tobacco tissue cultures. Physiol Plant 15:473–497
15. Gamborg OL, Miller RA, Ojima K (1968) Nutrient requirements of suspension cultures of soybean root cells. Exp Cell Res 50:151–158
16. Widholm JM (1972) The use of fluorescein diacetate and phenosafranine for determining viability of cultured plant cells. Stain Technol 47:189–194
17. Huang CN, Cornejo MJ, Bush DS, Jones RL (1986) Estimating viability of plant protoplasts using double and single staining. Protoplasma 135:80–87
18. Ohdaira Y, Kakegawa K, Amino S, Sugiyama M, Fukuda H (2002) Activity of cell-wall degradation associated with differentiation of isolated mesophyll cells of *Zinnia elegans* into tracheary elements. Planta 215:177–184
19. Siegel SM (1953) On the biosynthesis of lignin. Physiol Plant 6:134–139
20. Hogetsu T (1990) Detection of hemicelluloses specific to the cell wall of tracheary elements and phloem cells by fluorescein-conjugated lectins. Protoplasma 156:67–73
21. Obara K, Kuriyama H, Fukuda H (2001) Direct evidence of active and rapid nuclear degradation triggered by vacuole rupture during programmed cell death in *Zinnia*. Plant Physiol 125:615–626
22. Christou P (1988) Habituation in *in vitro* soybean cultures. Plant Physiol 87:809–812
23. Piscke MS, Huttlin EL, Hegeman AD, Sussman MR (2006) A transcriptome-based characterization of habituation in plant tissue culture. Plant Physiol 140:1255–1278
24. Chu CC, Wang CC, Sun CS, Hsu C, Yin KC, Chu CY, Bi BY (1975) Establishment of an efficient medium for anther culture of rice through comparative experiments on the nitrogen sources. Sci Sinica 18:659–668
25. Lloyd G, McCown B (1980) Commercially-feasible micropropagation of mountain laurel, *Kalmia latifolia*, by use of shoot-tip culture. Comb Proc Int Plant Prop Soc 30:421–427
26. Owen HR, Miller AR (1992) An examination and correction of plant tissue culture basal medium formulations. Plant Cell Tissue Organ Cult 28:147–150

Chapter 7

Protoplast Isolation and Manipulation in the Unicellular Model Plant *Penium margaritaceum*

David S. Domozych, Eleanore Ritter, Anna Lietz, Berke Tinaz, and Sandra C. Raimundo

Abstract

The unicellular freshwater green alga *Penium margaritaceum* has become a novel and valuable model organism for elucidating cell wall dynamics in plants. We describe a rapid and simple means for isolating protoplasts using commercial enzymes in a mannitol-based buffer. Protoplasts can be cultured and cell wall recovery can be monitored in sequentially diluted mannitol-based medium. We also describe an optimized protocol to prepare highly pure, organelle-free nuclei fractions from protoplasts using sucrose gradients. This technology provides a new and effective tool in *Penium* biology that can be used for analysis of cell wall polymer deposition, organelle isolation and characterization, and molecular research including genetic transformation and somatic hybridization.

Key words Protoplasts, Enzymes, Nucleus, *Penium margaritaceum*, Cell wall recovery

1 Introduction

The biosynthesis and extracellular release of cell wall polymers followed by their insertion into the wall's microarchitecture over precisely defined geographic and temporal conditions constitute essential life processes for plant cell development [1–3]. Cell wall deposition encompasses activation of specific gene sets that are directed by the cell's ontogenetic program and in response to prompts such as environmental stress and pathogen attack. Subsequently, mobilization of the endomembrane and cytoskeletal systems, activation of cross talk signal transduction cascades and plasma membrane-focused dynamics provide the machinery for cell wall development [4–7]. Elucidation of these dynamic subcellular events remains a major challenge for plant biologists. This is particularly true for multicellular systems where resolution of specific wall-related mechanisms in a single cell surrounded by neighboring cells within tissue is experimentally difficult [8]. This includes application of many of the currently employed

Zoë A. Popper (ed.), *The Plant Cell Wall: Methods and Protocols*, Methods in Molecular Biology, vol. 2149, https://doi.org/10.1007/978-1-0716-0621-6_7, © Springer Science+Business Media, LLC, part of Springer Nature 2020

technologies of cell wall research such as biochemical extraction/analysis and high resolution microscopy, including immunocytochemical protocols.

Over the past several decades, the charophyte green algae (i.e., the Charophycean green algae or basal Streptophyta [9–11]) have been shown to be the most closely related and ancestral group of green algae to land plants [12]. Recently, taxa of late divergent clades (Zygnematophyceae, Charophyceae, Coleochaetophyceae) were also found to have many of the cell wall constituents typically found in the cell walls of land plants including cellulose, pectins including homogalacturonan (HG) and rhamnogalacturonan I (RGI), xyloglucans, mixed-linkage glucans, and proteoglycans (incl. Arabinogalactan proteins and extensin) [13–17]. Their small size and simple phenotype have also made the charophytes efficacious tools for studying key cell wall events including expansion and morphogenesis [18–20]. Recently, the unicellular charophyte, *Penium margaritaceum* (i.e., a placoderm desmid; Zygnematophyceae) has emerged as a valuable tool in resolving cell wall developmental events [21, 22]. This alga displays a simple cylindrical shape, possesses clearly defined zones of wall polymer secretion during development and has a cell wall whose components can be easily extracted. *Penium* is also an outstanding experimental tool because it grows quickly, adapts well to high throughput screening of exogenous agents (e.g., subcellular inhibitors) and can be live-labeled with antibodies with specificity to particular epitopes present in cell wall polymers. All of these attributes allows rapid and quantitative analyses of wall deposition and related cell expansion mechanics.

We describe the methodologies involved in the isolation and subsequent manipulation of protoplasts of *Penium*. Protoplasts derived from land plants have become valuable tools in plant biochemistry, cell biology and molecular biology especially in the areas of transformation and genetic engineering, somatic hybridization, organelle isolation and monitoring cell wall dynamics [23, 24]. Furthermore, nuclei isolation from protoplasts is a common experimental derivation of protoplast technology that results in acquisition of high yields of nuclei and DNA for subsequent molecular studies [25, 26]. Although efficient techniques have been developed for cellular organelle-free nuclei fractions for plant tissues [27], similar procedures for walled *Penium* cells poses a problem because nuclei fractions are often difficult to separate from the chloroplast fractions. This becomes a significant concern if pure fractions of nuclear DNA need to be isolated (i.e., free of any plastid DNA). Here, we provide an optimized protocol for removing cell wall polymers, culturing protoplasts, isolating nuclei, and recovery of the cell wall.

2 Materials

2.1 Culturing and Maintenance of *Penium margaritaceum*

Prepare all solutions using ultrapure water (18 MΩ cm) and analytical grade reagents. Store stock solutions in a refrigerator unless otherwise stated. Follow all institutional and government waste disposal protocols for the laboratory.

1. *Penium margaritaceum* is available from the Skidmore College Algal Culture Collection (contact corresponding author of this chapter for the strain used in current studies) or the Coimbra Culture Collection of Algae (http://acoi.ci.uc.pt/index.php).
2. 200 mL non-coated Nunc tissue culture flasks.
3. 15 mL sterile centrifuge tubes.
4. 50 mL sterile centrifuge tubes.
5. Refrigerated tabletop centrifuge that accommodates 15 mL and 50 mL centrifuge tubes.
6. 10 mL sterile plastic pipettes (Fisher Scientific) and pipettor.
7. Soil Water Extract: deposit into a 1 L glass flask, 2.5 cm of garden soil (without pesticides or fertilizers) and cover with 800 mL deionized water. Place on a hot plate and heat until boiling. Turn heat down and let simmer for 6 h. Let cool overnight and repeat the heating process for the second day. On the third day, filter the extract through multiple layers of filter paper. The extract should be brown and clear. (*See* **Note 1**).
8. Woods Hole (WH) Medium or MBL medium: *See* http://www.marine.csiro.au/microalgae/methods/Media%20CMARC%20recipes.htm#MBL for recipe. The sodium silicate macronutrient may be eliminated when growing this species.
9. Woods Hole Soil (WHS) medium: 50 mL of Soil Water Extract (Subheading 2.1, **item 7**) is added to 900 mL of WH medium. Mix, adjust pH to 7.2 and bring volume to 1000 mL. The medium is dispensed in 500 mL glass bottles and autoclaved at 120 °C for 20 min (*see* **Note 2**).

2.2 Protoplast Isolation

1. 30 × 15 mm sterile plastic Petri dishes.
2. Petri dish shaker or rotator for continuous mixing.
3. Aluminum foil.
4. Vortex (Vortex Genie2).
5. pH meter.
6. Plastic syringes (5 and 10 mL).
7. 0.22 μm syringe filters.

8. Small glass beaker.
9. Mannitol 2× stock: Place 8.5 g of mannitol in a 50 mL centrifuge tube. Fill the tube to 45 mL with WHS and shake for 1 min. Place tube in a beaker containing hot water for 5 min and shake until the mannitol is dissolved. Bring volume up to 50 mL with WHS. This solution can be stored for 1 week in a refrigerator.
10. $CaCl_2$ 10× stock: Dissolve 2.19 g of $CaCl_2$ in 100 mL of deionized water. Dissolve and store in a refrigerator. This solution can be kept for 1 year.
11. Cellulase (*Trichoderma* cellulase; Sigma #8546): Place 15 mg of cellulase in a 15 mL tube prior to use. This can be preweighed and stored in a refrigerator until needed.
12. Pectate lyase (Megazyme; (# E-PCLYAN2).
13. Protoplast Buffer (PB): To a clean glass beaker, add 10 mL of mannitol 2× stock, 2 mL of 10× $CaCl_2$ stock and 8 mL of WHS. Adjust pH to 7.2. (*See* **Note 3**). Under a fume hood, use a plastic syringe with a 0.22 μm syringe filter to sterilize the solution into a sterile 50 mL centrifuge tube. This solution is the Protoplast Buffer (PB).
14. Protoplast Enzyme Solution (PES): Add 15 mg of cellulase, 20 μL of pectate lyase, and 3 mL of PB to a 15 mL centrifuge tube. Mix gently and under a fume hood, use a plastic syringe with a 0.22 μm syringe filter to sterilize the solution into a sterile 15 mL centrifuge tube. This solution may be stored in a refrigerator for up to 48 h before use.

2.3 Protoplast Recovery

1. PB-agarose: Add 0.2 g of low gelling agarose to a 15 mL centrifuge tube. Add 8 mL of PB and mix thoroughly. Microwave carefully for short time periods (20 s) to dissolve the agarose. Bring the volume to 10 mL with PB, mix and store in 45 °C incubator or oven (this solution may last for 1–2 weeks if sealed with Parafilm®).
2. Recovery medium (RM): Dissolve 0.2 g of glucose and 0.1 g of yeast extract in 20 mL of PB. Add 2 μL of the following stocks: Benzyladenine 1000× (500 mg/L), 1 μL 2–4,D (1,000 mg/L), and 1 μL NAA (500 mg/L). Adjust pH to 7.2 and filter-sterilize into 15 mL tube.
3. 80% Recovery medium (RM-80): Into a sterile 15 mL centrifuge tube mix 4 mL of RW and 1 mL of sterile WHS.
4. 60% Recovery medium (RM-60): Into a sterile 15 mL centrifuge tube mix 5 mL of RW and 2 mL of sterile WHS.
5. 40% Recovery medium (RM-40): Into a sterile 15 mL centrifuge tube mix 2 mL of RW and 3 mL of sterile WHS.

6. 20% Recovery medium (RM-20): Into a sterile 15 mL centrifuge tube mix 1 mL of RW and 4 mL of sterile WHS.
7. 30 × 15 mm Petri dishes
8. 1–200 μL and 100–2000 μL micropipettors with sterile pipette tips.
9. Sterile 22 × 22 mm glass coverslips. Individual coverslips are dipped in 100% ethanol and flame sterilized. After cooling, they can be collected and stored in a sterile 30 × 15 mm Petri dish.
10. Calcofluor 1000× stock: Dissolve 1 mg of Calcofluor in 1 mL of WHS. Keep refrigerated until use.
11. Single-well immunoslide.
12. Fine-tip forceps.
13. Alcohol lamp.
14. Beaker with 50 mL 100% ethanol.

2.4 Nuclei Isolation

1. Nuclei isolation buffer (NIB).
   (a) 10 mM MES-KOH (pH 5.4): Prepare a 10× stock solution by dissolving 1.95 g of MES in deionized water to a final volume of 100 mL. Adjust the pH to 5.4 with KOH.
   (b) 10 mM NaCl: Prepare a 10× stock solution by dissolving 0.58 g of NaCl in deionized water to a final volume of 100 mL.
   (c) 10 mM KCl: Prepare a 10× stock solution by dissolving 0.75 g of KCl in deionized water to a final volume of 100 mL.
   (d) 2.5 mM EDTA: Prepare a 10× stock solution by dissolving 0.93 g of EDTA in deionized water to a final volume of 100 mL.
   (e) 250 mM sucrose: Prepare a 4× stock solution by dissolving 34.23 g of sucrose in deionized water to a final volume of 100 mL.
   (f) 0.1 mM spermine: Prepare a 1000× stock solution by dissolving 20 mg of spermine (Sigma) in 1 mL of deionized water.
   (g) 0.5 mM spermidine: Prepare a 1000× stock solution by dissolving 72.6 mg of spermidine (Sigma) in 1 mL of deionized water.
   (h) 1 mM dithiothreitol (DTT): Prepare a 1000× stock solution by dissolving 154.3 mg of DTT (Sigma) in 1 mL of deionized water.

   These stock solutions should be stored at 4 °C until needed.

NIB solution: Add 12.5 mL of sucrose, 5 mL of NaCl, 5 mL of KCl, 5 mL of EDTA and 5 mL of MES stock solutions in a 50 mL Falcon tube. Add 50 μL of spermine, 50 μL of spermidine, and 50 μL of DTT and bring the final volume to 50 mL. Keep the solution on ice.

2. Nuclei isolation buffer with detergent: Add 10 mL of NIB to a 15 mL Falcon tube and add 100 μL of Triton X-100 (use a cut pipette tip). Vortex vigorously for 10 s. Keep the solution on ice.
3. Sucrose solutions.
   (a) 2.5 M sucrose: Add 42.8 g of sucrose to a 50 mL Falcon tube and add WHS to 50 mL. Leave in a rotator overnight at room temperature in order to dissolve the solution completely. Add WHS to a final volume of 50 mL.
   (b) 1.25 M sucrose: Add 20 mL of the 2.5 M sucrose solution and add WHS to a final volume of 40 mL.
   (c) 0.625 M sucrose: Add 20 ml of the 1.25 M sucrose solution and add WHS to a final volume of 40 mL.

   These solutions should be stored at 4 °C until needed.
4. DAPI solution (1000×): 1 mg/mL in d-H_2O.
5. 15 mL Falcon tubes
6. Styrofoam box or other container to hold ice.

2.5 Microscopy

Olympus BX60 (Olympus America Inc., Melville, NY, USA) equipped with an Olympus DP73 digital camera; Olympus IX70 Inverted Microscope equipped with an Olympus DP71 digital camera.

3 Methods

3.1 Culturing

Perform all culture handling in aseptic conditions and under a laminar flow hood at room temperature.

1. Fill a 200 mL Nunc tissue culture flask to the recommended volume as marked on its side with sterile WHS medium.
2. Obtain a subculture of *Penium* that is 7–14 days old (i.e., since its last transfer), shake and aseptically pipet 10 mL of cell suspension into a 200 mL flask.
3. *Penium* can be grown at 18–24 °C with a 16:8 light–dark cycle under cool white fluorescent light (74 μmol/m^2s Photosynthetic Photon Flux). Subcultures will "turn" green after a few days. Cell suspensions from log-phase culture (7–14 days old) are used for subsequent labeling and experiments.

3.2 Harvesting Cells for Protoplast Isolation

1. Aseptically collect 40 mL aliquots of 7–14 days old cell cultures and place in sterile 50 mL centrifuge tubes.
2. Centrifuge the cell suspensions for 2 min at 1000 × *g*. Discard the supernatant and resuspend the pelleted cells in 20 mL of fresh sterile WHS, cap the tube, vortex vigorously for 20 s and recentrifuge. Repeat this step two times. This procedure should be performed immediately before the protoplast isolation (*see* **Note 4**).

3.3 Protoplast Isolation

1. Resuspend the washed pellet obtained in Subheading 3.1, **step 2** in 5 mL of PB. Vortex and place on a rotator with gentle shaking for 20 min.
2. Centrifuge the cell suspension at 1000 × *g* for 2 min and discard the supernatant.
3. Resuspend the pellet in 3 mL of filter sterilized PES and vortex. Pour the cell suspension in a 30 × 15 mm sterile petri dish. Wrap in aluminum foil and place on a rotator with constant gentle rotation at room temperature.
4. Every hour, carefully remove the aluminum foil and check the progress of protoplast formation with an inverted microscope (Fig. 1). Typically, 90% of the cells will release protoplasts within 3 h (*see* **Note 5**).
5. Once sufficient protoplasts have been obtained, aseptically pour the suspension into a sterile 15 mL centrifuge tube. Centrifuge at 150 × *g* for 2 min. The pellet contains the protoplasts. Discard the supernatant. (*See* **Note 6**). Very gently, resuspend the pellet with 2 mL of PB and recentrifuge.
6. Repeat the washing step two times and discard the supernatant. The pellet contains the protoplasts that may be used for subsequent experiments. Protoplasts are typically used immediately but may be kept at 4 °C for 12 h before use (Fig. 2).

3.4 Cell Wall Recovery

1. Resuspend the protoplast pellet (Subheading 3.3, **step 6**) in 500 μL of RM. Aseptically remove a 50 μL drop of this suspension and place onto the top of a sterile coverslip placed in the well of a sterile 30 × 15 mm Petri dish. Add a 50 μL drop of the warm PB-agarose next to this drop. After 20 s, mix the two drops and rapidly swirl the Petri dish on the table top in order to create a thin gelled layer over the coverslip. The agarose gels within 20 s. Repeat this for several coverslips/Petri dishes.
2. Gently pour 5 mL of RM-80 over the coverslip. Seal the Petri dish and place under light conditions (Subheading 3.1, **step 3**).
3. After 24 h, observe the protoplasts with an inverted microscope. Check to see if a cell wall appears to be forming and/or the protoplast starts to change shape.

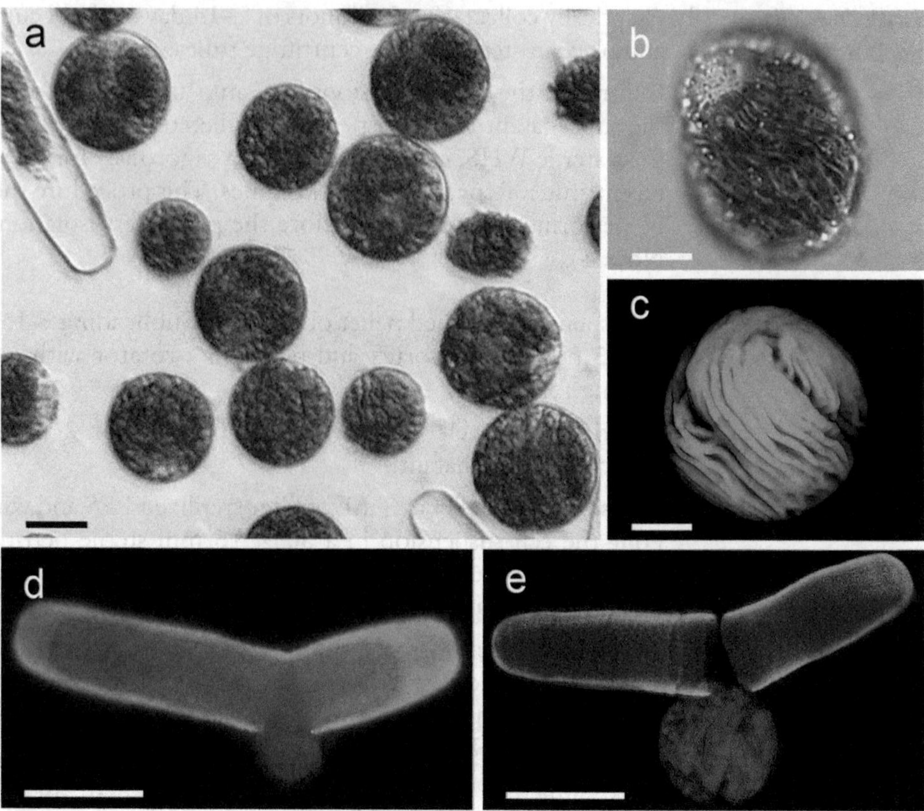

Fig. 1 Protoplasts are released from the cell walls after enzymatic treatment; after 3 h, the protoplast yield is typically around 90% (**a**). (**b**) DAPI staining of a protoplast showing the nucleus (blue) under both UV light and bright field illumination. (**c**) Chloroplast autofluorescence is visible using a FITC filter set, and is an accurate indicator of the protoplast's health. (**d**) JIM5-immunolabeled cell during protoplast formation shows the point where the cell wall (green) disrupts, and the protoplast (red) starts being released. (**e**) JIM5-immunolabeled cell shows the protoplast (red) completely released from the cell wall (green). *Scale bars*: (**a**) 50 μm, (**b**, **c**) 10 μm, and (**d**, **e**) 20 μm

4. To check for wall recovery, add 5 μL of Calcofluor 1000× stock to the dish containing the recovering protoplasts in RM-80. Mix, cover with aluminum foil and gently mix on a lab rotator for 5 min.
5. Remove the Calcofluor RM-80 from the Petri dish with a sterile pipette.
6. Add 2.5 mL of fresh sterile RM-80, mix gently for 1 min and discard the RM-80. Repeat this step once more. Add 2.5 mL of RM-80.
7. Remove the coverslip from the Petri dish and be sure that the agarose sheet containing the protoplasts is still adhered. Turn the coverslip over and place on the well of a single immunoslide containing 25 μL of RM-80. Observe the protoplasts with a

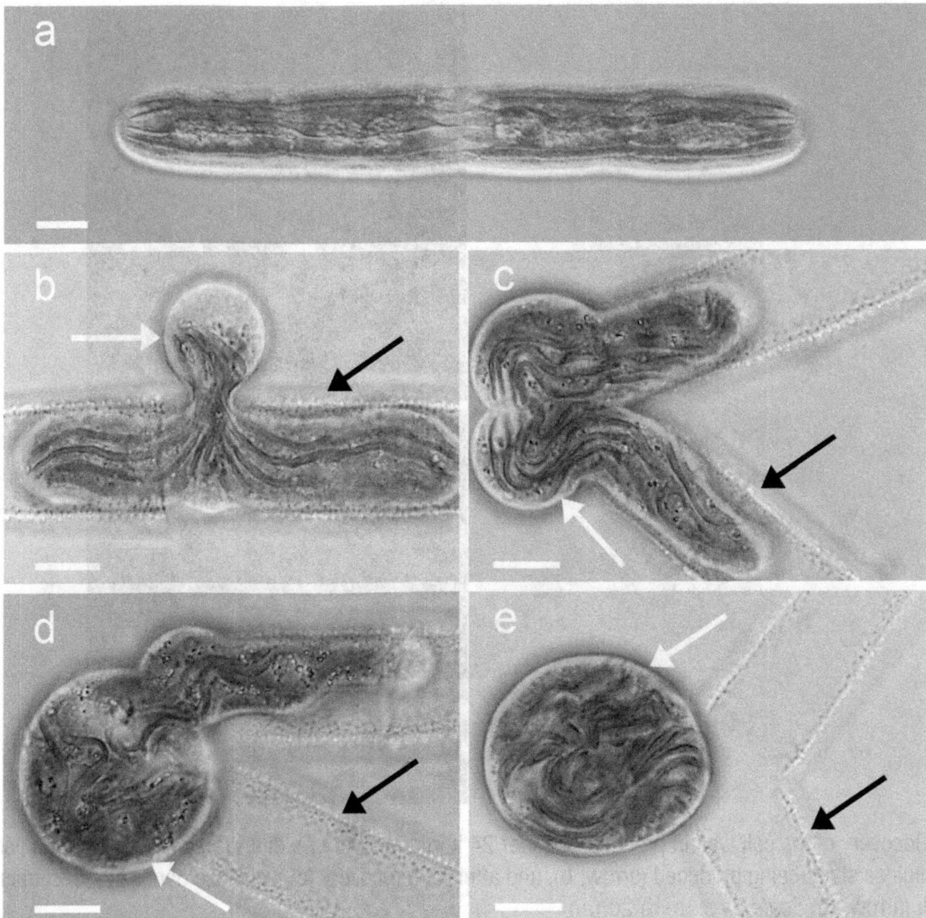

Fig. 2 Differential interference contrast (DIC) images of protoplast generation. (**a**) Typical morphology of *Penium* cell. Upon enzymatic degradation of cellulose and pectin, the wall integrity is compromised and disrupts at the isthmus zone (**b**), releasing the protoplast (**c**). Slowly, the protoplast emerges from the wall (**d**) and becomes detached from the cell wall (**e**). In (**b**)–(**e**), the black arrow highlights the cell wall and the white arrow highlights the protoplast. *Scale bars*: (**a**) 20 μm (**b**–**e**) 10 μm

fluorescence microscope using UV filter. β-Glucans such as cellulose will fluoresce on the cell surface (Fig. 3).

8. In order to monitor further events in process of wall recovery, repeat **steps 2**–7, protoplasts grown in RM-80 may be placed in RM-60 for 24 h. Repeat **steps 2**–7, for processing the protoplasts for microscopy.
9. After 24 h culturing in RM-60, protoplasts may be cultured in RM-40 for 24 h and then RM-20 for 24 h; check the cell wall formation with Calcofluor at each stage (**steps 2**–7).
10. If wall and shape recovery have occurred, WHS may be added to the petri dish and cells cultured as in Subheading 3.1, **steps 1**–**3**.

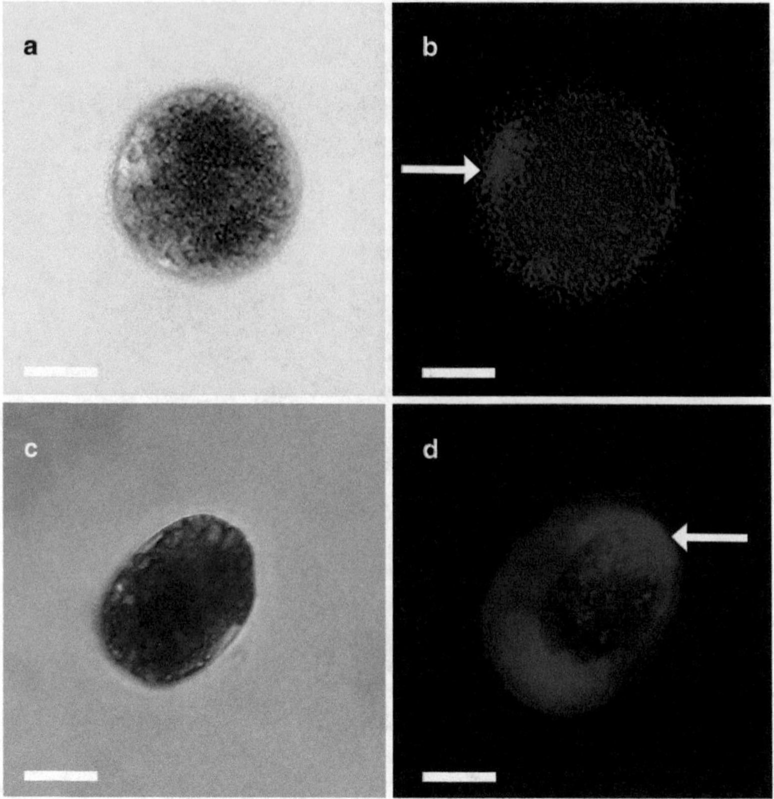

Fig. 3 Recovery of the cell wall in protoplasts. After 24 h in a recovery medium (**a**), calcofluor staining shows that cellulose starts being produced (arrow, **b**), and after 72 h recovery (**c**), cellulose deposition becomes very obvious (arrow, **d**). *Scale bars*: (**a–d**) 20 μm

3.5 Nuclei Isolation

Perform the procedure at room temperature and keep all solutions in a box with ice.

1. Resuspend the protoplast pellet (Subheading 3.3, **step 6**) in 4 mL of NIB with detergent. Leave the Falcon tube in a small box with ice and incubate with gentle shaking for 40 min.
2. Sucrose gradient: Add 3 mL of 2.5 M sucrose to a 15 mL Falcon tube. Add 3 mL of 1.25 M sucrose, gently discarding on the previous layer without disturbing it. Add 3 mL of the 0.625 M sucrose, without disturbing the previous layer. Do this procedure for four Falcon tubes.
3. Add 2 mL of the nuclei solution to the sucrose gradient to two of the tubes containing the sucrose gradient and centrifuge at $1000 \times g$ for 10 min (Fig. 4a).
4. Collect a 20 μL aliquot from each layer obtained, add 0.5 μL of DAPI and observe with the microscope under UV light for nuclei observation (Fig. 4).

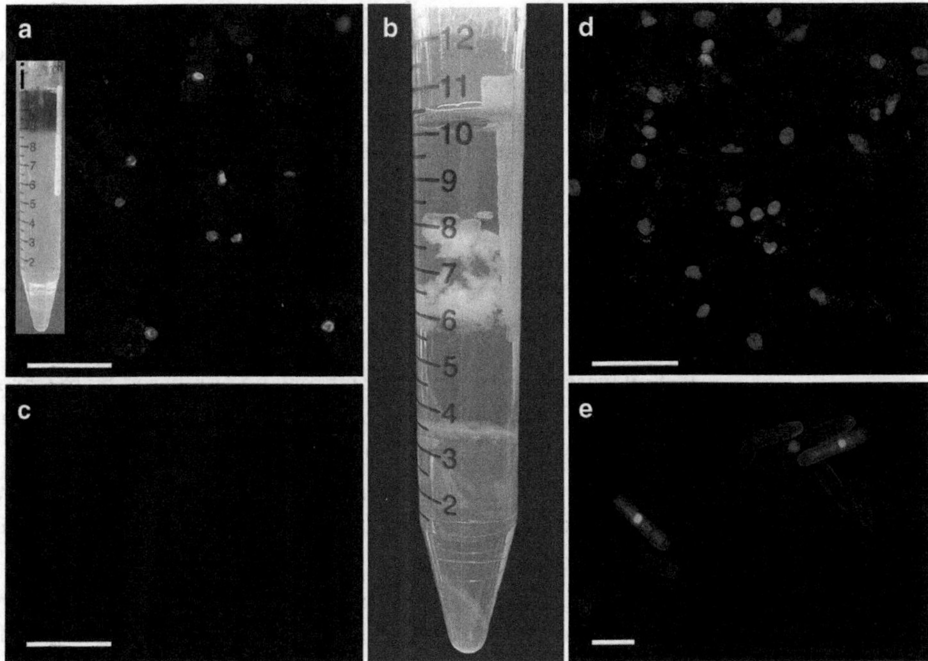

Fig. 4 Nuclei isolation from protoplasts. (**a**) DAPI staining of nucleic acids before centrifugation in the sucrose gradient (i) shows the nuclei (blue) attached to the cell content, namely the chloroplasts (red). After the second centrifugation in sucrose gradient (**b**), three layers are distinguishable: the green top layer contains chlorophyll and no nuclei are detected (**c**); the white fluffy middle layer, in the interface of the 0.625 M and 1.25 M sucrose, contains the pure nuclei fraction (**d**); the bottom layer, in the interface of the 1.25 M and 2.5 M sucrose, contains cell wall debris and cells that did not release protoplasts (**e**). *Scale bars*: 50 μm

5. Remove the top layer which contains chlorophyll, by pipetting (do not pour the liquid) (Fig. 4b, c) and discard. Collect the layer that contains the nuclei fraction (Fig. 4d) to a 15 mL Falcon tube and repeat **steps 1–3**, Subheading 3.5 (*see* **Note** 7). The third layer contains cell debris and should be discarded (Fig. 4e).
6. The layer that contains the nuclei can be collected with a pipette for DNA isolation or other experiments (*see* **Note 8**). Keep on ice until needed.

3.6 Conclusion

Penium is an efficacious model for studying cell wall deposition. With the production of protoplasts, the complete recovery of a cell wall with underlying subcellular mechanisms can be monitored. This protocol holds promise for understanding not only the initial cellulose synthesis but also the sequential deposition of the cell wall polymers, and the cytoskeleton's role in the coordination events that leads to wall polymer production and assembly, ultimately establishing cell shape and architectural complexity. Likewise, protoplasts are versatile tools in organelle isolation and here we show how nuclei may be isolated in a simple and efficacious way. The methods described here provide a rapid and practical means for future protocols in such areas as transformation technology, somatic hybridization, and isolation/characterization of subcellular components (e.g., endomembrane system organelles). Likewise, protoplast isolation and subsequent wall recovery will allow for detailed observation of specific cell wall macromolecules during wall development and concurrent morphogenesis.

4 Notes

1. Soil water supernatant extract may be purchased from Carolina Biological Supply Company (#153790) and can be used in the same way as the "homemade" soil extract.
2. Medium and stock reagents can be stored in a refrigerator (4 °C) for up to a year. If any stock turns cloudy, prepare a fresh supply.
3. Although usually protoplast protocols state that the pH of the protoplast buffer should be adjusted to 5.5, we successfully obtain protoplasts by adjusting the pH to 7.2, as used for the alga culturing conditions.
4. It is important to wash *Penium* cells with WHS before any experimental technique, in order to remove the extracellular polysaccharides this species produces, which can influence or prevent the action of the treatments/chemicals applied to the cells.
5. Sometimes the cells will take longer to release the protoplasts. Incubating the cells for 1 h at room temperature, followed by an overnight incubation at 4 °C with continuous gentle rotation, followed by 1 h incubation at room temperature proved to be very efficient to release the protoplasts.
6. The protoplasts can also be filtered through a 100 μm mesh filter.
7. The second incubation with NIB with detergent and centrifugation in sucrose cushion is required in order to completely separate the chloroplasts and cell wall remains from the nuclei. The fraction will be more pure in relation to the first isolation only.
8. The nuclei fraction has a very characteristic fluffy and sticky appearance, and is easy to collect with a cut pipette tip. If this layer is green, that means it still contains chloroplast debris. If this is inconvenient, perform another incubation with NIB with detergent and further centrifugation in sucrose gradient. Nonetheless, be aware that the DNA might show signs of degradation.

Acknowledgments

This work was supported by NSF grants NSF-MCB-RUI-1517345 and NSF-MCB-09919925.

References

1. Keegstra K (2010) Plant cell walls. Plant Physiol 154:483–486
2. Cosgrove DJ (2005) Growth of the plant cell wall. Nat Rev Mol Cell Biol 6:850–861
3. Cosgrove DJ, Jarvis MC (2012) Comparative structure and biomechanics of plant primary and secondary cell walls. Front Plant Sci 3:204. https://doi.org/10.3389/fpls/2012.00204
4. Ivakov A, Persson S (2013) Plant cell shape: modulators and measurements. Front Plant Sci 4:439. https://doi.org/10.3389/fpls.2013.00439
5. Li S, Lei L, Yingling YG, Gu Y (2015) Microtubules and cellulose biosynthesis: the emergence of new players. Curr Opin Plant Biol 28:76–82
6. Anderson CT (2016) We be jammin': an update on pectin biosynthesis, trafficking and dynamics. J Exp Bot 67:495–502
7. Cosgrove DJ (2016) Plant cell wall extensibility: connecting plant cell growth with cell wall structure, mechanics, and the action of wall-modifying enzymes. J Exp Bot 67:463–476
8. Domozych DS, Sorensen I, Popper ZA, Ochs J, Andreas A, Fangel JU, Pielach A, Sachs C, Brechka H, Ruisi-Besares P, Willats WGT, Rose JKC (2014) Pectin metabolism and assembly in the cell wall of the charophyte green alga *Penium margaritaceum*. Plant Physiol 165:105–118
9. Leliaert F, Smith DR, Moreau H, Herron MH, Verbruggen H, Delwiche CF, De Clerck O (2012) Phylogeny and molecular evolution of the green algae. Crit Rev Plant Sci 31:1–46
10. Lewis LA, McCourt RM (2004) Green algae and the origin of land plants. Am J Bot 91:1535–1556
11. Becker B, Marin B (2009) Streptophyte algae and the origin of embryophytes. Ann Bot 103:999–1004
12. Delwiche CF, Cooper ED (2015) The evolutionary origin of a terrestrial Flora. Curr Biol 25:R899–R910
13. Popper ZA, Ralet M-C, Domozych DS (2014) Plant and algal cell walls: diversity and functionality. Ann Bot 114:1043–1048
14. Popper ZA, Fry SC (2003) Primary cell wall composition of bryophytes and charophytes. Ann Bot 91:1–12
15. Mikkelsen MD, Harholt J, Ulvskov P, Johansen IE, Fangel JU, Doblin MS, Bacic A, Willats WGT (2014) Evidence for land plant cell wall biosynthetic mechanisms in charophyte green algae. Ann Bot 114:1217–1236
16. Domozych DS, Domozych CE (2014) Multicellularity in green algae: upsizing in a walled complex. Front Plant Sci 5:649. https://doi.org/10.3389/fpls.201400649
17. Sørensen I, Domozych D, Willats WGT (2010) How have plant cell walls evolved? Plant Phys 153:366–372
18. Ochs J, LaRue T, Tinaz B, Yongue C, Domozych DS (2014) The cortical cytoskeletal network and cell-wall dynamics in the unicellular charophycean green alga *Penium margaritaceum*. Ann Bot 114:1237–1249
19. Proseus TE, Boyer JS (2012) Calcium deprivation disrupts enlargement of *Chara corallina* cells: further evidence for the calcium pectate cycle. J Exp Bot 63:3953–3958
20. Meindl U (1993) *Micrasterias* cells as a model system for research on morphogenesis. Microbiol Rev 57:415–433
21. Domozych DS, Sørensen I, Sacks C et al (2014) Disruption of the microtubule network alters cellulose deposition and causes major changes in pectin distribution in the cell wall of the green alga, *Penium margaritaceum*. J Exp Bot 65:465–479
22. Domozych DS, Lambiasse L, Kiemle SN, Gretz MR (2009) Cell-wall development and bipolar growth in the desmid *Penium margaritaceum* (Zygnematophyceae, Streptophyta). Asymmetry in a symmetric world. J Phycol 45:879–893
23. Eeckhaut T, Laksshmanan PB, Deryckere D, Van Bockstaele E, Van Huylenbroeck J (2013) Progress in plant protoplast research. Planta 238:991–1003
24. Davey MR, Anthony P, Power JB, Lowe KC (2005) Plant protoplasts: status and biotechnological perspectives. Biotech Adv 23:131–171

25. Ohyama K, Lawrence EP, Horn D (1977) A rapid, simple method for nuclei isolation from plant protoplasts. Plant Phys 60:179–181
26. Saxena PK, Fowke LC, King J (1985) An efficient procedure for isolation of nuclei from plant protoplasts. Protoplasma 128:184–189
27. Sikorskaite S, Rajamäki M-L, Baniulis D, Stanys V, Valkonen JPT (2013) Protocol: optimized methodology for isolation of nuclei from leaves of species in the *Solanaceae* and *Rosaceae* families. Plant Methods 9:1–9

Chapter 8

Knocking Out the Wall: Revised Protocols for Gene Targeting in *Physcomitrella patens*

Alison W. Roberts, Christos S. Dimos, Michael J. Budziszek Jr, Chessa A. Goss, Virginia Lai, and Arielle M. Chaves

Abstract

The moss *Physcomitrella patens* has become established as a model for investigating plant gene function due to the feasibility of gene targeting. The chemical composition of the *P. patens* cell wall is similar to that of vascular plants and phylogenetic analyses of glycosyltransferase sequences from the *P. patens* genome have identified genes that putatively encode cell wall biosynthetic enzymes, providing a basis for investigating the evolution of cell wall polysaccharides and the enzymes that synthesize them. The protocols described in this chapter provide methods for targeted gene knockout in *P. patens*, from constructing vectors and maintaining cultures to transforming protoplasts and analysing the genotypes and phenotypes of the resulting transformed lines.

Key words Gene targeting, Immunocytochemistry, *Physcomitrella patens*, Gateway® cloning

1 Introduction

The moss *Physcomitrella patens* has become established as a model for investigating plant gene function due to its high rate of homologous recombination, which enables targeted gene modification [1]. Wild-type alleles can be knocked out by targeting with vectors containing upstream and downstream homologous sequences flanking a selection cassette [2]. Targeted mutagenesis can be accomplished by replacing genes with vectors carrying insertions, deletions, or point mutations [3]. Functional transgenes can be inserted using expression vectors targeted to intergenic regions [4]. Translational fusions can also be created at the native locus by gene targeting [5]. A further advantage of *P. patens* is that the dominant life cycle phase is haploid, which enables detection of mutant phenotypes in primary transformants and eliminates the need for backcrossing [6]. Genomic resources for *P. patens* include EST sequences and their corresponding full-length cDNA clones (http://moss.nibb.ac.jp/) [7, 8], microarrays [9–13], high-

Zoë A. Popper (ed.), *The Plant Cell Wall: Methods and Protocols*, Methods in Molecular Biology, vol. 2149, https://doi.org/10.1007/978-1-0716-0621-6_8,

throughput transcriptomes [14], and the annotated genome sequence [15–17]. Methods for gene silencing by RNAi [18] and miRNA [19], for genome editing [20–26], for auxotrophic selection [27] and for producing temperature sensitive mutants [28] have been developed. A complementation assay for rapid functional testing of engineered mutations has also been described [29].

Several aspects of *P. patens* morphology and development are advantageous for investigating cell wall biosynthesis and evolution. Development of *P. patens* from haploid spores, protoplasts, or fragmented tissue begins as photosynthetic chloronemal filaments that extend by apical division and tip growth [30]. Growth of chloronemal filaments can be maintained for several weeks on medium containing ammonium tartrate. Thus, in contrast to vascular plants, it is possible to produce tissue consisting of a single cell type (chloronemal filaments) from *P. patens*. This greatly simplifies the task of assigning changes in cell wall composition to alterations in the expression of specific genes and the activity of specific enzymes. In the absence of ammonium tartrate, the apical cells of the chloronemal filaments increase their growth rate to begin producing caulonemal filaments [6]. Caulonemal filaments produce buds that develop into the familiar leafy portion of the moss plant, the gametophore, which enlarges by diffuse growth. Zygotes derived from fusion of gametes, produced at the gametophore apex, develop into diploid sporophytes consisting of a short stalk and a sporangium [30].

Like those of vascular plants, *P. patens* cell walls contain cellulose, xyloglucan, mannan, xylan, pectins, callose, and arabinogalactan proteins [31–41]. Like other bryophytes, they lack lignin [42, 43]. Phylogenetic analyses of glycosyltransferase sequences from the *P. patens* genome have identified orthologs of genes that encode cell wall biosynthetic enzymes in other plant species [39, 44–50] and provide a basis for investigating the evolution of cell wall polysaccharides and the enzymes that synthesize them. Various aspects of *P. patens* biology have been reviewed [1, 6, 51–58] and protocols for culture and transformation have been described previously in print [59–63] and online (http://moss.nibb.ac.jp/; http://www.plant-biotech.net/; https://sites.dartmouth.edu/bezanillalab/moss-methods/). The following protocols provide a guide for targeted gene knock out in *P. patens*, from constructing a vector and maintaining cultures to transforming protoplasts and analysing the genotypes and phenotypes of the resulting transformed lines.

2 Materials

2.1 Vector Construction

1. Sequence of interest as a query for searching the *P. patens* genome.
2. Multisite Gateway® Pro 3.0 three-fragment or Multisite Gateway® Pro Plus flexible cloning kit (Invitrogen, Carlsbad, CA, USA).
3. Reagents for PCR.
4. Destination vector (available upon request from the authors).
5. Commercial plasmid DNA midi- or maxi-prep kit.

2.2 Culture of P. patens

1. Incubator set to 25 °C with constant illumination at 50–80 μmol/m^2/s.
2. BCDAT or BCD medium (*see* Table 1, **Note 1**).
3. Unvented Petri plates, 95 mm (Greiner Bio-One, Monroe, NC, USA, *see* **Note 2**).
4. Sterile cellophane disks (Type 325P, AA Packaging Ltd., Lancashire, UK) interleaved with circles of copy paper and autoclaved in glass Petri plates (*see* **Note 3**).
5. Sterile forceps.

Table 1
Media for culture and transformation of *P. patens* [60]

	BCD (per L)	BCDAT (per L)	PRMB (per L)	PRMT (per L)	PRML (per L)
$MgSO_4$ heptahydrate	0.25 g	0.25 g	0.25 g	0.25 g	0.25 g
KH_2PO_4	0.25 g	0.25 g	0.25 g	0.25 g	0.25 g
KNO_3	1.0 g	1.0 g	1.0 g	1.0 g	1.0 g
$FeSO_4$ septahydrate	12.5 mg	12.5 mg	12.5 mg	12.5 mg	12.5 mg
Diammonium tartrate	–	0.92 g	0.92 g	0.92 g	0.92 g
Trace element solution[a]	1 mL	1 mL	1 mL	1 mL	1 mL
Mannitol	–	–	60 g	80 g	80 g
Agar	7 g	7 g	8 g	4 g	–
1 M Calcium chloride solution 147 g/L $CaCl_2 \cdot 2H_2O$ (sterile) Add after autoclaving	1 mL	1 mL	10 mL	10 mL	10 mL

Dry ingredients are added to purified water, q.s. to 1 L, and sterilized by autoclaving. Calcium chloride solution is added after autoclaving to prevent precipitation

[a]55 mg/L cupric sulfate pentahydrate; 55 mg/L zinc sulfate heptahydrate; 614 mg/L boric acid; 389 mg/L manganous chloride tetrahydrate; 55 mg/L cobalt chloride hexahydrate; 28 mg/L potassium iodide; 25 mg/L sodium molybdate dehydrate [68]

6. Inoculum of *P. patens* (Gransden, available from Stefan Rensing, University of Marburg).
7. Sterile 15 mL disposable centrifuge tubes (polystyrene or polypropylene).
8. Sterile 10 mL serological pipettes.
9. Sterile purified water.
10. Rotor-stator type tissue homogenizer and sterile plastic tips for hard tissue (Omni International, Kenneshaw, GA, USA). Tips can be wrapped in aluminium foil, autoclaved, and reused.

2.3 Transformation and Selection

1. Materials for subculture of *P. patens* (see above).
2. Pipettors (20, 200, and 1000 μL capacity) and sterile tips.
3. Heated water bath with thermometer.
4. Orbital shaker.
5. Haemocytometer.
6. 50 μg vector DNA: linearized, 1 μg/μL in sterile purified water or 5 mM Tris-HCl, pH 7.5.
7. 8.5% (w/v) D-mannitol (sterilized by autoclaving).
8. 3 M solution: the following are added to 40 mL purified water: 4.55 g D-mannitol, 750 μL of 1 M $MgCl_2$, 5 mL of 1% MES-KOH, pH 5.6; q.s. to 50 mL; filter sterilized; stored at 4 °C for up to 6 months.
9. 2% Driselase solution: 1 g Driselase (Sigma-Aldrich, St. Louis, MO, USA, *see* **Note 4**) is dissolved in 50 mL 8.5% D-mannitol; stirred gently 30 min 21–25 °C; chilled 30 min 4 °C; stirred 5 min 21–25 °C; centrifuged 2500 × *g* 10 min; filter sterilized; stored as 3 mL aliquots in sterile 15 mL centrifuge tubes at –20 °C).
10. PEG solution: To prepare Part 1: 9 mL of 8.5% D-mannitol is combined with 1 mL of 1 M $Ca(NO_3)_2$ and 100 μl of 1 M Tris–HCl, pH 8.0, and filter sterilized; to prepare Part 2: 4 g PEG 8000 (Sigma-Aldrich) is melted in a sterile 50 mL disposable centrifuge tube by microwaving; Part 1 is added to Part 2 and vortexed until completely mixed; kept at 21–25 °C for 2 h before use; stored in 1 mL aliquots at –20 °C).
11. PRMB and PRMT media (*see* Table 1).
12. Sterile nylon filters (EASYstrainer, Greiner Bio-One, Monroe, NC, USA).
13. Sterile 50 mL disposable conical centrifuge tubes and 15 mL round-bottom culture tubes.
14. Antibiotic stock (e.g., 15 mg/mL hygromycin in purified water).

2.4 Genotype Analysis by PCR

1. Vector primers (Vector F = TGACAGATAGCTGGGCAATG, Vector R = TCCGAGGGCAAAGAAATAGA) and flanking primers (gene specific, see below).
2. Kontes Pellet Pestle® Micro Grinder (Kimble/Kontes, Vineland, NJ, USA).
3. DNA extraction buffer for PCR: 0.2 M Tris–HCl, pH 9.0, 0.4 M LiCl, 25 mM EDTA, 1% SDS.
4. Isopropanol.
5. TE buffer: 10 mM Tris–HCl, pH 8, 1 mM EDTA.

2.5 Genotype Analysis by Southern Blotting

1. Transformed *P. patens* lines cultured on BCDAT plates for 7 days (*see* Subheading 3.2).
2. DNA extraction buffer for Southern blotting: the following are combined in the order listed in a 15 mL disposable centrifuge tube: Solution 1: 350 mM sorbitol, 100 mM Tris–HCl, pH 7.5, 5 mM EDTA, sterilized by autoclaving (1 mL); 3.8 mg sodium bisulfite; Solution 2: 200 mM Tris–HCl, pH 7.5, 50 mM EDTA, 2% (w/v) CTAB, 2 M sodium chloride, sterilized by autoclaving (1 mL); 0.4 ml of 5% (w/v) N-lauroylsarcosine; 5 μL of 10 mg/mL RNaseA; prepared just before use.
3. Chloroform–octanol 24:1.
4. Cryocup® Grinder (Research Products International, Mt. Prospect, IL, USA).
5. Liquid nitrogen.
6. DNA precipitation buffer: 80 mL ethanol, 20 mL 1 M sodium acetate, pH 7.0.
7. 70% (v/v) ethanol.
8. Supplies and reagents for Southern blotting.

2.6 Immunofluorescent Cell Wall Labeling

1. 1 mL pipette tips, tips cut off with a razor blade at 1 cm, autoclaved.
2. Primary antibodies or carbohydrate binding modules (CBM) (*see* Plant Probes www.plantprobes.net; Biosupplies Australia www.biosupplies.com.au).
3. Secondary antibodies (ALEXA Fluor® 488–labeled goat anti-rat or anti-mouse) as appropriate for primary antibody, Invitrogen).
4. FlexiPERM® 12-well chambers (ISC BioExpress, Kaysville, UT, USA).
5. Poly-L-lysine–coated slides (Fisher Scientific).
6. Sterile nylon filters (*see* Subheading 2.3, **item 12**).
7. 15 and 50 mL disposable centrifuge tubes.

8. 2× fixative stock: 100 mM PIPES, pH 6.8, 5 mM magnesium sulfate, 10 mM EGTA.
9. Fixative: 500 μL 2× fixative stock, 270 μL purified water, 230 μL 16% formaldehyde (methanol-free, Polysciences Inc., Warrington, PA, USA), prepared just before use.
10. Phosphate-buffered saline (PBS): 137 mM NaCl, 2.7 mM KCl, 10 mM Na_2HPO_4, 1.4 mM KH_2PO_4, pH 7.4.
11. Blocking solution: 5% w/v nonfat dry milk in PBS, prepared just before use.
12. Coverslips (24 × 60 mm).
13. Antifade reagent (SlowFade® Gold, Invitrogen).

2.7 Morphological Analysis

See Subheading 2.2.

3 Methods

3.1 Preparation of Knockout Vectors

Replacement vectors for deleting target genes include homologous sequences upstream and downstream of the target gene coding sequence, separated by a selection cassette (Fig. 1). The selection cassette replaces the target gene when the vector is integrated into the genome by homologous recombination. Vectors are constructed using Gateway Multisite® Cloning (Invitrogen). The following describes methods for cloning homologous sequences upstream and downstream of the target gene into the appropriate pDONR vectors for cloning into pBHSNRG (Fig. 1a, *see* **Note 5**).

1. *P. patens* homologs of genes of interest are identified using the BLAST function available at CoGe (https://genomevolution.org/coge/CoGeBlast.pl). In the "Organism" box, choose "Physcomitrella patens." Paste your peptide or nucleic acid sequence into the box, choose the appropriate BLAST type (i.e., peptide or nucleic acid) and click "Run CoGe BLAST" at the bottom of the page. When your search is complete, click the "GenomeView" link for your chosen hit. When you find the desired gene model, zoom out to show at least 2 kb of flanking sequence on both ends of the coding sequence. Open the "Sequence" pull-down menu and choose "Save track data" and "Save" to save the text file.
2. Open the sequence text file in any sequence analysis program or text editor. After the start and stop codons are identified, primers are designed to amplify about 1 kb of homologous sequence upstream and 1 kb of homologous sequence downstream of the gene of interest. Sequences for cloning into pDONR P1-P4 (Element 1) are added to the upstream

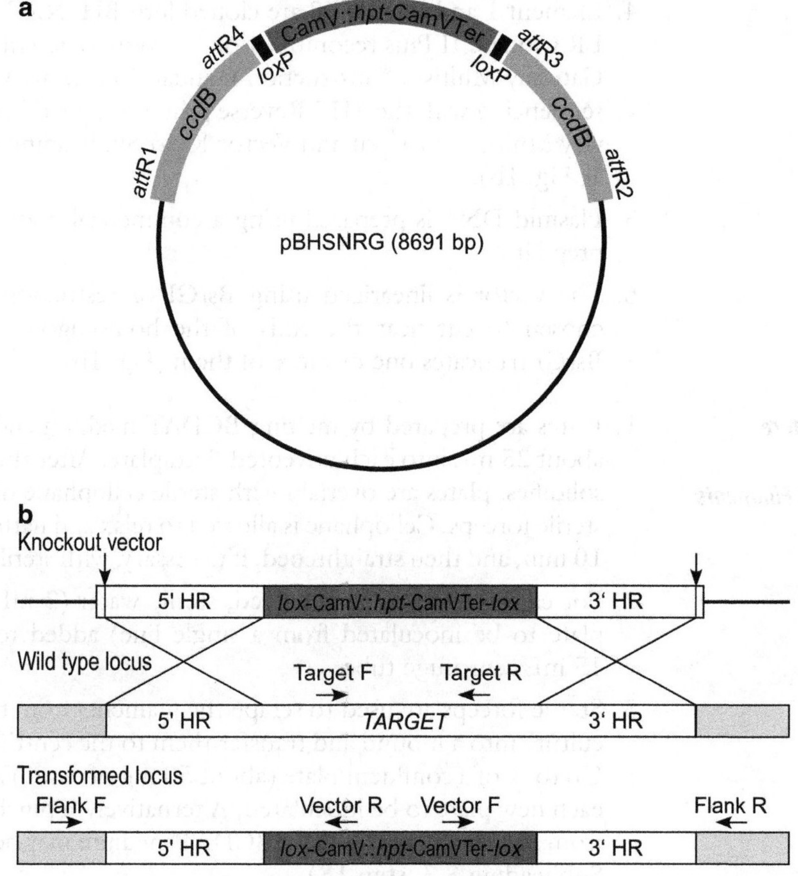

Fig. 1 (**a**) Multisite Gateway® destination vector pBHSNRG. Amplified 5′ and 3′ homologous regions cloned into pDONR P1–P4 and pDONR P3–P2, respectively, are inserted in the destination vector when the *att*R1/*att*R4 and *att*R3/*att*R2 sites of pBHSNRG recombine with the pDONR *att*L1/*att*L4 and *att*L3/*att*L2 sites to make the knockout vector. (**b**) Knockout vector, wild type target genomic locus and transformed target genomic locus. The vector is linearized using BsrGI or restriction enzymes chosen to cut near the ends of the homologous region (vertical arrows). The positions of primers used to test for targeted integration (Flank F/Vector R and Vector F/Flank R), deletion of the target gene (Target F/Target R) and absence of concatenated vector sequences (Vector F/Vector R) are shown

forward and reverse primers, and sequences for cloning into pDONR P3-P2 (Element 3) are added to the downstream forward and reverse primers (see Gateway Multisite® instruction manual).

3. Primers are used to amplify genomic DNA isolated from wild type *P. patens* (*see* Subheading 3.5) and the amplified fragments are cloned into their respective pDONR vectors using the BP Clonase II recombination reaction as described in the Gateway Multisite® instruction manual to construct Element 1 and Element 3.

4. Element 1 and Element 3 are cloned into BHSNRG using the LR Clonase II Plus recombination reaction as described in the Gateway Multisite® instruction manual. Clones are verified by sequencing with the M13 Reverse primer supplied in the Gateway Multisite Pro® kit and Vector R (*see* Subheading 2.4, **item 1**, Fig. 1b).
5. Plasmid DNA is prepared using a commercial midi- or maxi-prep kit.
6. The vector is linearized using BsrGI or restriction enzymes chosen to cut near the ends of the homologous regions if BsrGI truncates one or more of them (Fig. 1b).

3.2 Subculture of *P. patens* Chloronemal Filaments

1. Plates are prepared by melting BCDAT medium and pouring about 25 mL into each unvented Petri plate. After the medium solidifies, plates are overlain with sterile cellophane disks using sterile forceps. Cellophane is allowed to relax and flatten out for 10 min, and then straightened, if necessary, with sterile forceps.
2. For each line to be subcultured, sterile water (2 mL for each plate to be inoculated from a single line) added to a sterile 15 mL centrifuge tube.
3. Sterile forceps are used to scrape the filaments from the starter culture into a mound and transfer them to the centrifuge tube. Up to ¼ of a confluent plate (about 50 mg of tissue) is used for each new plate to be inoculated. Alternatively, a pinch of tissue from a colony maintained on BCDAT medium may be used (*see* Subheading 3.3, **step 18**).
4. Tissue is homogenized for approximately 5 s on medium setting using a tissue homogenizer with sterile plastic tip (*see* **Note 6**).
5. Approximately 2 mL of suspension are poured or transferred using a serological pipette to the surface of each plate and spread evenly.
6. Plates are incubated at 25 °C with constant illumination at 50–80 μmol/m^2/s (*see* **Note** 7). They are not sealed with Parafilm® or stacked.
7. After 1 week, tissue is subcultured or plates are sealed with Parafilm® and transferred to an incubator set to 10 °C with a 2 h photoperiod at 20 μmol/m^2/s for storage up to 1 year. Clones can also be conserved for several years by suspending about 50 mg of tissue in 1 mL of sterile distilled water in a sterile microcentrifuge tube and storing at 4 °C (*see* **Note 8**).

3.3 Transformation and Selection

1. 5–7 days before transformation, chloronemal tissue from the line to be transformed is subcultured from a fresh plate (*see* **Note 9**) on cellophane-overlain BCDAT plates and incubated described in Subheading 3.2.
2. Protoplast regeneration plates (three per transformation) are prepared with PRMB medium and overlain with sterile cellophane (*see* Subheading 3.2, **step 1**).
3. Before beginning protoplast preparation, all materials are made ready: water bath is equilibrated to 45 °C for heat shock; a 500 mL beaker containing 300 mL of water is equilibrated to room temperature (21–25 °C); PRMT medium is melted and equilibrated to 45 °C in the water bath; linearized vector DNA is ethanol precipitated and dissolved in sterile water at 1 μg/μL; one aliquot of Driselase solution (*see* **Note 4**) per line to be transformed and one aliquot of PEG solution per one to three vectors are thawed and completely redissolved.
4. Protoplast preparation is begun by pipetting 9 mL of 8.5% mannitol into a Petri plate and adding chloronemal filaments scraped from the plate using sterile forceps, followed by 3 mL of 2% Driselase solution. Petri plates are incubated for 60 min with shaking at 60 rpm on an orbital shaker at 21–25 °C.
5. Using a serological pipette, protoplast suspension is gently drawn from the Petri plate and passed through a nylon filter placed on top of a 50 mL disposal centrifuge tube.
6. Filtrate is transferred to a 15 mL disposable conical centrifuge tube and centrifuged at speed 2–3 in a clinical centrifuge (*see* **Note 10**) for 7 min; supernatant is discarded.
7. Protoplasts are resuspended in 10 mL 8.5% mannitol pipetted directly onto the pellet from a serological pipette. It is important to resuspend protoplasts gently and to avoid aspirating protoplasts into the pipette. The suspension is centrifuged for 7 min, speed 2–3; supernatant is discarded. This step is repeated.
8. During **step 7**, 15–30 μL of each vector is pipetted into a labeled 15 mL disposable round-bottomed centrifuge tube.
9. Protoplasts are resuspended in 10 mL of 8.5% mannitol. After 10 μL of suspension are removed and loaded into a haemocytometer, the suspension is centrifuged for 7 min, speed 2–3.
10. Intact protoplasts (*see* **Note 11**) are counted and the density is estimated; use the instructions supplied with your haemocytometer, 2–4 × 10^5 protoplasts/mL is typical.
11. Supernatant is discarded and protoplasts are resuspended in 3 M solution at 2 × 10^6 protoplasts/mL; 2–4 mL of suspension is typical.

12. Protoplast suspension (0.3 mL) and PEG solution (0.3 mL) are added to each tube containing vector DNA. The suspension is mixed gently, but thoroughly and incubated at 21–25 °C for 10 min.
13. Protoplasts are heat-shocked for 3 min at 45 °C and transferred immediately to a beaker of 21–25 °C water for 10 min. Important—the water bath heater is turned off just before submerging the tubes containing the protoplasts to prevent overheating should the heater turn on during incubation.
14. Protoplasts are resuspended in 5 mL PRMT held at 45 °C and 1.6 mL of suspension is spread on each of three cellophane-overlain PRMB plates. This should result in an inoculation rate of 2×10^5 protoplasts/plate (i.e., 2×10^6 protoplasts/mL $\times$ 0.3 mL/transformation $\div$ 3 plates/transformation).
15. Plates are incubated for 5 days at 25 °C with constant illumination at 50–80 μmol/m^2/s.
16. The regeneration rate can be estimated on day 5 as follows: using a mm ruler, determine the field area of your dissecting microscope on high power. Using the same magnification, count the number of regenerated protoplasts in three to five fields and calculate the average. The area of a plate is about 6400 mm^2. Estimate the number of regenerated protoplasts per plate by multiplying the average number per field by 6400 mm^2 and dividing by the field area in mm^2. Expect 30–70% regeneration (*see* **Note 12**).
17. Selection is initiated after 5 days. Selection plates are prepared by melting BCDAT medium, cooling to 55 °C, adding antibiotic (e.g., 15 μg/mL hygromycin), and pouring about 25 mL each into unvented petri plates. After the medium solidifies, cellophane disks are lifted from PRMB plates with sterile forceps and transferred to BCDAT/antibiotic plates, taking care to avoid trapping air bubbles under the cellophane. Plates are incubated for 7 days at 25 °C with constant illumination at 50–80 μmol/m^2/s.
18. Clones surviving after 7 days on selection consist of a mixture of both stable and unstable transformants. To select for stable transformants, cellophane disks are transferred to BCDAT plates for 7 days, then back to BCDAT/antibiotic plates for 7 days. Typically, clones that grow vigorously during the second round of selection are stable. When large enough, they are split and arrayed on duplicate BCDAT plates without cellophane and incubated for 7 days at 25 °C with constant illumination at 50–80 μmol/m^2/s, and then stored at 10 °C with a 2 h photoperiod at 20 μmol/m^2/s. A small amount of tissue (about 25 mm^3) is collected for genotype analysis.

3.4 Genotype Analysis by PCR

1. Genomic DNA for PCR is prepared by homogenizing tissue (about 25 mm^3) using a Pellet Pestle® Micro Grinder in a compatible 1.5 mL microcentrifuge tube and immediately adding 500 μL of DNA extraction buffer. Debris is pelleted in a microcentrifuge at high speed for 5 min and 350 μL of supernatant is transferred to a 1.5 mL microcentrifuge tube containing 350 μL of isopropanol. DNA is pelleted in a microcentrifuge at high speed for 15 min. The supernatant is poured off and the tube containing the pellet is dried upside down on a paper towel. The pellet is dissolved in 200 μL of TE buffer by shaking for 30 min at 21–25 °C.
2. PCR primers oriented outward from the selection cassette (Vector R/Vector F; *see* Subheading 2.4, **item 1**) are paired with primers designed to amplify inward from the genomic regions flanking the homologous sequences contained within the knockout vector (Flank F/Flank R) to test for 5′ and 3′ integration of the vector. Forward and reverse primers designed to amplify the target sequence (Target F/Target R) are used to test for deletion of the target gene (Fig. 1b). Flank and target sequence primers are gene-specific.
3. DNA extracted from each stable line (2–4 μL) is amplified with primers Flank F/Vector R (Fig. 1b) in a 25 μL PCR reaction to test for integration of the 5′ end of the vector (*see* **Note 13**).
4. For stable lines testing positive for 5′ integration, genomic DNA (2–4 μL) is amplified with primers Vector F/Flank R (Fig. 1b) in a 25 μL PCR reaction to test for integration of the 3′ end of the vector.
5. For stable lines testing positive for both 5′ and 3′ integration, genomic DNA (2–4 μL) is amplified with primers Target F/Target R (Fig. 1b) in a 25 μL PCR reaction to test for deletion of the target sequence.
6. For stable lines testing positive for 5′ and 3′ integration, and negative for the target sequence, genomic DNA (2–4 μL) is amplified with primers Vector F/Vector R (Fig. 1b) in a 25 μL PCR reaction to test for insertion of concatenated vector.

3.5 Genotype Analysis by Southern Blotting

Southern blots are performed to test for nonhomologous integration of the vector. The probe is synthesized using the selection cassette (amplified or restriction fragment) as a template. Genomic DNA (about 2–3 μg each) is digested with two to four restriction enzymes, each chosen to excise a 3–10 kb fragment containing the selection cassette and run on a 0.7% agarose gel at 1 V/cm for 18–24 h. DNA is transferred, hybridized, and developed using standard methods. Hybridization only to the expected fragments is evidence for integration of a single copy of the vector at the target locus. The following procedure yields about 20 μg of genomic DNA (enough for six to eight digests) and requires only a microcentrifuge:

1. Seven-day old chloronemal tissue is scraped from a plate and squeezed firmly between layers of filter paper to remove excess liquid.
2. Up to 240 mg of squeeze-dried tissue is ground in a liquid nitrogen-cooled Cryocup grinder. The resulting powder is transferred to a 15 mL disposable centrifuge tube containing 2.4 mL extraction buffer for Southern blotting.
3. Homogenate is incubated at 65 °C for 20 min with occasional mixing by inversion.
4. Homogenate is transferred to three microcentrifuge tubes (0.75 mL each). After adding 0.75 mL of chloroform–octanol, each tube is inverted six times to mix.
5. Phases are separated by microcentrifuging at high speed, 5 min. Upper aqueous phase is transferred into clean microcentrifuge tubes, 0.9 mL DNA precipitation buffer is added and each tube is inverted six times and chilled at 4 °C for 15 min.
6. DNA is pelleted in a microcentrifuge at high speed, 15 min.
7. Pellets are washed with of 1.5 mL 70% ethanol and microcentrifuged at high speed, 5 min.
8. Supernatant is removed and remaining ethanol is evaporated at 21–25 °C, 10 min. Pellets are dissolved in 100 μL TE buffer and combined.

3.6 Immunofluorescent Cell Wall Labeling

1. Protoplasts are prepared as described in Subheading 3.3, **steps 1–7** and plated in PRML (1 mL per plate) on cellophane-overlain PRMB at a density of about 15,000 per plate. The cellophanes are transferred to BCDAT plates after 2 days (*see* **Note 14**) and cultured on BCDAT for 3–5 days.
2. To collect colonies, the plate is flooded with 3–5 mL of sterile H_2O and a cutoff 1 mL pipette tip is used to pipet the suspension through a nylon filter placed over a 50 mL disposable centrifuge tube. This removes dead protoplasts, which create undesirable background labeling. The filter is placed upside down in a clean Petri plate and colonies are washed into the plate with 3–5 mL of purified water before pipetting them into a 15 mL tube.
3. Colonies are centrifuged for 7 min at speed 3 in a clinical centrifuge (speed is adjusted as necessary to collect tissue without damaging it). Water is removed and colonies are suspended in 1 mL of fixative for 20 min at 21–25 °C or overnight at 4 °C.

4. Colonies are centrifuged for 7 min at speed 3 in a clinical centrifuge. Supernatant is removed and colonies are suspended in 3 mL of PBS. This step is repeated two more times.
5. During the washes, the flexiPERM cell chamber is pressed onto a poly-L-lysine–coated slide. The slide is incubated on a slide warmer at 80 °C for 20 min to ensure adhesion, and then cooled to 21–25 °C.
6. Colonies are centrifuged for 7 min at speed 3 in a clinical centrifuge. Supernatant is removed and colonies are suspended in 3 mL of purified water. This step is repeated 2 more times, resuspending in 1 mL of purified water after the final wash. All salts must be removed as they interfere with adhesion of the tissue to the poly-L-lysine–coated slides.
7. A cut-off 1 mL pipette tip is used to pipet about 50 μL of suspension into each well (including an extra well for a negative control). Tissue sinks to the bottoms of the wells and should cover them completely. More suspension may be added, if necessary. After 20 min, water is carefully removed and the slide is allowed to sit for 20 min to maximize adhesion.
8. Following the schedule below, solutions are pipetted into each well, and then removed using care not to dislodge the tissue. A multichannel pipettor may be used, if available.
   (a) 200 μL of blocking solution, 10 min.
   (b) 50 μL of primary antibody or CBM (*see* **Note 15**) diluted to the recommended working concentration with blocking solution OR 50 μL of blocking solution for negative control wells, 1.5 h at 21–25 °C or overnight at 4 °C.
   (c) 200 μL of PBS, 5 min. Repeat for a total of three washes.
   (d) 50 μL of secondary antibody diluted 1:100 in blocking solution, 1.5 h at 21–25 °C in the dark.
   (e) 200 μL of PBS, 5 min. Repeat for a total of three washes.
   (f) After removal of the final PBS wash, the flexiPERM® is carefully removed, and slides are air-dried, 5–10 min. A few drops of antifade reagent and a coverslip are added and the slide is left in a dark dry place overnight.

3.7 Morphological Analysis

Cell wall defects are often manifested as alterations in the morphology of specific cells or tissues. The following methods can be used to test for alterations in various stages of *P. patens* growth and development.

3.7.1 Protonemal Morphology Assay

Detailed methods for quantitative analysis of protonemal morphology, including culture and image acquisition, have been published [64, 65]. An ImageJ-based macro is available from the authors [64].

3.7.2 Gametophore Development Assay

Protoplasts are prepared as described in Subheading 3.3, **steps 1**–**7** and plated in PRML on cellophane-overlain PRMB at a density of about 1000 per plate. The cellophanes are transferred to BCD plates after 2–4 days (*see* **Note 14**) and cultured for 10 days at 25 °C with constant illumination at 50–80 μmol/m^2/s. Gametophore buds appear 6 days after transfer to BCD and develop into leafy gametophores over the next 4 days [38].

3.7.3 Caulonemal Gravitropism Assay

Petri plates containing BCDAT with 1.2% agar and 0.5% sucrose [66, 67] are prepared. About seven small clumps of fresh chloronemal tissue are plated along the diameter of each plate. After incubation at 25 °C with constant illumination at 50–80 μmol/m^2/s, the plates are positioned vertically and incubated in the dark at 25 °C for 14 days. Pigmented caulonemal filaments are negatively gravitropic in wild type and may exceed 2 cm in length.

3.7.4 Rhizoid Development Assay

Chloronemal tissue is plated on BCD medium supplemented with 0.1–1.0 μM naphthyleneacetic acid and cultured at 25 °C with constant illumination at 50–80 μmol/m^2/s for 14 days. Leafless gametophores develop numerous rhizoids [5].

4 Notes

1. Some media formulas for *P. patens* use $Ca(NO_3)_2$ as a nitrate source [68]. However, calcium phosphate precipitates form during autoclaving and this is prevented by using KNO_3 as a nitrate source and adding $CaCl_2$ solution after autoclaving.
2. The use of unvented Petri plates reduces evaporation and contamination. Alternatively, vented Petri plates can be sealed with Micropore® tape (3 M Corporation, St. Paul, MN, USA).
3. We have also used roll cellophane (Research Products International, Mt. Prospect, IL, USA) cut in 8.5 × 11″ sheets, interleaved with copy paper, cut into 9 cm circles, and autoclaved in glass Petri plates. However, growth inhibition has been noted with some lots of this product.
4. We discovered that Driselase lot # SLBP0654V, available since the end of 2015 (Sigma-Aldrich), is ineffective for protoplast production from *P. patens.* However, a mixture of 21 units/mL of cellulase from *Trichoderma reesei* (Worthington Biochemical Corporation, Lakewood, NJ, USA) along with 0.4% Driselase from this batch can be used in place of 0.5% Driselase from previous batches.
5. Multisite Gateway® is a rapid method for producing knockout vectors. A destination vector (pBHSNRG) was constructed by inserting R1-R4 and R3-R2 Gateway® cassettes into the multiple cloning sites of pBHSNR (gift of D. Schaefer, University of

Lausanne, constructed by inserting the *hpt* gene driven by a double 35S promoter (SacI-NotI fragment from pCAMBIA) in reverse orientation between the two loxP sites of pBilox) [69]. The Gateway® Reading Frame Cassette A (Invitrogen) was modified by PCR (converting R2–R4 or R1–R3) and cloned into pGEM-T Easy (Promega Corp., Madison, WI, USA). The R1–R4 and R3–R2 cassettes were excised from pGEM-T Easy with SphI/SpeI and AvrII/NsiI, respectively, and ligated into the SphI/XbaI and SpeI/NsiI sites, respectively, of pBHSNR. pBHSNRG is available upon request. Vector construction methods using standard restriction digestion and ligation are available in print [60] and online (http://moss.nibb.ac.jp/). Detailed treatments of the effects of various vector parameters on recombination efficiency are also available [70, 71].

6. For subculturing large numbers of different lines, tissue can be placed in sterile tubes with 2 mL of sterile water and two sterile stainless steel beads (3.2 mm, Biospec Products, Bartlesville, OK, USA), capped tightly, clamped in a paint shaker (e.g., Tornado II, Blair Equipment Company, Flint, MI, USA), and shaken for several minutes until large clumps are broken up and small clumps of a few dozen cells remain. When older starter cultures are used, longer shaking is required. When subculturing fresh cultures, vortexing can be substituted for shaking. However, older cultures are often not sufficiently broken up by vortexing.
7. *P. patens* may be grown with continuous light or a long-day photoperiod (typically 18 h). The cell division cycle becomes synchronized to the long-day photoperiod [59].
8. Other methods for long-term storage, including cryopreservation have been reported [59].
9. Chloronemal tissue used to generate protoplasts must be in excellent condition. Dead filaments and cell wall debris cause clumping of protoplasts and reduce transformation efficiency. Tissue should be bright green and should be inspected with a dissecting microscope to confirm that no dead or brown filaments are present. Poor-quality tissue can be rejuvenated through two to three rounds of subculturing, starting with a small amount of inoculum to reduce the introduction of cell debris and using plates containing a generous amount of medium to prevent the cultures from drying out.
10. This protocol uses more gentle centrifugation (about 25 × *g*) compared to published protocols [59, 60], and the resulting pellet is easier to resuspend. The supernatant from the protoplast washes should be clear. A green tinge results from free chloroplasts, a sign of protoplast lysis.

11. Intact protoplasts are spherical and appear turgid with their chloroplasts pressed against the plasma membrane. Protoplasts with chloroplasts aggregated in the centre will not regenerate and should not be counted. Intact protoplasts should substantially outnumber damaged protoplasts.
12. The rate of protoplast regeneration after 5 days is an indicator of successful transformation. At this stage protoplasts should have divided several times and there should be more than 50,000 per plate. A successful transformation typically yields hundreds of unstable transformants after the first round of selection, and dozens of stable transformants after the second round of selection.
13. The following should be considered when interpreting the PCR genotyping results. The percentage of stable transformants in which the vector is integrated by homologous recombination at both the 5′ and 3′ ends can range from 25–100%. In some cases, several tandem and/or inverted copies of the vector may be integrated at the target locus. Protoplast fusion can occur during transformation, producing diploid clones carrying both the vector integrated by homologous recombination and the wild type gene. Refer to detailed treatments of integration mechanisms and resulting genotypes [70–73] for more information.
14. Developing protoplasts can be transferred from PRMB to antibiotic-free BCDAT after 2 days. However, the protoplasts require 4 days to develop antibiotic resistance.
15. The same protocol with modifications is used to label with CBM. CBM is substituted for primary antibody in Subheading 3.6, **step 8b**, and an antipolyhistidine incubation and three washes are added before the secondary antibody incubation.

Acknowledgments

We thank the members of the moss community for their collegiality and helpful suggestions. Magdalena Bezanilla and the members of her group provided invaluable advice and assistance. We thank Didier Schaefer for the gift of pBHSNR. Development of the first version of this protocol was supported by National Research Initiative Competitive Grant no. 2007-35318-18389 from the USDA National Institute of Food and Agriculture. Updates contributed by A. M. Chaves and included in the revised protocol were supported as part of The Center for LignoCellulose Structure and Formation, an Energy Frontier Research Center funded by the U.S. Department of Energy, Office of Science, Office of Basic Energy Sciences under Award Number DE-SC0001090.

References

1. Schaefer DG (2002) A new moss genetics: targeted mutagenesis in *Physcomitrella patens.* Annu Rev Plant Biol 53:477–501
2. Strepp R, Scholz S, Kruse S, Speth V, Reski R (1998) Plant nuclear gene knockout reveals a role in plastid division for the homolog of the bacterial cell division protein FtsZ, an ancestral tubulin. Proc Natl Acad Sci U S A 95:4368–4373
3. Schaefer D, Zryd J-P (2004) Principles of targeted mutagenesis in the moss *Physcomitrella patens.* In: Wood AJ, Oliver MJ, Cove DJ (eds) New frontiers in bryology: physiology, molecular biology and functional genomics. Kluwer, Dordrecht, pp 37–49
4. Perroud PF, Quatrano RS (2006) The role of ARPC4 in tip growth and alignment of the polar axis in filaments of *Physcomitrella patens.* Cell Motil Cytoskeleton 63:162–171
5. Sakakibara K, Nishiyama T, Sumikawa N, Kofuji R, Murata T, Hasebe M (2003) Involvement of auxin and a homeodomain-leucine zipper I gene in rhizoid development of the moss *Physcomitrella patens.* Development 130:4835–4846
6. Cove D (2005) The moss *Physcomitrella patens.* Annu Rev Genet 39:339–358
7. Nishiyama T, Fujita T, Shin I-T, Seki M, Nishide H, Uchiyama I, Kamiya A, Carninci P, Hayashizaki Y, Shinozaki K, Kohara Y, Hasebe M (2003) Comparative genomics of *Physcomitrella patens* gametophytic transcriptome and *Arabidopsis thaliana*: implication for land plant evolution. Proc Natl Acad Sci U S A 100:8007–8012
8. Fujita T, Nishiyama T, Hiwatashi Y, Hasebe M (2004) Gene tagging, gene- and enhancer-trapping, and full-length cDNA overexpression in *Physcomitrella patens.* In: Wood AJ, Oliver MJ, Cove D (eds) New frontiers in bryology. Physiology, molecular biology and functional genomics. Kluwer Academic Publishers, Dordrecht, pp 111–132
9. Hiss M, Laule O, Meskauskiene RM, Arif MA, Decker EL, Erxleben A, Frank W, Hanke ST, Lang D, Martin A, Neu C, Reski R, Richardt S, Schallenberg-Rudinger M, Szovenyi P, Tiko T, Wiedemann G, Wolf L, Zimmermann P, Rensing SA (2014) Large-scale gene expression profiling data for the model moss *Physcomitrella patens* aid understanding of developmental progression, culture and stress conditions. Plant J 79:530–539
10. Richardt S, Timmerhaus G, Lang D, Qudeimat E, Correa LG, Reski R, Rensing SA, Frank W (2010) Microarray analysis of the moss *Physcomitrella patens* reveals evolutionarily conserved transcriptional regulation of salt stress and abscisic acid signalling. Plant Mol Biol 72:27–45
11. Cuming AC, Cho SH, Kamisugi Y, Graham H, Quatrano RS (2007) Microarray analysis of transcriptional responses to abscisic acid and osmotic, salt, and drought stress in the moss, *Physcomitrella patens.* New Phytol 176:275–287
12. O'Donoghue MT, Chater C, Wallace S, Gray JE, Beerling DJ, Fleming AJ (2013) Genome-wide transcriptomic analysis of the sporophyte of the moss *Physcomitrella patens.* J Exp Bot 64:3567–3581
13. Wolf L, Rizzini L, Stracke R, Ulm R, Rensing SA (2010) The molecular and physiological responses of *Physcomitrella patens* to ultraviolet-B radiation. Plant Physiol 153:1123–1134
14. Khraiwesh B, Qudeimat E, Thimma M, Chaiboonchoe A, Jijakli K, Alzahmi A, Arnoux M, Salehi-Ashtiani K (2015) Genome-wide expression analysis offers new insights into the origin and evolution of *Physcomitrella patens* stress response. Sci Rep 5:17434
15. Rensing SA, Lang D, Zimmer AD, Terry A, Salamov A, Shapiro H, Nishiyama T, Perroud PF, Lindquist EA, Kamisugi Y, Tanahashi T, Sakakibara K, Fujita T, Oishi K, Shin IT, Kuroki Y, Toyoda A, Suzuki Y, Hashimoto S, Yamaguchi K, Sugano S, Kohara Y, Fujiyama A, Anterola A, Aoki S, Ashton N, Barbazuk WB, Barker E, Bennetzen JL, Blankenship R, Cho SH, Dutcher SK, Estelle M, Fawcett JA, Gundlach H, Hanada K, Heyl A, Hicks KA, Hughes J, Lohr M, Mayer K, Melkozernov A, Murata T, Nelson DR, Pils B, Prigge M, Reiss B, Renner T, Rombauts S, Rushton PJ, Sanderfoot A, Schween G, Shiu SH, Stueber K, Theodoulou FL, Tu H, Van de Peer Y, Verrier PJ, Waters E, Wood A, Yang L, Cove D, Cuming AC, Hasebe M, Lucas S, Mishler BD, Reski R, Grigoriev IV, Quatrano RS, Boore JL (2008) The *Physcomitrella* genome reveals evolutionary insights into the conquest of land by plants. Science 319:64–69
16. Zimmer AD, Lang D, Buchta K, Rombauts S, Nishiyama T, Hasebe M, Van de Peer Y, Rensing SA, Reski R (2013) Reannotation and extended community resources for the genome of the non-seed plant *Physcomitrella patens* provide insights into the evolution of

plant gene structures and functions. BMC Genomics 14:498

17. Lang D, Ullrich KK, Murat F, Fuchs J, Jenkins J, Haas FB, Piednoel M, Gundlach H, Van Bel M, Meyberg R, Vives C, Morata J, Symeonidi A, Hiss M, Muchero W, Kamisugi Y, Saleh O, Blanc G, Decker EL, van Gessel N, Grimwood J, Hayes RD, Graham SW, Gunter LE, McDaniel SF, Hoernstein SNW, Larsson A, Li FW, Perroud PF, Phillips J, Ranjan P, Rokshar DS, Rothfels CJ, Schneider L, Shu S, Stevenson DW, Thummler F, Tillich M, Villarreal Aguilar JC, Widiez T, Wong GK, Wymore A, Zhang Y, Zimmer AD, Quatrano RS, Mayer KFX, Goodstein D, Casacuberta JM, Vandepoele K, Reski R, Cuming AC, Tuskan GA, Maumus F, Salse J, Schmutz J, Rensing SA (2018) The Physcomitrella patens chromosome-scale assembly reveals moss genome structure and evolution. Plant J 93:515–533
18. Bezanilla M, Perroud PF, Pan A, Klueh P, Quatrano RS (2005) An RNAi system in *Physcomitrella patens* with an internal marker for silencing allows for rapid identification of loss of function phenotypes. Plant Biol 7:251–257
19. Fattash I, Khraiwesh B, Arif MA, Frank W (2012) Expression of artificial microRNAs in *Physcomitrella patens*. Methods Mol Biol 847:293–315
20. Lopez-Obando M, Hoffmann B, Gery C, Guyon-Debast A, Teoule E, Rameau C, Bonhomme S, Nogue F (2016) Simple and efficient targeting of multiple genes through CRISPR-Cas9 in *Physcomitrella patens*. G3 6:3647–3653
21. Mallet DR, Chang M, Cheng X, Bezanilla M (2019) Efficient and modular CRISPR-Cas9 vector system for *Physcomitrella patens*. Plant Direct 3:e00168
22. Ako AE, Perroud PF, Innocent J, Demko V, Olsen OA, Johansen W (2017) An intragenic mutagenesis strategy in *Physcomitrella patens* to preserve intron splicing. Sci Rep 7:5111
23. Collonnier C, Epert A, Mara K, Maclot F, Guyon-Debast A, Charlot F, White C, Schaefer DG, Nogue F (2016) CRISPR-Cas9-mediated efficient directed mutagenesis and RAD51-dependent and RAD51-independent gene targeting in the moss *Physcomitrella patens*. Plant Biotechnol J 15:122–131
24. Collonnier C, Guyon-Debast A, Maclot F, Mara K, Chalot F, Nogue F (2017) Towards mastering CRISPR-induced gene knock-in in plants: survey of key features and focus on the model *Physcomitrella patens*. Methods 121–122:103–107
25. Ermert AL, Nogue F, Stahl F, Gans T, Hughes J (2019) CRISPR/Cas9-mediated knockout of *Physcomitrella patens* phytochromes. Methods Mol Biol 2026:237–263
26. King BC, Vavitsas K, Ikram NK, Schroder J, Scharff LB, Bassard JE, Hamberger B, Jensen PE, Simonsen HT (2016) *In vivo* assembly of DNA-fragments in the moss, *Physcomitrella patens*. Sci Rep 6:25030
27. Ulfstedt M, Hu G-Z, Johansson M, Ronne H (2017) Testing of auxotrophic selection markers for use in the moss *Physcomitrella* provides new insights into the mechanisms of targeted recombination. Front Plant Sci 8:1850
28. Vidali L, Augustine RC, Fay SN, Franco P, Pattavina KA, Bezanilla M (2009) Rapid screening for temperature-sensitive alleles in plants. Plant Physiol 151:506–514
29. Scavuzzo-Duggan TR, Chaves AM, Roberts AW (2015) A complementation assay for *in vivo* protein structure/function analysis in *Physcomitrella patens* (Funariaceae). Appl Plant Sci 3:1500023
30. Schumaker KS, Dietrich MA (1998) Hormone-induced signaling during moss development. Annu Rev Plant Physiol Plant Mol Biol 49:501–523
31. Fu H, Yadav MP, Nothnagel EA (2007) *Physcomitrella patens* arabinogalactan proteins contain abundant terminal 3-*O*-methyl-L-rhamnosyl residues not found in angiosperms. Planta 226:1511–1524
32. Lawton MA, Saidasan H (2011) Cell wall genomics in the recombinogenic moss *Physcomitrella patens*. In: Buckeridge MS, Goldman GH (eds) Routes to cellulosic ethanol. Springer, Berlin, pp 241–261
33. Lee KJD, Sakata Y, Mau S-L, Pettolino F, Bacic A, Quatrano RS, Knight CD, Knox JP (2005) Arabinogalactan proteins are required for apical cell extension in the moss *Physcomitrella patens*. Plant Cell 17:3051–3065
34. Peña MJ, Darvill AG, Eberhard S, York WS, O'Neill MA (2008) Moss and liverwort xyloglucans contain galacturonic acid and are structurally distinct from the xyloglucans synthesized by hornworts and vascular plants. Glycobiology 18:891–904
35. Roberts AW, Roberts EM, Haigler CH (2012) Moss cell walls: structure and biosynthesis. Front Plant Sci 3:166
36. Moller I, Sørensen I, Bernal AJ, Blaukopf C, Lee K, Øbro J, Pettolino F, Roberts AW, Mikkelsen JD, Knox JP, Bacic A, Willats WGT (2007) High-throughput mapping of cell-wall polymers within and between plants using novel microarrays. Plant J 50:1118–1128
37. Berry EA, Tran ML, Dimos CS, Budziszek MJ Jr, Scavuzzo-Duggan TR, Roberts AW (2016) Immuno and affinity cytochemical analysis of

cell wall composition in the moss *Physcomitrella patens*. Front Plant Sci 7:248

38. Goss CA, Brockmann DJ, Bushoven JT, Roberts AW (2012) A *Cellulose Synthase* (*CESA*) gene essential for gametophore morphogenesis in the moss *Physcomitrella patens*. Planta 235:1355–1367
39. Schuette S, Wood AJ, Geisler M, Geisler-Lee J, Ligrone R, Renzaglia KS (2009) Novel localization of callose in the spores of *Physcomitrella patens* and phylogenomics of the callose synthase gene family. Ann Bot 103:749–756
40. Liepman AH, Nairn CJ, Willats WGT, Sørensen I, Roberts AW, Keegstra K (2007) Functional genomic analysis supports conservation of function among *Cellulose synthase-like A* gene family members and suggests diverse roles of mannans in plants. Plant Physiol 143:1881–1893
41. Kulkarni AR, Peña MJ, Avci U, Mazumder K, Urbanowicz BR, Pattathil S, Yin Y, O'Neill MA, Roberts AW, Hahn MG, Xu Y, Darvill AG, York WS (2012) The ability of land plants to synthesize glucuronoxylans predates the evolution of tracheophytes. Glycobiology 22:439–451
42. Graham LE, Cook ME, Busse JS (2000) The origin of plants: body plan changes contributing to a major evolutionary radiation. Proc Natl Acad Sci U S A 97:4535–4540
43. Niklas KJ (2004) The cell walls that bind the tree of life. Bioscience 54:831–841
44. Roberts AW, Bushoven JT (2007) The cellulose synthase (*CESA*) gene superfamily of the moss *Physcomitrella patens*. Plant Mol Biol 63:207–219
45. Harholt J, Sørensen I, Fangel J, Roberts A, Willats WGT, Scheller HV, Petersen BL, Banks JA, Ulvskov P (2012) The glycosyltransferase repertoire of the spikemoss *Selaginella moellendorffii* and a comparative study of its cell wall. PLoS One 7:e35846
46. McCarthy TW, Der JP, Honaas LA, dePamphilis CW, Anderson CT (2014) Phylogenetic analysis of pectin-related gene families in *Physcomitrella patens* and nine other plant species yields evolutionary insights into cell walls. BMC Plant Biol 14:79
47. Yin Y, Chen H, Hahn MG, Mohnen D, Xu Y (2010) Evolution and function of the plant cell wall synthesis-related glycosyltransferase family 8. Plant Physiol 153:1729–1746
48. Yin Y, Huang J, Xu Y (2009) The cellulose synthase superfamily in fully sequenced plants and algae. BMC Plant Biol 9:99
49. Hornblad E, Ulfstedt M, Ronne H, Marchant A (2013) Partial functional conservation of IRX10 homologs in *Physcomitrella patens* and *Arabidopsis thaliana* indicates an evolutionary step contributing to vascular formation in land plants. BMC Plant Biol 13:3
50. Jensen JK, Johnson NR, Wilkerson CG (2014) Arabidopsis thaliana IRX10 and two related proteins from *Psyllium* and *Physcomitrella patens* are xylan xylosyltransferases. Plant J 80:207–215
51. Cove D, Benzanilla M, Harries P, Quatrano R (2006) Mosses as model systems for the study of metabolism and development. Annu Rev Plant Biol 57:497–520
52. Decker EL, Frank W, Sarnighausen E, Reski R (2006) Moss systems biology en route: phytohormones in *Physcomitrella* development. Plant Biol 8:397–405
53. Lang D, Zimmer AD, Rensing SA, Reski R (2008) Exploring plant biodiversity: the *Physcomitrella* genome and beyond. Trends Plant Sci 13:542–549
54. Schaefer DG (2001) Gene targeting in *Physcomitrella patens*. Curr Opin Plant Biol 4:143–150
55. Schaefer DG, Zryd JP (2001) The moss *Physcomitrella patens*, now and then. Plant Physiol 127:1430–1438
56. Prigge MJ, Bezanilla M (2010) Evolutionary crossroads in developmental biology: *Physcomitrella patens*. Development 137:3535–3543
57. Kofuji R, Hasebe M (2014) Eight types of stem cells in the life cycle of the moss *Physcomitrella patens*. Curr Opin Plant Biol 17:13–21
58. Frank W, Decker EL, Reski R (2005) Molecular tools to study *Physcomitrella patens*. Plant Biol 7:220–227
59. Cove DJ, Perroud PF, Charron AJ, McDaniel SF, Khandelwal A, Quatrano RS (2009) The moss *Physcomitrella patens*. A novel model system for plant development and genomic studies. In: Behringer RR, Johnson AD, Krumlauf RE (eds) Emerging model organisms: a laboratory manual, vol 1. Cold Spring Harbor Laboratory Press, Cold Spring Harbor, pp 69–104
60. Knight CD, Cove DJ, Cuming AC, Quatrano RS (2002) Moss gene technology. In: Gilmartin PM, Bowler C (eds) Molecular plant biology—a practical approach, vol 2, Oxford Press. Oxford, New York, pp 285–301
61. Liu YC, Vidali L (2011) Efficient polyethylene glycol (PEG) mediated transformation of the moss *Physcomitrella patens*. J Vis Exp 50:2560
62. Bonhomme S, Nogue F, Rameau C, Schaefer DG (2013) Usefulness of *Physcomitrella patens* for studying plant organogenesis. Methods Mol Biol 959:21–43

63. Strotbek C, Krinninger S, Frank W (2013) The moss *Physcomitrella patens*: methods and tools from cultivation to targeted analysis of gene function. Int J Dev Biol 57:553–564
64. Bibeau JP, Vidali L (2014) Morphological analysis of cell growth mutants in *Physcomitrella*. Methods Mol Biol 1080:201–213
65. Vidali L, Augustine RC, Kleinman KP, Bezanilla M (2007) Profilin is essential for tip growth in the moss *Physcomitrella patens*. Plant Cell 19:3705–3722
66. Cove DJ, Quatrano RS (2004) The use of mosses for the study of cell polarity. In: Wood AJ, Oliver MJ, Cove DJ (eds) New frontiers in bryology: physiology, molecular biology and functional genomics. Kluwer Academic Publishers, Dordrecht, pp 189–203
67. Knight CD, Cove D (1991) The polarity of gravitropism in the moss *Physcomitrella patens* is reversed during mitosis and after growth on a clinostat. Plant Cell Environ 14:995–1001
68. Ashton NW, Grimsley NH, Cove DJ (1979) Analysis of gametophytic development in the moss, *Physcomitrella patens*, using auxin and cytokinin resistant mutants. Planta 144:427–435
69. Thelander M, Nilsson A, Olsson T, Johansson M, Girod PA, Schaefer DG, Zryd JP, Ronne H (2007) The moss genes PpSKI1 and PpSKI2 encode nuclear SnRK1 interacting proteins with homologues in vascular plants. Plant Mol Biol 64:559–573
70. Kamisugi Y, Cuming AC, Cove DJ (2005) Parameters determining the efficiency of gene targeting in the moss *Physcomitrella patens*. Nucleic Acids Res 33:e173
71. Kamisugi Y, Schlink K, Rensing SA, Schween G, von Stackelberg M, Cuming AC, Reski R, Cove DJ (2006) The mechanism of gene targeting in *Physcomitrella patens*: homologous recombination, concatenation and multiple integration. Nucleic Acids Res 34:6205–6214
72. Wendeler E, Zobell O, Chrost B, Reiss B (2015) Recombination products suggest the frequent occurrence of aberrant gene replacement in the moss *Physcomitrella patens*. Plant J 81:548–558
73. Schween G, Fleig S, Reski R (2002) High-throughput-PCR screen of 15,000 transgenic *Physcomitrella* plants. Plant Mol Biol Rep 20:43–47

Chapter 9

Expression of Cell Wall–Modifying Enzymes in Aspen for Improved Lignocellulose Processing

Marta Derba-Maceluch and Ewa J. Mellerowicz

Abstract

Wood is an important source of biomass for materials and chemicals, and a target for genetic engineering of its properties for different applications or for research. Wood properties can be altered by using different enzymes acting on cell wall polymers postsynthetically in cell walls. This approach allows for a precise polymer structure modification thanks to the specificity of enzymes used. Such enzymes can originate from all kinds of organisms, or even be modified in a desired way for novel attributes. Here we present a general strategy for expressing a microbial enzyme in aspen and targeting it to cell wall, using an example of fungal glucuronoyl esterase. We describe methods of vector cloning, plant transformation, transgenic line selection and multiplication, testing for the presence of enzymatic activity in different cell compartments, and finally the method of plant transferring from sterile culture to the greenhouse conditions.

Key words Fungal wood degrading enzymes, Protein targeting to cell wall, Gateway® cloning, Aspen transformation, Transgenic trees, *Populus tremula* × *tremuloides*, Vector design, In vitro culture

1 Introduction

The plant cell is surrounded by a unique, complex, and dynamic structure—the cell wall. It consists of a network of different polysaccharides, which can be combined with lignins, lipids, and proteins, depending on plant species, cell types, timing of biosynthesis during the growing season (exemplified by early and late wood), ontogeny, or exposure to biotic and abiotic stresses of the external environment [1, 2]. Secondary cell walls of the woody plants are particularly valuable as they are main source of energy in the form of heat and they constitute raw material in many different industries like pulping, construction, chemical industry and textile production. They are also considered as a raw material for biofuel production using different types of processing [3–5]. Today, the most common biofuel is ethanol made from starch- or sugar-based feedstocks like corn or sugarcane, whereas lignocellulosic feedstocks would be desirable, since they are not used for food and they may

Zoë A. Popper (ed.), *The Plant Cell Wall: Methods and Protocols*, Methods in Molecular Biology, vol. 2149,
https://doi.org/10.1007/978-1-0716-0621-6_9, © Springer Science+Business Media, LLC, part of Springer Nature 2020

come from short-rotation tree plantations, or from agricultural residues and industrial wastes. To make lignocellulose-based feedstocks more economically attractive for biofuel production, researchers try to modify cell wall to enhance accessibility of lignocellulose components to fermentation enzymes. Such modifications may affect not only lignocellulose processing properties, but also wood mechanical properties as well as tree growth and resistance to stresses [6].

One way to modify cell walls is manipulation of expression level of known native genes or transcription factors involved in biosynthesis of cell wall polymers. Unfortunately, many of these pathways have dramatic impact on plant growth and development and their manipulation results in growth defects that outweigh the gains from improved sugar release. For instance, global reduction of phenylpropanoid biosynthesis level results in growth defects [7]. Similarly, knocking out genes involved in xylan biosynthesis results in dwarfism and collapsed xylem vessels (irregular xylem phenotype) [8–13]. Mutants affected in pectin biosynthesis (for example, *quasimodo1*) exhibit dwarfism and defect in cellular adhesion [14]. Dwarf phenotype is observed when cellulose biosynthesis is compromised, for example in *cobra*, *radial swelling1*, or *korrigan1* [15, 16].One way to attenuate these adverse effects on growth while improving wood properties is to modify the structure of different polymers without affecting their amount. For example, lignin polymer composition has been engineered using foreign enzymes that can synthesize novel lignin monomers [17–19]. Such modified plants grew normally, but their lignocellulose exhibited superior processing properties.

Cell wall polymer structure can be tailored also postsynthetically, using polymer-modifying enzymes acting in cell walls. The polysaccharides are modified by different classes of carbohydrate active enzymes (CAZymes), such as hydrolases, transglycosylases, carbohydrate esterases, polysaccharide lyases, or oxidases. Some of these enzymes are expressed by plants themselves, and play a role in cell wall loosening, maintaining proper wall plasticity and flexibility as well as in mobilization of storage wall components. Taking into account the great variety of CAZymes in different organisms, many of which do not have homologs in plants, exploration of these enzymes considerably broadens possibilities of cell wall modification. Successful postsynthetic modification of cell wall polymer by foreign enzymes is exemplified by the reduction of xylan acetylation by expressing acetyl xylan esterase (AXE) from *Trichoderma reesei* in aspen [20], or from *Aspergillus niger* in *Arabidopsis thaliana* [21]. The latter resulted in more easily extractable xylan, higher enzymatic saccharification and much higher ethanol production during fermentation process, whereas the growth of plants was not affected. However, the manipulation of cell wall structure can trigger plant defense systems increasing their resistance to pathogens [22, 23], or triggering other stress symptoms [24, 25].

Remodeling of particular cell wall components can be useful in applications for bioenergy crops, improving lignocellulose quality and resulting in reduced costs during biofuels production. However, any modification of cell wall polymer structure can lead to desirable and nondesirable effects, and requires a custom-made approach to be effective, starting from a design of a vector, and ending with a very thorough evaluation of transgenic plants. Methods of generation of transgenic aspen lines with altered structure of secondary cell wall components by their postsynthetic modification, and basic line evaluation procedures are described in this chapter. These methods are described using an example of expressing fungal glucuronoyl esterase.

2 Materials

2.1 A DNA Clone

1. A DNA clone containing the coding sequence of an enzyme of interest, for example glucuronoyl esterase *Pc*GCE, (GenBank Accession: AFM93784.1) [25].

2.2 Cloning Materials

Cloning materials for the preparation of expression clone of the fungal gene cDNA (*Pc*GCE) fused with plant signal peptide.

1. Primers for signal peptide exchange between the native *Pc*GCE signal peptide and a hybrid aspen apoplastic protein (*Ptxt*GH9B3, GenBank accession AY660968.1; [26]):
   Fsp: 5′caccATGAGAAGGGGAGCTTCTTTCTGCCTCTT G3′,

   Rsp: 5′CGACTGGGCTTCTTTgttgtaattgggtttGGCTTGGA CAAAACC3′,

   Ffp: 5′aaacccaattacaacAAAGAAGCCCAGTCGTTTGGCTG CTCCACG3′,

   Rfp: 5′ctaATGAGACAACGTAGGGGTGGTCCAGTTG3′.
2. Reagents for PCR, preferably including a proofreading DNA polymerase.
3. pENTR™/D-TOPO® Cloning Kit (Invitrogen, Carlsbad, CA, USA).
4. Commercial kits for plasmid DNA isolation and agarose purification.

2.3 Materials for Cloning

Materials for cloning of the destination vector containing the "wood specific promoter" from *Populus trichocarpa* (*PtGT43B* promoter) [20].

1. *Populus trichocarpa* (genotype Nisqually 1) plants grown in vitro.
2. Commercial kits for plant genomic DNA isolation, plasmid DNA isolation and agarose purification.
3. Primers used for amplification of *PtGT43B* promoter:
   pGT43BSacF:
   5′CACC***GAGCTC***CACCTCACCTCACCTCACCT3′,
   pGT43BSpeR: 5′***ACTAGT***GCCCACCAAGCTAAACCC3′.
4. pENTR™/D-TOPO® Cloning Kit (Invitrogen, Carlsbad, CA, USA).
5. Restriction enzymes: SacI and SpeI.
6. T4DNA ligase.
7. Vector pK2GW7.0 [27].

2.4 Materials for the Preparation of Agrobacterium Clone

Materials for the preparation of *Agrobacterium* clone containing the expression clone.

1. Gateway®_LR Clonase™_II enzyme mix (Invitrogen, Carlsbad, CA, USA).
2. *Agrobacterium tumefaciens* GV3101-pMP90RK strain [28].

2.5 Materials for Aspen In Vitro Growing and Line Multiplication

1. Hybrid aspen (*Populus tremula* L × *tremuloides* Michx) clone T89.
2. Solidified MS6 medium: ½ strength MS, pH 5.6 (Murashige & Skoog medium including vitamins, Duchefa Biochemie; Haarlem, the Netherlands, M0222), 133 mL per jar (*see* Table 1 and **Note 1**).
3. Large jars—tissue culture vessels with a revolutionary breathing system OS140box/green filter (ROUND model, 140 mm H, Duchefa Biochemie, Haarlem, the Netherlands, E1674).
4. Chromatography paper: 3MM Chr (Whatman 3030 917).
5. Growth chamber: photoperiod 18 h day (25 °C)/6 h night (18 °C), light intensity max. 150 μmol/m^2 s; light source, warm white fluorescent tubes (color code 830).

2.6 Materials for Aspen Transformation and Selection

1. Chemicals
   BAP—6-benzylaminopurine (Sigma-Aldrich, B3408).
   IBA—Indole-3-butyric acid (Sigma-Aldrich, I5386).
   TDZ—Thidiazuron (Duchefa Biochemie, Haarlem, the Netherlands, T0916).
   Acetosyringone—3′,5′-dimethoxy-4′-hydroxy-acetophenone (Sigma-Aldrich, D134406).

Table 1
Media for aspen in vitro culture and transformation

Medium	Content	Purpose/notes
YEB	5 g/L beef extract, 1 g/L yeast extract, 5 g/L peptone, 5 g/L sucrose, 0.5 g/L $MgCl_2$ Solidifying agent: agar 20 g/L	Agrobacterium plate
LB	NaCl 10 g/L, tryptone 10 g/L, yeast extract 5 g/L	Mini and midi liquid culture of Agrobacterium
MS0	MS medium, 20 μM acetosyringone; pH 5.8	Transformation
MS1	MS medium, sucrose 2%, BAP 0.2 mg/L, IBA 0.1 mg/L, TDZ 0.01 mg/L; pH 5.8 Solidifying agent: plant agar 8.5 g/L	Coincubation, solid plate Washing—liquid medium with 500 mg/L cefotaxime Callus induction—solid plate with 500 mg/L cefotaxime and antibiotics (80 mg/L kanamycin or 25 mg/L hygromycin)
MS2	MS medium, sucrose 2%, BAP 0.2 mg/L, IBA 0.1 mg/L, 500 mg/L cefotaxime, pH 5.6 Solidifying agent: Phytagel 2.7 g/L	Shoot induction—small jars (100 mL of medium/jar)
MS6	½ MS, pH 5.6 Solidifying agent: Phytagel 2.7 g/L	Root induction/maintains of lines—big jars (100–133 mL of medium/jar)

MS (Murashige & Skoog medium including vitamins, Duchefa Biochemie, Haarlem, the Netherlands, M0222).

Cefotaxime—Cefotaxime sodium (Duchefa Biochemie, Haarlem, the Netherlands, C0111).

Kanamycin—Kanamycin sulfate monohydrate (Duchefa Biochemie, Haarlem, the Netherlands, K0126).

Hygromycin—Hygromycin B (Duchefa Biochemie, Haarlem, the Netherlands, H0192).

Plant Agar (Duchefa Biochemie, Haarlem, the Netherlands, P1001).

Phytagel (Sigma-Aldrich, P8169).

2. Media (*see* Table 1 and **Note 1**).
3. Small jars. Tissue culture vessels with a revolutionary breathing system ECO2box/green filter (OVAL model 80 mm H) (Duchefa Biochemie, Haarlem, the Netherlands, E1654).

4. Growth chamber. Photoperiod 18 h day (25 °C)/6 h night (18 °C), light intensity max 150 μmol/m^2 s; Light source, warm white fluorescent tubes (colour code 830).

2.7 Materials for Evaluation of the Transgene Expression Level by RT-PCR

1. Commercially available kits for plant DNA and RNA isolation.
2. Primers used for amplification of vector fragment, here kanamycin-resistance gene, reference gene, for example actin (GenBank accession XM_002308329.2) and the expressed transgene, here *Pc*GCE (GenBank accession AFM93784.1):
   KanaF1: 5′gatgtttcgcttggtggtc3′,
   KanaR2: 5′gcttgggtggagaggctatt3′,
   ACT11for: 5′tattgttctcagtggtggctct3′,
   ACT11rev: 5′ggactcatcatactctgccttt3′,
   PcGCEf: 5′ ctggtaacacaaccacgttc3′,
   PcGCEr: 5′aagaactggccgacgcct3′.
3. DNase enzyme or kit to remove DNA from RNA sample.
4. Reverse transcriptase and reverse primer used for reverse transcription reaction or reverse transcription kit.

2.8 Materials for the Greenhouse Experiment

Materials for the greenhouse experiment with transgenic aspen lines.

1. 3 L pots.
2. Soil (Hasselfors K-jord, Örebro, Sweden).
3. 5 L transparent foil bags.
4. Plastic sticks, 30 cm long (three to four per pot).
5. Greenhouse equipped with lamps (Lamps OSRAM POWERSTAR 400w/D PRO Daylight E 40). Light conditions: 18 h of day (22 °C) and 6 h of night (18 °C) and relative humidity 60–80%.
6. Fertilizer Rika-S (N/P/K 7:1:5; Weibulls Horto, Hammenhög, Sweden).

2.9 Materials for Protein Isolation

Materials for protein isolation from different cell compartments.

1. Stem segments of transgenic and wild type trees, freshly collected from greenhouse-grown trees.
2. Syringe, 5–20 mL, depending on the amount of used material.
3. Razor blades or scalpel.
4. Mortar, pestle and liquid nitrogen.
5. Protease inhibitor (here, c*O*mplete, Roche).
6. Apoplastic and soluble protein isolation buffer. In the case of *Pc*GCE it is composed of 50 mM sodium phosphate buffer, pH 6.0, 2 mM EDTA, 4% PVP mw 360,000, and supplemented with protease inhibitor cocktail.

7. Cell wall bound protein isolation buffer. In the case of *Pc*GCE it is composed of apoplastic and soluble protein isolation buffer (**item 6**) supplemented with 1 M NaCl.
8. Quick Start™ Bradford protein assay (Bio-Rad).

2.10 Materials for GCE Enzyme Activity Test

1. ELISA plate reader.
2. Flat bottom ELISA plates.
3. 50 mM sodium phosphate buffer pH 6.0.
4. The substrate. Commercially available benzyl D-glucuronate (Carbosynth, Compton, UK, MB07246) (*see* **Note 2**).
5. 2 M hydroxylamine hydrochloride.
6. 3.5 N NaOH.
7. HCl diluted with 2 volumes of water.
8. 0.37 M $FeCl_3.6H_2O$ dissolved in 0.1 N HCl.
9. Acetylcholine chloride for the standard curve (*see* **Note 3**).

3 Methods

3.1 Selection of the Enzyme

Most important general considerations for selecting the enzymes for cell wall modification are their specificities, pH, temperature, and cofactor dependence. The pH in cell wall is acidic, around 5.5, although it might be locally different. Polymers in cell wall might be not as easily accessible as the similar polymers in solution. Moreover, the enzymatic tests are frequently carried out using substrates which are not exactly mimicking the substrates in cell wall. For example, many native hemicellulosic polysaccharides in cell wall are acetylated and may be inaccessible to hydrolyzing enzymes, whereas hemicelluloses used for testing enzymatic activities in vitro might be deacetylated [29]. In such cases, coexpressing a hydrolase with an acetyl esterase could be a solution. Here we give an example of expressing *Pc*GCE (GenBank Accession: AFM93784.1) in aspen. Some sources for other cDNA are listed in **Note 4**.

3.2 Preparation of the Entry Vector Used as Entry Clone in Gateway® Cloning System, Which Contains the Fungal cDNA Fused with Plant Signal Peptide

The cloned cDNA of a fungal gene (*PcGCE*, Subheading 2.1) is used as a template to create the entry clone for further cloning in a Gateway® system. To express *Pc*GCE enzyme in cell walls, the native fungal signal peptide sequence is exchanged to the signal peptide sequence from *Ptxt*GH9B3 (alias *Ptt*Cel9B3) form hybrid aspen by PCR (GenBank accession AY660968.1 [26]) (*see* **Note 5**).

1. *PtxtCel9B3* signal peptide sequence is amplified using forward signal peptide primer (Fsp), in which the adaptor sequence for TOPO® cloning CACC is present on 5′ end. Reverse signal peptide primer (Rsp) contains 3′ end of *PtxtCel9B3* signal

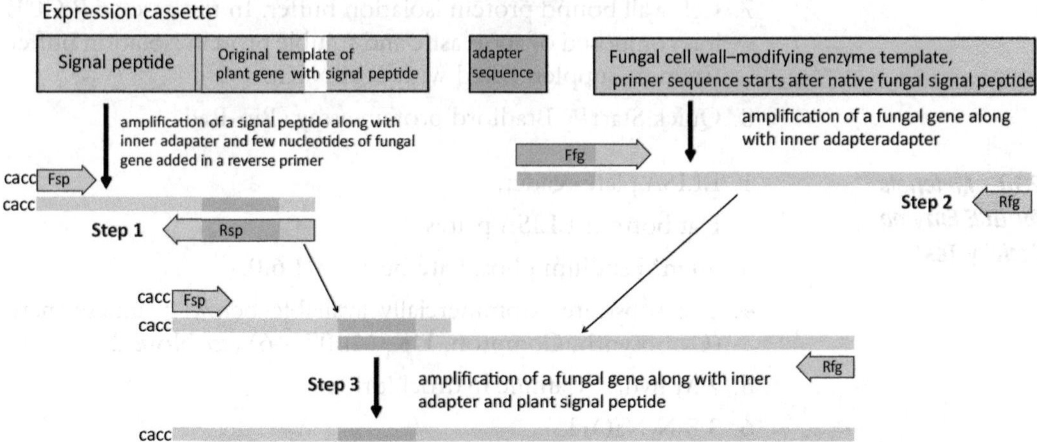

Fig. 1 Schematic representation of the vector cloning strategy. To target the enzyme to cell walls, we used the signal sequence of hybrid aspen secreted cellulase, *Ptxt*Cel9B3 and exchanged it with the native fungal signal peptide

peptide, adaptor sequence (15 nucleotides), and a beginning of fungal enzyme sequence starting after its native peptide signal. This way *PtxtCEL9B3* signal peptide sequence is amplified with the adaptor for TOPO® cloning, inner adaptor sequence (15 nt) and 5′ fragment of fungal gene (step 1 in Fig. 1).

2. Fungal gene is amplified using forward fungal primer (Ffp) which contains the same adaptor sequence (15 nucleotides) as described above followed by the fungal gene sequence starting after its native fungal signal peptide. Part of this fungal sequence was also present in Rsp primer. Reverse fungal primer (Rfp) includes the stop codon. Thus the obtained fragment contains the 15 nucleotides adapter, and the fungal enzyme without signal peptide sequence, including the stop codon (step 2 in Fig. 1).
3. In a final PCR reaction, the diluted PCR products obtained in **steps 1** and **2** are used in a molar ratio 1:3 (fungal gene amplicon: signal peptide amplicon) forming a hybrid template (*see* **Note 6**). For a final hybrid product amplification (step 3 in Fig. 1), primers Fsp and Rfp are used so that the product contains the TOPO® cloning adapter (CACC), *PtxtCel9B3* signal peptide followed by the inner adaptor sequence (15 nucleotides) and fungal enzyme cDNA sequence.
4. Agarose gel electrophoresis is used to identify and purify full-length product obtained in **step 3**, using commercially available agarose gel purification kit.

5. Purified the hybrid sequence obtained in **step 4** is introduced into *pENTR™/D-TOPO®* vector using pENTR™/D-TOPO® Cloning Kit, to obtain the entry clone (Gateway® Technology protocol). Finally the vector is cloned in *E. coli* by the standard procedures.
6. Several colonies containing the entry clone are sequenced using M13 forward and reverse primers to exclude any clones with mutations and the clones with correct construct sequence are saved.

3.3 Preparation of the Destination Vector Containing the "Wood Specific" Promoter

Gateway® cloning system facilitates rapid and efficient cloning into many different destination vectors. Efficient cell wall modification requires proper tissue-localization of the expressed enzyme, which depends on the promoter. We have therefore replaced the commonly used the CaMV 35S promoter for ectopic expression from the Gateway® destination vector, that is, pK2GW7.0 [27] with a tissue specific promoter, *PtGT43B*, active in cells forming secondary cell walls [20] (*see* **Note** 7).

1. Genomic DNA from young leaves of in vitro–grown *Populus trichocarpa* (genotype Nisqually 1) is extracted using a commercially available kit for gDNA isolation.
2. *PtGT43B* promoter is amplified with primers pGT43BSacF and pGT43BSpeR. On the 5′ end of the forward primer there is a specific sequence obligatory for TOPO® cloning (CACC) followed by the SacI restriction site. On the 3′ end of the reverse primer, the SpeI restriction site is included. Thus, the product sequence starts at the 5′ end with TOPO® cloning adaptor, followed by the Sac I restriction site, then the *PtGT43B* promoter sequence, and finally the SpeI restriction site.
3. The PCR product obtained in the previous step is introduced into *pENTR™/D-TOPO®* vector using pENTR™/D-TOPO® Cloning Kit and cloned to *E. coli* by the standard procedures. Several bacterial colonies containing the entry clone with the *PtGT43B* promoter are used for plasmid sequencing using M13 forward and reverse primers and clones with correct sequence are saved.
4. The cloning cassette is cut out by SacI and SpeI restriction enzymes from a correct *pENTR™/D-TOPO®* vector containing *PtGT43B* promoter (Subheading 3.3, **step 3**). Destination vector *pK2GW7.0* is digested with the same restriction enzymes resulting in removal of the 35S promoter. Agarose gel electrophoresis is run to identify and purify the digestion products from the two reactions: the *pK2GW7.0* vector without 35S promoter and the *PtGT43B* promoter cut out from the *pENTR™/D-TOPO®* vector. The fragments are purified with commercially available agarose gel purification kit.

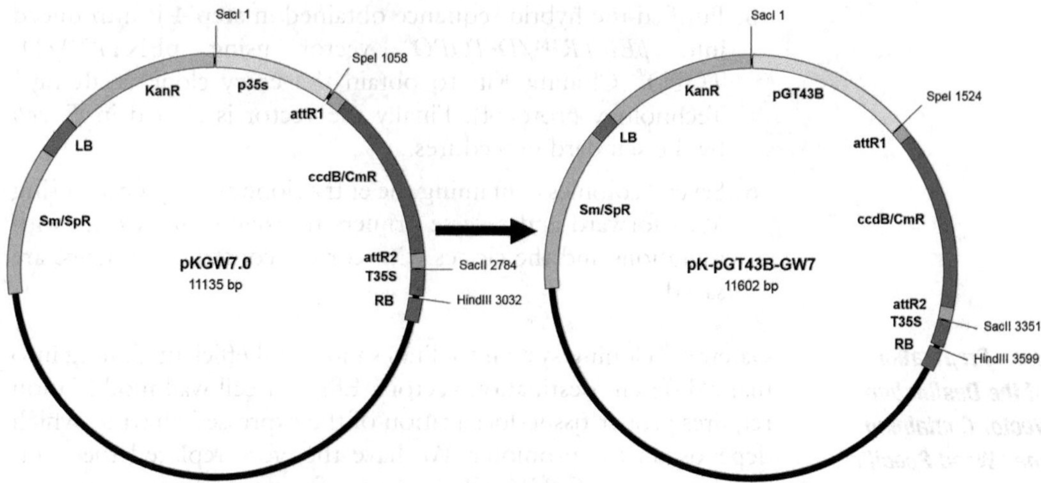

Fig. 2 Destination vector *pK-pGT43B-GW7* carrying the *PtGT43B* promoter driving the expression in cells depositing secondary cell walls. The *pK-pGT43B-GW7* vector was described in Ratke et al. [20], and it originated from Gateway® destination vector *pK2GW7.0*. The *pK-pGT43B-GW7* vector is used together with an entry vector in the LR Clonase ™ reaction as described in Gateway® Technology, to obtain the final expression vector

5. The fragment containing the *PtGT43B* promoter is ligated with the linearized *pK2GW7.0* vector using T4 DNA ligase. This step generates the destination vector *pK-pGT43B-GW7.0* [20] (Fig. 2) which can be used as a regular destination vector in Gateway® cloning system. This particular destination vector possesses the spectinomycin resistance gene for bacterial selection and the kanamycin resistance gene for in planta selection. The vector is sequenced to confirm correct orientation of the *PtGT43B* promoter.

3.4 Preparation of the Expression Vector and Its Insertion to the Agrobacterium

1. Destination vector obtained in the previous step (Subheading 3.3, **step 5**) and the entry clone containing the fungal enzyme sequence obtained in Subheading 3.2, **step 6**. are used in LR Clonase™ reaction as described in Gateway® Technology manual (https://tools.thermofisher.com/content/sfs/manuals/gatewayman.pdf) to generate the expression vector (called expression clone in Gateway® manual). Since the expression vector is created in *pK-GT43B-GW7.0* it possesses the spectinomycin resistance gene for bacterial selection, and kanamycin resistance gene for in planta selection.
2. Multiplied expression vector is introduced into *Agrobacterium tumefaciens* strain GV3101-pMP90RK using either the electroporation, the freeze/thaw method of transformation, or triparental mating [28, 30]. Positive colonies are verified by PCR using Fsp and Rfp primers.

3.5 Aspen Subculturing and Multiplication

The clone T89 of hybrid aspen has been maintained in vitro for several years, and it is easy to transform, multiply, adapt to soil culture and to greenhouse conditions. All manipulations described below are done in a sterile hood using the sterile technique.

1. Hybrid aspen plants (clone T89) are aseptically grown on MS6 medium in big jars (Fig. 3j). Each big jar has room for growing up to four rooted cuttings which can grow continuously for a maximum of 12 months in a controlled environment chamber, as described in Subheading 2.6, **item 4**.
2. When the rooted cuttings reach the lid of the jar, they stop elongate their leaders and start producing many side shoots. These side shoots are used for the line multiplication. Explants about 3–4 cm long of each side shoot including its apical meristem and two to three leaves are cut out and placed in the fresh MS6 medium in a big jar. (*see* **Note 8**) (Fig. 3h). Jars are carefully closed, covered with one sheet of white paper (Subheading 2.5, **item 4**) and grown in a controlled environment chamber (Subheading 2.6, **step 4**). After 1 week, the paper is removed. The cuttings produce roots and may be kept for the next 12 months as a stock, or for 3–4 weeks before being transformed, or for 5–6 weeks before being transferred to soil.

Fig. 3 Different stages of in vitro hybrid aspen culture (**a**–**j**) and its greenhouse cultivation following the in vitro multiplication (**k**, **l**). (**a**) explants after transformation; (**b**–**d**) explants with developing callus resistant to the antibiotics in the medium; (**e**) callus removed from explants and grown on plates; (**f**, **g**) stem development; (**h**, **i**) root development; (**j**) fully developed aspen rooted cuttings growing in the jar; (**k**) 5-week-old plants freshly transferred to pots and covered with plastic bags; (**l**) 3-month-old aspen trees in the greenhouse

3.6 Aspen Transformation and Selection

The transformation protocol described below is optimized for T89 clone and is based on the one described by Nilsson et al. [31]. All manipulations are done in a sterile hood and using the sterile technique (Fig. 3).

1. Agrobacterium strain GV3101-pMP90RK containing the binary vector with the cloned fungal gene of interest (Subheading 3.4, **step 2**) is grown on YEB-plates (*see* Table 1) supplemented with antibiotics (100 mg/L rifampicin, 50 mg/L gentamycin, 50 mg/L kanamycin, and 100 mg/L spectinomycin) for 48 h at 28 °C. The final antibiotic depends on the bacterial resistance gene present in the binary vector.
2. Single colonies are picked and inoculated into 10 mL LB medium (*see* Table 1) supplemented with antibiotics (as above) and incubated at 28 °C for 48 h with shaking to obtain mini cultures of Agrobacterium.
3. To obtain midi cultures of Agrobacterium, 500 μL of mini culture is transferred into 50 mL LB supplemented with antibiotics (as above) and incubated at 28 °C for about 24 h with shaking, until optical density OD_{600} reaches 0.6–0.8. The optical density cannot exceed 1.
4. Midi cultures are centrifuged at 3000 × *g* for 10 min at RT. The supernatant is decanted and the pellet is suspended in MS0 medium (without sugars, pH 5.8 and supplemented with 20 μM acetosyringone, *see* Table 1) to the final density corresponding to OD_{600} of around 0.5.
5. Bacterial suspensions are incubated on Petri plates at 25 °C for about 30 min with gentle shaking (25 rpm) (*see* **Note 9**).
6. Explants from 4–5-week-old hybrid aspen (clone T89) grown in vitro on MS6 medium as described in Subheading 3.5, **step 2** should be used for transformation. Petioles are cut into segments around 10 mm long and stems (internodes) are cut to segments between 5 mm and 10 mm long (*see* **Note 10**). Cutting needs to be done within 10 min to prevent the drying of the material, or it can be done in sterile water in a Petri dish.
7. Directly after cutting, each explant is transferred into MS0 medium (*see* Table 1) with agrobacterium (Subheading 3.6, **step 5**) and incubated for 1 h in the light with gentle shaking (25 rpm).
8. Explants are transferred to a plate with MS1 medium, and the plate wrapped with aluminum foil is incubated for 48 h in the growth chamber (Subheading 2.6, **step 4**).
9. After 48 h, the explants are washed twice for 20 min in sterile MS medium without any antibiotics, pH 5.8) supplemented with 500 mg/L cefotaxime.

10. After that the explants are transferred to plates with MS1 medium (*see* Table 1) (supplemented with 500 mg/L cefotaxime and 80 mg/L kanamycin (selection is determined by the selection in planta of the used binary vector) and incubated for 2 weeks (Fig. 3a). During the first week the plates with explants are covered with one sheet of white paper and then the explants are exposed to light (*see* **Note 11**). Every 2 weeks the explants are transferred onto the fresh plate with MS1 medium (Table 1) with antibiotics as above and incubated in the growth chamber (Subheading 2.6, **step 4**) until green callus appears (Fig. 3b). When the callus is big enough to be handled, it is transferred onto the fresh medium without the parent explant (Fig. 3e). Callus tissue is continuously being transferred to plates with MS1 medium with antibiotics as long as it does not reach the lid.
11. Larger callus are transferred into small jars with MS2 medium (*see* Table 1) to induce shoot production (Fig. 3f, g).
12. When shoots are 4 cm long, they are cut close to the medium and transferred to big jars with MS6 medium (*see* Table 1) (Fig. 3h, i). At least 24 independent lines are prepared per construct. An independent line should originate from a single transformation event that produces callus (*see* **Note 12**).

3.7 Transgene Expression Level and the Presence of Transgene in the Selected Lines

After transformation, all obtained lines need to be screened for successful transformation and also for the best expression. The three to five best lines are usually selected based on this screening and maintained in vitro. These lines can be multiplied as described in Subheading 3.5 when needed, for planting in soil or other experiments.

1. Green parts of rooted cuttings left in the big jars after subculturing are homogenized in liquid nitrogen and DNA is isolated with any commercially available kit suitable for plant DNA extraction. DNA from each line is used as a template for PCR. PCR primers amplifying the selection gene, that is, kanamycin resistance, or any other vector part of interest, can be used to identify the transgenic lines.
2. The RNA for RT-PCR is prepared from desired parts of rooted cuttings by homogenizing them in liquid nitrogen. RNA is isolated with any commercially available kit suitable for plant RNA isolation. DNA is removed by DNase treatment and reverse transcription is done using any kit or the selected reverse transcriptase and selected reverse primer, for instance, the random hexamers. Two sets of PCR primers are needed. One to amplify the fragment of expressed transgene and the second to amplify a reference gene, for example actin. Semi-quantitative analysis of expression level of introduced gene is

done based on relative analysis with the reference gene. Alternatively, the quantitative PCR can be used instead of semiquantitative RT-PCR.

3.8 Greenhouse Experiment

There is a considerable variability among trees, and therefore as many replicates as possible should be planned in the greenhouse experiment. Typically, ten trees from each independent line should be grown and compared with at least ten wild-type T89 trees, used as a control. If transgenic control is used, at least two lines of such controls need to be produced to be able to detect any effects caused by gene disruption caused by random insertion.

1. Best lines are multiplied in vitro (Subheading 3.5). Five 6-week-old rooted cuttings (note that the age may be variable among the lines, if transgenic effect is manifesting already during in vitro culture) grown in the big jars on MS6 medium are transferred to 3 L pots filled with watered soil. Approx. 200 mL water is poured to each pot before planting, and 200 mL is added after planting (*see* **Note 13**).
2. Plants need to be covered with transparent foil bags immediately after planting. To support the bags, three to four sticks are inserted in soil around each plant (Fig. 3k).
3. Pots are transferred to the table in a greenhouse (Subheading 2.8, **item 5**) and plants are grown for 1 week under bags. Twice a day, the condensed moisture is removed from the bags by shaking them.
4. After 1 week, two corners of each bag are cut and plants are grown for another week. Then the bags are removed.
5. After the bags are removed, the plans are watered twice a day and fertilized once a week with 150 mL of 1% Rika-S.
6. Plants are initially grown on the tables, and later are moved to the floor. They can be grown for 3–6 months until they reach the desired height (Fig. 3l).

3.9 Protein Isolation from Different Cell Compartments

One of the most important considerations when evaluating the effects of transgenes is a proper localization of the expressed protein. There are many methods of protein identification, depending on the available equipment and materials. Here we present a relatively quick and reliable method of protein identification in one of the three compartments: apoplastic fluid, soluble fraction, and cell wall ionically bound fraction. These isolated protein fractions can be used in numerous experiments, for example, western blot, immunoprecipitation, or in enzyme activity assays. The latter confirms not only that the gene is expressed but also that the enzyme is active in the cell wall compartment, and thus expected to act on the substrates localized in cell walls.

1. Apoplastic fluid preparation according to Pogorelko et al. [32]. Green aerial parts of 6-week-old hybrid aspen growing in vitro in big jars or desired organs from the greenhouse grown plants can be used. The organs are cut with a scalpel into 5 mm long pieces.
2. The collected material is transferred to a syringe without a plunger, having an appropriate size to fit the material. The tip of the syringe is sealed with a Parafilm® and the syringe is filled with a suitable buffer for expressed protein activity. In case of *Pc*GCE, 50 mM sodium phosphate buffer, pH 6.0, containing 2 mM EDTA, 4% PVP mw 360,000, and protease inhibitor cocktail is used.
3. The syringe with plant material immersed in buffer without a plunger is placed in a vacuum for 2 × 30 min with a 5 min break in between.
4. Then the Parafilm® is carefully removed and the buffer is drained out.
5. The syringe barrel is then placed in a 50 mL Falcon tube (suitable for 5, 10 and 20 mL syringe) and centrifuged at 4000 × *g* for 10 min at 4 °C. The liquid collected at the bottom of the flacon tube contains the apoplastic fluid and can be used for the protein activity test.
6. Material placed in a syringe is taken out and ground in a mortar in liquid nitrogen. Part of the ground material is transferred to a fresh tube.
7. The same buffer as in Subheading 3.9, **step 2** is added to the material at 1:1 volume ratio, and then incubated with shaking at 4 °C for 1 h.
8. The tubes are centrifuged at 20,000 × *g* for 10 min at 4 °C and the supernatant containing the soluble protein fraction is collected.
9. To the remaining pellet, the same buffer as in Subheading 3.9, **step 2**, supplemented with 1 M NaCl, is added, the material is suspended and incubated with shaking for 1 h at 4 °C.
10. After centrifugation at 20,000 × *g* for 10 min at 4 °C, the supernatant containing ionically bound proteins is collected.
11. Concentrations of proteins in all three fractions (apoplastic fluid, soluble and ionically bound proteins) are determined in a Bradford protein assay and used for activity analyses.

3.10 GCE Enzyme Activity Test

1. Reaction is carried out in ELISA plate in a volume of 40 μL.
2. 5 μg of extracted proteins from the three fractions are incubated in 50 mM sodium phosphate buffer, pH 6.0, containing 5 mM substrate (Subheading 2.10, **item 4**, *see* **Note 2**). As a control, thermally denatured extracted proteins are used.

3. The plate is incubated for 1 h at 30 °C.
4. After that time, ester bonds remaining in the substrate are detected according to Hestrin [33] as follows:
5. 80 μL of a mixture of 2 M hydroxylamine hydrochloride: 3.5 N NaOH (1:1; vol:vol) is added. The solutions are mixed and left for 1 h.
6. 40 μL of diluted HCl (Subheading 2.10, **item** 7) is added and mixed.
7. 40 μL of $FeCl_3.6H_2O$ (Subheading 2.10, **item 8**) is added and mixed.
8. Optical density at 540 nm is measured and compared to the standard curve prepared with acetylcholine chloride (*see* **Note 3**). The purple-brown color indicates the ester bonds.

4 Notes

1. The pH of media is critical for all in vitro cultures and it is known that medium pH can be slightly changing during autoclaving or extended heat exposure. With our experience, all media can be supplemented with 0.07% MES, pH 5.75 before autoclaving to stabilize the pH. Volume of MS6 medium in the big jars depends on the duration of cultivation. For the long-term cultivation (up to 12 months) ~133 mL per jar is used, but for the short cultivation (up to 6 weeks), 100 mL per jar is enough.
2. Other substrates [34] and methods for glucuronoyl esterase activity assays are available including thin layer chromatography, HPLC [35] and spectrophotometric analysis [36].
3. For the colorimetric measurement of esterase activity by identification of ferric–acetohydroxamic acid complex, the standard curve needs to be prepared. When one ester bond is present in a substrate then the absorbance is a direct function of concentration. If the amount of substrate is limited, acetylcholine chloride, which contains one ester bond, can be used to prepare the standard curve [33].
4. Many different CAZymes and lignin-acting enzymes can be used for improvement of wood processing in different applications. Many fungal enzymes are available as clones in DNA stocks (http://www.fgsc.net/). If a gene sequence is known, a DNA synthesis can be purchased from biotech companies and codon optimization can be carried out at the same time, for an efficient protein synthesis in planta.
5. Effective expression of enzymes requires their proper targeting to their site of action. To target enzymes to cell walls, we used the signal sequence present in hybrid aspen secreted cellulase,

*Ptxt*Cel9B3 [37]. Other signal peptides with known organelle-specific targeting can be used as well. For instance, 5′ fusion of β-expansin signal peptide efficiently targets the enzymes to cell wall [23]. Native fungal signal peptides can be also effective *in planta* [21]. In some cases, wall-degrading enzymes targeted to apoplast can have a negative impact on plant development, which can be avoided by targeting such enzymes to other subcellular compartments. For example, an endoglucanase from *Trichoderma reesei* expressed in apoplast of tobacco resulted in reduced content of crystalline cellulose and inhibited growth but its expression in ER only altered the matrix polysaccharide composition without visible effects on plant development [38].

6. Molar ratio depends on amplicon length. As more similar in length two amplicons are as closer to 1 molar ratio is used.
7. *Pt*GT43B promoter was found effective for genetic engineering of properties of wood and fibers [30]. It is specific to cells forming secondary cell walls. Other promoters successfully used for modification of wood include: *C4H* promoter active in cells producing monolignols, or CesA8 and FRA8 promoters active in similar cell types as *PtGT43B* promoter [12, 18, 39]. Utilization cell or tissue specific promoter to express or silence wood-associated gene expression ensures that wood cells properties can be specifically changed while other cells remain similar to the wild type. This strategy probably allows avoiding many unintended phenotypic changes.
8. Few leaves [2–4] should be visible on the explant that is used for rooting. Phenotypes could appear at this stage and fully developed leaves may not always can be identified (*see* Fig. 3i) but even if reduced they need to be present on the stem segment for successful rooting.
9. Agrobacterium is incubated in MS0 medium during the preparation of plant explants. If this preparation requires a longer time then 30 min it is possible to extend the Agrobacterium incubation as well, but it should not take longer than 60 min. If longer preparation time is needed, we recommend to start preparing explants before transferring Agrobacterium to MS0 medium.
10. Agrobacterium strain GV3101 is rifampicin resistant, and it contains Ti-plasmid pMP90RK with gentamycin and kanamycin resistance genes [28]. Binary vector should be carefully chosen because of limited bacterial selection possibilities. Also plant selection need to be considered. In the described method, kanamycin and hygromycin selection in planta works best. If hygromycin selection is used, internodes used as explants give better transformation efficiency than petioles.

11. Each time after transferring explants onto a fresh medium (in plates or in jars) one sheet of white paper needs to be placed on the culture containers for the first week. During the incubation of explants on a Petri dish, the temperature control is crucial because of relatively small volume of medium and air in a dish. If the incubation takes place on a shelf in the growth chamber, consider that lights installed under the shelf can significantly increase local temperature on this shelf.
12. For over 20 years of using this transformation protocol, we have not observed any nontransgenic escapes. Presence of selective antibiotics in MS1 medium during callus induction is enough to prevent the growth of nontransformed tissue. All trees, even after many years of cultivation in vitro, were able to grow in soil and they contained the introduced transgenes.
13. Plants grown in vitro have very little cuticle developed and thus they are extremely sensitive to drought stress during planting. Taking them out from the jars, planting and covering them with foil bags needs to be done as quickly as possible. Bags should be tightly attached to pots so humidity can reach a similar level as in the jars.

Acknowledgments

We thank Dr. Emma Master and Dr. Satoshi Endo for their advice on the vector cloning strategy and Veronica Bourquin for sharing her experience in plant tissue culture techniques.

References

1. Mellerowicz EJ, Gorshkova TA (2012) Tensional stress generation in gelatinous fibres: a review and possible mechanism based on cell-wall structure and composition. J Exp Bot 63 (2):551–565. https://doi.org/10.1093/jxb/err339
2. Scheller HV, Ulvskov P (2010) Hemicelluloses. Annu Rev Plant Biol 61:263–289. https://doi.org/10.1146/annurev-arplant-042809-112315
3. Fenning TM, Gershenzon J (2002) Where will the wood come from? Plantation forests and the role of biotechnology. Trends Biotechnol 20(7):291–296
4. Pauly M, Keegstra K (2008) Cell-wall carbohydrates and their modification as a resource for biofuels. Plant J 54(4):559–568. https://doi.org/10.1111/j.1365-313X.2008.03463.x
5. Pauly M, Keegstra K (2010) Plant cell wall polymers as precursors for biofuels. Curr Opin Plant Biol 13(3):305–312. https://doi.org/10.1016/j.pbi.2009.12.009
6. Leboreiro J, Hilaly AK (2011) Biomass transportation model and optimum plant size for the production of ethanol. Bioresour Technol 102(3):2712–2723. https://doi.org/10.1016/j.biortech.2010.10.144
7. Bonawitz ND, Chapple C (2013) Can genetic engineering of lignin deposition be accomplished without an unacceptable yield penalty? Curr Opin Biotechnol 24(2):336–343. https://doi.org/10.1016/j.copbio.2012.11.004
8. Brown DM, Goubet F, Wong VW, Goodacre R, Stephens E, Dupree P, Turner SR (2007) Comparison of five xylan synthesis mutants reveals new insight into the mechanisms of xylan synthesis. Plant J 52 (6):1154–1168. https://doi.org/10.1111/j.1365-313X.2007.03307.x

9. Brown DM, Zeef LA, Ellis J, Goodacre R, Turner SR (2005) Identification of novel genes in Arabidopsis involved in secondary cell wall formation using expression profiling and reverse genetics. Plant Cell 17 (8):2281–2295. https://doi.org/10.1105/tpc.105.031542
10. Lee C, O'Neill MA, Tsumuraya Y, Darvill AG, Ye ZH (2007) The irregular xylem9 mutant is deficient in xylan xylosyltransferase activity. Plant Cell Physiol 48(11):1624–1634. https://doi.org/10.1093/pcp/pcm135
11. Lee C, Zhong R, Ye ZH (2012) Arabidopsis family GT43 members are xylan xylosyltransferases required for the elongation of the xylan backbone. Plant Cell Physiol 53(1):135–143. https://doi.org/10.1093/pcp/pcr158
12. Pena MJ, Zhong R, Zhou GK, Richardson EA, O'Neill MA, Darvill AG, York WS, Ye ZH (2007) Arabidopsis irregular xylem8 and irregular xylem9: implications for the complexity of glucuronoxylan biosynthesis. Plant Cell 19 (2):549–563. https://doi.org/10.1105/tpc.106.049320
13. Wu AM, Hornblad E, Voxeur A, Gerber L, Rihouey C, Lerouge P, Marchant A (2010) Analysis of the Arabidopsis IRX9/IRX9-L and IRX14/IRX14-L pairs of glycosyltransferase genes reveals critical contributions to biosynthesis of the hemicellulose glucuronoxylan. Plant Physiol 153(2):542–554. https://doi.org/10.1104/pp.110.154971
14. Bouton S, Leboeuf E, Mouille G, Leydecker MT, Talbotec J, Granier F, Lahaye M, Hofte H, Truong HN (2002) QUASIMODO1 encodes a putative membrane-bound glycosyltransferase required for normal pectin synthesis and cell adhesion in Arabidopsis. Plant Cell 14(10):2577–2590
15. Schindelman G, Morikami A, Jung J, Baskin TI, Carpita NC, Derbyshire P, McCann MC, Benfey PN (2001) COBRA encodes a putative GPI-anchored protein, which is polarly localized and necessary for oriented cell expansion in Arabidopsis. Genes Dev 15(9):1115–1127. https://doi.org/10.1101/gad.879101
16. Turner SR, Somerville CR (1997) Collapsed xylem phenotype of Arabidopsis identifies mutants deficient in cellulose deposition in the secondary cell wall. Plant Cell 9 (5):689–701. https://doi.org/10.1105/tpc.9.5.689
17. Eudes A, Sathitsuksanoh N, Baidoo EE, George A, Liang Y, Yang F, Singh S, Keasling JD, Simmons BA, Loque D (2015) Expression of a bacterial 3-dehydroshikimate dehydratase reduces lignin content and improves biomass saccharification efficiency. Plant Biotechnol J 13(9):1241–1250. https://doi.org/10.1111/pbi.12310
18. Wilkerson CG, Mansfield SD, Lu F, Withers S, Park JY, Karlen SD, Gonzales-Vigil E, Padmakshan D, Unda F, Rencoret J, Ralph J (2014) Monolignol ferulate transferase introduces chemically labile linkages into the lignin backbone. Science 344(6179):90–93. https://doi.org/10.1126/science.1250161
19. Zhang K, Bhuiya MW, Pazo JR, Miao Y, Kim H, Ralph J, Liu CJ (2012) An engineered monolignol 4-*O*-methyltransferase depresses lignin biosynthesis and confers novel metabolic capability in Arabidopsis. Plant Cell 24 (7):3135–3152. https://doi.org/10.1105/tpc.112.101287
20. Ratke C, Pawar PM, Balasubramanian VK, Naumann M, Duncranz ML, Derba-Maceluch M, Gorzsas A, Endo S, Ezcurra I, Mellerowicz EJ (2015) Populus GT43 family members group into distinct sets required for primary and secondary wall xylan biosynthesis and include useful promoters for wood modification. Plant Biotechnol 13(1):26–37. https://doi.org/10.1111/pbi.12232
21. Pawar PM, Derba-Maceluch M, Chong SL, Gomez LD, Miedes E, Banasiak A, Ratke C, Gaertner C, Mouille G, McQueen-Mason SJ, Molina A, Sellstedt A, Tenkanen M, Mellerowicz EJ (2016) Expression of fungal acetyl xylan esterase in *Arabidopsis thaliana* improves saccharification of stem lignocellulose. Plant Biotechnol 14(1):387–397. https://doi.org/10.1111/pbi.12393
22. Ferrari S, Galletti R, Pontiggia D, Manfredini C, Lionetti V, Bellincampi D, Cervone F, De Lorenzo G (2008) Transgenic expression of a fungal endo-polygalacturonase increases plant resistance to pathogens and reduces auxin sensitivity. Plant Physiol 146 (2):669–681. https://doi.org/10.1104/pp.107.109686
23. Pogorelko G, Lionetti V, Fursova O, Sundaram RM, Qi M, Whitham SA, Bogdanove AJ, Bellincampi D, Zabotina OA (2013) Arabidopsis and *Brachypodium distachyon* transgenic plants expressing *Aspergillus nidulans* acetylesterases have decreased degree of polysaccharide acetylation and increased resistance to pathogens. Plant Physiol 162(1):9–23. https://doi.org/10.1104/pp.113.214460
24. Latha Gandla M, Derba-Maceluch M, Liu X, Gerber L, Master ER, Mellerowicz EJ, Jonsson LJ (2015) Expression of a fungal glucuronoyl esterase in *Populus*: effects on wood properties and saccharification efficiency. Phytochemistry 112:210–220. https://doi.org/10.1016/j.phytochem.2014.06.002

25. Tsai AY, Canam T, Gorzsas A, Mellerowicz EJ, Campbell MM, Master ER (2012) Constitutive expression of a fungal glucuronoyl esterase in Arabidopsis reveals altered cell wall composition and structure. Plant Biotechnol 10 (9):1077–1087. https://doi.org/10.1111/j.1467-7652.2012.00735.x
26. Rudsander U, Denman S, Raza S, Teeri TT (2003) Molecular features of Family GH9 cellulases in hybrid Aspen and the filamentous fungus *Phanerochaete chrysosporium*. J Appl Glycosci 50(2):253–256. https://doi.org/10.5458/jag.50.253
27. Karimi M, Inze D, Depicker A (2002) GATEWAY vectors for Agrobacterium-mediated plant transformation. Trends Plant Sci 7 (5):193–195
28. Koncz C, Schell J (1986) The promoter of TL-DNA gene 5 controls the tissue-specific expression of chimaeric genes carried by a novel type of Agrobacterium binary vector. Mol Gen Genet 204(3):383–396. https://doi.org/10.1007/BF00331014
29. Chong SL, Derba-Maceluch M, Koutaniemi S, Gomez LD, McQueen-Mason SJ, Tenkanen M, Mellerowicz EJ (2015) Active fungal GH115 alpha-glucuronidase produced in *Arabidopsis thaliana* affects only the UX1-reactive glucuronate decorations on native glucuronoxylans. BMC Biotechnol 15:56. https://doi.org/10.1186/s12896-015-0154-8
30. Wise AA, Liu Z, Binns AN (2006) Three methods for the introduction of foreign DNA into Agrobacterium. In: Wang K (ed) Agrobacterium protocols. Humana Press, Totowa, NJ, pp 43–54. https://doi.org/10.1385/1-59745-130-4:43
31. Nilsson O, Aldén T, Sitbon F, Little CHA, Chalupa V, Sandberg G, Olsson O (1992) Spatial pattern of cauliflower mosaic virus 35S promoter-luciferase expression in transgenic hybrid aspen trees monitored by enzymatic assay and non-destructive imaging. Transgenic Res 1(5):209–220. https://doi.org/10.1007/bf02524751
32. Pogorelko G, Fursova O, Lin M, Pyle E, Jass J, Zabotina OA (2011) Post-synthetic modification of plant cell walls by expression of microbial hydrolases in the apoplast. Plant Mol Biol 77(4–5):433–445. https://doi.org/10.1007/s11103-011-9822-9
33. Hestrin S (1949) The reaction of acetylcholine and other carboxylic acid derivatives with hydroxylamine, and its analytical application. J Biol Chem 180(1):249–261
34. Spanikova S, Biely P (2006) Glucuronoyl esterase—novel carbohydrate esterase produced by *Schizophyllum commune*. FEBS Lett 580 (19):4597–4601. https://doi.org/10.1016/j.febslet.2006.07.033
35. Biely P, Mastihubova M, Cote GL, Greene RV (2003) Mode of action of acetylxylan esterase from *Streptomyces lividans*: a study with deoxy and deoxy-fluoro analogues of acetylated methyl beta-D-xylopyranoside. Biochim Biophys Acta 1622(2):82–88
36. Sunner H, Charavgi MD, Olsson L, Topakas E, Christakopoulos P (2015) Glucuronoyl esterase screening and characterization assays utilizing commercially available benzyl glucuronic acid ester. Molecules 20(10):17807–17817. https://doi.org/10.3390/molecules201017807
37. Takahashi J, Rudsander UJ, Hedenstrom M, Banasiak A, Harholt J, Amelot N, Immerzeel P, Ryden P, Endo S, Ibatullin FM, Brumer H, del Campillo E, Master ER, Scheller HV, Sundberg B, Teeri TT, Mellerowicz EJ (2009) KORRIGAN1 and its aspen homolog PttCel9A1 decrease cellulose crystallinity in Arabidopsis stems. Plant Cell Physiol 50 (6):1099–1115. https://doi.org/10.1093/pcp/pcp062
38. Klose H, Gunl M, Usadel B, Fischer R, Commandeur U (2015) Cell wall modification in tobacco by differential targeting of recombinant endoglucanase from *Trichoderma reesei*. BMC Plant Biol 15:54. https://doi.org/10.1186/s12870-015-0443-3
39. Huntley SK, Ellis D, Gilbert M, Chapple C, Mansfield SD (2003) Significant increases in pulping efficiency in C4H-F5H-transformed poplars: improved chemical savings and reduced environmental toxins. J Agric Food Chem 51(21):6178–6183. https://doi.org/10.1021/jf034320o

Chapter 10

Activity and Action of Cell-Wall Transglycanases

Lenka Franková and Stephen C. Fry

Abstract

Transglycanases (endotransglycosylases) are enzymes that "cut and paste" polysaccharide chains. Several transglycanase activities have been discovered which can cut (i.e., use as donor substrate) each of the major hemicelluloses [xyloglucan, mannans, xylans, and mixed-linkage β-glucan (MLG)], and, as a recent addition, cellulose. These enzymes may play interesting roles in adjusting the wall's physical properties, influencing cell expansion, stem strengthening, and fruit softening.

Activities discussed include the homotransglycanases XET (xyloglucan endotransglucosylase, i.e., xyloglucan–xyloglucan endotransglycosylase), trans-β-mannanase (mannan–mannan endotransglycosylase), and trans-β-xylanase (xylan–xylan endotransglucosylase), plus the heterotransglycanases MXE (MLG–xyloglucan endotransglucosylase) and CXE (cellulose–xyloglucan endotransglucosylase).

Transglycanases acting on polysaccharide donor substrates can utilize small, labeled oligosaccharides as acceptor substrates, generating easily recognizable polymeric labeled products. We present methods for extracting transglycanases from plant tissues and assaying them in vitro, either quantitatively in solution assays or by high-throughput dot-blot screens. Both radioactively and fluorescently labeled substrates are mentioned. A general procedure (glass-fiber blotting) is illustrated by which proposed novel transglycanase activities can be tested for.

In addition, we describe strategies for detecting transglycanase action in vivo. These methods enable the quantification of, separately, XET and MXE action in *Equisetum* stems. Related methods enable the tissue distribution of transglycanase action to be visualized cytologically.

Key words Cellulose, Histochemistry, Dot blots, Enzyme action, Enzyme activity, Glass-fiber blotting, Mannans, Mixed-linkage glucan (MLG), Transglycosylation, Xylans, Xyloglucan

1 Introduction

Transglycanases (endotransglycosylases) are enzymes that cleave a polysaccharide (donor substrate) in mid-chain, then transfer a portion of it onto another poly- or oligosaccharide molecule (acceptor substrate), conserving the energy of the cleaved bond (Fig. 1) [1]. Several such enzymes occur in the plant cell wall and are likely to contribute to wall-polysaccharide assembly, restructuring and disassembly during plant growth and development. The study of these enzymes is a relatively recent endeavor, and we present here

Zoë A. Popper (ed.), *The Plant Cell Wall: Methods and Protocols*, Methods in Molecular Biology, vol. 2149, https://doi.org/10.1007/978-1-0716-0621-6_10,

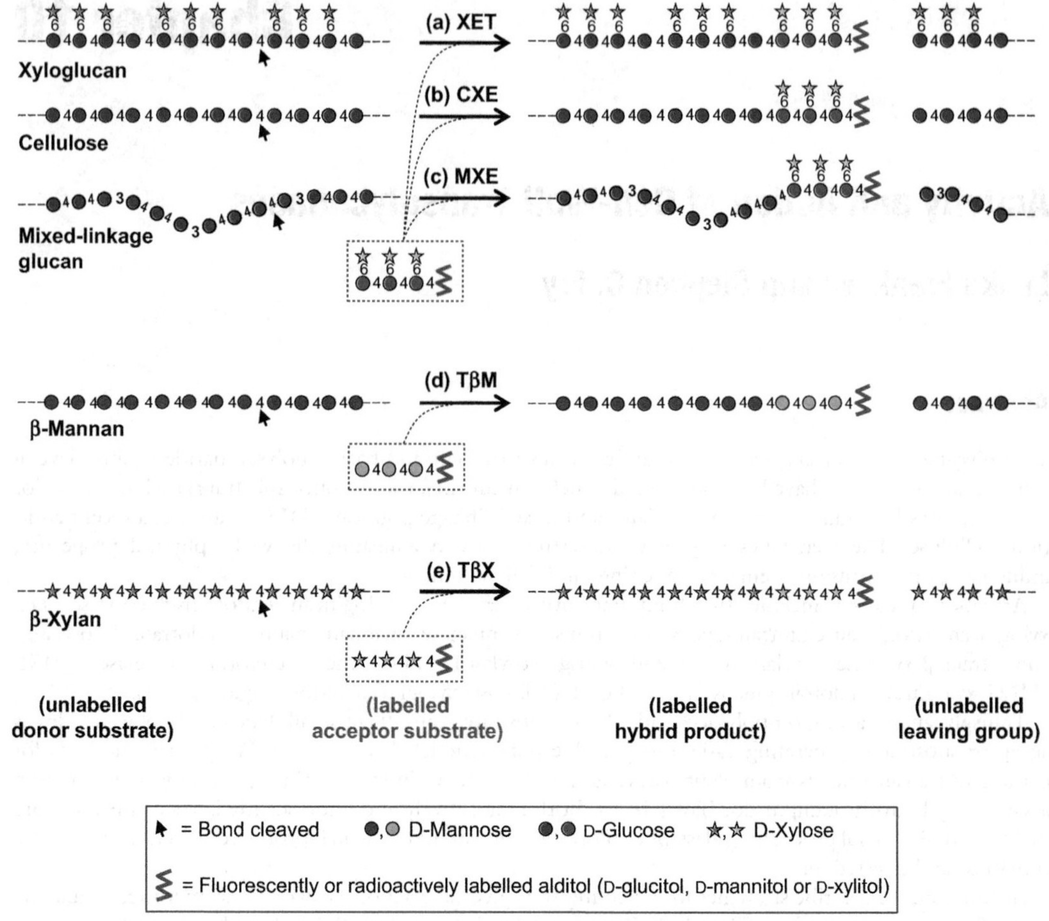

Fig. 1 Five transglycanase reactions using labeled, low-M_r model acceptor substrates. In each case, a high-M_r labeled reaction product is formed. The "label" is either radioactive (e.g., ^3H) or fluorescent (e.g., sulforhodamine; SR). Symbols used for sugar residues generally follow a standard system (glucose, blue circle; mannose, green circle; xylose, orange star; http:/www.functionalglycomics.org/static/consortium/Nomenclature.shtml) except that lighter colors have been used in the acceptor substrates to distinguish these from the donors. The D-xylose residues in xyloglucan are α- whereas those in xylan are β-; all D-glucose and D-mannose residues are β-. Pink zigzags represent labeled moieties: either ^3H-alditols or SR-conjugated 1-amino-1-deoxyalditols. Dashes (-----) indicate continuation of a long polysaccharide chain. (Diagram adapted from [17])

some of the techniques for investigating their activity in vitro and action in vivo.

The morphology of plant cells, and ultimately whole plants, is determined by the cell wall—a complex fabric, some of whose structural polysaccharides, including xyloglucan, mixed-linkage (1 → 3,1 → 4)-β-D-glucan (MLG), mannans, xylans, and cellulose, can be remodeled by enzymes such as hydrolases, transglycanases and transglycosidases. The best-known transglycanase activity is xyloglucan endotransglucosylase (XET; Fig. 1a; Enzyme Commission no. E.C. 2.4.1.207; belonging to CAZy class GH16), which grafts part of one xyloglucan chain onto another [2–6]. Plant proteins that possess XET activity closely resemble those that have xyloglucan endohydrolase (XEH) activity, together described as xyloglucan endotransglucosylase/hydrolases (XTHs) [7]. The *Arabidopsis* genome has 33 *XTH* sequences [8] of which only two appear to encode XEH-active proteins [9, 10].

In addition to XTHs, plants possess trans-β-mannanase (TβM; Fig. 1d; GH5) [11, 12] and trans-β-xylanase (TβX; Fig. 1e; GH10; also known as xylan endotransglycosylase) [13–15] activities, which catalyze mannan-to-mannan and xylan-to-xylan grafting respectively. XET, TβM and TβX are termed homotransglycanase activities because the donor is qualitatively similar to the acceptor in each case. Recently, a heterotransglycanase activity, MLG–xyloglucan endotransglucosylase (MXE; Fig. 1c; E.C. 2.4.1.B52; GH16), was discovered that preferentially grafts segments of MLG onto a qualitatively different polysaccharide, xyloglucan [16]. Interestingly, the MXE-active protein can also utilize cellulose as its donor substrate, exhibiting cellulose–xyloglucan endotransglucosylase activity (CXE; Fig. 1b), and is therefore termed hetero-trans-β-glucanase (HTG). In contrast to conventional XTHs, HTG has limited XET activity [17].

Plant cell walls also contain transglycosidase (exotransglycosylase) activities, which transfer a terminal monosaccharide residue from one poly- or oligosaccharide molecule to another. Examples that act on xyloglucan include trans-α-xylosidase and trans--β-galactosidase [18, 19]. However, transglycosidases will not be discussed here.

HTG is of special interest because it is the first reported transglycanase to target cellulose [17], the major insoluble, microfibrillar component of plant cell walls, as distinct from artificial soluble derivatives of cellulose such as hydroxyethylcellulose [20], carboxymethylcellulose, or H_2SO_4-swollen "cellulose" [21] (which is a cellulose sulfate ester). Remarkably, HTG appears to be confined to *Equisetum*, an evolutionarily isolated genus of fern-like plants; there is a 370-million-year gap between *Equisetum* and its closest living relatives. Exciting opportunities now exist to engineer this enzyme into major crops. HTG is most abundant in mature *Equisetum* stems, and possibly serves to strengthen *Equisetum* plants by covalently bonding cellulose and/or MLG in the secondary cell wall to xyloglucan chains in the primary wall.

Although HTG may be confined to *Equisetum*, its substrates (cellulose, MLG and xyloglucan) are widespread in crop plants and in "waste" biomass. HTG may therefore be valuable agriculturally and industrially. For example, an HTG transgene might have potential for modifying crop plants' mechanical properties, for example, strengthening cereal straw and dicot fibers. HTG could also be used postharvest in "green" technologies to enhance the properties of biomass—for example, functionalizing cellulose.

XTHs are ubiquitous in land plants, and XET activity has also been detected in several charophytic algae [16], even though most such algae possess no detectable xyloglucan [22]; it seems likely that their preferred natural substrates in charophytes are not xyloglucans. In contrast to HTG, XTHs usually peak during periods of rapid cell expansion [7], and the cutting and pasting of xyloglucan chains by their XET activity probably contributes to wall assembly and restructuring during primary growth [23, 24]. TβM and TβX activities have been found in a wide range of plants and charophytic algae; they seem to be particularly active in lower land plants and higher charophytes [1]. The biological significance of hemicellulose restructuring by TβM and TβX awaits further exploration.

Some enzymes with transglycanase activity catalyze very predominantly transglycosylation, others also catalyze hydrolysis. In the latter cases, the transglycosylation–hydrolysis ratio is governed by the concentration of the acceptor substrate (relative to water, the substrate for hydrolysis). Since polysaccharide concentrations in a cell wall are very high [25, 26], transglycosylation may sometimes be a significant function of certain enzymes that are traditionally thought of as glycanases (i.e., hydrolases). Sometimes a newly sequenced enzyme, classified in the CAZy system as a predicted glycosylhydrolase (GH), may have a more significant biological function in transglycosylation, differing from the prediction. Many interesting transglycanase (and transglycosidase) activities would have been overlooked if a reliable "wet biochemistry" approach of the type described in the present chapter had not been applied.

Transglycanases often preferentially utilize polysaccharides as the donor substrate, and may be unable to cleave oligosaccharides [5, 6, 12, 16]; nevertheless, they will readily use oligosaccharides as (model) acceptor substrates. This ability is the basis of simple, quantitative transglycanase assays—during the reaction, a labeled oligosaccharide receives a large segment of polysaccharide, such that the labeled molecule greatly increases in M_r (Fig. 1). It should be remembered, however, that these assays are based on "model" reactions, and the major acceptor substrates in vivo are probably polysaccharides rather than oligosaccharides. This basic principle is used in vitro in radiochemical [5] and fluorescent dot-blot assays [20, 27] and in muro [28, 29].

In the case of XET, MXE and CXE activities, there is a very simple and effective way of separating the labeled polymeric reaction-product from the remaining unreacted labeled oligosaccharide: the former binds avidly to filter paper (or indeed is filter paper) whereas the latter readily washes off [5, 16, 17]. This strategy is less successful for TβM and TβX activities, because (a) the polysaccharides (mannans and xylans) do not remain 100% bound to filter paper when washed in water, and (b) the acceptor substrates (manno- and xylo-oligosaccharides) themselves have some affinity for filter-paper and cannot be efficiently washed off in solvents preventing elution of the polymeric product [1]. In these cases, we recommend a novel glass-fiber blotting method that not only allows the assay of TβM and TβX but also offers a general means of screening for novel transglycanases that may become of interest in the future.

In this chapter we distinguish between enzyme activity (assayed in vitro, under optimized conditions, usually after solubilization of the enzyme) and enzyme action (usually monitored in vivo in the "rough and tumble" of the walls of living plant cells) [30]. An enzyme that exhibits high activity in vitro might fail to act in vivo, for example because the enzyme is not physically in contact with the substrate(s) (colocalized; [28]), or because of the presence on the endogenous acceptor substrate's reducing terminus of an "end-cap" such as the xyloglucan building block V [α-D-Xyl*p*-(1 → 4)-α-D-Xyl*p*-(1 → 6)-D-Glc, added by trans-α-xylosidase] [18, 19], or because of a nonoptimal pH or redox potential, or because the enzyme has been taken out of play by immobilization to structural wall components [31]. Conversely, an enzyme might sometimes exhibit higher action in vivo than predicted by its measured in-vitro activity, for example because the intact cell wall contains beneficial factors that the enzymologist did not consider when designing the in-vitro assay, such as the presence of arabinogalactan proteins or inorganic cations [32].

We describe here methods for extracting transglycanases from plant tissues, procedures for assaying XET, MXE, CXE, TβM, and TβX activities in vitro (Fig. 2a–c), and strategies for quantifying and localizing the action of transglycanases on their endogenous donor substrates in vivo. The "action" assays described here deal with two endogenous players (endogenous enzyme acting on endogenous donor substrate), although the third player (a labeled acceptor substrate) is exogenous. The methodology thus demonstrates that the enzyme cleaves its donor substrate in vivo, presumably forming a polysaccharide–enzyme intermediary covalent complex, but not necessarily that a transglycosylation reaction is consummated by the covalent transfer of the polysaccharide segment onto an endogenous acceptor substrate. Evidence for transglycanase action with all three players (enzyme, polymeric donor, and polymeric acceptor) being endogenous can be obtained by a dual-labeling strategy (radiolabeling with ^{3}H and density labeling with ^{2}H and/or ^{13}C) [23], which is not included in this chapter.

Note on nomenclature: Transglycanase (endotransglycosylase) and glycanase are general terms used to denote an enzyme that cleaves the backbone of a polysaccharide without specifying which particular sugar residue is attacked. The terms transglucanase (endotransglucosylase) and glucanase on the other hand specify that the backbone is broken at a glucose residue rather than mannose, xylose, and so on.

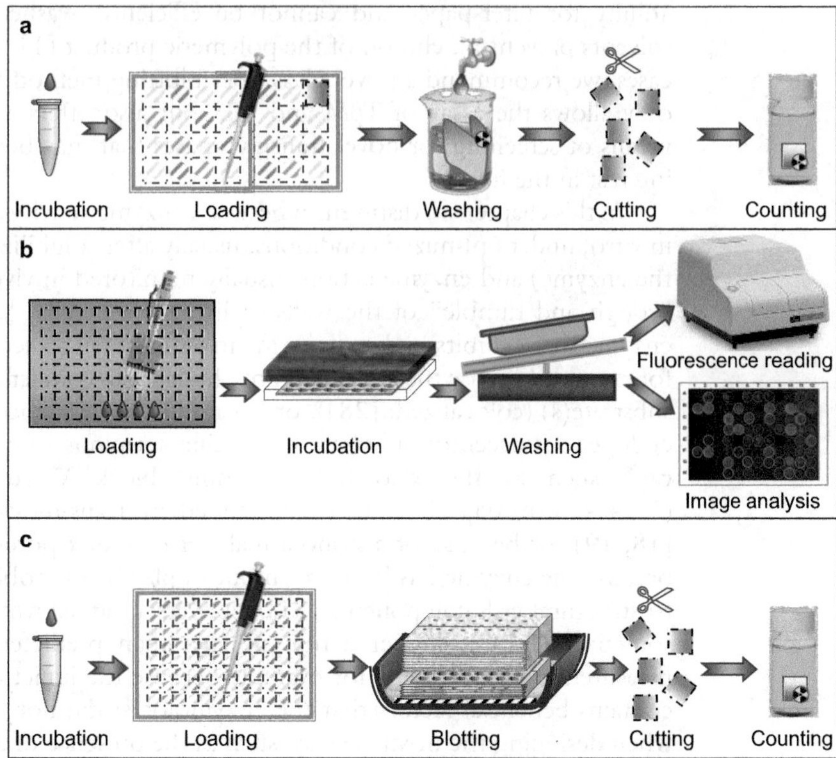

Fig. 2 in-vitro transglycanase activity assays: three experimental approaches. (**a**) Radiochemical filter paper binding assay for XET and MXE activities. (**b**) Fluorescent dot-blot screen for XET and MXE activities. (**c**) Radiochemical glass fiber blotting method for any transglycanase activity (e.g., trans-β-mannanase or trans-β-xylanase). (Diagram adapted from [1])

2 Materials

Note all solutions are aqueous, unless otherwise stated.

2.1 General Materials

2.1.1 Oligosaccharides

1. Xyloglucan oligosaccharides (XGOs) are available commercially, or can be prepared from tamarind xyloglucan by digestion with a xyloglucan-digesting cellulase (e.g., Megazyme's EGII; from *Trichoderma longibrachiatum*) and purified chromatographically. Cellotetraose, mannohexaose, and xylohexaose are available commercially.
2. Tritium-labeled oligosaccharides. The oligosaccharides (e.g., XXXG, XLLG, or mannohexaose) are reductively tritiated with NaB^3H_4 [33], yielding $[^3H]$XXXGol, $[^3H]$XLLGol, or $[^3H]$mannohexaitol respectively, which are then purified chromatographically (for example by thin-layer chromatography (TLC) in the system described in Subheading 3.2.1 followed by elution with water). These and other reductively tritiated oligosaccharides are available commercially (EDIPOS; http://fry.bio.ed.ac.uk//edipos.html).

3. Sulforhodamine (SR)-labeled oligosaccharides (e.g., XGO-SRs). The oligosaccharides (e.g., XGOs and cellotetraose) are reductively aminated with ammonium carbonate plus $NaCNBH_3$, then derivatized with lissamine rhodamine sulfonyl chloride, as described [34]. These and other SR-labeled oligosaccharides are available commercially (EDIPOS).

2.1.2 Hemicelluloses

1. Tamarind seed xyloglucan, medium-viscosity barley MLG ("-β-glucan"), konjac glucomannan, ivory-nut mannan, and arabinoxylan were from Megazyme Inc., Ireland; methylglucuronoxylan and birchwood xylan from Sigma.
2. Dissolving hemicelluloses. If the enzyme solution to be used contains sufficient buffer for the assay, dissolve the hemicellulose at 1% (w/v) in 0.5% (w/v) chlorobutanol (1,1,1-trichloro-2-methylpropan-2-ol; a volatile, antimicrobial agent). For use in the radiochemical XET and MXE assays described, dissolve the polysaccharide in 1.5 × strength buffer "A" instead of plain 0.5% chlorobutanol. In either case, suspend 1 g of the polysaccharide in 100 mL of the aqueous solution, autoclave at 121 °C for 15 min. If necessary (*see* **Note 1**), stir on a magnetic stirrer overnight (starting while still hot); if necessary, autoclave and stir again until a clear solution has formed. Centrifuge the solution (10 min at 2000 × *g*) to remove traces of insoluble material (e.g., contaminating proteins that have been denatured by heating). Store the polysaccharide solutions frozen in aliquots in plastic tubes. After thawing, heat the plastic tube in a beaker of boiling water for a few minutes to ensure complete redissolution. If no autoclave is available, boil the polysaccharide suspension 2× for ~15 min with occasional vortexing. Ivory-nut mannan is not soluble upon heating but can be solubilized in 2.5 M NaOH and then neutralized immediately prior the assay—as described by suppliers.

2.1.3 Cellulose

1. Whatman No. 3 and No. 1 filter paper (chromatography paper) are almost pure cotton cellulose and can be used for most purposes without further treatment.
2. For use in CXE assays, pretreat paper in alkali to increase the accessibility of the cellulose chains. Bathe a piece (e.g., 12 × 8 cm) of Whatman No. 1 paper for 16 h with very gentle rocking in 50 mL of 6 M NaOH in a closed polypropylene sandwich box, rinse the paper in water until neutral to litmus paper, rinse again with 5% (v/v) acetic acid, then again in water, and dry. A prescribed weight of the paper can conveniently be estimated from its area (basis weight ≈ 8 mg cm^{-2}) (*see* **Note 2**).

2.1.4 Buffers, Acids and Alkalis

1. Buffer "A," containing (final concentrations) 100 mM succinate (Na^+, pH 5.5), 5 mM $CaCl_2$, chlorobutanol and optionally 0.1% bovine serum albumin (BSA), is suitable for most transglycanase assays (*see* **Note 3**). Prepare triple-strength buffer "A" as follows: dissolve 35 g succinic acid and 2.2 g $CaCl_2.2H_2O$ in ~800 mL 0.5% chlorobutanol, then adjust the pH to 5.5 by adding 6 M NaOH at the pH meter, and add BSA to 0.3% w/v if desired. Adjust the volume to 1 L with water.
2. 90% (w/w) formic acid: commercially available at this concentration (caution—highly corrosive; avoid skin contact).
3. 6 M NaOH: weigh out 24 g solid NaOH (caution—highly caustic; avoid skin contact) and dissolve with constant stirring in about 75 mL water. Adjust final volume to 100 mL with more water. Store in a stoppered bottle.

2.1.5 Scintillation Counting

1. Radioactive samples are assayed for ^3H in 22-mL plastic vials in a scintillation counter.
2. Dried papers (e.g., 4 × 4 cm squares from the XET ^3H assay) or glass-fiber squares are assayed for ^3H in 2–4 mL of water-immiscible scintillation fluid (e.g., Meridian Gold Star "O").
3. Water-containing solutions (or papers that have been soaked in 1–2 mL water to increase the counting efficiency) are thoroughly mixed with 10 volumes of a water-miscible scintillant (e.g., Fisher ScintiSafe 3).
4. Scintillation counter.

2.1.6 Equipment

1. Centrifugal evaporator (e.g., Savant SpeedVac).
2. Freeze drier (lyophilizer).
3. TLC tank.
4. Magnetic stirring plate and stirring bar.
5. Microcentrifuge.
6. Bench centrifuge capable of 2000 × *g*.

2.2 Materials for Specific Procedures

2.2.1 Extraction of Enzymes from Plants and Algae

1. Buffer B: 200 mM succinate (Na^+), pH 5.5, supplemented with 3% polyvinylpolypyrrolidone (PVPP) (for enzyme extraction).
2. Buffer C: 300 mM succinate (Na^+), pH 5.5, supplemented with 15% glycerol, 20 mM ascorbate, and 3% polyvinylpolypyrrolidone (PVPP) (for enzyme extraction).
3. Buffer D: 100 mM succinate (Na^+), pH 5.5, supplemented with 1 M NaCl and 0.05% Triton X-100 (for enzyme extraction) (*see* **Note 4**).
4. Buffer E: 40 mM succinate (Na^+) pH 5.5 (for dialysis).
5. Buffer F: 40 mM succinate (Na^+) pH 5.5 supplemented with 2% glycerol and 2.6 mM ascorbate (for dialysis).

6. Nylon cloth.
7. Liquid nitrogen or dry ice (solid CO_2).
8. Polystyrene ice-box.
9. Hand-mixer.
10. Pestle and mortar.
11. Refrigerated centrifuge (capable of 10,500 × *g*).
12. Dialysis tubing (Medicell).

2.2.2 Radiochemical Assay of Extracted XET and MXE with [³H]XXXGol

1. Ethanolic or aqueous solution of [^{3}H]XGO, 100 kBq/mL, optionally also containing 0.1% (w/v) Orange G.
2. Nonradioactive XGO, chemically identical to the [^{3}H]XGO (optional).
3. 90% (w/w) formic acid (commercially available at this concentration).
4. Whatman No. 3 chromatography paper, marked out into 4 × 4 cm squares in pencil.
5. Scintillation vials (22 mL).
6. Scintillation fluid (water-immiscible) (e.g., Meridian Gold Star "O").

2.2.3 Dot-Blot Assay of Extracted XET and MXE Activity with XXXG–SR on Paper

1. XET and MXE test-paper [27].
   - Place 1% xyloglucan or MLG solution in a tray, and quickly pass a sheet of Whatman No. 1 paper over the surface of the polysaccharide solution so that only one side of the paper comes into contact with the solution. Do this while holding the paper along one edge, which is kept dry.
   - Hang the paper using clips along its dry edge, and leave to dry (*see* **Note 5**).
   - Dip the hemicellulose-impregnated paper through a 5 μM solution of XGO-SR in 75% (v/v) acetone (dilute the solution in 75% acetone until the absorbance at 567 nm is about 0.4) and redry (*see* **Note 6**).
   - Attach rectangles (typically 8 × 12 cm, if a 96-well format is to be used) of the resulting test-paper on to sheets of cellulose acetate (smooth overhead projector transparencies), using a thin layer of adhesive (spray-gun, supplied by art shops for wall-mounting).
2. Access to cold room (e.g., 4 °C).
3. Ethanol–formic acid–water (1:1:1 by vol.) (optional).
4. Acetone (optional).
5. Ultraviolet lamp emitting at 254 nm.
6. Ultraviolet eye protection (suitable goggles).

7. Camera set up for recording fluorescent spots on paper (e.g., MultiDoc-It, Ultra-Violet Products Ltd., Cambridge).
8. Dilute acetic acid: 5% (v/v).

2.2.4 Radiochemical CXE Assay

1. Ethanolic or aqueous solution of [^{3}H]XXXGol, 100 kBq/mL (optionally containing 0.01% Orange G).
2. Nonradioactive XXXGol (optional).
3. Triple-strength buffer “A,” with 0.3% w/v BSA.
4. 10% formic acid: 1 mL of 90% (w/w) formic acid added to 8 ml water.
5. Scintillation vials (22 mL).
6. Scintillation fluid (water-immiscible) (e.g., Meridian Gold Star “O”).

2.2.5 Trans-β-Mannanase and Trans-β-Xylanase Assay by Glass-Fiber Method

1. Donor substrate: 1% xylan or glucomannan, unbuffered.
2. Radiolabeled acceptor substrate: 1 MBq/mL [^{3}H]mannohexaitol or [^{3}H]xylohexaitol; final concentration in the reaction mixture ~0.2 μM.
3. Microplate shaker.
4. PCR strip tubes (Sarstedt).
5. 4 sheets of coarse polypropylene gauze (or dense plastic mesh)
6. Plastic tray (25 × 25 cm).
7. Blotting stratum: 20 × 20 cm glass-fiber sheets (GF/A grade; Whatman) with 2.5 × 2.5 cm pencil-marked grid layout (*see* **Note** 7).
8. 21 × 21 cm paper towels (enough to produce ~3-cm-thick layer when pressed; e.g., 65 sheets of Kimberly-Clark towels)
9. Blotting solvent: 75% (v/v) ethanol in 0.5% acetic acid (~1.5 L).
10. Flat weight (e.g., two ~5-cm-thick catalogs).
11. Glass plate.

2.2.6 In-Situ Detection of XET and/or MXE Action on Endogenous Donor Substrates with [^{3}H]XXXGol

1. Ethanol–formic acid (20:1), freshly prepared.
2. 15-mL plastic centrifuge tubes (e.g., Corning CentriStar).
3. 75% ethanol (ethanol–water, 3:1).
4. Benchtop tube rotator (“wheel”; e.g., Stuart blood tube rotator SB2 or SB3).
5. 6 M NaOH: see above.
6. Acetic acid (“glacial,” which is approximately 17 M).
7. Dialysis tubing (Medicell International, 12–14 kDa molecular weight cutoff, $^{18}/_{32}$ in.).

8. Pyridine–acetic acid–water (1:1:98 by vol.) containing 0.5% (w/v) chlorobutanol.
9. Xyloglucan-digesting cellulase: 50 units/mL in 1% BSA (e.g., Megazyme's EGII, from *Trichoderma longibrachiatum*).
10. Lichenase: 50 units/mL in 1% BSA (*Bacillus* enzyme, from Megazyme).
11. Plastic backed, 20 × 20 cm TLC plates (e.g., Merck silica-gel 60; with or without fluorescent indicator).
12. TLC solvent "BAW": 40 mL butan-1-ol, 20 mL acetic acid, 20 mL water; freshly prepared.
13. Fluor (organic scintillator): 7% (w/v) 2,5-diphenyloxazole (PPO) dissolved in diethyl ether.
14. Photographic flash gun.
15. Film (e.g., Kodak Biomax XAR). Preflash the film with a flash gun at such a distance that the background of the film will be slightly fogged when developed.
16. Dark room.

2.2.7 Localization of XET Action on Its Endogenous Donor Substrate with XXXG–SR

1. Single-edged razor blade.
2. 12 μM XGO-SR solution (and as a control, cellotetraose-SR).
3. Flat-bottomed glass tubes (e.g., 5 × 1 cm).
4. Muslin (cheesecloth).
5. Waterproof tape (e.g., autoclave tape).
6. Dilute formic acid (~0.1 M or 0.5% v/v; this has a pH of about 2.4).
7. Ethanol–formic acid–water (15:1:4 by vol.).
8. 75% ethanol (ethanol–water, 3:1).
9. Acetone.
10. Fluorescence microscope with excitation ~315 or 540–570 nm, emission 615 nm.

3 Methods

3.1 Activity of Extracted Enzymes, Assayed In Vitro

3.1.1 Extraction of Enzymes from Plants and Algae [1, 13]

Usually, the plant material is homogenized with 3–5 volumes of an extraction buffer (i.e., 1 g fresh weight in 3–5 mL extractant), according to the toughness and water content of the tissue. All the procedures are performed at ~4 °C; buffers are prechilled on ice.

Extraction of Enzymes from Land Plants

1. Chop the plant into small segments.
2. For small amounts of tissue, homogenize it in liquid nitrogen using a pestle and mortar, then add three quarters of chilled buffer B and grind for 5 min.
3. In the case of bulk material, place the plant material in a hand mixer without liquid nitrogen, add three quarters of the required volume of buffer B and blend for 5 min.
4. Transfer the homogenate with the final quarter of the buffer into a beaker and gently stir for another 3 h to extract the enzyme.
5. Filter the homogenate through three layers of nylon cloth; collect the filtrate.
6. Centrifuge at 10,500 × *g* for 45 min.
7. Gently pipet out the supernatant and divide it into small aliquots (*see* **Note 8**).
8. Store the aliquots at −80 °C.

Extraction of Enzymes from Charophytic Algae

1. Wash the charophyte in deionized water. Allow the alga to sediment (centrifuge briefly if necessary) and discard the supernatant (*see* **Note 9**).
2. Homogenize with extraction buffer C or D by pestle and mortar. Use 3 mL per g fresh weight or 6 mL per g dry (lyophilized) weight (*see* **Note 10**).
3. Centrifuge the extract at 10,500 × *g* for 45 min.
4. Pipet out the supernatant, and store in small aliquots (e.g., in 200-μL PCR tubes) (*see* **Note 8**).

Partial Purification of Enzymes

Crude, dilute extracts can be fractionated and concentrated by ammonium sulfate precipitation (e.g., up to 30% or 70% saturation for HTG or TβM and TβX respectively). This step is advantageous especially if the fluorescent dot-blot method (Subheading 3.2.1) is to be used for algal transglycanases.

1. While stirring, gradually add powdered ammonium sulfate (176 g/L or 472 g/L for 30% or 70% saturation respectively) into a crude extract. Stir for ~1 h until all crystals dissolve.
2. Centrifuge at 12,500 × *g* for 45 min.
3. Gently discard the supernatant. Resuspend the pellet in a small volume of extraction buffer and desalt by dialyzing 2 × 12 h against deionized water followed by 1 × 12 h against dialysis buffer E or F (depending whether extraction buffer B or C was used) (*see* **Note 11**).

4. Centrifuge the contents of the dialysis sac in a microcentrifuge at 2000 × *g* for 2 min.
5. Collect the supernatant and store in small aliquots at −80 °C (*see* **Note 12**).

3.1.2 Assay of Extracted XET and MXE Activity

Radiochemical Assay of Extracted XET and MXE [5, 16]

1. The methods for assaying XET [5] and MXE [16] are essentially identical, except that xyloglucan and MLG, respectively, are used as the donor substrates. The procedure is summarized in Fig. 2a.
2. Pipet 10 μL of an aqueous or ethanolic solution containing 1 kBq of [^{3}H]XXXGol or [^{3}H]XLLGol into each of a series of microfuge tubes. Dry off the solvent, for example in a SpeedVac (*see* **Note 13**).
3. Add 20 μL of donor substrate solution (either 0.5% tamarind xyloglucan or 1.0% barley MLG, dissolved in triple-strength buffer "A") (*see* **Notes 14** and **15**).
4. Mix thoroughly and allow sufficient time for the [^{3}H]XGO to redissolve (*see* **Note 16**).
5. Add 10 μL of putative enzyme solution (supplemented with BSA to a final concentration of 0.1%, w/v), and mix thoroughly. Cap the tube and incubate at 20 °C.
6. After the desired time period, add 10 μL 90% formic acid (*see* **Note 17**) and mix well, to stop the enzymic reaction. Depending on the activity of the enzyme, this may be anything from 1 min to several days; we find that these enzymes are extremely durable, and an extremely slow rate may continue at an almost constant rate for a week or more (Fig. 3) (*see* **Note 18**).
7. Dry the sample onto a 4 × 4 cm square, marked in pencil on Whatman No. 3 paper (*see* **Note 19**).
8. Wash the paper in running tap water for 16–48 h (*see* **Note 20**).
9. Dry the paper, cut out the individual 4 × 4 cm squares and place each in a 22-mL scintillation vial (roll the paper, with the loaded face outward) and add 2–4 mL of scintillation fluid. Assay ^{3}H, representing the polymeric reaction product, polysaccharide–[^{3}H]XGO, in a scintillation counter (*see* **Note 21**).
10. If the acceptor XGO had been adjusted to a concentration close to the enzyme's K_M, the reaction rate can be expressed in katals (or as pmol polysaccharide product formed h^{-1}) per mg enzyme. If the XGO concentration was much lower than the K_M (e.g., if carrier-free ^{3}H-labeled XGO was used), then the rate can be usefully expressed as Bq/kBq/h/mg (Bq incorporated into polymeric product, per kBq [^{3}H]XGO supplied, per h, per mg enzyme). The data obtained in the latter case are valuable for comparing different enzyme

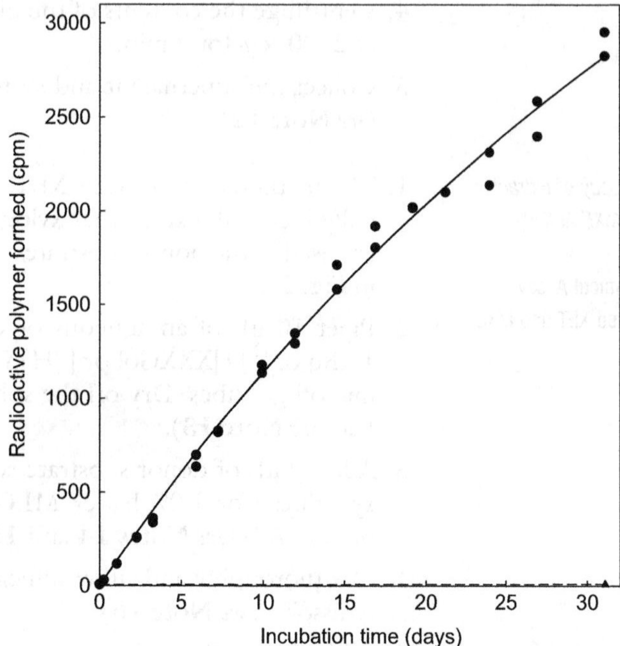

Fig. 3 Prolonged time-course of an enzyme preparation with low MXE activity. A sample of *Equisetum* HTG, heterologously produced in the yeast *Pichia* (Simmons et al., 2015), was monitored by the radiochemical assay at 23 °C. The graph shows the extreme durability of the enzyme, which continued to generate radioactive polymeric products for at least a month. Donor substrate, barley MLG; acceptor substrate, [^{3}H]XXXGol. The control (– – –) lacked enzyme. (Authors' unpublished work)

preparations, different donor substrates, different acceptor substrates, and different pH values, temperatures, inhibitors, all other variables being kept constant.

Fluorescent Dot-Blot Assay of Extracted XET and MXE Activity [27, 35]

Transglycanase activity test papers provide a rapid means of screening numerous enzyme extracts for XET or MXE activity (Figs. 2b and 4). Other applications of the technique include the screening of a collection of substances for inhibitory effects on a single enzyme preparation [35]. The same test-papers can also be used for "tissue printing" [27, 36], whereby a cut surface of a plant organ is pressed onto the test paper and a pattern showing the distribution of XET activity is generated; it should be noted, however, that this approach only reveals enzymes that readily leach from the cut surface, and not firmly wall-bound enzymes.

1. Attach an 8 × 12 cm piece of test-paper [impregnated with XGO-SR plus either xyloglucan or MLG (or neither polysaccharide)] to an inert, water-repellent surface (e.g., an overhead-projector acetate sheet) (*see* **Note 22**).

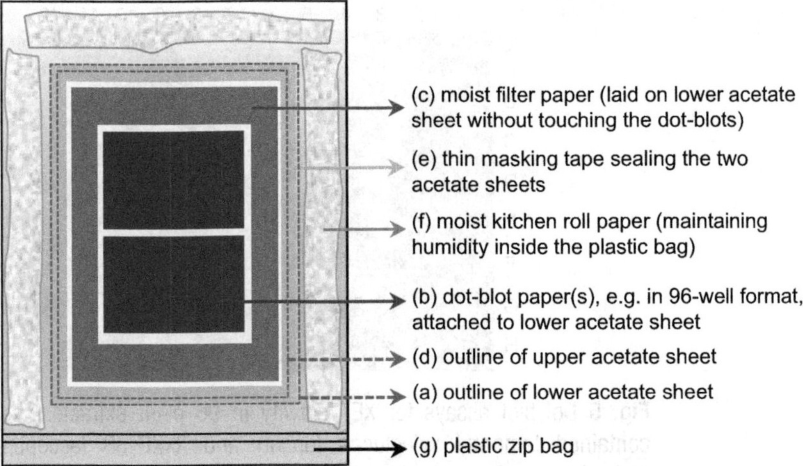

Fig. 4 Method for dot-blot screening for XET or MXE activity. Plant enzyme extracts are quickly loaded (preferably in 96-well plate format) on to transglycanase test paper(s) (b) previously mounted onto a transparency acetate sheet (a), then surrounded by a moist cut-out of Whatman No. 1 paper (c). The dot-blot is then overlaid with another sheet of acetate (d) and the two acetates are quickly sealed with thin masking tape (e). The whole sandwich is placed into a zip bag (g) with a thin layer of wet tissue paper (f) to maintain humidity, especially during prolonged incubations, and pressed under a heavy load. After the desired time of incubation, the sandwich is opened and the dot-blot dried before being washed

2. Working in a cold room, quickly pipet 5 μL of putative enzyme solution (e.g., in 1× strength buffer "A," containing 0.1% BSA and 0.01% (w/v) Orange G, and also containing any substances being tested as promoters, inhibitors etc.) onto each square. Include a few enzyme-free controls (*see* **Note 23**).
3. Add a cut-out of moist Whatman No. 1 paper as shown in Fig. 4, then cover the papers with a second acetate sheet, seal the two sheets together with thin masking tape, insert the acetate–paper–acetate sandwich into a polythene zip bag, and place the bag on a slightly soft surface (e.g., newspaper) under a heavy weight, for example a pile of books.
4. Incubate the assembly at room temperature for 1–12 h.
5. Remove and dry the paper. Check uniformity of loading of Orange G spots (scan the dried paper to provide a permanent record of any errors if required).
6. Wash the paper, either in running tap water overnight or in ethanol–formic acid–water (1:1:1 by vol) for 1 h (in the latter case, followed by a rinse in acetone). The Orange G (and unreacted background XGO-SR) should now have disappeared.
7. Dry the paper and document the orange-fluorescing spots of polysaccharide–XGO-SR. A handheld UV lamp emitting at 254 or 315 nm is adequate (wear eye protection!), or the

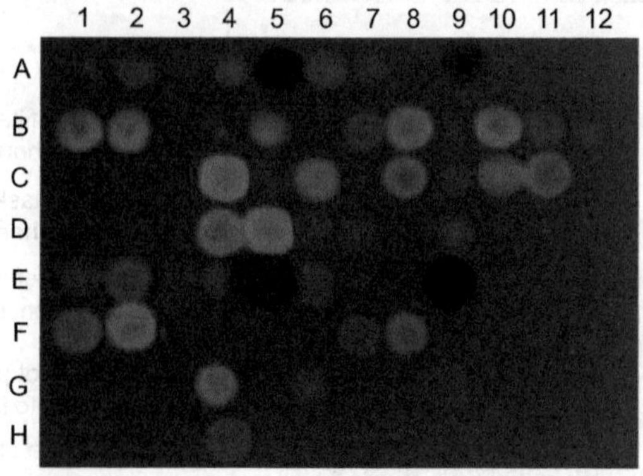

Fig. 5 Dot-blot assays for XET activity in 96 plant extracts. The test-paper contained tamarind xyloglucan (donor) and XXXG-SR (acceptor), and the paper-binding fluorescent product (xyloglucan-SR) remaining after prolonged washing in water was documented in a MultiDoc-It (excitation 254 nm). Authors' unpublished results

spots can be photographically documented in a MultiDoc-It (Fig. 5). Alternatively, epi-excitation with a green laser (560–570 nm) and measurement of emitted orange fluorescence at 600–615 nm may give an improved signal–noise ratio (*see* **Note 24**) when quantified by a multilabel plate reader (Fig. 2b).

8. If the presence of CXE activity is suspected (orange-fluorescing spots on the control test-paper lacking xyloglucan and MLG), this can be checked by subsequently bathing the paper in 6 M NaOH for 16 h (in a closed plastic sandwich box, with very gentle rocking) and then rinsing copiously with, sequentially, gently running tap water (1 h), 5% acetic acid (15 min, gentle rocking) and gently running tap water (15 min), before redrying. Any remaining orange fluorescence is likely to be cellulose–XGO-SR, generated by CXE activity [17].

3.1.3 CXE Assay [17]

1. Pipet 10 μL of an aqueous or ethanolic solution containing 1 kBq of [^{3}H]XXXGol into each of a series of microfuge tubes. Dry off the solvent (*see* **Note 25**).
2. Add 10 μL of triple-strength buffer "A" containing 0.3% BSA.
3. Mix thoroughly, and allow sufficient time for the [^{3}H]XGO to redissolve (*see* **Note 26**).
4. Add 20 μL of putative enzyme solution, and mix thoroughly.
5. To the whole 30 μL add 36 mg of dry, NaOH-pretreated Whatman No. 1 paper (~20.8 × 20.8 mm).
6. Cap the tube and incubate at 20 °C for 1–72 h.

7. At the desired time-point, stop the reaction by adding 200 μL of 10% formic acid.
8. Wash the paper in copious volumes of water (*see* **Note 27**).
9. Dry the paper, place in a 22-mL scintillation vial and add 2–4 mL of scintillation fluid. Assay ^{3}H, representing the polymeric reaction product, cellulose–[^{3}H]XXXGol, in a scintillation counter (*see* **Note 28**).
10. If the enzyme contained traces of hemicelluloses, these may have served as donor substrates for XET or MXE activity during the CXE assay, leaving a deposit of xyloglucan–[^{3}H]XXXGol hydrogen-bonded to the paper. To check for this artifact, wash the 36 mg paper (*see* **Note 29**) in 10 mL 6 M NaOH (in a 15-mL closed centrifuge tube, on a tube rotator for 16 h), then pellet the cellulose by centrifugation at 2000 × *g* for 5 min (reject the supernatant) and wash the pellet with, sequentially, water (several times, until the supernatant is almost neutral to litmus paper) then 5% acetic acid (once). Finally redry and measure the ^{3}H. Any remaining ^{3}H is likely to be cellulose–[^{3}H]XXXGol, generated by CXE activity.

3.1.4 Trans-β-Mannanase and Trans-β-Xylanase Assay by Glass-Fiber Method [1]

In contrast to the filter paper-binding methods used for XET and MXE (Fig. 2a), the glass-fiber blotting method (Fig. 2c) is preferred in the cases of TβM and TβX for reasons explained in the Introduction.

1. Dry the enzyme acceptor substrate [4 kBq (typically ~4 pmol) of [^{3}H]mannohexaitol or [^{3}H]xylohexaitol) for TβM and TβX assay respectively] in a 200-μL PCR strip tube.
2. Add 5 μL 1% polymeric donor substrate (konjac glucomannan or birchwood xylan) (*see* **Note 30**).
3. Initiate the reaction by adding 15 μL enzyme (crude extract or dialysate).
4. Vortex, spin down for few seconds and finally incubate on a microplate shaker for 1–24 h at 20 °C. Stop the reaction by adding 6 μL 90% formic acid.
5. Dry the entire 20 μL reaction mixture on to a 20 × 20 cm glass fiber (GF/A) sheet with 2.5 × 2.5 cm layout (one sample centered on each square).
6. Blot the GF sheet sandwiched in a blotting assembly for 2 × 48 h (Fig. 6).
7. Place 4 sheets of coarse polypropylene gauze in the tray and overlay with 25 sheets of paper toweling (forming a ~1 cm-thick layer when compressed).
8. Place GF/A blotting stratum, with the dried samples facing up, on top of the towels.

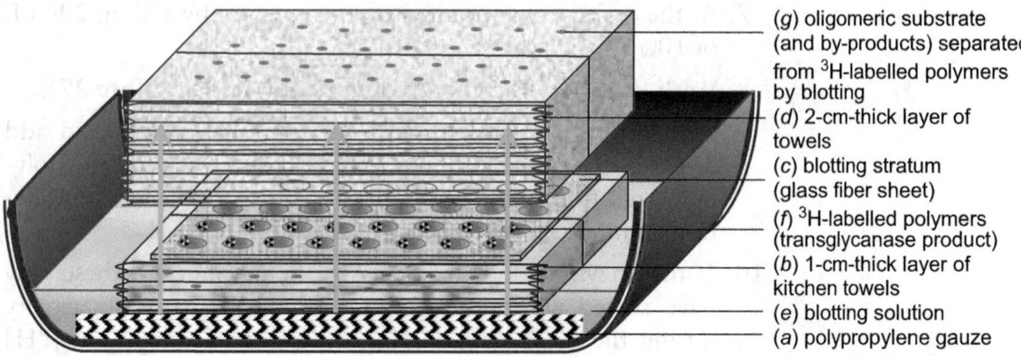

Fig. 6 A general in-vitro assay for known and potential novel transglycanase activities extracted from plants and algae. The strategy (*see* Fig. 2c) is given in detail for TβM and TβX. Into a tray are placed (a) polypropylene gauze, (b) kitchen towels, (c) a glass fiber sheet (loaded with a radioactive reaction mixture), (d) more towels, and finally (not shown here) a glass plate. Ethanolic blotting solution is added (e), and when the paper is saturated the glass plate is removed. Evaporation causes the ethanol to percolate through the radioactive reaction mixture (f) on the glass fiber, such that the oligomeric ^{3}H-substrate (and any smaller hydrolysis products) are carried up to the surface (g), leaving the polymeric radiolabeled transglycanase products in the glass fiber sheet. (Diagram adapted from [1])

9. Overlay with another layer of paper towels (40 sheets; ~2 cm thick when compressed).
10. Cover with a thick glass plate and compress with a heavy load (e.g., catalogs).
11. Gently pour the ethanolic blotting solution into both sides of the tray and allow to be absorbed by the blotting papers until the solvent reaches the top of the assembly (*see* **Note 31**).
12. When the solvent stops being absorbed by towels, remove the heavy load and glass plate so that the evaporation flow process can start.
13. At the first and 12th hour of blotting, add ~100 mL fresh solvent to maintain an evaporative flow.
14. After 12 h of blotting, place the blotting tray in a fume hood to allow the solvent to gradually evaporate from the top layer of paper towels.
15. Once blotted, remove the upper paper toweling, then remove the wet GF sheet; gently transfer the latter on to aluminum foil and air-dry.
16. Reblot for another 48 h in a new blotting sandwich.
17. Cut the dried GF sheet into squares. Place each square (potentially containing ^{3}H-labeled polymeric reaction products) in 2 mL of scintillation fluid and quantify by scintillation counting (Fig. 2c).

3.2 Action of Endogenous Enzymes on Endogenous Donor Substrates, Assayed In-Situ

3.2.1 In Situ Radiochemical Detection of XET and/or MXE Action [29]

1. Slice the tissue of interest (e.g., *Equisetum* stems or barley leaves) into ~0.5-mm-thick strips with a razor blade. Keep the sections moist.
2. Immediately suspend 50 mg of sliced tissue in 250 μL water containing 25 kBq [^{3}H]XXXGol (concentration <5 μM) in a 1.5-mL tube, and incubate for 1–16 h at 20 °C.
3. Add 750 μL ethanol–formic acid (20:1) to stop the reaction, precipitate all polysaccharides and extract the unreacted [^{3}H] XGO. The samples will be stable in the acidified ethanol at room temperature for at least several weeks, and can be analyzed later.
4. Transfer the tissue into a 15-mL plastic centrifuge tube. Use 75% ethanol to ensure complete transfer.
5. Top up to the 14-mL mark with 75% ethanol. Incubate with effective movement, for example on a benchtop tube rotator, for at least 1 h; then bench-centrifuge (5 min at 2000 × *g*) and reject the supernatant.
6. Repeat the previous step several times, until the supernatant is nonradioactive (*see* **Note 32**). If the supernatant is still not free of [^{3}H]XGO, wash with a series of more dilute ethanol solutions (e.g., 66% and 50%). However, concentrations lower than 50% are not recommended unless only firmly wall-bound polysaccharides are of interest.
7. Dry the thoroughly washed pellet (alcohol-insoluble residue; AIR) (*see* **Note 33**) then incubate it in 1.35 mL 6 M NaOH at 37 °C for 16 h to solubilize the hemicelluloses. A second or third NaOH treatment may be beneficial to ensure complete solubilization of the ^{3}H-hemicelluloses (check by scintillation-counting a small portion), in which case the volumes of acetic acid in the following steps will need increasing.
8. Centrifuge as before and keep the supernatant.
9. Resuspend the pellet in 1 mL water and recentrifuge; pool the new supernatant with the first supernatant.
10. Repeat the previous step.
11. To the pooled solutions (total volume ~3.25 mL), add enough pure (glacial) acetic acid to take the pH down to ~5–6 (750 μL should be sufficient; check the final pH with indicator paper).
12. Dialyse the hemicellulose solution against several changes of 0.5% chlorobutanol (each 1 L), then remove the solution from the dialysis sac and freeze-dry it.
13. Redissolve the hemicellulose in 500 μL of volatile pH 4.7 buffer (pyridine–acetic acid–water, 1:1:98, containing 0.5% chlorobutanol) by heating in a capped tube in a boiling water bath for 30 min.

14. To one 100-μL aliquot of the hemicellulose solution, add 10 μL 1% (w/v) BSA containing 50 μg (or 0.5 units) of xyloglucan-acting cellulase (e.g., Megazyme EGII cellulase [from *Trichoderma longibrachiatum*]) (*see* **Note 34**).
15. To a second 100-μL aliquot add 10 μL 1% (w/v) BSA containing 0.5 units lichenase (Megazyme).
16. To a third 100-μL aliquot add 10 μL water.
17. Incubate the three aliquots at 20 °C for 1.5 h, to digest respectively both polysaccharides (xyloglucan and MLG), MLG only, and neither.
18. Add 300 μL ethanol–formic acid (20:1), mix well, incubate overnight at room temperature, then centrifuge (~12,000 × *g*) in a microfuge for 10 min to pellet undigested hemicelluloses.
19. Collect 360 μL of the clear ethanolic supernatant (containing oligosaccharides generated by the cellulase and lichenase), and dry it (e.g., in a SpeedVac).
20. Redissolve the dried oligosaccharides in 10 μL 67% ethanol, and load all three solutions on to a silica-gel TLC plate in the form of 2-cm streaks (*see* **Note 35**).
21. When the loadings are thoroughly dry, develop the plate in freshly prepared butan-1-ol–acetic acid–water (2:1:1). It takes ~8 h for the solvent to reach the top of the plate (*see* **Note 35**).
22. Dry the plate (good draft in a fume cupboard for ~1 h), then redevelop the plate in the same solvent (and same dimension) for a second 8-h period to improve the resolution of the oligosaccharides (*see* **Note 36**).
23. Redry the plate until it no longer smells of acetic acid.
24. Fluorograph the plate (*see* Subheading 3.7.2 of Chapter 1) to reveal the radioactive oligosaccharides on film (*see* **Note 37**).
25. Interpretation (*see* Fig. 7): a spot of [^{3}H]XXXGol in the first aliquot (cellulase-digested) indicates that in-situ XET and/or MXE action had occurred in the plant material; a spot of [^{3}H]GGXXXGol in the second aliquot (lichenase-digested) indicates in-situ MXE action. [^{3}H]XXXGol and GG[^{3}H]XXXGol approximately comigrate with maltohexaose and maltononaose respectively. No chromatographically mobile material should be seen in the third (undigested) aliquot; if [^{3}H]XXXGol is nevertheless detected in it, this indicates that the ethanol washing steps were incomplete.

3.2.2 In-Situ Fluorescent Localization of XET Action on Its Endogenous Donor Substrate [28]

This procedure can demonstrate the colocalization of endogenous XTH with its endogenous donor substrate, xyloglucan, in a botanical specimen (Fig. 8b).

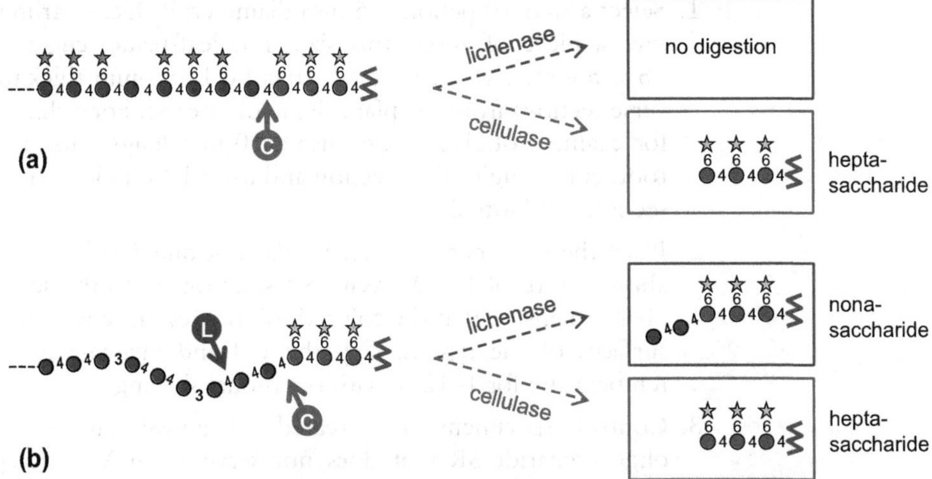

Fig. 7 Distinguishing the products of XET and MXE action when either or both may have been formed in vivo, for example by *Equisetum* tissue. Aliquots of the alkali-extracted hemicelluloses, previously radiolabeled in situ as shown in Fig. 1a, c, are treated with lichenase (**L**) or xyloglucan-acting cellulase (**C**), which attack at the points indicated by bold arrows to give the oligosaccharides shown. These can then be separated by TLC. (**a**) XET product. (**b**) MXE product. Symbols as in Fig. 1

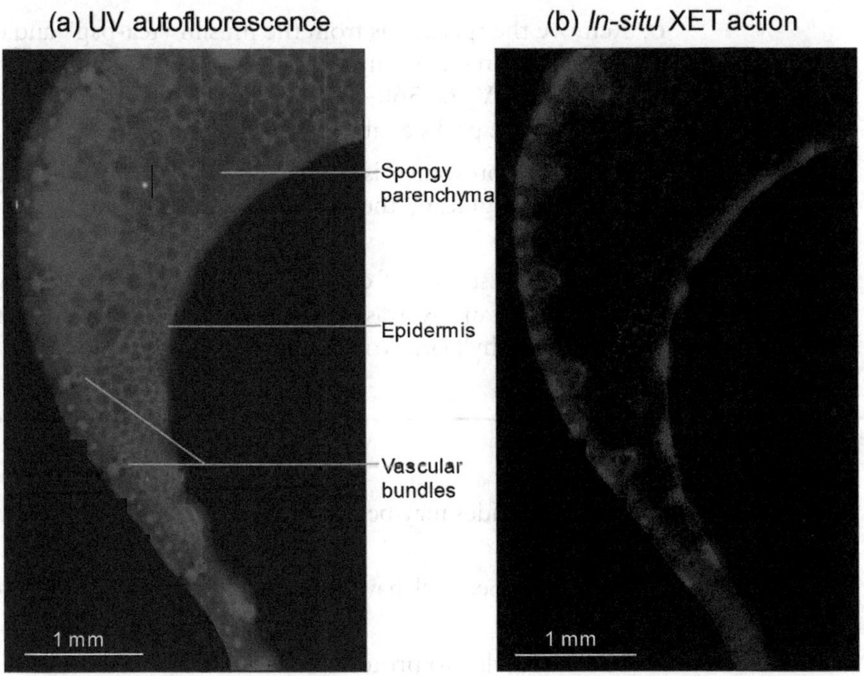

Fig. 8 In-situ XET action, visualized in a transverse section of a maize leaf. (**a**) Autofluorescence, revealing all cell walls. (**b**) The orange fluorescence shows in-situ XET action: colocalization of endogenous enzyme with accessible endogenous donor substrate molecules (xyloglucan). (Authors' unpublished results)

1. Select a stem or petiole <6 mm diameter; if thicker, trim it to give a piece of about this size. For leaf-blade, cut a piece ~5 × 5 mm. Use a razor blade to cut 0.1–0.5-mm-thick transverse sections from the plant tissue of interest. For a thin root, for example of *Arabidopsis*, use a 10-mm length; for a thick root, cut a longitudinal section and use a 10-mm length of the section (*see* **Note 38**).
2. Place the specimens in a small, flat-bottomed vial, and pipet about 50 μL of 12 μM XGO-SR solution on to the sections. Shake to ensure that the colored solution evenly covers the cut surfaces of the tissues. Cap the vial and incubate at room temperature for 1–12 h, with occasional shaking.
3. Control specimens are treated as above but with an oligosaccharide-SR that does not serve as an XET acceptor substrate (e.g., cellotetraose-SR (*see* **Note 39**).
4. Place the specimens in a small muslin envelope, formed like a tea-bag, tightly sealed round the edges with waterproof tape (e.g., autoclave tape). Place the envelope in a flask containing 1 L of 0.1 M formic acid and leave stirring overnight (*see* **Notes 40** and **41**).
5. Use a final wash with pure water or 75% ethanol to remove the acid.
6. Remove the specimens from the muslin "tea-bag" and observe under a fluorescence microscope (Fig. 8b). Excitation can be at 315 nm (UV) or 560–570 nm (green); the emitted (orange) fluorescence peaks at about 600–615 nm.
7. Orange fluorescence is due to sulforhodamine-labeled, wall-bound xyloglucan, the product formed by XET action (*see* **Note 42**).
8. If the fluorescence of chlorophyll interferes in observation of the xyloglucan-SR, wash the specimen with acetone to remove the chlorophyll (*see* **Note 43**).

4 Notes

1. Polysaccharides may become highly viscous, and some are slow to dissolve.
2. But the paper will have irreversibly shrunk during the alkali treatment.
3. The BSA helps to protect the enzymes from denaturation and tends to prevent them binding to the walls of the vial or (in the case of CXE assays) to cellulose. Addition of BSA is particularly important for maintaining a high CXE reaction rate [17].

4. Buffer D is usually used for extracting firmly bonded cell wall proteins, for example from algae.
5. The loading of polysaccharide on the paper achieved by this method is typically ~2.5 g m^{-2}.
6. The amount of XGO-SR thereby impregnated into the paper is ~1 pmol m^{-2}.
7. As GF sheets contain traces of polysaccharides that might interfere with the assay, they were preheated in the oven at 350 °C for 24 h. After this treatment, GF sheets become more fragile and should be handled carefully.
8. This preparation, rich in cell wall enzymes, will be termed "crude extract."
9. Alternatively, freeze-dry the sample to eliminate excess liquid.
10. Since algae are aquatic, their wall enzymes tend to be firmly wall-bound; thus buffer D, with a high salt concentration and containing a surfactant, may be preferred for solubilization of these enzymes.
11. For de-salting, an alternative to dialysis (recommended if the proteins were ammonium sulfate–precipitated from an extract containing buffer D), is removal of both surfactant and salt on a mini-desalting column such as PD 10 (Sephadex G-25) or detergent removing Extracti-Gel D (Pierce). However, there may be reasons to omit the step since some wall enzymes tend to bind dialysis membrane or chromatography column resins, which may affect enzyme activity recovery.
12. This preparation will be termed "dialysate."
13. Different ^{3}H-labeled XGOs can be used if substrate specificity is of interest. Generally, [^{3}H]XLLGol is optimal for conventional XTHs, for example from dicots; [^{3}H]XXXGol is best for HTG. For some purposes it is valuable to know the final concentration of the XGO used in the assay. If so, the [^{3}H] XGO should be supplemented with the same oligosaccharide in nonradioactive form, using a weight which when finally dissolved in the 30-μL reaction mixture will give a concentration close to the enzyme's K_M. The K_M of a classical XTH is typically 50–200 μM XXXGol; that of *Equisetum* HTG is in the order of 1 μM and the final XGO concentration in this case should not exceed ~10 μM because higher concentrations are inhibitory.
14. BSA, if used, should be added after the polysaccharide has been heated in the buffer.
15. These polysaccharide solutions are rather viscous and difficult to pipet accurately; a positive-displacement pipette (e.g., Microman M; Gilson) is recommended.

16. Since the dried [^{3}H]XGO is practically invisible, checking for its dissolution may be difficult (and dissolution in the viscous polysaccharide solution may be slow); it is helpful if the [^{3}H]XGO had been codried with 1 μL of 0.1% (w/v) Orange G, an inert colored substance. When the [^{3}H]XGO has been successfully redissolved, you should have a uniformly orange solution.
17. Care—avoid skin contact!
18. The aim should be to continue the incubation until in the order of 10% of the radioactive substrate has been converted to product; the yield should not be allowed to exceed 20% so that an initial, linear rate can be determined.
19. The Orange G provides a visual check of uniform loading.
20. Unreacted [^{3}H]XGO will be removed and disposed of in the drains: check that this is permitted in your jurisdiction. The Orange G will also be removed.
21. Counting efficiency for the XET reaction product is typically ~25%, in which case complete conversion of oligosaccharide substrate to polysaccharide product would give 15,000 cpm.
22. Use a thin layer of adhesive or masking tape to mount the test paper onto an acetate sheet; mark out 9 × 9 mm squares with a pencil or a laser printer. It is a wise precaution to include a control test-paper impregnated with XGO-SR but no hemicellulose: a positive result with this would indicate that the enzyme extract either contained significant amounts of a hemicellulosic donor substrate (typically xyloglucan or MLG) or included CXE activity capable of using the cellulose of the paper as its donor substrate.
23. Work quickly so that the 5-μL spots of enzyme solution do not start drying on the paper.
24. If crude enzyme preparations, or any additives under investigation, are found to have interfered in the detection of the orange-fluorescing polysaccharide–XGO-SR spots, a wide range of additional washes can subsequently be used to try to clear the interfering substances, for example, 1% SDS (100 °C), phenol–acetic acid–water (2:1:1; 70 °C—care, avoid skin contact!), 1 M H_2SO_4 (20 °C), 1 M NH_4OH (20 °C), DMSO, and acetone. After any of these washes, rinse the paper thoroughly in water, and redry. Drying is expedited if the final wash is in acetone.
25. Ideally, use a centrifugal evaporator. Diverse ^{3}H-labeled XGOs can be used if substrate specificity is of interest.
26. Since 1 kBq of dried [^{3}H]XGO is practically invisible, checking for its dissolution may be difficult; it is helpful if the [^{3}H]XGO had been codried with Orange G, an inert colored substance. When the [^{3}H]XGO has been successfully redissolved, you should have a uniformly orange solution.

27. If the paper is still present as a coherent piece, labeled in pencil, washing can conveniently be done in slowly running tap water overnight; however, if the paper is disintegrating, it should be washed in a 15-mL centrifuge tube by repeated resuspension in water and pelleting at 2000 × *g* for 5 min. In the latter case, continue the washing until a 2-mL sample of the supernatant is nonradioactive when scintillation counted in 20 mL of water-miscible scintillant.
28. The counting efficiency of the CXE product is typically ~7%, in which case complete conversion of oligosaccharide substrate to polysaccharide product would give 4200 cpm.
29. This check can be done on the same sample after a preliminary scintillation count: remove the paper from the scintillation fluid with forceps, wash it thoroughly in acetone and redry, then proceed with the NaOH treatment as described.
30. There is no need to add the buffer since the enzyme extract itself contains sufficient buffering compounds.
31. Never allow the free blotting solution to exceed the level indicated by the red line in Fig. 6.
32. Assay a 2-mL sample of supernatant in 20 mL water-miscible scintillant to check for remaining radioactivity.
33. Drying the pellet is expedited if the final wash is in acetone.
34. Xyloglucan endoglucanase (XEG), as used by [29] on the basis that it was thought not to digest the MXE product, is not recommended because we have recently found that it does eventually digest MLG–[^{3}H]XXXGol to yield a mixture of [^{3}H]XXXGol, [^{3}H]GXXXGol, and [^{3}H]GGXXXGol.
35. On the same plate, also load a suitable external marker such as 1 kBq [^{3}H]XXXGol. A common mistake is not to dry the loadings adequately; appearance can be deceptive. Use warm air from a hair drier, or leave the plate to dry overnight.
36. A third 8-h run will further improve it, but is not essential.
37. Alternatively, cut each track of the (plastic-backed) TLC plate into 1-cm strips, and assay these for ^{3}H by scintillation counting; or monitor the ^{3}H profile of each track with a radioisotope TLC analyzer (e.g., LabLogic AR2000).
38. For demonstration purposes, a young celery petiole, a maize leaf-blade, an *Equisetum* stem or the apical 10 mm of an *Arabidopsis* root gives clear results. It should be borne in mind that the fluorescent substrate may not penetrate the cuticle of the shoot or the endodermis of the root—hence the need for cutting.

39. An alternative control is to use the 12 μM XGO-SR as above, supplemented with 1 mM nonlabeled XGO as a competing acceptor substrate. Either of these controls will indicate any nonspecific (XET-independent) binding of the sulforhodamine group to the specimen.
40. The acid will inactivate enzymes, preventing further XET action. It will also wash out the remaining unreacted XGO–SR. However, it will not wash out wall-bound xyloglucan-SR produced by XET action.
41. Some of the reaction product may not be wall-bound and may thus be washed out by the aqueous acid. If soluble polymeric products are of interest, use ethanol–formic acid–water (15:1:4 by vol.) instead of aqueous acid for the washing step; this will precipitate and immobilize any water-soluble polysaccharides in situ.
42. In *Equisetum*, which possesses HTG, the observed fluorescence does not distinguish between MXE, CXE, and XET action. Further treatment of the section (e.g., with lichenase or 6 M NaOH) may allow investigation of which endogenous polysaccharide has been used as donor substrate by the endogenous enzyme.
43. The use of specific green-excitation fluorescence filter with narrow nm range of emission will also help to eliminate signal from any autofluorescing chlorophyll.

References

1. Franková L, Fry SC (2015) A general method for assaying homo- and hetero-transglycanase activities that act on plant cell-wall polysaccharides. J Integr Plant Biol 57:411–428
2. Baydoun EA-H, Fry SC (1989) *In vivo* degradation and extracellular polymer-binding of xyloglucan nonasaccharide, a natural anti-auxin. J Plant Physiol 134:453–459
3. Smith RC, Fry SC (1991) Endotransglycosylation of xyloglucans in plant cell-suspension cultures. Biochem J 279:529–535
4. Farkaš V, Sulová Z, Stratilová E, Hanna R, Maclachlan G (1992) Cleavage of xyloglucan by nasturtium seed xyloglucanase and transglycosylation to xyloglucan subunit oligosaccharides. Arch Biochem Biophys 298:365–370
5. Fry SC, Smith RC, Renwick KF, Martin DJ, Hodge SK, Matthews KJ (1992) Xyloglucan endotransglycosylase, a new wall-loosening enzyme activity from plants. Biochem J 282:821–828
6. Nishitani K, Tominaga T (1992) Endo-xyloglucan transferase, a novel class of glycosyltransferase that catalyses transfer of a segment of xyloglucan molecule to another xyloglucan molecule. J Biol Chem 267:21058–21064
7. Rose JKC, Braam J, Fry SC, Nishitani K (2002) The XTH family of enzymes involved in xyloglucan endotransglucosylation and endohydrolysis: current perspectives and a new unifying nomenclature. Plant Cell Physiol 43:1421–1435
8. Nishitani K (2005) Division of roles among members of the XTH gene family in plants. Plant Biosyst 139:98–101
9. Zhu XF, Shi YZ, Lei GJ et al (2012) *XTH31*, encoding an in-vitro XEH/XET-active enzyme, controls Al sensitivity by modulating *in-vivo* XET action, cell wall xyloglucan content and Al binding capacity in Arabidopsis. Plant Cell 24:4731–4747
10. Kaewthai N, Gendre D, Eklöf JM, Ibatullin FM, Ezcurra I, Bhalerao R, Brumer H (2013) Group III-A XTH genes of Arabidopsis thaliana encode predominant xyloglucan endo-

hydrolases that are dispensable for normal growth. Plant Physiol 161:440–454

11. Schröder R, Wegrzyn TF, Bolitho KM, Redgwell RJ (2004) Mannan transglycosylase: a novel enzyme activity in cell walls of higher plants. Planta 219:590–600
12. Schröder R, Wegrzyn TF, Sharma NN, Atkinson RG (2006) LeMAN4 endo-β-mannanase from ripe tomato fruit can act as a mannan transglycosylase or hydrolase. Planta 224:1091–1102
13. Franková L, Fry SC (2011) Phylogenetic variation in glycosidases and glycanases acting on plant cell wall polysaccharides, and the detection of transglycosidase and trans-β-xylanase activities. Plant J 67:662–681
14. Johnston SL, Prakash R, Chen NJ, Kumagai MH, Turano HM, Cooney JM, Atkinson RG, Paull RE, Cheetamun R, Bacic A, Brummell DA, Schröder R (2013) An enzyme activity capable of endotransglycosylation of heteroxylan polysaccharides is present in plant primary cell walls. Planta 237:173–187
15. Derba-Maceluch M, Awano T, Takahashi J, Lucenius J, Ratke C, Kontro I, Busse-Wicher-M, Kosík O, Tanaka R, Winzéll A, Kallas Å, Leśniewska J, Berthold F, Immerzeel P, Teeri TT, Ezcurra I, Dupree P, Serimaa R, Mellerowicz EJ (2015) Suppression of a xylan transglycosylase PtxtXyn10A affects cellulose microfibril angle in secondary wall in aspen wood. New Phytol 205:666–681
16. Fry SC, Mohler KE, Nesselrode BHWA, Franková L (2008) Mixed-linkage β-glucan: xyloglucan endotransglucosylase, a novel wall-remodelling enzyme from *Equisetum* (horsetails) and charophytic algae. Plant J 55:240–252
17. Simmons TJ, Mohler KE, Holland C, Goubet F, Franková L, Houston DR, Hudson AD, Meulewaeter F, Fry SC (2015) Hetero-trans-β-glucanase, an enzyme unique to *Equisetum* plants, functionalises cellulose. Plant J 83:753–769
18. Franková L, Fry SC (2012a) Trans–αxylosidase and trans-β-galactosidase activities, widespread in plants, modify and stabilise xyloglucan structures. Plant J 71:45–60
19. Franková L, Fry SC (2012b) Trans--α-xylosidase, a widespread enzyme activity in plants, introduces (1→4)-α-D-xylobiose sidechains into xyloglucan structures. Phytochemistry 78:29–43
20. Kosík O, Auburn RP, Russell S, Stratilová E, Garajová S, Hrmová M, Farkaš V (2010) Polysaccharide microarrays for high-throughput screening of transglycosylase activities in plant extracts. Glycoconj J 27:79–87
21. Hrmova M, Farkaš V, Lahnstein J, Fincher GB (2007) A barley xyloglucan xyloglucosyl transferase covalently links xyloglucan, cellulosic substrates, and (1,3;1,4)-β-D-glucans. J Biol Chem 282:12951–12962
22. Popper ZA, Fry SC (2003) Primary cell wall composition of bryophytes and charophytes. Ann Bot 91:1–12
23. Thompson JE, Fry SC (2001) Restructuring of wall-bound xyloglucan by transglycosylation in living plant cells. Plant J 26:23–34
24. Carpita NC, Campbell M, Tierney M (eds) (2001) Plant cell walls. Springer Netherlands, 342 pp, [ISBN 978-94-010-3861-4]
25. Bacic A, Harris PJ, Stone BA (1988) Structure and function of plant cell walls. In: Priess J (ed) The biochemistry of plants. Academic Press, London, pp 297–371
26. Varner JE, Lin LS (1989) Plant cell wall architecture. Cell 56:231–239
27. Fry SC (1997) Novel 'dot-blot' assays for glycosyltransferases and glycosylhydrolases: optimization for xyloglucan endotransglycosylase (XET) activity. Plant J 11:1141–1150
28. Vissenberg K, Martinez-Vilchez IM, Verbelen J-P, Miller JG, Fry SC (2000) *In-vivo* colocalization of xyloglucan endotransglycosylase activity and its donor substrate in the elongation zone of *Arabidopsis* roots. Plant Cell 12:1229–1238
29. Mohler KE, Simmons TJ, Fry SC (2013) Mixed-linkage glucan:xyloglucan endotransglucosylase (MXE) re-models hemicelluloses in *Equisetum* shoots but not in barley shoots or *Equisetum* callus. New Phytol 197:111–122
30. Fry SC (2004) Tansley review: primary cell wall metabolism: tracking the careers of wall polymers in living plant cells. New Phytol 161:641–675
31. Franková L, Fry SC (2013) Biochemistry and physiological roles of enzymes that 'cut and paste' plant cell-wall polysaccharides. (Darwin review). J Exp Bot 64:3519–3550
32. Takeda T, Fry SC (2004) Control of xyloglucan endotransglucosylase activity by salts and anionic polymers. Planta 219:722–732
33. Hetherington PR, Fry SC (1993) Xyloglucan endotransglycosylase activity in carrot cell suspensions during cell elongation and somatic embryogenesis. Plant Physiol 103:987–992
34. Miller JG, Farkaš V, Sharples SC, Fry SC (2007) *O*-Oligosaccharidyl-1-amino-1-deoxyalditols as intermediates for fluorescent

labelling of oligosaccharides. Carbohydr Res 342:44–54

35. Chormova D, Franková L, Defries A, Cutler SR, Fry SC (2015) Discovery of small molecule inhibitors of xyloglucan endotransglucosylase (XET) activity by high-throughput screening. Phytochemistry 117:220–236

36. Van Sandt VST, Guisez Y, Verbelen J-P, Vissenberg K (2006) Analysis of a xyloglucan endotransglycosylase/hydrolase (XTH) from the lycopodiophyte Selaginella kraussiana suggests that XTH sequence characteristics and function are highly conserved during the evolution of vascular plants. J Exp Bot 57:2909–2922

Chapter 11

Screening for Cellulolytic Plant Enzymes Using Colorimetric and Fluorescence Methods

Kirsten Krause and Stian Olsen

Abstract

Cellulolytic activity can be measured using a variety of methods, the choice of which depends on the raw material and goals. An inexpensive, rapid, and reliable method, suitable for plants and other sources alike, is based on digestion of the easily degradable soluble cellulose derivative carboxymethylcellulose (CMC). Direct detection of CMC digestion by cellulolytic activity is based on the "negative staining principle," where undigested CMC is stained with appropriate colorimetric or fluorescent stains, while CMC exposed to digestion by cellulase shows a reduction in staining intensity. The reduction is proportional to the enzyme activity and is not influenced by endogenous levels of glucose in the sample, making this method applicable for a wide variety of samples, including plant material.

Key words Calcofluor White Stain, Carboxymethylcellulose, Cellulase, Cellulose, Gram's iodine stain

1 Introduction

The plant cell wall component cellulose (β-1,4-glucan) is among the most abundant biopolymers on earth. Its monomers, glucose molecules, are joined to form long linear chains that in turn use hydrogen bonds to align to stiff microfibrils that exhibit a high tensile strength. In plants these microfibrils confer strength and rigidity, providing a counterforce for the cell turgor. Enzymes that can break this rigid structure are found in plants [1] as well as in organisms feeding on living or dead plant material (e.g., [2–4]). In microbes, fungi and parasitic plants, cellulolytic enzymes facilitate plant infection.

The breakdown of cellulose is cooperatively mediated by endocellulases that randomly cleave internal bonds in the cellulose chains and by exocellulases that progressively cleave sugar units from the ends of the chains. Several methods have been developed for estimating the total saccharifying activity of crude cellulolytic extracts, mostly for the purpose of characterizing the capacity of soil-borne microorganisms to decompose cellulosic matter [4–

Zoë A. Popper (ed.), *The Plant Cell Wall: Methods and Protocols*, Methods in Molecular Biology, vol. 2149,
https://doi.org/10.1007/978-1-0716-0621-6_11,

6]. The most popular ones can be roughly divided into two types: qualitative petri dish-based assays and quantitative assays. In both cases, cellulosic substrates that are more readily soluble in water than pure cellulose (e.g., carboxymethylcellulose [CMC]), have become the substrate of choice. In quantitative assays, the release of reducing sugars that can be monitored using spectrophotometric, fluorometric or chromatographic methods is used as a measure for cellulolytic activity. While this is a very valuable method for purified enzymes or partially purified extracts, it is prone to produce artifacts when sugars are naturally present at higher levels (such as in crude plant extracts).

In petri dish assays, the detection of the cellulolytic activity is achieved by staining of undigested CMC in areas not subjected to cellulose decomposition [7]. Typically, agar is used as an inert carrier substance into which the substrate is mixed and the staining is performed using Gram's iodine stain, Safranin or Congo Red [8]. The visual results of CMC cleavage are clear halos surrounding the source of the enzyme [9], making this in theory a rapid, visual and inexpensive method. However, Zitomer and Eveleigh in 1987 [10] and more recently Johnsen and Krause [11] have advocated caution in the use of this method in connection with Gram's iodine stain, as commercial agars were shown to produce artifacts with this dye. An even simpler method omitting the agar and exploiting the natural gelling capacity of the CMC substrate proved to be a good alternative. Moreover, when 96-well plates are used, a larger number of samples can be processed simultaneously with a minimal use of reagents [11]. Limitations, unfortunately, were still observed with crude plant extracts due to the interference with the spectrophotometric quantification by photosynthetic pigments and other extract discolorations. We have therefore tested a fluorescent dye as an alternative and found that this allows measuring crude plant extracts with little or no interference of endogenous sample components. This staining alternative is described alongside the colorimetric stain-based method that was published earlier [11].

2 Materials

In the following, the materials needed for one 96-well plate assay are given.

2.1 CMC Solution

Carboxymethylcellulose is commercially available in different viscosity grades. We use medium viscosity CMC (e.g., Sigma-Aldrich, Catalog number C4888) at a concentration of 7% in demineralized water, at which it has the consistency of very thick syrup and is difficult to pipet.

1. CMC powder, medium viscosity grade.
2. Demineralized water.
3. Microwave oven.
4. Duran glass bottle, magnetic stirrer, and magnetic bar.

2.2 Colorimetric and Fluorescent Dyes

1. For colorimetric detection: 20 mL Gram's iodine stain (0.133 g KI and 0.067 g Iodine dissolved in 20 mL demineralized water).
2. For fluorescence detection: 1 mL Calcofluor White (CFW) Stain (Sigma-Aldrich, Catalog number 18909) composed of 1 g/L Calcofluor White M2R (alias Fluorescent Brightener 28) and 0.5 g/L Evans Blue.

2.3 Cellulase Calibration Solution and Extraction/Dilution Buffer

Cellulase Onozuka R-10 from *Trichoderma viride* (Duchefa, Catalog number C8001) at a stock concentration of 20 mg/mL in demineralized water. Dilutions of the stock for digestion were made in extraction/dilution buffer (Table 1). The buffer is also used for preparing enzyme extracts from the samples (*see* **Note 1**).

Chemicals are purchased from standard suppliers including Sigma, Merck, Fluka, and so on, and are of analytical grade.

2.4 Miscellaneous Equipment

1. 96-well plates
2. Multipette (e.g., from Eppendorf) and matching Combitips for dispensing 200 μL volumes.
3. Aluminum foil.
4. Filter paper.
5. 96% v/v Ethanol.
6. Demineralized water.

Table 1
Extraction/dilution buffer

Chemical	Concentration	Needed for 100 mL buffer
Sodium acetate, pH 5.5	50 mM	0.41 g (or 10 ml of a 500 mM stock)
Sodium chloride	300 mM	1.75 g
Sodium ascorbate (add fresh)	20 mM	0.4 g
Calcium chloride, anhydrous	10 mM	0.11 g
Glycerol, 85%	15% (v/v)	17.6 mL
Polyvinylpyrrolidone (PVP-40)	3% (w/v)	3 g

7. For fluorescence detection: Molecular Imager with UV-transillumination source for excitation and a 530/28 emission filter for blue fluorophore detection.
8. For colorimetric detection: plate reader for spectrophotometric absorbance measurements.

3 Methods

3.1 Preparation of CMC Solution

1. Heat 25 mL demineralized water in a 100 mL Duran glass bottle briefly in a microwave oven without actually bringing it to boil.
2. Place the bottle on a magnetic stirrer and stir moderately with a magnetic bar.
3. Add 1.75 g of CMC powder slowly in small portions to the hot water, carefully avoiding clumping of the CMC.

The solution increases in viscosity as the concentration increases and as the solution cools down to room temperature. It should finally have a syrupy consistence, at which it can still be aspired and dispensed with a Multipette.

3.2 CMC-CFW Stain Solution

For fluorescence detection, the CFW stain can be added as soon as the CMC gel has cooled down to room temperature. Leave the bottle on the magnetic stirrer and add 1 mL CFW stain to 25 mL CMC. Wrap the bottle in aluminum foil and stir until the blue color (from Evans Blue) is evenly distributed.

3.3 Preparing a Dilution Series of Cellulase for Calibration

1. Prepare stock solution by dissolving 20 mg of cellulase Onozuka R10 (specific activity >1000 U/mg) in 1 mL of sterile demineralized water.
2. Dilute 50 μL stock solution 1:10 with 450 μL extraction/dilution buffer.
3. Dilute 100 μL of the first dilution (above) 1:5 with 400 μL extraction/dilution buffer.
4. Dilute 100 μL of the second dilution 1:5 with 400 μL extraction/dilution buffer.
5. Repeat dilutions in the same way until desired amount of dilution steps are reached.

3.4 Sample Extract Preparation

1. Grind 1 g of plant material using mortar and pestle and a pinch of sea sand in 1 mL ice-cold extraction/dilution buffer.
2. Incubate for 2 h on ice (*see* **Note 2**).
3. Centrifuge the homogenate for 5 min at 12000 × *g*.
4. Use the supernatant undiluted or diluted in extraction/dilution buffer for subsequent enzyme assays.

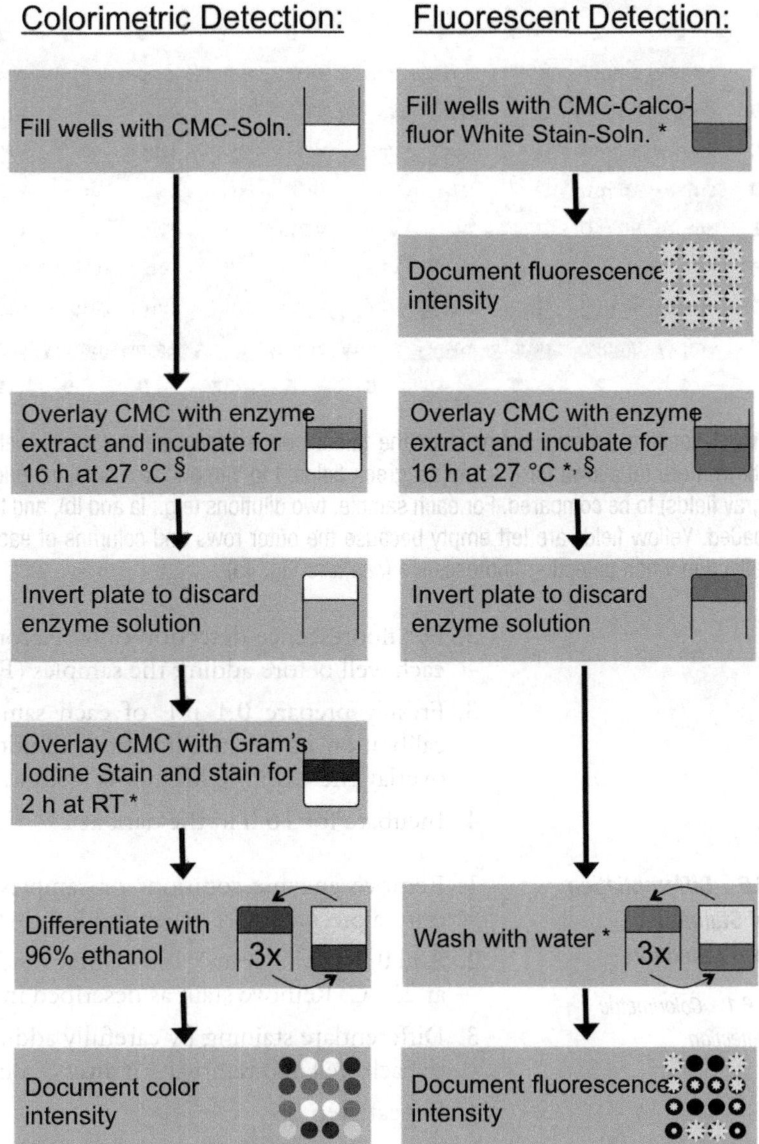

Fig. 1 Flow diagram showing the key steps of the two alternative detection methods. ∗: Protect from light. §: Optimal incubation temperature may vary with the type of sample

3.5 Preparation of Plates

The steps below are schematically depicted in Fig. 1. The colorimetric assay was described by Johnsen and Krause [11]. Figure 2 shows an example of a sample loading scheme.

1. Dispense 0.2 mL CMC solution (for colorimetric detection) or CMC-CFW solution (for fluorescence detection) into each well of a 96-well plate using a Multipette dispenser.

	1	2	3	4	5	6	7	8	9	10	11	12	
A	empty	empty	empty	empty	empty	empty	empty	empty	empty	empty	empty	empty	A
B	empty	0.64 mU	Ia	IIIa	Va	VIIa	Ib	IIIb	Vb	VIIb	2000 mU	empty	B
C	empty	3.2 mU	Ia	IIIa	Va	VIIa	Ib	IIIb	Vb	VIIb	400 mU	empty	C
D	empty	16 mU	Ia	IIIa	Va	VIIa	Ib	IIIb	Vb	VIIb	80 mU	empty	D
E	empty	80 mU	IIa	IVa	VIa	VIIIa	IIb	IVb	VIb	VIIIb	16 mU	empty	E
F	empty	400 mU	IIa	IVa	VIa	VIIIa	IIb	IVb	VIb	VIIIb	3.2 mU	empty	F
G	empty	2000 mU	IIa	IVa	VIa	VIIIa	IIb	IVb	VIb	VIIIb	0.64 mU	empty	G
H	empty	empty	empty	empty	empty	empty	empty	empty	empty	empty	empty	empty	H
	1	2	3	4	5	6	7	8	9	10	11	12	

Fig. 2 Loading scheme example for the fluorescence assay method in 96-well plate format. The scheme shown here (a) allows for 8 samples (green fields, I to VIII) and two dilution series of the calibration enzyme (gray fields) to be compared. For each sample, two dilutions (e.g., Ia and Ib), and three technical replicates are loaded. Yellow fields are left empty because the outer rows and columns of each plate tend to show higher reflection when detecting fluorescence (*see also* Fig. 4a)

2. For fluorescence detection only: Record fluorescence levels in each well before adding the samples (Fig. 3a) (*see* **Note 3**).
3. Freshly prepare 0.1 mL of each sample or enzyme for the calibration curve in extraction/dilution buffer and carefully overlay the CMC/CMC-CFW with it.
4. Incubate for 16 h in the dark at 27 °C.

3.6 Differentiation of Staining and Detection

3.6.1 Colorimetric Detection

1. Remove enzyme solutions or samples by inverting the plate onto a piece of filter paper and lightly tapping it.
2. Add 0.2 mL of Gram's Iodine Stain and stain for 2 h in the dark at 27 °C. Remove stain as described in **step 1**.
3. Differentiate staining by carefully adding 0.2 mL 96% Ethanol to each well and removing it after 5 min as described in **step 1**.
4. Repeat **step 3**.
5. Repeat **step 4**, but wait for 15 min before removing the ethanol.
6. Record the absorbance at 575 nm.
7. Calculate enzyme activity using the calibration curve (see examples shown in Fig. 3c, d) or select extracts of interest based on visual inspection of the results.

3.6.2 Fluorescence Detection

1. Remove enzyme solutions or samples by inverting plate onto a piece of filter paper and lightly tapping it.
2. Differentiate staining by carefully adding 0.2 mL demineralized water to each well and removing it after 15 min as described in **step 1**.

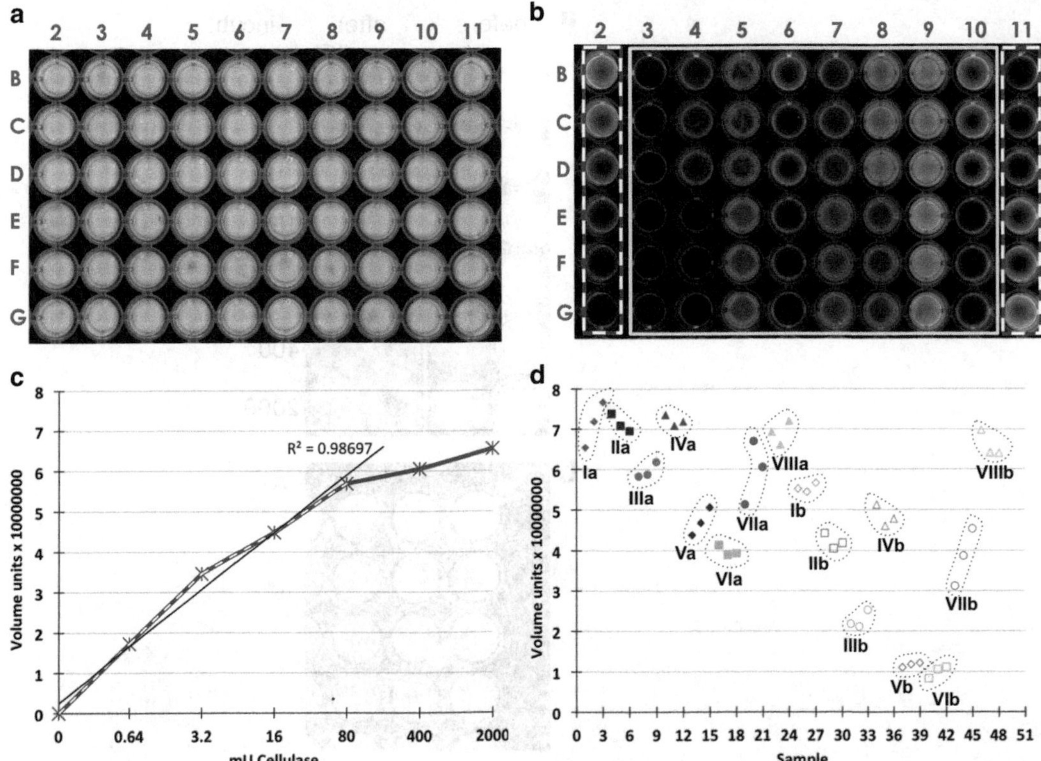

Fig. 3 Example for evaluation of CMC digestion by cellulolytic action. (**a**, **b**) Fluorescence of CMC-CFW is detected before (**a**) and after (**b**) incubation with the samples. The labeling of rows (B-G) and columns (2–11) corresponds to the scheme shown in Fig. 2. (**c**) A calibration curve was calculated from the difference in fluorescence values of wells containing defined amounts of cellulose before and after incubation with the enzyme (squares with broken blue-white lines in (**b**)). The difference increases with increasing units of cellulase. The linear range of activity detection is between 0 and ~100 milliunits (mU) enzyme. (**d**) The difference in fluorescence values before and after incubation is plotted for 48 data points (outlined in yellow in (**b**)), corresponding to eight samples, each in two dilutions and in technical triplicates (**d**). Filled symbols represent undiluted samples, open symbols represent the same samples after a 1:5 dilution

3. Repeat **step 2**, but wait for 30 min before removing the water as described in **step 1**.
4. Repeat **step 2**, but wait for 90 min before removing the water as described in **step 1**.
5. Record staining intensity using a Molecular Imager. Excitation of fluorescence is carried out using UV light (UV table) and emission is recorded at 530 nm (Fig. 3b).
6. Calculate enzyme activity using the calibration curve (see examples shown in Fig. 3c, d) or select extracts of interest based on visual inspection of the results.

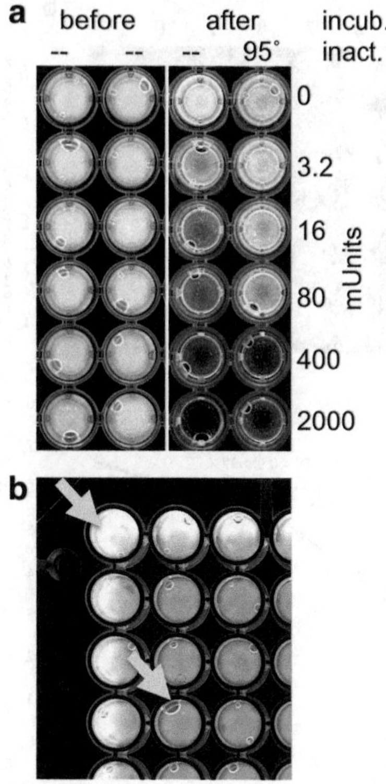

Fig. 4 Potential caveats. (**a**) Heat inactivation at 95 °C for 15 min is effective in the linear range of enzyme activity detection, between 0 and ~100 milliunits (*see* Fig. 3). (**b**) Reflection in the outer rows and columns of each plate (upper arrow) and from air bubbles (lower arrow) in the CMC matrix can give false results when detecting fluorescence (*see* **Notes 4** and **5**)

4 Notes

1. Unused extracts and ascorbate-containing buffer can be stored at −20 °C but may lose activity if freeze-thawed frequently or stored over a longer period.
2. For negative controls, enzyme extracts can be inactivated by heating to 95 °C.
3. Recording of the fluorescence before incubation can be omitted if sample readings are related to control wells without enzyme. However, this step gives the possibility to correct for a priori CMC volume differences due to pipetting errors. Given the gelatinous nature of the CMC solution, such differences are difficult to avoid completely.

4. The outermost rows and columns of the 96-well plates show strong reflection during illumination on a UV table. (Fig. 4b) and should not be used for samples or calibration curves.
5. Bubbles in the wells (Fig. 4b) show reflection and can interfere with measuring accuracy. They should thus be avoided.

Acknowledgments

The authors thank Tromsø Research Foundation (Mohn Foundation) and the Faculty of Biosciences, Fisheries and Economics at UiT for providing the funds for this work. Marie Cooper (UiT) is thanked for proofreading.

References

1. Popper ZA et al (2011) Evolution and diversity of plant cell walls: from algae to flowering plants. Annu Rev Plant Biol 62:567–590
2. Kudo T (2009) Termite-microbe symbiotic system and its efficient degradation of lignocellulose. Biosci Biotechnol Biochem 73 (12):2561–2567
3. Cai S et al (2010) Cellulosilyticum ruminicola, a newly described rumen bacterium that possesses redundant fibrolytic-protein-encoding genes and degrades lignocellulose with multiple carbohydrate- borne fibrolytic enzymes. Appl Environ Microbiol 76(12):3818–3824
4. Dashtban M et al (2010) Fungal biodegradation and enzymatic modification of lignin. Int J Biochem Mol Biol 1(1):36–50
5. Vancov T, Keen B (2009) Rapid isolation and high-throughput determination of cellulase and laminarinase activity in soils. J Microbiol Methods 79(2):174–177
6. Zhang YH, Hong J, Ye X (2009) Cellulase assays. Methods Mol Biol 581:213–231
7. Dashtban M, Schraft H, Qin W (2009) Fungal bioconversion of lignocellulosic residues; opportunities & perspectives. Int J Biol Sci 5 (6):578–595
8. Gohel HR et al (2014) A comparative study of various staining techniques for determination of extracellular cellulase activity on carboxy methyl cellulose (CMC) agar plates. Int J Curr Microbiol Appl Sci 3(5):261–266
9. Kasana RC et al (2008) A rapid and easy method for the detection of microbial cellulases on agar plates using gram's iodine. Curr Microbiol 57(5):503–507
10. Zitomer SW, Eveleigh DE (1987) Cellulase screening by iodine staining: an artefact. Enzym Microb Technol 9(4):214–216
11. Johnsen HR, Krause K (2014) Cellulase activity screening using pure carboxymethylcellulose: application to soluble cellulolytic samples and to plant tissue prints. Int J Mol Sci 15 (1):830–838

Chapter 12

Using Solid-State ^{13}C NMR Spectroscopy to Study the Molecular Organization of Primary Plant Cell Walls

Stefan J. Hill, Tracey J. Bell, Laurence D. Melton, and Philip J. Harris

Dedicated to the memory of Roger H. Newman (1 March 1949-26 July 2014) who was a much-valued colleague and pioneer in the application of solid-state ^{13}C NMR to the study of plant cell walls.

Abstract

A knowledge of the mobilities of the polysaccharides or parts of polysaccharides in a cell-wall preparation provides information about possible molecular interactions among the polysaccharides in the cell wall and the relative locations of polysaccharides within the cell wall. A number of solid-state ^{13}C NMR techniques have been developed that can be used to investigate different types of polysaccharide mobilities: rigid, semirigid, mobile, and highly mobile. In this chapter techniques are described for obtaining spectra from primary cell-wall preparations using CP/MAS, proton-rotating frame, proton spin–spin, spin–echo relaxation spectra and single-pulse excitation. We also describe how proton spin relaxation editing can be used to obtain subspectra for cell-wall polysaccharides of different mobilities, and how 2D and 3D solid-state NMR experiments have recently been applied to plant cell walls.

Key words Primary cell walls, Polysaccharide mobility, Solid-state ^{13}C NMR, Proton-spin relaxation editing (PRSE), Single-pulse excitation NMR, 2D and 3D NMR, TEM, X-ray diffraction

1 Introduction

Primary walls surround growing plant cells and are composed of rigid cellulose microfibrils embedded in a gel-like matrix of noncellulosic polysaccharides with a range of different structures [1–3]. The major noncellulosic, matrix polysaccharides of the primary cell walls of eudicotyledons and noncommelinid monocotyledons are xyloglucans (XG) and pectic polysaccharides [2, 3]. The pectic polysaccharides consist mainly of homogalacturonan (HG) and rhamnogalacturonan I (RG-I), with smaller proportions of the substituted galacturonans rhamnogalacturonan II (RG-II) and xylogalacturonan (XGA). Arabinans, galactans and arabinogalactans are attached as side chains to RG-I [4, 5]. The major polysaccharides of the primary cell walls of most commelinid

Zoë A. Popper (ed.), *The Plant Cell Wall: Methods and Protocols*, Methods in Molecular Biology, vol. 2149, https://doi.org/10.1007/978-1-0716-0621-6_12, © Springer Science+Business Media, LLC, part of Springer Nature 2020

monocotyledons are glucuronoarabinoxylans (GAXs) with smaller proportions of pectic polysaccharides and XGs [2, 6, 7]. Variable proportions of $(1 \rightarrow 3),(1 \rightarrow 4)$-β-glucans also occur in the primary walls of grasses and cereals (Poaceae) and related families [8–10].

Solid-state ^{13}C nuclear magnetic resonance (NMR) spectroscopy is one of several techniques used to study the molecular organization of such walls. In contrast to TEM, this technique can be used on cell walls that have never been dried, so minimizing possible chemical and physical modifications. Moreover, in contrast to X-ray diffraction, this technique does not require multiple planes of ordered molecules. Solid-state ^{13}C NMR is sensitive to ordering over relatively short dimensions and measurements of chemical shifts can indicate differences in the conformations of backbone chains or side chains [11]. Measurements of spin relaxation time constants can indicate differences in the short-range environment, specifically the nature and dynamics of neighboring molecules. In studies of the cellulose in walls, solid-state ^{13}C NMR can be used to distinguish between molecules on the surfaces of crystallites and those in the interior, since their molecular conformations differ [12–14]. This information can be used to estimate the lateral dimensions of cellulose crystallites. In this application, solid-state ^{13}C NMR is particularly informative when the crystallites are of such a small cross-section that there are as many chains exposed on surfaces as contained in the interior, as occurs in primary-wall cellulose [11, 15]. TEM and X-ray diffraction are more informative if the cellulose crystallites are of a relatively large cross section [13]. However, with the development of synchrotron based X-ray techniques, complementary cellulose crystallite dimensions can be obtained and combined with solid-state ^{13}C data to better understand the detailed structure [16, 17].

Solid-state ^{13}C NMR spectroscopy can also be used to distinguish cell-wall polymers that have different mobilities because of their locations and interactions with other molecules within the cell wall [18–20]. A summary of how this type of mobility-resolved NMR may help to identify polysaccharide domains and interactions within the cell wall is shown in Fig. 1. For example, the $(1 \rightarrow 4)$-β-D-glucan backbones of XG molecules, or parts of molecules, adsorbed on to the surface of cellulose microfibrils in muro are predicted to adopt a rigid, flattened conformation, rather than the twisted-backbone conformation of free XG [21]. These two different conformations can be detected by solid-state ^{13}C NMR spectroscopy [22, 23] (Fig. 1). Therefore, not only is solid-state ^{13}C NMR spectroscopy useful for determining the mobilities of different components in plant cell walls, but it also effectively makes in situ investigations [24].

Two main types of mobility-resolved solid-state ^{13}C NMR experiments are commonly used to investigate primary cell walls, cross-polarization (CP) NMR and single-pulse excitation (SPE)

RIGID DOMAIN

MOBILE DOMAIN

VERY MOBILE DOMAIN

Polysaccharides that respond to CP/MAS NMR

Polysaccharides that separate into PSRE subspectrum A with long $T_{1\rho}$(H) and short T_2(H)

Inherently rigid polysaccharides e.g. cellulose

Also, more mobile polysaccharide structures affected by proton-spin diffusion from rigid polysaccharides eg. Glc and Xyl of XG adsorbed on to cellulose

Polysaccharides that separate into PSRE subspectrum A with short $T_{1\rho}$(H) and long T_2(H)

Mobile polysaccharides

Semi-rigid polysaccharide structures affected by proton-spin diffusion from the mobile polysaccharides

Polysaccharides that do not respond to CP/MAS NMR

Very mobile polysaccharides e.g. arabinans, galactans, arabinogalactans

Polysaccharides that respond to SPE/MAS (recovery delay 1 ms) NMR

Polysaccharides that respond to SE-SPE/MAS NMR

Polysaccharides that separate into subspectrum A of SE-PSRE

Long T_2(C)

Rigid polysaccharides e.g. cellulose

Polysaccharides that separate into subspectrum B of SE-PSRE

§Short T_2(C)

Mobile polysaccharide structures e.g. XG-Xyl of XG adsorbed to cellulose

Intermediate T_2(C)

Semi-rigid polysaccharide structures e.g. XG-Glc in XG cross-links between cellulose microfibrils

Short T_2(C)

Mobile polysaccharides e.g. pectic homogalacturonan

Very long T_2(C)

Very mobile polysaccharides e.g. arabinans, galactans and arabinogalactans

Fig. 1 Solid-state ^{13}C NMR, proton- and ^{13}C-spin relaxation techniques described in this chapter and used to identify polysaccharides of different mobilities in cell-wall preparations. § These mobile polysaccharides separate into the same $T_{1\rho}$(H) and T_2(H) subspectra as cellulose due to proton spin diffusion from adjacent cellulose but separate into the mobile T_2(C) subspectra as these structures are inherently mobile

NMR. Both are generally used in combination with MAS, hence they are referred to as CP/MAS NMR and SPE/MAS NMR [25]. CP/MAS NMR suppresses signals from relatively mobile molecules and SPE/MAS NMR suppresses signals from relatively rigid molecules [26].

Cross-polarization NMR combines both the magnetic spins of protons and ^{13}C nuclei. By transferring the magnetization from protons to ^{13}C, the range of dispersion of the chemical shifts is increased compared to the range of chemical shift values for proton NMR, thus increasing the resolution of the signals from solid-state samples [27]. CP/MAS NMR also allows the indirect measurement of proton NMR relaxation by variations in the strengths of selected ^{13}C signals [28]. The proton relaxation process is complicated by spin diffusion where neighboring nuclei exchange spin information [29]. Proton spin diffusion may occur over distances of

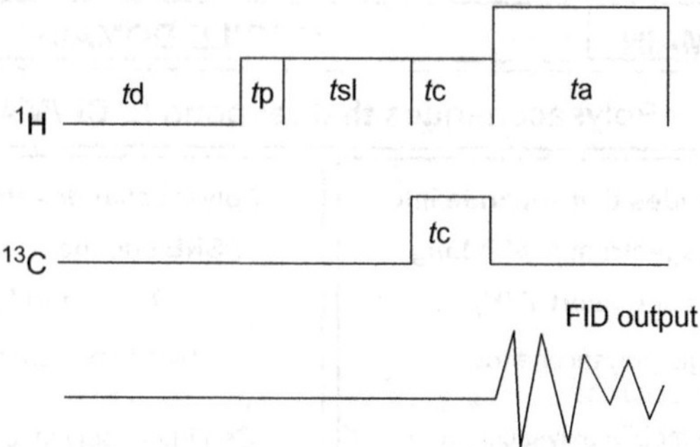

Fig. 2 Schematic representation of a proton rotating-frame spin relaxation pulse sequence. *tp* preparation pulse, *tsl* spin-locking pulse, *tc* cross-polarization contact time, *ta* data acquisition time, *td* recovery delay. In each case, *t* = time. (Adapted from Newman and Hemmingson [57])

nanometers and over time scales of milliseconds [30]. Therefore, proton spin relaxation time constants are mean values for all protons within a finite volume, irrespective of the ^{13}C-signal used to monitor the proton relaxation.

Two main approaches using CP/MAS NMR are used to investigate polysaccharide mobility in primary cell walls. Both approaches exploit the fact that the relaxation of spins of magnetic nuclei for polysaccharides in cell walls can be sensitive to the molecular conformations of those polysaccharides. First, PSRE can be used to separate NMR subspectra containing signals from polysaccharides in rigid and mobile domains of the cell wall [12]. To do this, a proton spin relaxation event is introduced prior to the CP contact time. A typical PSRE pulse sequence is illustrated in Fig. 2. Linear combinations, generated from spectra obtained under the normal CP/MAS and the PSRE conditions, can then be used to edit the CP/MAS spectra [26]. Second, a number of PSRE experiments may be carried out each with different recovery delays [31]. From these spectra, full relaxation curves for component signals may be constructed. The shape of the relaxation curve and the time constants associated with the proton relaxations can be used to distinguish between polysaccharides in rigid and mobile domains in the cell wall, remembering that the proton spin relaxation time constants of a particular polysaccharide will be similar to those of the polysaccharides in its immediate vicinity.

Segmental motion within a polysaccharide can be investigated by measuring relaxation of the ^{13}C-nuclei by using a spin–echo (SE) NMR pulse sequence. The rarity of the ^{13}C-isotope (1.1% of carbon) means the diffusion of ^{13}C spin information is relatively slow [32]. Therefore, unlike proton spin relaxation, ^{13}C relaxation is sensitive to segmental motion of the polysaccharide at the site of the ^{13}C nucleus and is much less sensitive to the motion of more distant segments of the same polysaccharide chain, or other polysaccharide chains.

There are three main spin-relaxation processes that have been used to investigate primary cell walls: spin–lattice relaxation with time constants T_1(H) and T_1(C), rotating-frame relaxation with time constants $T_{1\rho}$(H) and $T_{1\rho}$(C), and spin–spin relaxation with time constants T_2(H) and T_2(C) [11, 27]. Newman et al. [11] found that the most useful relaxation processes for investigating proton relaxations of polysaccharides in primary cell walls were the rotating-frame and spin–spin experiments.

SPE/MAS NMR is an excellent complementary technique to CP/MAS NMR [31] for investigating mobile polysaccharides in cell-wall preparations. The CP/MAS NMR spectra of the cell-wall preparations are generally dominated by signals assigned to cellulose, and the broader underlying signals are assigned to noncellulosic polysaccharides, along with, in primary cell-wall preparations containing predominantly pectic polysaccharides, signals at 173, 54, and 22 ppm assigned to carboxylic acid, methoxyl and acetyl carbons, respectively. As cellulose generally accounts for less than half of the primary cell wall, a portion of the noncellulosic material must be too mobile to respond to the cross-polarization pulse sequence. Although SPE/MAS can potentially detect all components in the cell-wall preparations [25], the highly mobile cell-wall components, which are less responsive to CP/MAS NMR [33, 34], are particularly responsive to SPE/MAS. By using a short recovery delay in the SPE/MAS pulse sequence (1 s), signals which have long T_1(C) relaxations, such as the rigid cellulose molecules, are suppressed [25, 34] (Fig. 1).

Studies of polysaccharide mobility in an isolated cell-wall preparation should use not only the relaxation properties of the proton or ^{13}C of that moiety, as indicated in Table 1, but also take into account alterations in the chemical shift for that polysaccharide or monosaccharide or part of the polysaccharide. Generally, polysaccharides investigated using solution NMR will have lower chemical shifts than in solid-state NMR (Table 1). These different chemical shifts may indicate differences in molecular conformations, for example, extended chains rather than coiled chains [23]. Chemical shifts can therefore provide insights into the reasons for reduced mobility within the cell wall, particularly when combined with the relaxation studies described above (Table 1).

One of the major drawbacks of both solution and solid-state NMR spectroscopic techniques is their relative lack of sensitively, with ^{13}C being only approximately 1% naturally abundant [50]. Solid-state NMR also suffers from being, on the whole, not suitable for 1H spectroscopy due to homonuclear dipolar coupling causing signal broadening [51]. Because of these two factors, 2D or 3D NMR experiments, routinely used in solution state NMR, have not been widely applied to solid-state NMR experiments. Specifically gaining information on the direct coupling of carbons in cell walls is difficult as it requires adjacent ^{13}C atoms, which statically

Table 1
Cited solid-state ^{13}C NMR assignments for the main primary cell-wall polysaccharides of eudicotyledons and noncommelinid monocotyledons

	Assignment	Chemical shift (ppm)	References
C-1 Glc	Cellulose I_{α} (interior)	105.6	[35]
	Cellulose I_{β} (interior)	106.5 and 104.4	[35]
	Cellulose II	107.6	[36]
C-4 Glc	Cellulose I_{α} (interior)	90.3 and 89.4	[37, 38]
	Cellulose I_{β} (interior)	89.4 and 88.5	[37, 38]
	Cellulose I (surface)	84.8 and 83.9	[39]
	Cellulose II	89.4 and 88.2	[40]
C-6 Glc	Cellulose I_{α} (interior)	65.2	[36]
	Cellulose I_{β} (interior)	66.3 and 65.2	[36]
	Cellulose I (surface)	63.1 and 62.1	[36]
	Cellulose II	63.5 and 62.5	[38]
C-1 XG-Xyl	Cyclamen seed XG	99.8	[41]
	Tamarind XG	99.8	[42]
	Arabidopsis cell walls (chemically extracted to remove pectic polysaccharides)	100.0	[43]
	Tamarind/BC composite	100.3	[44]
	Rubus suspension cells	100.4[a]	[45]
	Tamarind XG	99.1–100[a]	[46]
	Cyclamen seed XG	100.2[a]	[41]
C-4 XG-Xyl	Tamarind XG	70.1–70.4[a]	[46]
	Cyclamen seed XG	70.8[a]	[41]
C-1 XG-Glc	Tamarind XG	103.6	[42]
	Rubus suspension cells	103.9[a]	[45]
	Tamarind XG	103.1–103.7[a]	[46]
	Cyclamen seed XG	103.5[a]	[41]
C-4 Glc	Cyclamen seed XG	82–85	[41]
	Tamarind XG	70.3-80.5[a]	[46]
	Cyclamen seed XG	79.5-80.8[a]	[41]
C-1	5-Ara Arabinans	108.4[a]	[47]
	5-Ara Arabinans	108.5[a]	[48]
	4-Gal Galactans	105.2[a]	[47]
	4-GalA HG	101.1	[49]
	4-GalA HG	99.8[a]	[48]
C-4	4-Ara Arabinans	83.2[a]	[47]
	4-GalA HG	79–80	[49]
	4-Gal Galactans	78.5[a]	[47]
	4-Ara Arabinans	77.7[a]	[47]
	t-Gal Galactans	61.8[a]	[47]
$-COOCH_3$	HG	52.8–53.7	[49]

BC bacterial cellulose
[a]Solution NMR spectroscopy; all other values were obtained with CP/MAS of solid material

would occur for only 0.01% of molecules. This low probability means natural abundant ^{13}C experiments are almost impossible to carry out successfully for complex cell walls, although such experiments have been used for simpler systems such as individual amino acids [52].

The methods previously discussed based on proton mobility are indirect approaches required due to the low natural abundance of ^{13}C in nonenriched samples. This limitation has recently been overcome by growing plant material in the dark in liquid culture with uniformly labeled ^{13}C-glucose as the sole carbon source. By enhancing the abundance of ^{13}C a selection of 2D and 3D correlation experiments were made possible and resulted in a detailed study of atomic connectivities and molecular [53, 54].

2 Materials

1. Cell wall isolation buffer: HEPES-KOH buffer (20 mM, pH 6.7) containing 10 mM dithiothreitol (DTT) (*see* **Note 1**).
2. Ponceau 2R solution: Make up an aqueous solution of Ponceau 2 R (CI 16150) (0.2% w/v; 100 mL). Add 2 drops of 18 M H_2SO_4. Store solution at 4 °C.
3. Polydimethylsilane was from Hüls America Incorporated, Cincinnati, Ohio, USA (*see* **Note 2**).
4. Polytrichlorofluoroethylene grease was from Halocarbon Products Corporation, River Edge, New Jersey, USA (*see* **Note 3**).

3 Methods

3.1 Preparation of Cell Walls

A cell-wall preparation should be free of cytoplasmic contents as NMR signals from proteins and lipid can interfere with the NMR signals from polysaccharides. To achieve this, cell walls are isolated by mechanically breaking the cells open and washing out the cells contents with cold aqueous buffer [55, 56]. All procedures are carried out at 4 °C.

1. Plant tissue (approx. 20 g wet weight) is homogenized in 100 mL of cell wall isolation buffer, using a Polytron blender (Model PT10-35, Kinematica, Luzern, Switzerland) on full power (three times for 20 s).
2. The tissue can be further homogenized using a Tenbroeck ground-glass homogenizer (15 mL; Kontes Glass Company, Vineland, NJ, USA). Breakage of the cells is monitored using bright-field microscopy after staining with Ponceau 2R solution, which stains protein red [55].

3. The homogenate is centrifuged (250 × *g*, 10 min), and the pellet washed three times by centrifugation with buffer (with DTT omitted), followed by six times with water.
4. The pellet is resuspended in water, washed onto nylon mesh (pore size 11 μm), and the residue on the mesh washed with water (500 mL).
5. The water content of the cell-wall preparation should be reduced but not completely eliminated for solid-state NMR (*see* **Note 4**). Two methods can be used to do this:
   (a) The preparation is washed three times by centrifugation with 80% (v/v) ethanol and the final suspension kept at 4 °C in 80% (v/v) ethanol until NMR spectroscopy can be done (*see* **Note 5**). Preparations in 80% (v/v) aqueous ethanol are prepared for NMR by filtering portions of the suspensions onto a glass-fibre filter (GF/C, Whatman Scientific, Maidstone, Kent, UK) and part-drying in air to a water-content of approximately 40% (w/w). This weight is estimated from the dry weight of an aliquot of preparation that is removed and freeze-dried.
   (b) An aliquot of the preparation is removed and freeze-dried to estimate the water. The remaining preparation is frozen and dried, carefully, under vacuum to a water-content of approximately 40% (w/w) based on the dry weight of the sub-sample. If this method is used, NMR spectroscopy must be done immediately to avoid possible sample degradation. The exact moisture content of the preparation can be determined by oven drying to constant weight (105 °C) following the completion of the NMR experiments.

3.2 Solid-State ^{13}C-NMR Spectroscopy

1. The never-dried cell-wall preparation is packed in a 7 mm-diameter cylindrical silicon nitride rotor, and retained with Vespel end caps. An internal standard of polydimethylsilane can be added to the centre of the sample during the packing of the rotor. Polydimethylsilane contributes a ^{13}C signal at −1.96 ppm (*see* **Note 2**). As the cell-wall preparations are partly hydrated, polytrichlorofluoroethylene grease (*see* **Note 3**) is used to ensure a water-tight seal between the cylinder and the end caps. The grease should be spread on the internal surface of the rotor, not on the cap, to avoid expelling grease when the cap is inserted. It is important that there should be no grease on the external surfaces of the rotor or caps, otherwise the NMR stator will become contaminated and sample spinning will be impeded.
2. The rotor is spun at 4 kHz in a magic-angle spinning probe for ^{13}C NMR spectroscopy at 50.3 MHz. In the worked example, the probe was supplied by Doty Scientific (Columbia, SC,

USA) and the Inova-200 spectrometer was supplied by Varian (Palo Alto, CA, USA). Slower spinning can cause interference from spinning-sideband signals, while faster spinning can cause physical degradation of the cell-walls because of the centrifugal forces generated.

3.3 CP/MAS Spectroscopy

In cross-polarization (CP) NMR experiments, each 90° proton preparation pulse is followed by a 1 ms CP contact time, 51 ms of data acquisition and a recovery delay of 1.0 s before the sequence was repeated (*see* **Note 6**). The correct duration of the 90° proton preparation pulse can be measured by using trial values and selecting the value that provides the best signal strength. A typical value is 6 μs, as used to illustrate this chapter. Both proton and ^{13}C transmitters are left on for the duration of the contact time. It is also important that the power levels are adjusted for a Hartmann–Hahn match, that is, the measurement of a 90° pulse should give the same value for both nuclei. Measurement of a 90° pulse for ^{13}C is discussed below, under the heading "Single pulse excitation/ magic angle spinning (SPE/MAS) NMR." The proton transmitter output is increased during data acquisition, to provide adequate power for spin decoupling, that is, a target power level corresponding to a precession frequency > 40 kHz and typically between 53 and 59 kHz as in the experiments used to illustrate this chapter. Lower power levels will cause noticeable broadening of the NMR signals.

3.4 Proton Spin Relaxation Experiments

Proton rotating-frame spin relaxation with time constant $T_{1\rho}(\mathrm{H})$, and proton spin–spin relaxation with time constant $T_2(\mathrm{H})$ are characterized by inserting relaxation intervals of duration t_1 or t_2, respectively, between the proton preparation pulse and the CP contact time (Fig. 2). The values for t_1 and t_2 are chosen to be within the ranges of values for $T_{1\rho}(\mathrm{H})$ and $T_2(\mathrm{H})$, respectively, to optimize the signal-to-noise ratios in separate subspectra [57]. Protons are spin-locked during t_1, but the proton transmitter is switched off during t_2.

PSRE NMR subspectra are generated by combining spectra labeled S, S′, and S″, where S is obtained by the normal cross-polarization pulse sequence, S′ with $t_1 = 4$ ms and S″ with $t_2 = 15$ μs. The experimental spectra are obtained in the order CP/MAS (S), proton-rotating frame experiment (S′), and proton spin–spin experiment (S″).

The number of transients of the pulse sequences required to obtain adequate signal-to-noise ratios from averaged spectra needed for PSRE editing varies, but 40,000 up to 100,000 transients is usual. This corresponds to 30–80 h of data accumulation time for a full PSRE experiment.

3.5 Separating the Proton-Spin Relaxation Edited (PSRE) Subspectra

The principles behind PSRE NMR have been described in Newman [17] and are summarized in the Introduction. In the simplest case, a spectrum S is the sum of subspectra A and B from two distinct types of domains.

A partly relaxed spectrum S′ is then.

$$S' = f_a A + f_b B \quad (1)$$

where f_a and f_b are signal suppression factors.

The subspectra can then be separated by computing:

$$A = kS + k'S' \quad (2)$$

$$B = (1 - k)S - k'S' \quad (3)$$

where

$$k = f_b / (f_b - f_a) \quad (4)$$

$$k' = -1 / (f_b - f_a) \quad (5)$$

In the context of a cell-wall preparation, subspectra A and B usually contain signals from the cellulose and noncellulosic polysaccharides, respectively.

Two NMR signals, characteristic of the two mobility domains, are selected for proton rotating-frame relaxation experiments. The signal at 89 ppm (assigned to C-4 of cellulose crystallite-interior) is selected as representative of cellulose crystallites, appearing at a chemical shift for which there is little overlap with signals from other polysaccharides. A signal at 69 ppm (assigned to C-2, C-3, and C-5 of GalA residues in pectic homogalacturonans) [49, 58] is selected as representative of mobile polysaccharides for those primary walls containing high proportions of pectic polysaccharides [1, 2]. [NB. *See* Ref. 34 for appropriate editing signals for GAX rich primary walls].

Because the $T_{1\rho}(H)$ relaxation time constants for the 69 ppm signal were not greatly different, it was not possible to achieve total elimination of signals from noncellulosic polysaccharides without also suppressing signals from cellulose, so linear combinations of S and S′ are generated to enhance signal suppression (Table 2).

1. Initial estimates of f_a and f_b are calculated from signal heights as illustrated in Fig. 3.
2. The values for f_a and f_b are used to determine the linear combinations used to separate the subspectra from the CP/MAS data, using Eqs. (4) and (5) then Eqs. (2) and (3).
3. Values for f_a and f_b are then adjusted until the signal at 89 ppm is eliminated from subspectrum B and signal at 69 ppm, assigned to pectic polysaccharides, are suppressed in subspectrum A without allowing any signals to become inverted. For

Table 2
Suppression factors, linear combinations and the corresponding relaxation time constants used to generate PSRE subspectra A and B for mung bean cell walls

Suppression factors		
$T_{1\rho}$(H) ms	f_a	0.69
	f_b	0.27
T_2(H) μs	f_a	0.30
	f_b	0.60
Linear combinations		
$T_{1\rho}$(H) ms	Subspectrum A	−0.64S + 2.38S′
	Subspectrum B	1.64S + 2.38S′
T_2(H) μs	Subspectrum A	2.00S - 3.33S″
	Subspectrum B	−1.00S + 3.33S″
Relaxation time constants		
$T_{1\rho}$(H) ms	Subspectrum A	10.8
	Subspectrum B	3.1
T_2(H) μs	Subspectrum A	9.7
	Subspectrum B	14.8

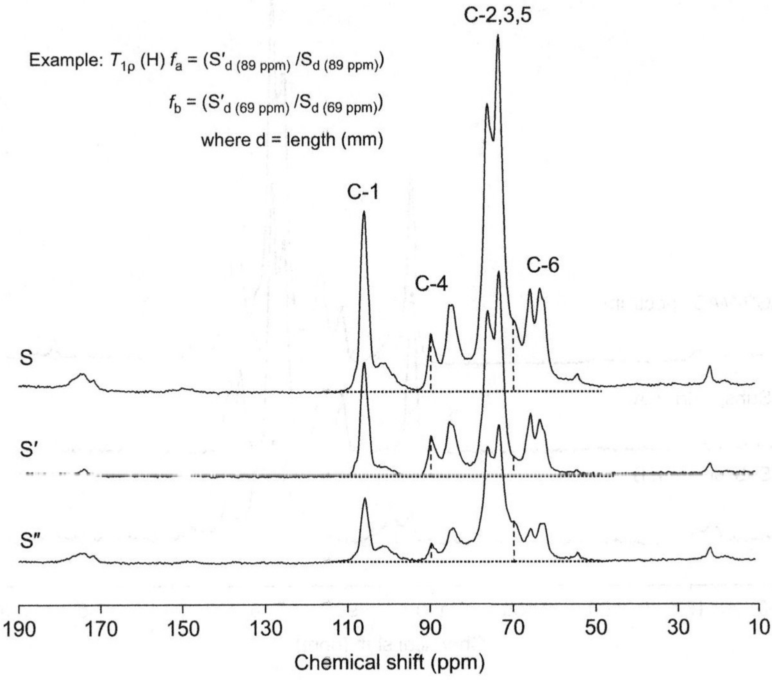

Fig. 3 Calculating suppression factors from the CP/MAS and PSRE spectra obtained of a mung bean cell-wall preparation. *S* obtained by the normal CP/MAS pulse sequence, *S*′ with 4 ms of proton rotating-frame spin relaxation, *S*″ with 15 μs of proton spin–spin relaxation. Carbon numbers refer to the Glc residues of cellulose

example, for the spectra from mung bean (*Vigna radiata*) cell walls shown in Fig. 3, the final suppression factors from the proton rotating frame experiment were:

$$f_a = 0.69 \text{ and } f_b = 0.27, \text{therefore; } k = -0.64 \text{ and } k' = -2.38$$

In the worked example, the linear combinations used to separate subspectra A and B from the CP/MAS NMR data were:

$$\text{Subspectrum A} = -0.64S + 2.38S' \text{ and Subspectrum B} = 1.64S + 2.38S'$$

The separation is successful when signals assigned to cellulose and mobile noncellulosic polysaccharides appeared in subspectra A and B, respectively. The resulting separated subspectra are shown in Fig. 4.

4. The final values of f_a and f_b can then be used to calculate improved values of the proton spin relaxation time constants. For example, if the spin relaxation process is exponential then:

$$f_a = \exp\left[-t_1\,(\text{ms})/T_{1\rho}(\text{H})\right]$$

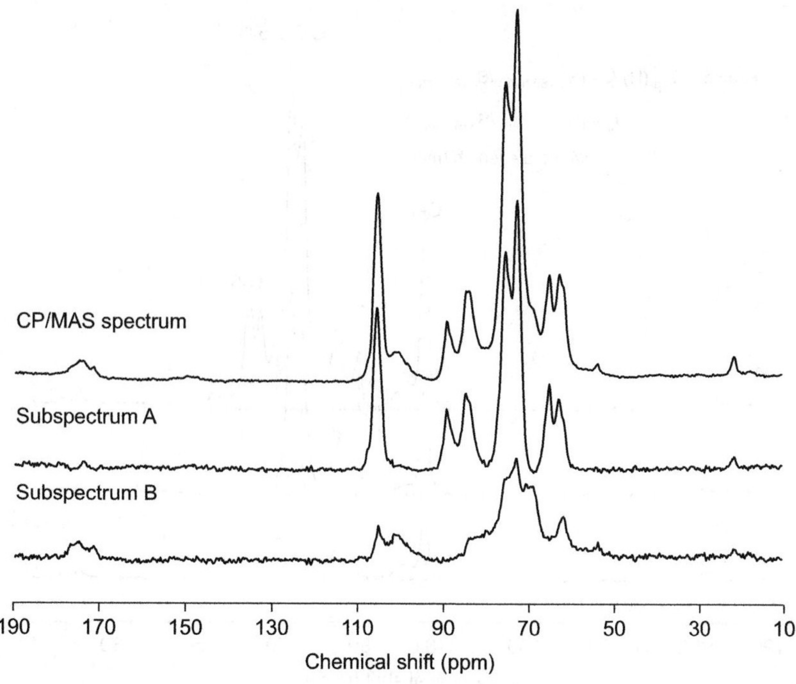

Fig. 4 Normal CP/MAS spectrum and separated PSRE subspectra, of a cell-wall preparation of the hypocotyls from mung bean seedlings. Subspectra are obtained by exploiting the differences in proton rotating-frame spin relaxation. Subspectra A and B display signals assigned primarily to cellulose and the noncellulosic matrix, respectively

Taking natural logarithms of both sides:

$$\ln f_a = -t_1(\text{ms})/T_{1\rho}(\text{H})$$

This can be rearranged to:

$$T_{1\rho}(\text{H}) = -t_1(\text{ms})/\ln f_a$$

If f_a is 0.69 and the t_1 value for the $T_{1\rho}(\text{H})$ experiment is 4 ms, then the estimated $T_{1\rho}(\text{H})$ value for subspectrum A in the proton rotating-frame experiments is 10.8 ms. The relaxation time constant for subspectrum B using f_b can be calculated using the same equations.

5. The editing process is repeated to separate the $T_2(\text{H})$ subspectra. This spin relaxation process suppresses cellulose signals more than the other signals, but does not eliminate them entirely. Linear combinations can again be generated to enhance the amount of suppression. Like the $T_{1\rho}(\text{H})$ relaxation, the separation is successful in that signals assigned to cellulose and noncellulosic polysaccharides appear in A and B, respectively.
6. As for $T_{1\rho}(\text{H})$ relaxation, the final values of f_a and f_b can be used to calculate improved values of the spin–spin relaxation time constants for the two subspectra. However, the relaxation curves for rigid solids, such as crystalline cellulose and more mobile polysaccharides, such as pectic homogalacturonans, are described by different functions [11]:

$$f_a = \exp\left[-\{t_2\ (\text{ms})/T_2(\text{H})\}^2/2\right]$$

$$f_b = \exp\left[-t_2\ (\text{ms})/T_2(\text{H})\right]$$

Table 2 shows the adjusted suppression factors, linear combinations and relaxation time constants from the PSRE experiments obtained for a preparation of mung bean cell walls.

3.5.1 Spin Echo NMR with PSRE (SE-PSRE)

As discussed above, unlike $T_{1\rho}(\text{H})$ and $T_2(\text{H})$, $T_2(\text{C})$ is sensitive to segmental motion of the polysaccharide at the site of the ^{13}C nucleus and is insensitive to the motion of more distant polysaccharides. Therefore, in SE-PSRE NMR experiments, the $T_2(\text{C})$ values for a polysaccharide will indicate the mobility of that particular polysaccharide, whereas the $T_{1\rho}(\text{H})$ and $T_2(\text{H})$ relaxation values will reflect the averaged molecular motion of the surrounding polysaccharides. For example, if a relatively rigid XG chain extends through a domain containing relatively flexible pectic polysaccharides, values of $T_{1\rho}(\text{H})$ and $T_2(\text{H})$ measured from XG signals will reflect molecular motion in the pectic polysaccharide environment. In this example, the values of $T_2(\text{C})$ measured from xyloglucan signals will reflect the rigidity of the XG chain and not the mobility of the pectic polysaccharide environment [18].

1. T_2(C) relaxation is characterized by a spin–echo sequence in which a delay of duration t_2 is inserted between the CP contact time and commencement of data acquisition [11]. A 180° refocusing pulse is applied halfway through t_2.
2. Multiple values of t_2 can be chosen so that $0.5t_2$ is always a multiple of the rotor rotation period, and protons are decoupled with an attenuated power output corresponding to a precession frequency of 43 kHz throughout t_2.
3. Proton spin relaxation edited (PSRE) NMR subspectra are generated for each of the T_2(C) spectra using the same S and S′ values used for editing the proton spin-relaxation spectra.
4. Relaxation time constants for T_2(C) can be calculated as described in Subheading 3.6. As indicated above, these relaxation values will reflect the mobility of that particular polysaccharide in the cell-wall preparation.

3.5.2 Single Pulse Excitation/Magic Angle Spinning (SPE/MAS) NMR

As discussed in the Introduction, SPE/MAS is a useful technique for investigating the mobility of the very mobile polysaccharides in the cell wall, such as the pectic polysaccharide side chains on RG-I.

1. Single pulse excitation/magic angle spinning (SPE/MAS) spectra are obtained with a pulse sequence in which each 90° ^{13}C excitation pulse is followed by 51 ms of data acquisition time and a 1 s recovery delay [34]. The correct duration of the 90° ^{13}C pulse can be measured by using trial values and selecting the value that provides the best signal strength. A typical value is 6 μs, as used to illustrate in this chapter. The 1.0 s delay is used to maximize the response from mobile components and minimize the response from rigid components of the cell walls [34]. The proton decoupler transmitter power during data acquisition should correspond to a precession frequency is increased to provide radiofrequency field strengths >40 kHz and preferably between 53 and 59 kHz.
2. As with all NMR experiments, the signal-to-noise ratio will improve with increased number of experimental transients. However, care must be taken to avoid sample degradation during long experiments (*see* **Note 7**). The SPE/MAS spectrum for a cell-wall preparation of mung bean hypocotyls, shown in Fig. 5, was acquired by the averaging of 23,931 experiment transients. This equated to approximately 6.7 h of data accumulation time.

3.5.3 Spin Echo with SPE/MAS NMR (SE-SPE/MAS)

The SE-SPE/MAS pulse sequence is similar to the SE-PSRE sequence except it is run with SPE/MAS and not PSRE and spectra are not separated into mobility domains. The SE-SPE/MAS experiments will provide mobility information about a particular monosaccharide component within the polysaccharide, provided the signal is relatively free from other overlapping signals.

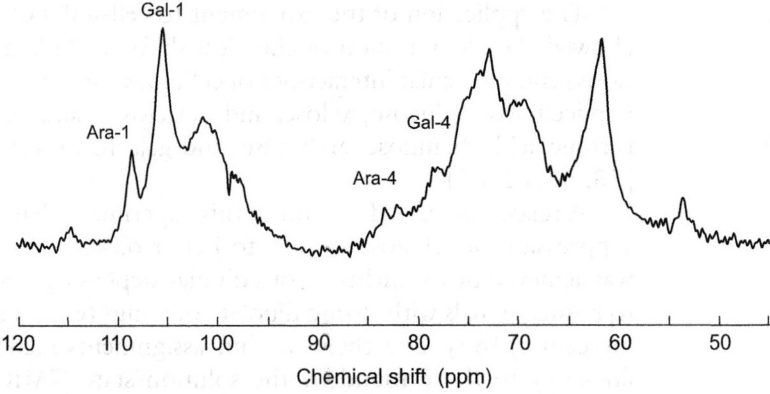

Fig. 5 SPE/MAS NMR spectrum of cell-wall preparation from hypocotyls of mung bean seedlings. Assignments refer to the carbon numbers of Ara or Gal in arabinans or galactans (*see* Table 1)

Each 90° ^{13}C excitation pulse is followed by a t_2 delay then 51 ms of data acquisition time and a 1 s recovery delay. A 180° refocusing pulse is applied halfway through t_2, and values of t_2 are chosen so that $0.5t_2$ is always a multiple of the rotor rotation period, and protons are decoupled with an attenuated power output corresponding to a precession frequency > 40 kHz throughout t_2. The durations of the 90° and 180° ^{13}C pulses are typically 6 μs and 12 μs, respectively.

3.6 Applications of 2D and 3D Solid-State NMR Experiments

3.6.1 Correlation Experiments

Carbon correlation experiments allow the through bond or through space connection of carbon spin systems to be determined. This means that carbons that are either directly bonded to each other or only a few bonds away from each other are observed as cross-peaks in a two-dimensional spectrum. The use of higher data dimensions (e.g., 2D or 3D) simplifies the NMR spectra by removing noncorrelated peaks and reducing peak overlap. By using such a strategy the complexity of the cell wall can be probed without the need to isolate only certain components prior to analysis.

2D J-INADEQUATE (Incredible Natural Abundance Double Quantum Transfer Experiment) Sequence Based Experiment

This experiment, originally developed for solution-state NMR, correlates natural abundance ^{13}C–^{13}C spin couplings by suppressing the spinning sidebands and other weak long range couplings and by selecting a double-quantum coherence pathway [59]. This type of experiment has been used for solid-state NMR using both fully and non ^{13}C labeled L-isoleucine [52] and a modified version has been used with fully labeled L-alanine, sucrose, and glycine [60]. In this experiment the usual cross-polarization to carbon from protons is followed by magnetization evolution under homonuclear coupling and refocusing. This is followed by the creation of a double-quantum coherence to produce a 2D ^{13}C–^{13}C spectrum where one axis has the expected carbon frequency and the other double the frequency.

The application of the experiment to cell-wall polysaccharides allowed the identification of chemical shifts (including conformations) and molecular interactions of cellulose (surface and interior), hemicellulose (glucose, xylose, and galactose), and pectin (galacturonic acid, rhamnose, arabinose, and galactose) related residues [53, 54, 61–65].

A relaxation edited version of this experiment that allowed the suppression of cellulose signals, to better observe hemicelluloses, was achieved by the inclusion of a dipolar-dephasing step that acted to reduce signals with strong dipolar couplings (e.g., rigid cellulose molecules) [66]. The chemical shift assignments made compared favorably to those found by the solution state NMR analysis of similar samples ball milled and extract into d_6-DMSO [67].

3.6.2 Spin Diffusion Experiments

1D/2D PDSD (Proton Driven Spin Diffusion) and DARR (Dipolar Assisted Rotational Resonance) Sequence Based Experiments

These experiments are based on the diffusion of magnetization from protons prior to transfer to carbons under MAS conditions. The time of diffusion can then be correlated to a distance as determined by the diffusion coefficient. By measuring the spin diffusion times, hence distances, constraints on possible molecular arrangements can allow a structure to be developed [68]. Such constraints provide through-space information about the *closeness* of various cell-wall components by altering the mixing-times during the experiment, allowing a range of molecular-scales to be investigated, such as water-polysaccharide interactions [69, 70], protein-polysaccharide interactions [63, 71], and polysaccharide–polysaccharide interactions [61, 64, 72, 73].

3D ^{13}C–^{13}C–^{13}C Sequence Based Experiments

Adding a 3D to solid-state NMR data provides a way to develop a concept of the 3-dimensional structure within molecules and between molecules. This approach uses a combination of two PDSD/DARR 2D NMR pulse sequences: one with a short mixing time to select for intramolecular correlations and the other with a longer mixing time that provides information on intermolecular correlations [64, 74]. When applied to the study of cell walls, this experiment allowed the unique assignment of NMR peaks for the major polysaccharides present in the cell wall and how these interact on the molecular scale. This has allowed insight into the dynamic nature of the cell wall where a number of different interactions between components drive the observed physical properties of the cell wall [53].

3.7 Ultrahigh Spinning Speed Solid-State NMR

Although increasing the magnetic field strength results in improvements in spectral resolution by a factor of about three compared with low to mid field strengths [73], further improvements can be gained by increasing the rate of spinning of the sample. Recent advances in solid-state probe design has allowed sample spinning speeds to be in the order of 100 kHz, with the potential to improve proton signal resolution and opening up the new area of inverse detected experiments [75]. However, the application of fast magic-angle sample spinning to plant cell-wall research has not been reported so far.

4 Notes

1. The reducing agent DTT is added to the isolation buffer to prevent the oxidation of phenols to quinones [55, 76]. Quinones can polymerize to form red/brown products and may also form covalent bonds with proteins resulting in insoluble precipitates [55]. Alternatively, buffer containing 10 mM 2-mercaptoethanol may replace DTT.
2. The negative chemical shift (−1.96 ppm) of polydimethylsilane is outside the range normally seen for cell-wall polysaccharides, and therefore does not interfere with signals from cell-wall material. An alternative standard could be polydimethylsiloxane, showing a ^{13}C signal at 1.50 ppm, which is closer to the range associated with cell wall material [77]. Polydimethylsiloxane is readily available from Sigma-Aldrich, and polydimethylsilane can be obtained from Meryer Chemical Technology Co., Ltd., Shanghai, China.
3. The polytrichlorofluoroethylene grease from Halocarbon is thickened with silica and provides a water tight seal for the Vespel end caps. Neither the grease nor the thickener contributes signal strength to the cross-polarization ^{13}C NMR spectra.
4. Moisture is essential for distinguishing polysaccharides in different molecular environments in a cell-wall preparation, for example, cellulose molecules at the surface of the crystallite [12]. However, over-drying or drying and rehydrating of the preparation can result in irreversible changes in the molecular order of the polysaccharides [78, 79]. Excessively high moisture contents dilute down the amount of carbon that can be packed into the rotor and therefore diminish the NMR signal. The ideal water content is between 30% and 50% w/w in many cases.
5. Exposure of cell-wall preparations to ethanol may, in principle physically alter polysaccharides [78], as well as denature and precipitate proteins. Although we have not seen evidence in the NMR spectra of such changes, we recommend caution.
6. Preliminary T_1(H) experiments indicated that a 1.0 s delay was adequate for the recovery of proton magnetization in relatively mobile segments of polysaccharides, as was also shown in experiments on other cell walls [11, 12].

7. It is desirable to check for sample degradation using SPE experiments. This can be achieved by breaking the experiment into several periods of data accumulation and comparing the spectra to test for changes. If the spectra are all similar, they may be added together to improve the signal-to-noise ratio.

Acknowledgments

The authors would like to dedicate this work to the late Dr. Roger H. Newman who was a pioneer in the application of solid-state NMR spectroscopy to plant cell walls. His techniques are still the basis for many of the experiments used today, and his work is still highly cited. Although Roger is missed, his science legacy is to be found in every solid-state NMR spectrum of a plant cell wall.

References

1. Harris PJ (2005) Diversity in plant cell walls. In: Henry RJ (ed) Plant diversity and evolution: genotypic and phenotypic variation in higher plants. CAB International, Wallingford, Oxon, pp 201–227
2. Harris PJ, Stone BA (2008) Chemistry and molecular organization of plant cell walls. In: Himmel ME (ed) Biomass recalcitrance: deconstructing the plant cell wall for bioenergy. Blackwell Publishing, Oxford, pp 61–93
3. Fry SC (2011) Cell wall polysaccharide composition and covalent crosslinking. Ann Plant Rev 41:1–42
4. Mohnen D (2008) Pectin structure and biosynthesis. Curr Opin Plant Biol 11:266–277
5. Atmodjo MA, Hao Z, Mohnen D (2013) Evolving views of pectin biosynthesis. Annu Rev Plant Biol 64:747–779
6. Hsieh YSY, Harris PJ (2009) Xyloglucans of monocotyledons have diverse structures. Mol Plant 2:943–965
7. Scheller HV, Ulvskov P (2010) Hemicelluloses. Annu Rev Plant Biol 61:263–289
8. Trethewey JAK, Campbell LM, Harris PJ (2005) (1→3),(1→4)-β-D-glucans in the cell walls of the Poales (*sensu lato*): an immunogold labelling study using a monoclonal antibody. Am J Bot 92:1660–1674
9. Harris PJ, Fincher GB (2009) Distribution, fine structure and function of (1,3;1,4)-β-glucans in the grasses and other taxa. In: Bacic A, Fincher GB, Stone BA (eds) Chemistry, biochemistry, and biology of (1→3)-β-glucans and related polysaccharides. Academic Press, Elsevier Inc., San Diego, pp 621–654
10. Burton RA, Fincher GB (2014) Evolution and development of cell walls in cereal grains. Front Plant Sci 5:456. https://doi.org/10.3389/fpls.2014.00456
11. Newman RH, Davies LM, Harris PJ (1996) Solid-state ^{13}C nuclear magnetic resonance characterisation of cellulose in the cell walls of *Arabidopsis thaliana* leaves. Plant Physiol 111:475–485
12. Newman RH, Ha M-A, Melton LD (1994) Solid-state ^{13}C NMR investigation of molecular ordering in the cellulose of apple cell walls. J Agric Food Chem 42:1402–1406
13. Newman RH (1999) Estimation of the lateral dimensions of cellulose crystallites using ^{13}C NMR signal strengths. Solid State Nucl Magn Reson 15:21–29
14. Newman RH, Davidson TC (2004) Molecular conformations at the cellulose-water interface. Cellulose 11:23–32
15. Bootten TJ, Harris PJ, Melton LD, Newman RH (2008) WAXS and ^{13}C-NMR study of *Gluconoacetobacter xylinus* cellulose in composites with tamarind xyloglucan. Carbohydr Res 343:221–229
16. Hill SJ, Kirby NM, Mudie ST, Hawley AM, Ingham B, Franich RA, Newman RH (2010) Effect of drying and rewetting of wood on cellulose molecular packing. Holzforschung 64:421–427
17. Newman RH, Hill SJ, Harris PJ (2013) Wide-angle X-ray scattering and solid-state nuclear magnetic resonance data combined to test models for cellulose microfibrils in mung bean cell walls. Plant Physiol 163:1558–1567

18. Bootten TJ, Harris PJ, Melton LD, Newman RH (2004) Solid-state ^{13}C-NMR spectroscopy shows that the xyloglucans in the primary cell walls of mung bean (*Vigna radiata* L.) occur in different domains: a new model for xyloglucan-cellulose interactions in the cell wall. J Exp Bot 55:571–583
19. Ng JK, Zujovic ZD, Smith BG, Johnston JW, Schröder R, Melton LD (2014) Solid-state ^{13}C NMR study of the mobility of polysaccharides in the cell walls of two apple cultivars of different firmness. Carbohydr Res 386:1–6
20. Zujovic Z, Chen D, Melton LD (2016) Comparison of celery (*Apium graveolens* L.) collenchyma and parenchyma cell walls enabled by solid-state ^{13}C NMR. Carbohydr Res 420:51–57
21. Levy S, Maclachlan G, Staehelin LA (1997) Xyloglucan sidechains modulate binding to cellulose during in vitro binding assays as predicted by conformational dynamics simulations. Plant J 11:373–386
22. Horii F, Hirai A, Kitamaru R (1984) Cross-polarization/magic angle spinning ^{13}C-NMR study. Molecular chain conformations of native and regenerated cellulose. In: Polymers for fibers and elastomers. American Chemical Society, pp 27–42
23. Jarvis MC (1994) Relationship of chemical shift to glycosidic conformation in the solid state ^{13}C NMR spectra of (1→4)-linked glucose polymers and oligomers: anomeric and related effects. Carbohydr Res 259:311–318
24. Jarvis MC, Apperley DC (1990) Direct observation of cell wall structure in living plant tissues by solid-state ^{13}C NMR spectroscopy. Plant Physiol 92:61–65
25. Tang H, Belton PS, Ng A, Ryden P (1999) ^{13}C MAS NMR studies of the effects of hydration on the cell walls of potatoes and Chinese water chestnuts. J Agric Food Chem 47:510–517
26. Newman RH (1999) Editing the information in solid-state carbon-13 NMR spectra of food. In: Belton PS, Hills BP, Webb GA (eds) Advances in magnetic resonance in food science. The Royal Society of Chemistry, Cambridge, pp 144–157
27. Newman RH (1992) Solid-state carbon-13 NMR spectroscopy of multiphase biomaterials. In: Glasser WG, Hatakeyama H (eds) Viscoelasticity of biomaterials. American Chemical Society, Washington, DC, pp 311–319
28. Tekely P, Vignon MR (1987) Proton T_1 and T_2 relaxation times of wood components using ^{13}C CP/MAS NMR. J Polym Sci Part C Polym Lett 25:257–261
29. Hediger S, Emsley L, Fischer M (1999) Solid-state NMR characterization of hydration on polymer mobility in onion cell-wall material. Carbohydr Res 322:102–112
30. Zumbulyadis N (1983) Selective carbon excitation and the detection of spatial heterogeneity in cross-polarization magic-angle-spinning NMR. J Magn Reson 53:486–494
31. Tang H, Hills BP (2003) Use of ^{13}C MAS NMR to study domain structure and dynamics of polysaccharides in the native starch granules. Biomacromolecules 4:1269–1276
32. VanderHart DL (1987) Natural-abundance ^{13}C-^{13}C spin exchange in rigid crystalline solids. J Magn Reson 72:13–47
33. Foster TJ, Ablett S, McCann MC, Gidley MJ (1996) Mobility-resolved ^{13}C-NMR spectroscopy of primary plant cell walls. Biopolymers 39:51–66
34. Smith BG, Harris PJ, Melton LD, Newman RH (1998) The range of mobility of the non-cellulosic polysaccharides is similar in primary cell walls with different polysaccharide compositions. Physiol Plant 103:233–246
35. Newman RH (1997) Crystalline forms of cellulose in the silver tree fern *Cyathea dealbata*. Cellulose 4:269–278
36. Newman RH, Redgwell RJ (2002) Cell wall changes in ripening kiwifruit: ^{13}C solid state NMR characterisation of relatively rigid cell wall polymers. Carbohydr Polym 49:121–129
37. Atalla RH, Vanderhart DL (1984) Native cellulose: a composite of two distinct crystalline forms. Science 223:283–285
38. Newman RH, Hemmingson JA (1995) Carbon-13 NMR distinction between categories of molecular order and disorder in cellulose. Cellulose 2:95–110
39. Newman RH (1998) Evidence for assignment of ^{13}C NMR signals to cellulose crystallite surfaces in wood, pulp and isolated celluloses. Holzforschung 52:157–159
40. Hirai A, Horii F, Kitamaru R (1990) Carbon-13 spin-lattice relaxation behaviour of the crystalline and non-crystalline components of native and regenerated celluloses. Cellulose Chem Technol 24:703–711
41. Braccini I, Hervé du NB, Penhoat C, Michon V, Goldberg R, Clochard M, Jarvis MC, Huang Z-H, Gage DA (1995) Structural analysis of cyclamen seed xyloglucan oligosaccharides using cellulase digestion and spectroscopic methods. Carbohydr Res 276:167–181
42. Gidley MJ, Lillford PJ, Rowlands DW, Lang P, Dentini M, Crescenzi V, Edwards M, Fanutti C, Reid JSG (1991) Structure and

solution properties of tamarind-seed polysaccharide. Carbohydr Res 214:299–314

43. Davies LM, Harris PJ, Newman RH (2002) Molecular ordering of cellulose after extraction of polysaccharides from primary cell walls of *Arabidopsis thaliana*: a solid-state CP/MAS ^{13}C NMR study. Carbohydr Res 337:587–593

44. Whitney SEC, Brigham JE, Darke AH, Reid JSG, Gidley MJ (1995) In vitro assembly of cellulose/xyloglucan networks: ultrastructural and molecular aspects. Plant J 8:491–504

45. Joseleau JP, Cartier N, Chambat G, Faik A, Ruel K (1992) Structural features and biological activity of xyloglucans from suspension-cultured plant cells. Biochemie 74:81–88

46. York WS, Harvey LK, Guillen R, Albersheim P, Darvill AG (1993) Structural analysis of tamarind seed xyloglucan oligosaccharides using β-galactosidase digestion and spectroscopic methods. Carbohydr Res 248:285–301

47. Ryden P, Colquhoun IJ, Selvendran RR (1989) Investigation of structural features of the pectic polysaccharides of onion by ^{13}C-N.M.R. spectroscopy. Carbohydr Res 185:233–237

48. Saulnier L, Brillouet J-M, Joseleau J-P (1988) Structural studies of pectic substances from the pulp of grape berries. Carbohydr Res 182:63–78

49. Sinitsya A, Čopíková J, Pavlíková H (1998) ^{13}C CP/MAS NMR spectroscopy in the analysis of pectins. J Carbohydr Chem 17:279–292

50. Levitt MH (2008) Spin dynamics: basics of nuclear magnetic resonance, 2nd edn. Wiley, Chichester, West Sussex

51. Schmidt-Rohr K, Spiess HW (1994) Multidimensional solid-state NMR and polymers. Academic Press, London

52. Lesage A, Auger C, Caldarelli S, Emsley L (1997) Determination of through-bond carbon-carbon connectivities is solid-state NMR using the INADEQUATE experiment. J Am Chem Soc 119:7867–7868

53. Dick-Perez M, Zhang Y, Hayes J, Salazar A, Zabotina OA, Hong M (2011) Structure and interactions of plant cell-wall polysaccharides by two- and three-dimensional magic-angle-spinning solid-state NMR. Biochemistry 50:989–1000

54. Wang T, Hong M (2016) Solid-state NMR investigations of cellulose structure and interactions with matrix polysaccharides in plant primary cell walls. J Exp Bot 6:503–514

55. Harris PJ (1983) Cell walls. In: Hall JL, Moore AL (eds) Isolation of membranes and organelles from plant cell walls. Academic Press, London, pp 25–53

56. Melton LD, Smith BG (2005) Isolation of plant cell walls and fractionation of cell wall polysaccharides. In: Wrolstad RE (ed) Handbook of food analytical chemistry: water, proteins, enzymes, lipids and carbohydrates. Wiley, Hoboken, NJ, pp 697–719

57. Newman RH, Hemmingson JA (1990) Determination of the degree of cellulose crystallinity in wood by carbon-13 nuclear magnetic resonance spectroscopy. Holzforschung 44:351–355

58. Jarvis MC, Apperley DC (1995) Chain conformation in concentrated pectic gels: evidence from ^{13}C NMR. Carbohydr Res 275:131–145

59. Bax A, Freeman R, Kempsell SP (1980) Natural abundance ^{13}C-^{13}C coupling observed via double-quantum coherence. J Am Chem Soc 102:4849–4851

60. Hohwy M, Rienstra CM, Jaroniec CP, Griffin RG (1999) Fivefold symmetric homonuclear dipolar recoupling in rotating solids: application to double quantum spectroscopy. J Chem Phys 110:7983–7992

61. Dick-Perez M, Wang T, Salazar A, Zabotina OA, Hong M (2012) Multidimensional solid-state NMR studies of the structure and dynamics of pectic polysaccharides in uniformly ^{13}C-labeled *Arabidopsis* primary cell walls. Magn Reson Chem 50:539–550

62. Harris DM, Corbin K, Wang T, Gutierrez R, Bertolo AL, Petti C, Smilgies D-M, Estevez JM, Bonetta DB, Urbanowicz BR, Ehrhardt DW, Somerville CR, Rose JKC, Hong M, DeBolt S (2012) Cellulose microfibril crystallinity is reduced by mutating C-terminal transmembrane region residues CESA1^{A903V} and CESA3^{T942I} of cellulose synthase. Proc Natl Acad Sci U S A 109:4098–4103

63. Wang T, Salazar A, Zabotina OA, Hong M (2014) Structure and dynamics of *Brachypodium* primary cell wall polysaccharides from two-dimensional ^{13}C solid-state nuclear magnetic resonance spectroscopy. Biochemistry 53:2840–2854

64. Dupree R, Simmons TJ, Mortimer JC, Patel D, Iuga D, Brown SP, Dupree P (2015) Probing the molecular architecture of *Arabidopsis thaliana* secondary cell walls using two- and three-dimensional ^{13}C solid state nuclear magnetic resonance spectroscopy. Biochemistry 54:2335–2345

65. Simmons TJ, Mortimer JC, Bernardinelli OD, Pöppler A-C, Brown SP, deAzevedo ER, Dupree R, Dupree P (2016) Folding of xylan onto cellulose fibrils in plant cell walls revealed

by solid-state NMR. Nat Commun 7:13902. https://doi.org/10.1038/ncomms13902

66. Komatsu T, Kikuchi J (2013) Selective signal detection in solid-state NMR using rotor-synchronized dipolar dephasing for the analysis of hemicellulose in lignocellulosic biomass. J Phys Chem Lett 4:2279–2283
67. Komatsu T, Kikuchi J (2013) Comprehensive signal assignment of ^{13}C-labelled lignocellulose using multidimensional solution state NMR and ^{13}C chemical shift comparison with solid state NMR. Anal Chem 85:8857–8865
68. Grommek A, Meier BH, Ernst M (2006) Distance information from proton-driven spin diffusion under MAS. Chem Phys Lett 427:404–409
69. Hill SJ (2010) Water in *Pinus radiata* wood secondary cell walls: an investigation using nuclear magnetic resonance and synchrotron x-ray diffraction. PhD thesis. Victoria University of Wellington, Wellington, New Zealand
70. White PB, Wang T, Park YB, Cosgrove DJ, Hong M (2014) Water-polysaccharide interactions in the primary cell wall of *Arabidopsis thaliana* from polarization transfer solid-state NMR. J Am Chem Soc 136:10399–10409
71. Wang T, Williams JK, Schmidt-Rohr K, Hong M (2015) Relaxation-compensated difference spin diffusion NMR for detecting ^{13}C-^{13}C long-range correlations in proteins and polysaccharides. J Biomol NMR 61:97–107
72. Wang T, Zabotina OA, Hong M (2012) Pectin-cellulose interactions in the *Arabidopsis* primary cell wall from two-dimensional magic-angle-spinning solid-state nuclear magnetic resonance. Biochemistry 51:9846–9856
73. Wang T, Phyo P, Hong M (2016) Multidimensional solid-state NMR spectroscopy of plant cell walls. Solid State Nucl Magn Reson 78:56–63
74. Li S, Zhang Y, Hong M (2010) 3D ^{13}C-^{13}C-^{13}C correlation NMR for *de novo* distance determination of solid proteins and application to a human α-defensin. J Magn Reson 202:203–210
75. Nishiyama Y (2016) Fast magic-angle sample spinning solid-state NMR at 60-100 kHz for natural abundance samples. Solid State Nucl Magn Reson 78:24–36
76. Bootten TJ, Harris PJ, Melton LD, Newman RH (2009) A solid-state ^{13}C-NMR study of a composite of tobacco xyloglucan and *Gluconacetobacter xylinus* cellulose: molecular interactions between the component polysaccharides. Biomacromolecules 10:2961–2967
77. Jelinski LW, Melchior MT (1996) High-resolution NMR of solids. In: Bruch MD (ed) NMR spectroscopy techniques, Practical spectroscopy series, vol 2. Marcel Dekker, New York, pp 417–486
78. Thimm JC, Burritt DJ, Ducker WA, Melton LD (2000) Celery (*Apium graveolens* L.) parenchyma cell walls examined by atomic force microscopy. Planta 212:25–32
79. Newman RH (2004) Carbon-13 NMR evidence for cocrystallization of cellulose as a mechanism for hornification of bleached kraft pulp. Cellulose 11:45–52

Chapter 13

High-Resolution Imaging of Cellulose Organization in Cell Walls by Field Emission Scanning Electron Microscopy

Yunzhen Zheng, Gang Ning, and Daniel J. Cosgrove

Abstract

Field emission scanning electron microscopy (FESEM) is a powerful tool for analyzing surface structures of biological and nonbiological samples. However, when it is used to study fine structures of nanometer-sized microfibrils of epidermal cell walls, one often encounters tremendous challenges to acquire clear and undistorted images because of two major issues: (1) Preparation of samples suitable for high resolution imaging; due to the delicateness of some plant materials, such as onion epidermal cell walls, many things can happen during sample processing, which subsequently result in damaged samples or introduce artifacts. (2) Difficulties to acquire clear images of samples which are electron-beam sensitive and prone to charging artifacts at magnifications over 100,000×. In this chapter we described detailed procedures for sample preparation and conditions for high-resolution FESEM imaging of onion epidermal cell walls. The methods can be readily adapted for other wall materials.

Key words High-resolution imaging, Scanning electron microscopy, Onion scales, Epidermal cell walls, Cellulose microfibrils

1 Introduction

Many microscopy techniques have been used to study the structures of plant cells and tissues to detect various features at different resolutions. Examples of optical microscopy that are useful tools to study plant cells and cell walls include: bright field with staining for specific components; differential interference contrast, phase contrast, darkfield, and polarized light microscopy for nonstained samples; and both widefield and confocal fluorescence microscopy for fluorescent samples. However, with a resolving power limited to ~200 nm due to diffraction of light [1, 2], these forms of microscopy cannot directly visualize cellulose microfibrils in the cell walls, which are approximately 3 nm in diameter [3]. In recent years a new optical microscopy called super-resolution microscopy, represented by numerous variations [4–7], exceeds the diffraction limit to improve resolution to values as small as 50 nm, which is still more

Zoë A. Popper (ed.), *The Plant Cell Wall: Methods and Protocols*, Methods in Molecular Biology, vol. 2149, https://doi.org/10.1007/978-1-0716-0621-6_13, © Springer Science+Business Media, LLC, part of Springer Nature 2020

than an order of magnitude too large for visualizing individual cellulose microfibrils. With its superb resolution, transmission electron microscopy (TEM) has been commonly used to study ultrastructure of plant cells and cells walls [8], but samples generally need special preparation and contrast-enhancement for high-resolution imaging.

Electron microscopy and atomic force microscopy (AFM) are two major methods used for high-resolution imaging of cell wall microfibrils, and quite often they complement each other to provide distinctive information of fine structure and cellulose organization [9–13]. Of various forms of electron microscopy, FESEM is often used today to study cellulose microfibrils because images obtained with it reveal surface texture and topography in a three-dimensional view [14–18]. FESEM has largely replaced an earlier technique based on TEM which made use of shadowed replicas of rapidly frozen, deep-etched cell walls [19–21].

While there are many reports making use of FESEM to study cell walls, major technical problems come from two aspects: sample preparation and microscopic imaging. Plant materials, especially the most recently deposited surfaces of epidermal cell walls, are very delicate. Samples can be damaged during various steps of the process, including dissection, fixation, washing, dehydration, drying, mounting, and sputter coating. Cell wall matrix polymers, such as pectin, can overlie and obscure the microfibrils. Therefore, extraction of these materials is often essential for obtaining clear microfibril images at high resolution. Furthermore, artifacts can be introduced during the process; in particular, drying can cause sample deformation, particularly when combined with pectin removal, and sputter coating can thicken the appearance of microfibrils and can obscure microfibril details; for example, two adjacent microfibrils can appear to be one. It can be a frustrating day when one spends hours on the microscope without being able to capture even a single clear image of microfibrils which are sensitive to electron beam damage at magnifications over 100,000×. While low voltage microscopy and environmental and variable pressure microscopy are often recommended as approaches to reduce charging and radiation damage [22–27], we found that they were unable to acquire clear high-resolution images of cellulose microfibrils in plant cell walls.

In this chapter we summarize the methods that we have developed in our lab for such imaging, with step-by-step procedures and some tricks and tips.

2 Materials

Use analytical grade reagents and be particularly scrupulous to maintain cleanliness and avoid contamination of solutions and samples with dust or other particulates that can ruin the imaging of cell wall samples.

2.1 Preparation of Epidermal Cell Wall Strips

1. Plant material: fresh white onions (*Allium cepa* L.) from local grocery store.
2. HEPES buffer: 20 mM (4-(2-hydroxyethyl)-1-piperazineethanesulfonic acid adjusted to pH 6.8 with 1 M NaOH, with 0.1% Tween 20.
3. TRIS buffer: 20 mM Tris (Tris(hydroxymethyl)aminomethane), adjusted to pH 8.5.
4. PL: Pectate lyase (Megazyme, Cat#E-PCLYCJ), 10 μg/mL in TRIS buffer with 2 mM $CaCl_2$ and 2 mM NaN_3.
5. Ethanol: 100%.
6. Ultrapure water.
7. Lab shaker.
8. Fine-tip forceps such as Dumont Biological #5, for handling onion scales (*see* Fig. 1 and **Note 1**).
9. Plastic fine-tip pipette (Transfer Pipette Extended Fine Tip, Small Bulb, SKU GSTP050 from www.pipettes.com).

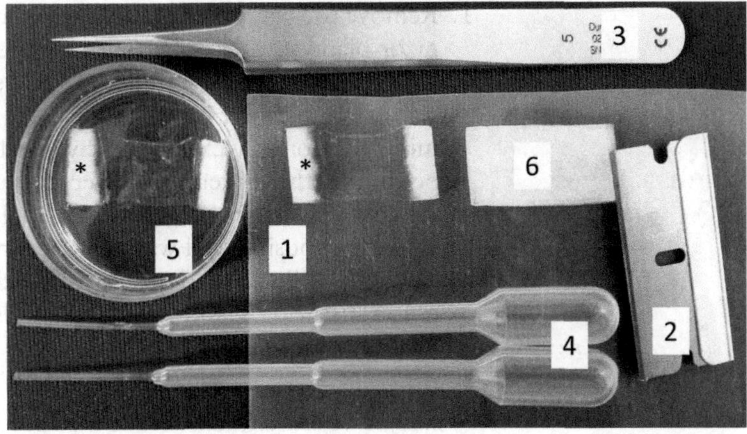

Fig. 1 Tools for handling onion samples. (#1) a Parafilm® sheet on cutting board, (#2) razor blade, (#3) fine-tip tweezers (Dumont Biological #5), (#4) fine-tip plastic transfer pipettes for transferring onion samples, (#5) Petri dish. An onion scale (#6) and onion epidermal strip samples (∗) are also shown

2.2 Materials for FESEM Sample Preparation

1. Automatic critical point dryer (e.g., Leica EM CPD300).
2. Razor blade: Personna GEM, single-edge, 0.009″/0.23 mm (Ted Pella Cat# 62–0178, Redding, CA).
3. Petri dish (35 × 10 mm) (Fisher, Cat. No. FB0875711YZ).
4. Sample cassette for Paraffin® embedding (Histosette I, Simport Prod # 27158-5B, Beloeil, QC).
5. Sponge pad (Ted Pella, Product No. 27154-1, Redding, CA).
6. Iridium sputter coater (Emitech K575X, Fall River, MA).
7. Round double-sided carbon tape (diameter: 12 mm) (Electron Microscopy Sciences, Cat. # 77824-12, Hatfield, PA).
8. Specimen Mount: Single pin SEM stub, aluminum, 1/2″ slotted head, 1/8″ Pin (Ted Pella, Product No. 16111, Redding, CA).

2.3 Image Acquisition

1. FESEM: Zeiss Sigma FESEM (Carl Zeiss, Germany).
2. Sample holder: Zeiss 9-hole standard holder.

3 Methods

3.1 Plant Material

3.1.1 Onion Epidermis Cell Walls Strips

A white onion of ~10 cm in diameter consists of about 12 concentric scales. We use the middle scales because they are relatively easy to peel off and handle compared with other layers which are either too thin and soft or too tough [3]. The onion bulbs should be kept at 4 °C in a refrigerator and taken out immediately before use.

The following steps are all carried out at room temperature.

1. Remove the dry, papery outer scales from the onion. Use a new razor blade to excise the onion in the vertical axis and take a slice of ~1 cm wide from the fifth scale; then take a thin, transparent strip of epidermal cell wall from the abaxial epidermis (outer or convex surface) by gently breaking the scale and peeling off the epidermis. The epidermal cells split open, resulting in a large sheet of outer epidermal cell walls with the most recently deposited cell wall surface exposed for imaging. *Note*: leave the onion epidermal strip attached to a small piece of onion tissue (~3 mm) at both ends to facilitate handling (Fig. 2).
2. With fine forceps hold the onion strip at one end of the peeled sample to lift and transfer it into HEPES buffer in a 35 mm Petri dish. Allow the sample to float in the buffer with inner (cell-facing) surface downward and immersed in the buffer.

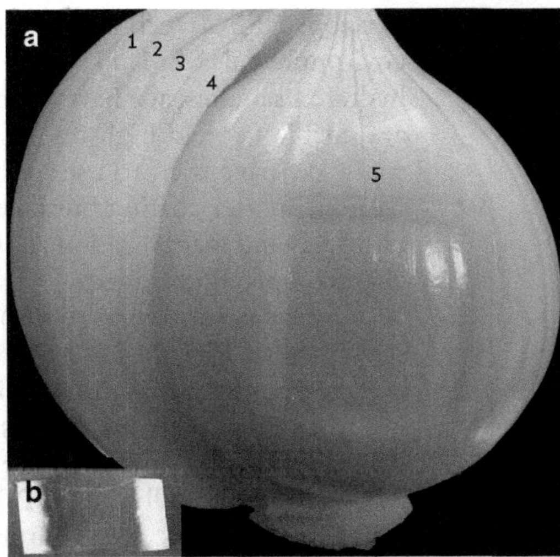

Fig. 2 (**a**) An onion is dissected to take samples from the fifth layer. (**b**) A freshly dissected onion epidermal strip sample. *Note* a piece of onion tissue is left attached at both ends for handling of the sample

3.1.2 Pectin Extraction and Wall Fixation/Dehydration

Matrix polysaccharides such as pectin and hemicellulose normally fill in the space between microfibrils within the cell walls [28, 29]; cellular debris and residues of plasma membrane may also remain on the surface of the cell wall sample, obscuring cellulose microfibrils. To fully expose the microfibrils for imaging at high resolution, the samples need to be thoroughly washed and pectin usually needs to be partially extracted prior to dehydration.

1. Move the Petri dish containing samples onto a lab shaker. Incubate in HEPES buffer with gently horizontal movement (~1 cycle/s) for 2 h to remove plasma membrane and cytoplasmic residues. After the incubation, remove the buffer from the dish and wash the samples with ultrapure water five times, 3 min each.
2. Transfer samples very carefully with two fine-tip plastic transfer pipettes to a clean Petri dish. Float samples in the PL solution for 40 h at room temperature to remove pectin (*see* **Note 2**). Wrap the Petri dish with Parafilm to prevent evaporation and protect from dust and dirt.
3. Wash by floating the samples in TRIS buffer three times, 5 min each, and then in ultrapure water five times, 3 min each.
4. Replace solution with 100% ethanol for one-step fixation/dehydration for 30 min (*see* **Note 3**).

3.2 Sample Drying, Mounting, and Sputter Coating

Biological samples must be dried and sputter-coated prior to FESEM imaging because these samples are electrically nonconductive and are prone to damage by high electron beam bombardment inside the SEM chamber under high vacuum. Although recently developed SEM techniques such as variable pressure SEM and environmental SEM [13–20] allow for direct observation of hydrated samples, we have found that these microscopy methods are unable to yield high-resolution images of microfibrils and cell walls. We have tried to use fresh and partially dehydrated samples without sputter coating but failed to obtain clear images of microfibrils beyond magnification at 1000× before the samples were fully charged up and damaged.

Critical point drying (CPD) is the method used for drying the specimen for FESEM [24]. Air drying and related methods create large surface tensions at the liquid–gaseous phase boundaries, leading to severe distortion, collapse, and shrinkage of delicate specimens; thus these methods are not suitable for high-resolution imaging of plant cell walls. CPD makes use of the critical point of a fluid at a specific temperature and pressure that avoids phase boundaries and circumvents distortions by surface tension as the fluid is removed [30].

After CPD, samples are mounted on a specimen stab, usually made of aluminum, with the surface of interest facing up and then they are sputter coated to deposit a thin layer of metal to the specimen surface to enhance electrical conductivity of the surface.

3.2.1 Critical Point Drying

1. Use two fine-tip transfer pipette to move samples onto a histocassette overlaid with a sponge pad (each cassette holds several pieces of tissue) (Fig. 3). Close and immerse the histocassette in 100% ethanol.

Fig. 3 An onion epidermal strip (∗) placed on blue sponge pad inside a histocassette immersed in ethanol

2. Transfer the histocassette into the CPD chamber with fresh 100% ethanol, completely covering the sample cassettes (*see* **Notes 4** and **5**). Use an Automated Critical Point Dryer system (*Leica EM CPD300* in our case) for slow and gentle cycles (*see* **Notes 6** and 7). This process usually takes 3–5 h to complete.

3.2.2 Sample Mounting

Carefully stick a double-sided carbon tape to a SEM stub; remove the protecting paper from the tape. Avoid introducing any wrinkles when doing this. Cut a piece of 3 mm × 3 mm onion tissue from the center portion of the sample; with a pair of fine forceps, hold the sample sheet to allow one corner of the sheet to land and stick on the carbon tape, and then slowly spread the entire sheet over the carbon tape; make sure cuticle side is downward (Fig. 4).

3.2.3 Sputter Coating

Sputter-coat samples with iridium (*see* **Note 8**) on a rotation stage for 3 s, with Argon gas tank pressure = 7 psi and sputter current = 40 mA.

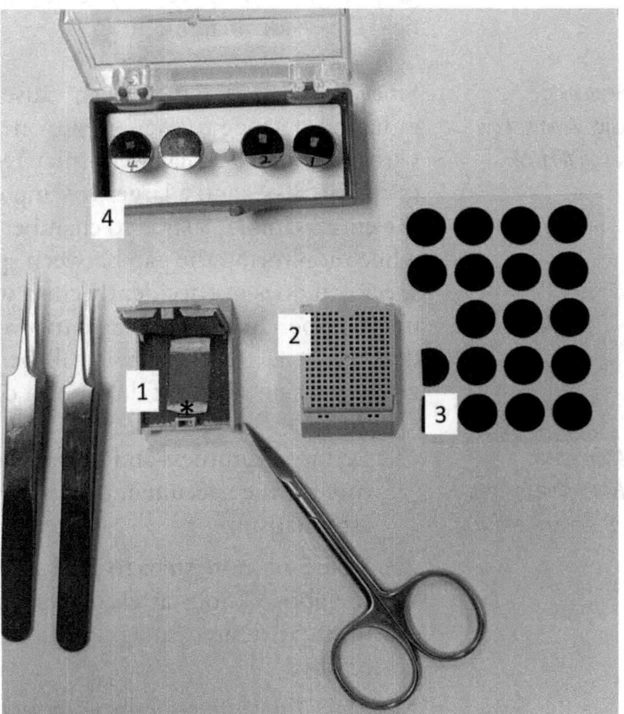

Fig. 4 Mounting onion tissue after CPD. (#1) Onion tissue (∗) laying on a blue sponge pad inside a histocassette after CPD. (#2) A histocassette. (#3) double-sided tapes for adhesive mounting of the sample. (#4) Samples (white squares in the center of black carbon tape) mounted on stubs in a storage box

3.3 Image Acquisition by FESEM

To obtain clear images of nm-scale cellulose microfibrils, one has to use an FESEM because a SEM with a tungsten filament cannot obtain the resolution needed for imaging individual microfibrils at magnifications over 100,000×. In our case, we used a *Zeiss Sigma* FESEM which is equipped with four detectors, that is, secondary electron (Everhart-Thornley) detector (SE2), in-lens detector, variable pressure detector, and backscattered electron detector. Only the in-lens detector is suited to high-resolution imaging. The in-lens detector is a high efficiency detector for high-resolution secondary electron imaging. It is located above the objective lens and detects directly in the beam path. The efficiency of the in-lens detector is mainly determined by the electric field of the electrostatic lens, which decreases exponentially with the distance [31]. Thus, the working distance (WD) is one of the most important factors affecting the signal-to-noise ratio of the images. The conditions to obtain reproducible images are as follows: working distant = 2–4 mm, voltage = 10 kV or above, aperture size = 30 μm (standard aperture), resolution = 1024 × 768, operation mode = normal, beam current = 80.0 μA, noise reduction = Pixel Avg for focusing sample; noise reduction = Line Int for photographing. Scan Speed = 7–8 for focusing (with a reduced view field), 6–7 for imaging.

3.3.1 Searching for a "Good" Area at Low Magnification (Fig. 5)

Since the in-lens detector can cause more specimen charging and radiation damage to the sample, one should start searching images with SE2 (The Everhart-Thornley Detector) at a low magnification (e.g., 20–30×) and a large working distance of 20–30 mm to view the entire sample. Switch to chamber scope to slowly raise the stage, while monitoring the gap between specimen and column to ensure specimen does not accidently crash with column cap to damage the microscope. Bring the stage to ~3.5 mm and switch to in-lens detector (the magnification will be about 50×). Find an area of interest which is free of debris and damage.

3.3.2 Acquisition of High-Resolution Images (See Fig. 6, Notes 9 and 10)

1. Reduce brightness and adjust contrast, focus at a specific structure on the specimen, center aperture by wobbler and correct astigmatism.
2. Take 5 or more steps to gradually raise magnification and adjust the above knobs at each step with scan speed at 4 or 5 (fast) until a desired magnification, for instance 150,000×, is reached.
3. Make final adjustments with focus, stigmator, wobbler, brightness, and contrast, quickly move the stage to a nearby area of specimen that has not been exposed to the beam, press to take photograph with moderate scan speed (Speed 6 or 7 on Zeiss Sigma which has 15 speed with 1 the fastest and 15 the slowest).

Fig. 5 A low magnification (30×) image is obtained with SE2 (The Everhart-Thornley Detector) and a large working distance (22.5 mm) at the beginning of the microscope session to view the entire sample and search areas for high magnification imaging

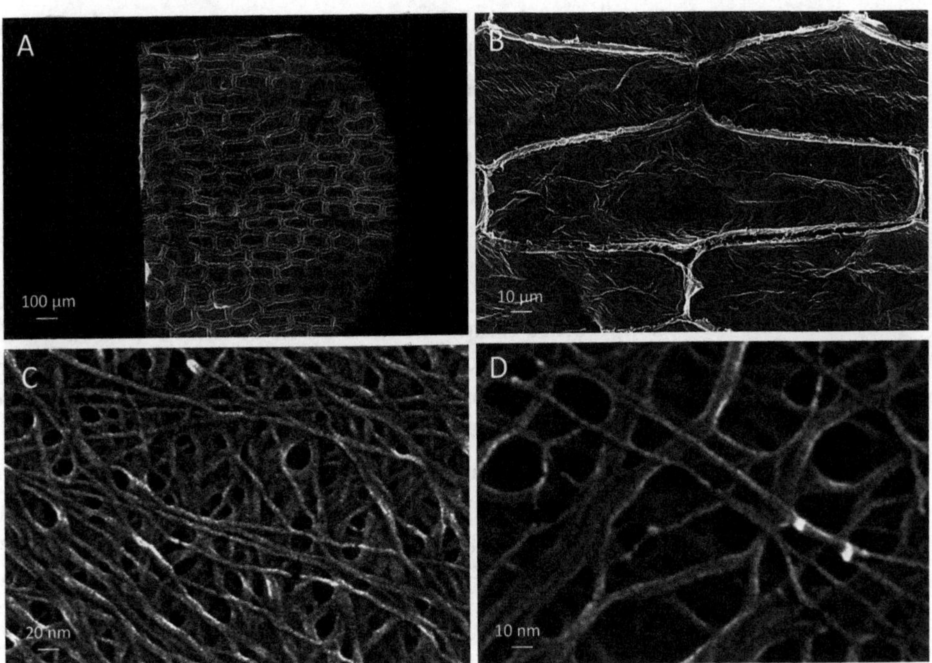

Fig. 6 Representative images taken with in-lens detector showing from low magnification (**a** at 50× and **b** at 500×) to ultrahigh magnification (**c** at 200,000× and **d** at 400,000×). The detailed organization of cellulose microfibrils in single microfibril and bundled forms is clearly seen

Final comments: In this chapter we used onion epidermal cell walls as a model to describe methods for high-resolution FESEM imaging of cellulose microfibrils. The method described here may not necessarily be applicable to high-resolution FESEM imaging of every type of cell wall tissue because these procedures are very much tissue-specific and equipment dependent. However, the methods described here can be used as a starting point for an experimentalist to obtain quality images in a relatively short time. Our imaging method starts with careful handling of tissue, followed by extraction of pectins and thoroughly washing off debris, one-step fixation/dehydration in pure ethanol, critical point drying in a programmed CPD at slow steps, short time sputter coating with iridium, and finally imaging at a moderate HV (10 kV) with in-lens secondary electrons.

4 Notes

1. Onion scale handling tools: we use fine-tip tweezers and plastic transfer pipettes (*see* picture in Fig. 1) to transfer and manipulate the onion skin because these plastic fine tips are flexible so that they can be bent in solutions to allow cell wall pieces to float away from the tips without damaging and trapping them during transfer.
2. Extraction: It is often important to extract pectin to expose cellulose microfibrils for clear images at high resolution. A two-step treatment, that is, EDTA followed by PL is commonly used but we found using PL alone can generate better results.
3. Samples chemically fixed with glutaraldehyde before dehydration were also tested, but the microfibrils looked distorted and flattened. We also used high pressure freezing to cryo-fix the samples followed by freeze substitution in an automatic freeze-substitution system but the results were not better than room temperature treatment with 100% ethanol followed by CPD. *Note* that because of the thinness and polysaccharide nature of the cell wall, standard methods of cell fixation and gradual dehydration in an ethanol series are not necessary.
4. Histocassettes and sponge pad for paraffin histology are commonly used in histology labs to restrain sample movement during the CPD process and to protect samples from mechanical damage while allowing exchange of liquids. We found the standard tissue basket and CPD holders are not suitable because cell wall samples are either lost or damaged after CPD.

5. During fixation/dehydration and prior to CPD, always keep samples immersed in pure ethanol; never allow samples to dry.
6. Make sure there is enough carbon dioxide (dehydrant) for the process by checking the pressure of the CO_2 tank connected to the CPD system. The following settings were used in our practice: Cooling Temperature: 15 °C; Heating Temperature: 40 °C; CO_2 In Step: Slow; Exchange Step: Speed 3 (medium) and 19 Cycles: Gas Out Step: Slow 90%.
7. It is important to use an automated CPD system for drying the samples because it allows for a gentle drying process and well controlled pressure change, while pressure is usually difficult to control with a manual system, especially for the last step (outgassing). We also tried manual CPD (Bal-Tek CPD030) to process samples but obtained wrinkled and deformed samples which were not suitable for high-resolution imaging.
8. We chose iridium over gold-palladium for sputter coating because coating with iridium generates a thinner coating layer to avoid artifacts in high-resolution imaging, in addition to providing better electrical conductivity over gold-palladium to reduce charging during image acquisition. The thickness of the coating is approx. 1.2 nm (estimated by equation $T = ma \times C$ (kV) (min) where $T =$ thickness (nm), $ma =$ current in mA, kV = voltage, min = time in minute, $C =$ constant for Argon gas. In our condition of sputtering, $T = 40$ mA (0.16 kV) (3/60 min) (3.6) = 1.2 nm). Wear gloves and use forceps when handling the mounted specimens.
9. Because of high sensitivity of these cell wall samples to electron beam damage, acquisition of high-resolution images during microscope operation is technically demanding, requiring the microscope operator to be very familiar with the microscope and to act quickly during image capture. The scanning speed usually is set at highest to search the field and at moderate speed to obtain clear pictures. Beam damage can be produced when relatively slow scan speeds are used for imaging (Fig. 7).
10. Facility room conditions are very important for high-resolution FESEM imaging. In our case, the Zeiss Sigma FESEM is placed in a "quiet area" specially designed for housing acoustically sensitive equipment in the basement of the building, where the resolution of the microscope is 1.14 nm, superior to the manufacturer's specifications of 1.2 nm for this instrument. When imaging is in progress, movement or traffic around the equipment is avoided to minimize noise; even loud music or speech can cause slight blurring of the images.

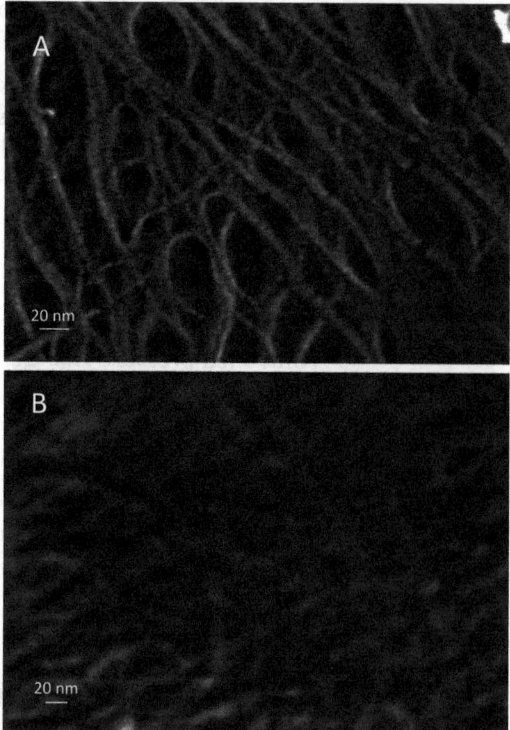

Fig. 7 Radiation damage can happen during imaging. In (**a**) the lower right corner of the picture shows a fuzzy area without microfibril detail. This is an area damaged by exposure to the beam before the image was taken at 300,000×. Image (**b**) shows a sample that was damaged by a single scan at slow scan speed. The image needs to be captured in an area of the sample that was not previously exposed to the beam; the adjacent area used for focusing cannot be used for detailed images

Acknowledgments

This work was supported as part of the Center for Lignocellulose Structure and Formation, an Energy Frontier Research Center funded by the US Department of Energy, Office of Science, Basic Energy Sciences under Award # DE-SC0001090.

References

1. Neice A (2010) Chapter 3—Methods and limitations of subwavelength imaging. In: Peter WH (ed) Advances in imaging and electron physics. Elsevier, Amsterdam, pp 117–140
2. Kuznetsova Y, Neumann A, Brueck SRJ (2007) Imaging interferometric microscopy–approaching the linear systems limits of optical resolution. Opt Express 15:6651–6663
3. Zhang T, Zheng Y, Cosgrove DJ (2016) Spatial organization of cellulose microfibrils and matrix polysaccharides in primary plant cell walls as imaged by multichannel atomic force microscopy. Plant J 85:179–192
4. Hell SW, Wichmann J (1994) Breaking the diffraction resolution limit by stimulated emission: stimulated-emission-depletion fluorescence microscopy. Opt Lett 19:780–782
5. Huang B, Bates M, Zhuang X (2009) Super-resolution fluorescence microscopy. Annu Rev Biochem 78:993–1016

6. Huang B, Babcock H, Zhuang X (2010) Breaking the diffraction barrier: super-resolution imaging of cells. Cell 143:1047–1058
7. Betzig E, Patterson GH, Sougrat R, Lindwasser OW, Olenych S, Bonifacino JS, Davidson MW, Lippincott-Schwartz J, Hess HF (2006) Imaging intracellular fluorescent proteins at nanometer resolution. Science 313:1642–1645
8. Fowke LC (1995) Transmission and scanning electron microscopy for plant protoplasts, cultured cells and tissues. In: Gamborg OL, Phillips GC (eds) Plant cell, tissue and organ culture: fundamental methods. Springer Berlin Heidelberg, Berlin, Heidelberg, pp 229–238
9. Zhang T, Mahgsoudy-Louyeh S, Tittmann B, Cosgrove DJ (2014) Visualization of the nanoscale pattern of recently-deposited cellulose microfibrils and matrix materials in never-dried primary walls of the onion epidermis. Cellulose 21:853–862
10. Marga F, Grandbois M, Cosgrove DJ, Baskin TI (2005) Cell wall extension results in the coordinate separation of parallel microfibrils: evidence from scanning electron microscopy and atomic force microscopy. Plant J 43:181–190
11. Baskin TI, Marga F, Grandbois M (2005) A comparison of atomic force microscopy and field-emission scanning electron microscopy for imaging the plant cell wall. Microsc Microanal 11:1130–1131
12. Ding S-Y, Zhao S, Zeng Y (2014) Size, shape, and arrangement of native cellulose fibrils in maize cell walls. Cellulose 21:863–871
13. Thimm JC, Burritt DJ, Ducker WA, Melton LD (2009) Pectins influence microfibril aggregation in celery cell walls: an atomic force microscopy study. J Struct Biol 168:337–344
14. Fujita M, Wasteneys GO (2014) A survey of cellulose microfibril patterns in dividing, expanding, and differentiating cells of *Arabidopsis thaliana*. Protoplasma 251:687–698
15. Zhu C, Ganguly A, Baskin TI, McClosky DD, Anderson CT, Foster C, Meunier KA, Okamoto R, Berg H, Dixit R (2015) The fragile fiber1 kinesin contributes to cortical microtubule-mediated trafficking of cell wall components. Plant Physiol 167:780–792
16. Xiao C, Zhang T, Zheng Y, Cosgrove DJ, Anderson CT (2016) Xyloglucan deficiency disrupts microtubule stability and cellulose biosynthesis in arabidopsis, altering cell growth and morphogenesis. Plant Physiol 170:234–249
17. Carpita NC, Defernez M, Findlay K, Wells B, Shoue DA, Catchpole G, Wilson RH, McCann MC (2001) Cell wall architecture of the elongating maize coleoptile. Plant Physiol 127:551–565
18. Domozych DS, Sorensen I, Popper ZA, Ochs J, Andreas A, Fangel JU, Pielach A, Sacks C, Brechka H, Ruisi-Besares P, Willats WG, Rose JK (2014) Pectin metabolism and assembly in the cell wall of the charophyte green alga *Penium margaritaceum*. Plant Physiol 165:105–118
19. McCann MC, Wells B, Roberts K (1990) Direct visualization of cross-links in the primary plant cell wall. J Cell Sci 96:323–334
20. Emons AMC (1988) Methods for visualizing cell wall texture. Acta Bot Neerl 37:31–38
21. Goodenough UW, Heuser JE (1985) The chlamydomonas cell-wall and its constituent glycoproteins analyzed by the quick-freeze, deep-etch technique. J Cell Biol 101:1550–1568
22. Liu J (2000) High-resolution and low-voltage fe-sem imaging and microanalysis in materials characterization. Mater Charact 44:353–363
23. Pawley JB, Schatten H (2008) Biological low-voltage scanning electron microscopy. Springer Verlag, New York, Berlin
24. Donald AM (2003) The use of environmental scanning electron microscopy for imaging wet and insulating materials. Nat Mater 2:511–516
25. Muscariello L, Rosso F, Marino G, Giordano A, Barbarisi M, Cafiero G, Barbarisi A (2005) A critical overview of esem applications in the biological field. J Cell Physiol 205:328–334
26. Griffin BJ (2007) Variable pressure and environmental scanning electron microscopy: imaging of biological samples. Methods Mol Biol 369:467–495
27. Kirk SE, Skepper JN, Donald AM (2009) Application of environmental scanning electron microscopy to determine biological surface structure. J Microsc 233:205–224
28. Cosgrove DJ (2014) Re-constructing our models of cellulose and primary cell wall assembly. Curr Opin Plant Biol 22C:122–131
29. Cosgrove DJ (2005) Growth of the plant cell wall. Nat Rev Mol Cell Biol 6:850–861
30. Bray D (2000) Critical point drying of biological specimens for scanning electron microscopy. In: Williams JR, Clifford AA (eds) Supercritical fluid methods and protocols. Humana Press, Totowa, NJ, pp 235–243
31. Griffin BJ (2011) A comparison of conventional Everhart-Thornley style and in-lens secondary electron detectors: a further variable in scanning electron microscopy. Scanning 33:162–173

Chapter 14

Analyzing Plant Cell Wall Ultrastructure by Scanning Near-Field Optical Microscopy (SNOM)

Tobias Keplinger and Ingo Burgert

Abstract

The importance of lignocellulosic materials in various research areas including for example biorefinery processes and new biomaterials is constantly rising. Therefore, a detailed knowledge on the macromolecular assembly of plant cell walls is needed. However, despite the tremendous progress in the structural and chemical analysis of plant cell walls there are still uncertainties concerning the ultrastructure. This is either due to Rayleigh criterion based limitations of the analytical methods or the relatively low chemical information gained with high spatial resolution techniques. In this chapter scanning near field optical microscopy (SNOM) is presented as a powerful tool for the subdiffraction limited chemical and structural characterization of plant cell walls.

Key words Scanning near field optical microscopy (SNOM), Plant secondary cell walls, Subdiffraction limited, Lignification

1 Introduction

The nanostructural and chemical analysis of plant cell walls is of great relevance for a better understanding of the spatial organization and the specific interactions of cell wall constituents. This has impact on cell growth processes, mechanical properties or degradability of cell walls for instance for biofuel production [1, 2]. In particular, genetic or chemical modifications result in changes in cell wall structure and composition, which require in-depth analysis, in order to evaluate their success, efficiency, and consequences [3, 4]. For the cell wall characterization various techniques exist, which allow for extracting valuable information. However, all of them face certain limitations. High-resolution electron microscopy techniques enable to study structural features but do not provide information on chemical composition, or the chemical composition is even impeding a detailed structural study (cellulose fibrils masked by lignin) [5]. The chemical composition can be investigated by spectroscopic methods such as fluorescence- or Raman

Zoë A. Popper (ed.), *The Plant Cell Wall: Methods and Protocols*, Methods in Molecular Biology, vol. 2149, https://doi.org/10.1007/978-1-0716-0621-6_14, © Springer Science+Business Media, LLC, part of Springer Nature 2020

spectroscopy in great detail while retaining the structure, but these techniques are limited by the so-called diffraction limit [6–9].

Plasmonic effects as in tip enhanced Raman spectroscopy (TERS) or surface enhanced Raman spectroscopy (SERS) can be utilized to circumvent this limit, but so far these techniques are highly challenging for the analysis of plant cell walls [10, 11]. Another option to overcome the diffraction limit for in-depth structural analysis is to bring an aperture in very close vicinity to the surface of the plant cell wall for performing near field optical microscopy (SNOM) measurements.

The basic considerations on near-field microscopy date back to 1928, when it was suggested to generate microscopic images of very high resolution, by illuminating an opaque plate with a sub-wavelength big aperture and collect and optical signal through that aperture [12]. However, due to technical limitations regarding signal collection efficiency and illumination intensity as well as issues in adjusting the appropriate distance of the aperture very close to the surface it lasted until 1984 for the first near field measurements to be conducted [13, 14]. Since then SNOM was used for the characterization of various biological samples such as proteins, lipids, polysaccharides, DNA, and entire cells [15, 16]; however, there are only a few reports which deal with the analysis of intact plant cell walls yet [17].

In the following, the basic measurement principle will be explained in more detail. In the material and methods parts a specific focus will be laid on the sample preparation, followed by a step-by-step guide of the measurement procedure. Selected images from a study on different plant species will be shown, that illustrate the structural and chemical information which can be extracted. Finally a critical evaluation of the influencing factors and the accordant interpretation of the obtained data will be provided which emphasizes the importance of control measurements on model systems.

2 Measurement Principle

After the first considerations on circumventing the Rayleigh criterion in microscopy by E. H. Synge in 1927/1928 and the first proof of principle with microwaves by E. A. Ash et al. in 1972 it took until 1984 that the first works, by two independent groups at IBM Zurich and the Cornell university, on the practical applications of scanning near field optical microscopy were published [12–14, 18]. This is explicable by the breakthrough of scanning probe microscopy in the 1980s that helped to deliver the needed prerequisite of bringing an aperture very close to the surface without touching it and also the possibility of scanning with this aperture

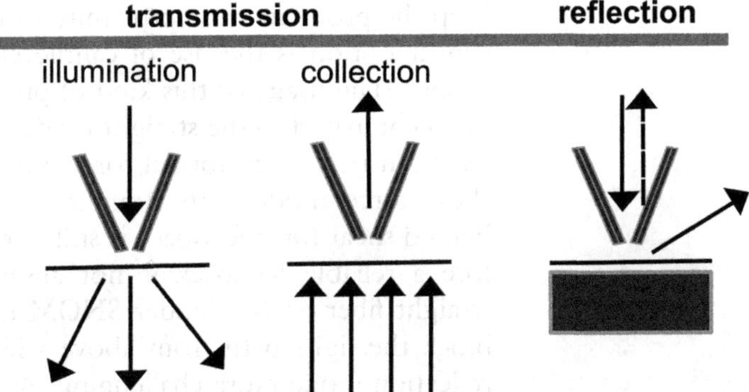

Fig. 1 Measurement modes in aperture based near field scanning optical microscopy (**a**) transmission (illumination or collection) and (**b**) reflection mode

across the sample. In close vicinity of the aperture to the sample the evanescent waves present in the near field can be used and consequently a sub-diffraction limited resolution is achieved, before the diffraction pattern in the far field of classical microscopy come into play [15, 19, 20].

Different measurement configurations are possible in SNOM and depending on the properties of the sample and the measurement environment needed they can be divided into the following three main options (Fig. 1):

1. Illumination mode imaging (transmission): in illumination mode the sample is illuminated through a point light source and the signal is collected from beneath of the sample.
2. Collection mode imaging (transmission): in collection mode the sample is illuminated from the far- field and the signal is collected through the NSOM probe.
3. Reflection mode imaging: for opaque samples the sample is illuminated through the NSOM probe and the signal is collected in the far field or again through the aperture.

Besides the measurement configuration described before, there are two main parts, closely interconnected to each other, that differ substantially between the various types of NSOM instruments. On the one hand they can be distinguished by the type of probes that are used and on the other hand by the feedback mechanism. Both of them are shortly described in the following paragraph. For a more detailed description the reader is referred to the literature [15, 20, 21].

The most common probes used in SNOM are straight, metal coated glass fiber probes that are mounted on one tine of a tuning fork. The tuning fork is modulated and if the fork with the attached probe approaches the sample surface the amplitude and the phase are affected and therefore can be used as a feedback mechanism to keep the probe in close proximity to the sample surface. Another similar option is the use of cantilevered metal coated glass fiber probes. The usage of this kind of probes possesses various advantages compared to the straight model. With these probes the feedback can be run in normal force mode in contrast to the needed shear force mode with straight fibers probes. The mechanism behind shear force feedback is still not fully understood and therefore a reliable feedback is not always guaranteed. Additionally straight fiber probes hinder SNOM measurements in liquids and block the light path from above which makes measurements in reflection mode more challenging. As a third option silicon based AFM cantilevers, known from beam-bounce AFM technology, are used. To implement a subwavelength aperture into the cantilever different techniques including for example microfabrication approaches or forming the aperture with a focused ion beam (FIB) are possible [15, 16].

The achievable spatial resolution of NSOM measurement is mainly dependent on the size of the aperture. However, it is also important to note that it is not feasible to fabricate infinitely small apertures due to fabrication limitations and the fact that the light throughput dramatically decreases as the aperture becomes smaller [20].

3 Materials

1. Plant samples from different plant organs can be used; number of replicates depends on the research question.
2. AFM stainless steel disks.
3. Instant glue.
4. Rotary or sliding microtome.
5. Ultramicrotome.
6. Diamond knife.
7. Disposable microtome blades.
8. Brushes.
9. Razor blades.
10. Tweezers.
11. Optical microscope.
12. NSOM microscope (e.g., Nanonics MultiView 2000).
13. Software (e.g., WSXM) for data treatment.

4 Methods

4.1 Sample Preparation

1. Cut plant samples with the dimensions of ca. $5 \times 5 \times 5$ mm^3.
2. Prepare a conical shape of the wood cube in order to decrease the area to be cut with the diamond knife (area size ca. 2×2 mm^2).
3. Glue the wood cube onto an AFM stainless steel disk with instant glue.
4. Fixation of the sample in the sample holder of a rotary microtome; as an alternative to a rotary microtome also a sliding microtome is suitable.
5. Start cutting sections with a thickness of ca. 5 μm until whole sections are obtained and then gradually decrease the thickness to about 1 μm (step size ca. 1 μm).
6. Put the precut sample cube into the sample holder of an ultramicrotome equipped with a diamond knife; start cutting section with a thickness of ca. 250 nm and gradually decrease the thickness until 60 nm are reached.
7. After cutting several sections with the lowest thickness check the quality of the cut block surface with a light microscope at high magnification.
8. If the surface is free of visible cutting artefacts the sample is ready for SNOM measurements; if not redo the ultramicrotome cutting step.

Depending on the sample properties (e.g., very soft plant materials) alternatives to the above described preparation procedure including for example cryosectioning or embedding of the samples in epoxy or PEG before cutting might be needed. For a detailed description of the procedures the reader is referred to a recently published protocol by Gierlinger et al. [6].

4.2 Conducting SNOM Measurements

The following step-by-step guide for performing SNOM measurements is based on a standard procedure for a MultiView2000 instrument (Nanonics Imaging, Jerusalem, Israel) in reflection configuration and an Avalanche photodiode as a detector (potential other detectors include PMTs or CCD). A scheme of the measurement setup is shown in Fig. 2.

Step-by-Step Guide

1. Turn on the instrument and wait for about 30 min for a stable performance of the instrument.
2. Mount a glass fiber probe with an aperture size between 50 and 150 nm with a typical resonant frequency of the tuning fork of around 35 kHz.

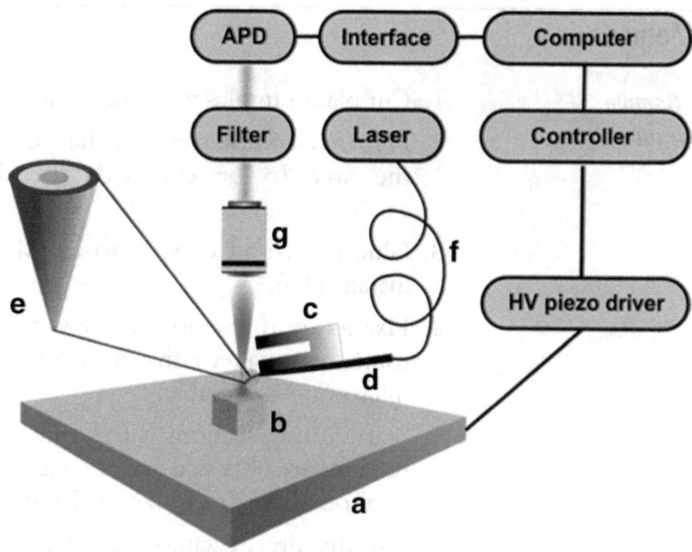

Fig. 2 Scheme of the SNOM-setup showing the arrangement of the main components. The sample (**a**) is placed on a piezo-scanning stage (**b**) and scanned with a tuning fork SNOM probe (**c**) through the SNOM tip zoom showing the inner Au/Cr coating the laser light coupled out of an optical fiber is coming in contact with the sample and reflected back through an objective and filter before the signal is detected with an APD [17]

3. Ensure a stable connection of the electronic contacts between the tuning fork and the probe holder for a stable feedback and fix the glass fiber with a tape on the scanner;
4. Check the resonance frequency of the tuning fork to guarantee a proper mounting of the probe.
5. Put the end of the glass fiber through the fiber holder and afterwards strip the fiber end; clean the stripped part with methanol and freshly cut the fiber with a fiber cutting tool;
6. Turn on the laser and the power supply for the detector and check if the laser shutter is closed.
7. Place the sample into the AFM; be aware that NSOM probes are longer than conventional AFM probes and therefore enough distance must be kept between the probe and the sample.
8. Approach the tuning fork by manually turning of the distance screw until the sample surface is 2 mm away from the probe tip.
9. Mount the enclosure.
10. Start with the tuning fork calibration procedure—determination of the resonant frequency and the setpoint.
11. Approach the probe to the surface with the stepper motor until feedback is achieved.

12. Center the NSOM probe exactly in the cross hair of the microscope, by adjusting the position of the probe with the piezo scanners.
13. Slightly retract the probe.
14. Reset the detector and open the detector shutter at the microscope.
15. Start scanning without being in contact over a flat sample surface area.
16. Open laser shutter.
17. Center the laser focus spot in the glass fiber (distance to the end app. 1.5–2 mm) in order to achieve an illumination of the entire glass fiber.
18. Adjust the detector position to the maximum detector signal; the minimum target value after alignment is depending on the size of the aperture.
19. Optimize the detector signal by repeating the procedure a couple of times.
20. When the maximum signal is achieved close the laser shutter and stop scanning.
21. Close the detector shutter and check the position of your probe in light microscope prior to approaching the probe.
22. Approach the probe to surface and start scanning and adjust the z-position of the microscope until maximal signal at the detector is achieved.
23. After adjusting restart the scan.
24. After the scan immediately close the laser shutter to avoid sample damage.

5 Images of Selected Examples

Secondary cell walls of wood species are characterized by a sandwich like structure consisting of three layers with a dominant middle layer (S2). These layers can be distinguished based on the different orientation of the parallel-aligned cellulose fibrils (microfibril angles (MFA)). In contrast for example bamboo exhibits a multilayered structure of alternating thick and thin layers with differences in the chemistry and the MFA between the layers. These structure are well understood and have been intensively investigated over the past years [22, 23]. Nevertheless there still exist uncertainties concerning the spatial arrangement of the constituents within the cell wall. In this respect several cell wall models have been proposed dealing with the distribution of the macromolecules within the cell wall. Kerr and Goring proposed a concentric

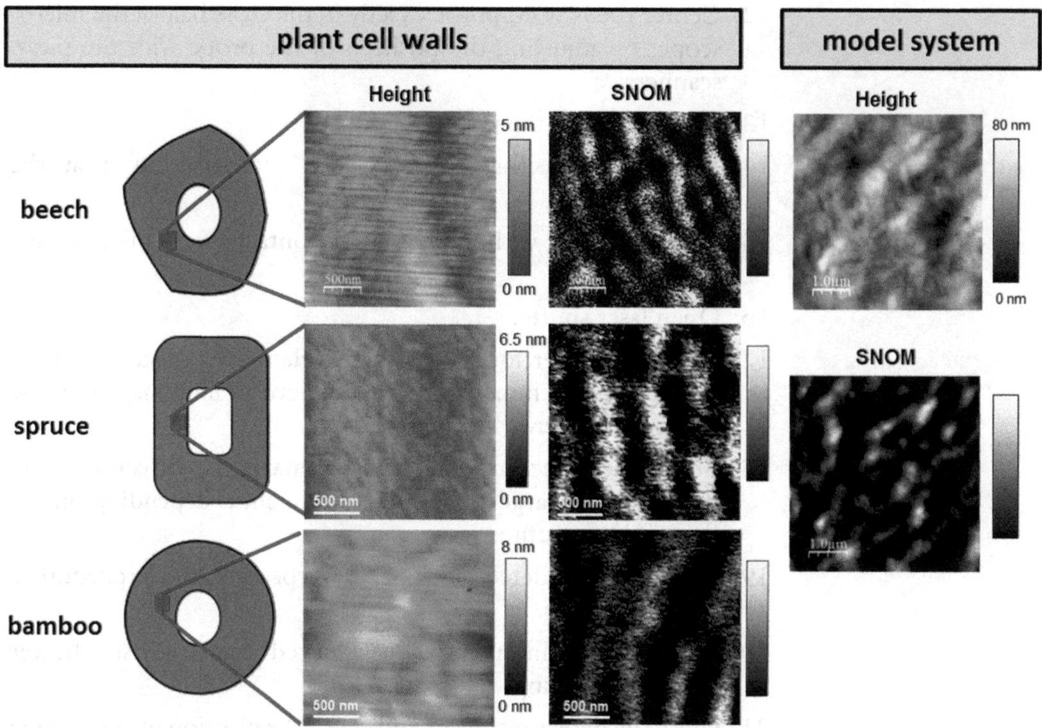

Fig. 3 SNOM and height images of plant cell walls (beech, spruce, and bamboo) and a cellulose–lignin model system (after [17])

arrangement of regions of higher cellulose and lignin content resulting in a segmented lamellar structure. Other models reveal a radial agglomeration or even a random texture of the macromolecules [22–26].

In Fig. 3 the results of Scanning Near Field Optical Microscopy measurements on the transverse section of different plant samples including beech wood, spruce wood and bamboo are shown. In addition to the SNOM signal also the corresponding height information of the scan is displayed. In the SNOM signal of the spruce, beech and bamboo samples a circumferential lamellar structure with strip like features with different sizes is visible. As this pattern is visible for all species measured, it seems that this might be a common secondary cell wall feature [17].

Based on considerations on the typical light–lignin interactions (fluorescence; resonance effects) it was concluded that the signal received is a result of changes in the amount or the lignin structure. In order to confirm this conclusion a comparative study on a bilayer system of cellulose nanocrystals and spin-coated DHP (dehydrogenation polymer) lignin on top of it was performed. Clear differences between the topography image and the SNOM image are visible. In the AFM image for lignin typical globular structures and the cellulose nanocrystals are visible. In contrast the NSOM image mainly reveals the agglomerations of lignin. These results support the conclusion drawn by the results of the measurements that predominately differences in the lignin content or structure are responsible for the contrast within the NSOM measurements of the secondary cell walls [17].

The obtained results are a strong indication for a superimposed circumferential pattern of the cellulose lignin architecture as it is proposed by the Kerr and Goring model. This structural pattern can be interpreted as a result of the cellulose fibril spinning process by the cellulose synthase complexes and the subsequent lignification process into the formed cellulose scaffold. As the lignification process represents a radical polymerization progress it is likely that regions with higher and lower degrees of lignin or differences in the structure arise [27–29].

6 Data and Image Interpretation

The contrast achieved in Scanning Near Field Optical Microscopy images can be of different origin including birefringence, fluorescence, Raman effect or differences in the refractivity index, reflectivity/transparency, and polarization. This broad spectrum of possible signal origin in combination with complex structured samples, as for example plant cell walls, makes it quite challenging to correlate structural or chemical features of the sample with a specific signal.

One possibility to circumvent these difficulties is to facilitate the measurement system by using model systems where the exact composition and structure is known and therefore the achieved signal can be directly correlated with the underlying chemistry or structure. The gained information on the signal origin can then be transferred to the actual sample of interest and helps to interpret the data, as it was demonstrated in the paragraphs before on the example of plant cell walls and a lignin-cellulose model system.

Another difficulty with regard to near field measurements is that they are quite prone to different types of artifacts. One main type is the so-called topography related artifacts. The SNOM signal is very sensitive to abrupt changes in the topography resulting sometimes in a nonreal SNOM signal. This can be easily followed from the fact that the near field only occurs very close to the surface. By using very flat surfaces (good sample preparation) the probability of topography artefacts can be reduced. The extent and the type of artefacts occurring also depend on the feedback mechanism of the used instrument. For a detailed description of the different types of artifacts, the reader is referred to the literature [30].

Acknowledgments

We thank the SNF for funding in the framework of NRP 66 as well BAFU (Bundesamt für Umwelt) and Lignum, Switzerland for the support of the Wood Materials Science Group at ETH Zurich.

References

1. Dixon RA (2013) Microbiology—break down the walls. Nature 493:36–37
2. Himmel ME, Ding SY, Johnson DK, Adney WS, Nimlos MR, Brady JW, Foust TD (2007) Biomass recalcitrance: engineering plants and enzymes for biofuels production. Science 315:804–807
3. Bjurhager I, Olsson A-M, Zhang B, Gerber L, Kumar M, Berglund LA, Burgert I, Sundberg B, Salmen L (2010) Ultrastructure and mechanical properties of Populus wood with reduced lignin content caused by transgenic down-regulation of Cinnamate 4-hydroxylase. Biomacromolecules 11:2359–2365
4. Burgert I, Cabane E, Zollfrank C, Berglund L (2015) Bio-inspired functional wood-based materials—hybrids and replicates. Int Mater Rev 60:431–450
5. Reza M, Kontturi E, Jaaskelainen A-S, Vuorinen T, Ruokolainen J (2015) Transmission electron microscopy for wood and fiber analysis—a review. Bioresources 10:6230–6261
6. Gierlinger N, Keplinger T, Harrington M (2012) Imaging of plant cell walls by confocal Raman microscopy. Nat Protoc 7:1694–1708
7. Gierlinger N, Keplinger T, Harrington M, Schwanninger M (2013) Raman imaging of lignocellulosic feedstock. In: Van de Ven T, Kadla J (eds) Cellulose - biomass conversion. InTech, London
8. Gierlinger N, Schwanninger M (2006) Chemical imaging of poplar wood cell walls by confocal Raman microscopy. Plant Physiol 140:1246–1254
9. Donaldson LA, Radotic K (2013) Fluorescence lifetime imaging of lignin autofluorescence in normal and compression wood. J Microsc 251:178–187
10. Schmid T, Opilik L, Blum C, Zenobi R (2013) Nanoscale chemical imaging using tip-enhanced Raman spectroscopy: a critical review. Angew Chem Int Ed Engl 52:5940–5954
11. Burgert I, Keplinger T (2013) Plant micro- and nanomechanics: experimental techniques for plant cell-wall analysis. J Exp Bot 64:4635–4649
12. Synge EH (1928) A suggested method for extending microscopic resolution into the ultra-microscopic region. Philos Mag 6:356–362
13. Pohl DW, Denk W, Lanz M (1984) Optical stethoscopy - image recording with resolution lambda/20. Appl Phys Lett 44:651–653
14. Lewis A, Isaacson M, Harootunian A, Muray A (1984) Development of a 500-a spatial resolution light microscope.1. Light is efficently transmitted through Gamma-16 diameter apertures. Ultramicroscopy 13:227–231
15. Lewis A, Taha H, Strinkovski A, Manevitch A, Khatchatouriants A, Dekhter R, Ammann E (2003) Near-field optics: from subwavelength illumination to nanometric shadowing. Nat Biotechnol 21:1377–1386
16. Dunn RC (1999) Near-field scanning optical microscopy. Chem Rev 99:2891
17. Keplinger T, Konnerth J, Aguie-Beghin V, Rueggeberg M, Gierlinger N, Burgert I (2014) A zoom into the nanoscale texture of secondary cell walls. Plant Methods 10:1
18. Ash EA, Nicholls G (1972) Super-resolution aperture scanning microscope. Nature 237:510–512
19. Hecht B, Sick B, Wild UP, Deckert V, Zenobi R, Martin OJF, Pohl DW (2000) Scanning near-field optical microscopy with aperture probes: fundamentals and applications. J Chem Phys 112:7761–7774
20. Novotny L, Stranick SJ (2006) Near-field optical microscopy and spectroscopy with pointed probes. Annu Rev Phys Chem 57:303–331
21. Zayats A, Richards D (2009) Nano-optics and near-field optical microscopy. Artech House, Norwood
22. Salmen L, Burgert I (2009) Cell wall features with regard to mechanical performance. A review COST action E35 2004-2008: wood machining - micromechanics and fracture. Holzforschung 63:121–129
23. Wang X, Keplinger T, Gierlinger N, Burgert I (2014) Plant material features responsible for bamboo's excellent mechanical performance: a

comparison of tensile properties of bamboo and spruce at the tissue, fibre and cell wall levels. Ann Bot 114:1627–1635

24. Zimmermann T, Thommen V, Reimann P, Hug HJ (2006) Ultrastructural appearance of embedded and polished wood cell walls as revealed by atomic force microscopy. J Struct Biol 156:363–369

25. Donaldson LA (2001) A three-dimensional computer model of the tracheid cell wall as a tool for interpretation of wood cell wall ultrastructure. IAWA J 22:213–233

26. Zimmermann T, Sell J, Eckstein D (1994) SEM studies on tension-fracture surfaces of spruce samples. Holz Roh Werkst 52:223–229

27. Cosgrove DJ (2005) Growth of the plant cell wall. Nat Rev Mol Cell Biol 6:850–861

28. Joseleau JP, Ruel K (1997) Study of lignification by noninvasive techniques in growing maize internodes—an investigation by Fourier transform infrared cross-polarization magic angle spinning C-13-nuclear magnetic resonance spectroscopy and immunocytochemical transmission electron microscopy. Plant Physiol 114:1123–1133

29. Terashima N, Fukushima K (1988) Heterogeneity in formation of lignin.11. An autoradiographic study on the heterogeneous formation and structure of pine lignin. Wood Sci Technol 22:259–270

30. Hecht B, Bielefeldt H, Inouye Y, Pohl DW, Novotny L (1997) Facts and artifacts in near-field optical microscopy. J Appl Phys 81:2492–2498

Chapter 15

Raman Imaging of Plant Cell Walls

Batirtze Prats Mateu, Peter Bock, and Notburga Gierlinger

Abstract

Raman imaging is a microspectroscopic approach revealing the chemistry and structure of plant cell walls in situ on the micro- and nanoscale. The method is based on the Raman effect (inelastic scattering) that takes place when monochromatic laser light interacts with matter. The scattered light conveys a change in energy that is inherent of the involved molecule vibrations. The Raman spectra are thus characteristic for the chemical structure of the molecules and can be recorded spatially ordered with a lateral resolution of about 300 nm. Based on thousands of acquired Raman spectra, images can be assessed using univariate as well as multivariate data analysis approaches. One advantage compared to staining or labeling techniques is that not only one image is obtained as a result but different components and characteristics can be displayed in several images. Furthermore, as every pixel corresponds to a Raman spectrum, which is a kind of "molecular fingerprint," the imaging results should always be evaluated and further details revealed by analysis (e.g., band assignment) of extracted spectra. In this chapter, the basic theoretical background of the technique and instrumentation are described together with sample preparation requirements and tips for high-quality plant tissue sections and successful Raman measurements. Typical Raman spectra of the different plant cell wall components are shown as well as an exemplified analysis of Raman data acquired on the model plant Arabidopsis. Important preprocessing methods of the spectra are included as well as single component image generation (univariate) and spectral unmixing by means of multivariate approaches (e.g., vertex component analysis).

Key words Arabidopsis, Chemical mapping, Confocal Raman microscopy, Plant cell wall, Plant microsections, Microtomy, Hyperspectral image analysis, Vertex component analysis

Abbreviations

AsLS	Asymmetric least squares
CCD	Charge-coupled detector
CRM	Confocal Raman microscopy
HCA	Hierarchical cluster analysis
MCR-ALS	Multivariate curve resolution alternating least squares
NMF	Nonnegative matrix factorization
PCA	Principal component analysis
PEG	Polyethylene glycol
S/G ratio	Syringyl–guaiacyl ratio
S/N ratio	Signal-to-noise ratio
VCA	Vertex component analysis

Zoë A. Popper (ed.), *The Plant Cell Wall: Methods and Protocols*, Methods in Molecular Biology, vol. 2149,
https://doi.org/10.1007/978-1-0716-0621-6_15,

1 Introduction

Life is not reductionist. Life and biotic materials cannot be explained by single atoms. However, they are based on combining these atoms in many different ways to form molecules. These may build polymers in different forms (e.g., amorphous, crystalline) and structures (e.g., fibrils), which again might interact with each other. The complexity is expressed at different hierarchical levels: from the nanoscale (atoms, molecules, fibrils) to the microscale (cell form, content, and wall thickness) and macroscale (tissues, organs). Plants are organized and compartmentalized at different hierarchical levels in order to specialize and accomplish all functions and requirements, such as limitation of light, mechanical stimuli (wind, slopes), drought, or viral infections. This individualization of organs and tissues makes plants a highly variable system to analyze [1]. Each plant is different from others and different with regards to its several organs, tissues or cells. Even within one cell the composition may vary at the nano- and microscale [2]. The plant cell wall has evolved to provide plant cells a dynamic support: at the beginning the primary cell wall is elastic and flexible to allow expansion and growth, while after cell death the cell wall lignifies and thickens in order to increase its stiffness [3], avoid buckling [4], and improve water transport [5] and biotic resistance [6]. From a materials viewpoint, plant cell walls are nanocomposites made of cellulose nanofibrils, whose arrangement and interaction with other plant cell wall polymers influence their performance [7]. Within this complicated setup, scientists studying this variable material ask themselves, ***How can this chemical and structural variability then be analyzed?***

Most of the characterization techniques (e.g., wet chemistry) approach the problem by analyzing the whole sample without spatial differentiation. The structure of the sample is broken down for sampling and is therefore already modified by the process before investigation. The result is a concentration of signal coming from the whole part of the analyzed plant—spatial information is lost in translation. Thus if chemistry has to be revealed in connection with structure this becomes a challenge.

Different microspectroscopic techniques have evolved to overcome the "bulk" problem and offer a spatially resolved information [8, 9]. Confocal Raman microscopy (CRM) non-destructively records spectra carrying chemical and structural information (molecular fingerprint) with high spatial resolution defined by the diffraction limit of light [10–12]. Furthermore, the weak sensitivity to water is advantageous for investigating the plant in-situ in its native, wet condition without prior drying or extraction.

In this book chapter the basics of Raman spectroscopy, sampling and sample preparation (and techniques available), state-of-the-art instrumentation, and different approaches in data analysis

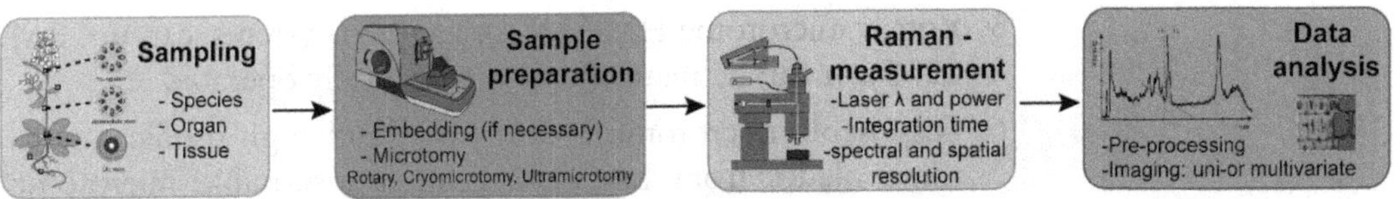

Fig. 1 Flow chart of the experimental procedure. After sampling the part of interest, thin sections or plane surfaces are cut (with or without embedding) using a microtome. Sections or blocks of the sample are measured/scanned with defined spatial and spectral resolution by adjusting different instrument parameters. A large amount of data is generated and analyzed by univariate (e.g., band integration) or multivariate methods

are described (Fig. 1). Several working steps and imaging results are shown for the most investigated plant model organism *Arabidopsis thaliana*. An overview of the technique as well as important advice on experimentation and troubleshooting is provided. Furthermore, the chapter provides the reader with a basic understanding of the hyperspectral image analysis methods and interpretation of Raman spectra, particularly of plant cell walls.

2 Materials

2.1 Reagents

1. Distilled water.
2. Deuterium oxide (D_2O; 99.9% D).
3. Immersion oil (Zeiss, low fluorescence).
4. Plant sample (if possible use fresh, never-dried-out material). Samples from all plant organs (stem, roots, and leaves) can be investigated, and replicates depend on the research question, but there should be at least three of these.

2.1.1 Polyethylene Glycol (PEG) Embedding

1. PEG 2000 melted at 60 °C (can be kept in the oven during embedding procedure).
2. PEG 2000 (60 °C)-deionized water (50%:50%).

2.1.2 Cryosectioning

1. Liquid nitrogen

2.2 Equipment

2.2.1 PEG Embedding and Sectioning

1. Plastic sample containers (minimum diameter 3 cm, minimum height 3 cm; are inexpensive and easy-to-cut plastic in order to remove the embedded sample block). Ice cube trays are suitable for small samples.
2. Blocks of wood (~1 × 1 × 1 cm) for fixing the embedded sample in the microtome (if it cannot be fixed alone).
3. Razor blades.
4. Tweezers (different sizes for handling the sample, microcuttings, and coverslips).

5. Rotary microtome HM 325 (Leica Microsystems, cat. no. RM 2255). It can be manually or automatically operated.
6. Disposable microtome blades (low- and high-profile microtome blades from Leica, or Feather disposable microtome blades (e.g., NH35R)). The blade choice can be optimized for tissue hardness.

   CAUTION! Blades are extremely sharp.
7. Toothpicks.

2.2.2 Cryosectioning

1. Superfrost microscope slides (Carl Roth, cat. no. 1879.1 or VWR, cat. no. 631-0909).
2. Corning 50-ml tubes (Corning, cat. no. 430829).
3. Disposable base molds (Fisher Scientific, cat. no. 22–363-552).
4. Leica CM15105 Cryostat (Leica Microsystems, cat. no. CM 15105).
5. Leica 819 low-profile microtome blades (Leica Microsystems, cat. no. D69226). The choice depends also on sectioning temperature and tissue hardness.

   CAUTION! Blades are extremely sharp.
6. Fine-nose forceps.
7. Razor blades.

2.2.3 Ultramicrotomy and Cryo-ultramicrotomy

1. Diatome diamond knifes: Histo for room temperature sectioning of PEG embedded or fresh samples. Cryo-35° dry for cryosectioning of hydrated material.
2. SuperFrost Plus Gold Adhesion microscope slides (Carl Roth, cat. no. ET09.2).

2.2.4 Sample Preparation

1. Brushes (any kind; different sizes for carrying the microsections, e.g., size 0).
2. Microscope slides (Carl Roth, cat. no. 0656.1).
3. Coverslips (Carl Roth; thickness according the microscope objective correction, usually 0.17 mm; size depends on the sample size).
4. Kimwipes (Fisher Scientific).
5. Nail polish (any brand). Old nail polish with higher viscosity works best.
6. Brush-On Nail glue (Fingers Europe AG; other nail glues or super glue might also work but has to be tested separately for temperature resistance).

2.2.5 Spectra Acquisition and Analysis

1. Raman microscope (WITec300, WITec) equipped with a 532-nm Nd-Yag laser, WITec UHTS 300 spectrometer, Andor, DV 401 BV CCD camera, and oil-immersion objective from Nikon (×100, NA = 1.4) for maximal resolution.
2. Software.
3. WITec Control 4.1 (or higher) for spectra acquisition, WITec Project 4.1 (or higher) and WITec ProjectPlus for data analysis.
4. OPUS (Bruker) for single spectrum analysis.
5. Epina Image Lab (for VCA and image preprocessing).
6. Matlab and GUI 2.O for MCR-ALS.

3 Methods

3.1 Sample Preparation

Depending on the research question, available sample amount, and/or sampling circumstances, either fresh, dry, or treated samples can be measured. Nevertheless, experience has shown that fresh and wet plant materials give the best results due to minor fluorescence (and sample burning) during spectra acquisition. For imaging a microscopic setup is needed and therefore all the restrictions for optical microscopes apply.

As the technique is based on scattering, no optimal sample thickness is required. Nevertheless, for good imaging results, plane surfaces are needed. This is achieved by micro-, ultra- or cryomicrotomy of native or embedded plant samples. Embedding is preferred in media which can be washed out after sectioning (e.g. polyethylene glycol (PEG)) or at least having no Raman bands overlapping with bands of interest. Cutting frozen samples is usually performed on so-called cryomicrotomes, although freezing a sample might damage its structure due to formation of ice crystals; the use of cryoprotectants such as sucrose might be helpful.

3.1.1 Embedding a Sample in PEG

PEG (polyethylene glycol) is a water-soluble polymer which is available in different polymeric lengths (i.e. molecular weight, given by the number next to it, e.g., PEG 1500). The higher the molecular weight, the stiffer the substance is. PEG-embedding is chosen when samples are otherwise too soft to be cut (*see* **Note 2**).

1. Cut your sample to a size that will fit into the embedding vials such that the height of the sample is half of the height of the embedding form.
2. Melt the PEG 2000 at 60 °C.
3. Add the same amount of Milli-Q® to achieve a 50% PEG solution; seal it, put the mixture into the oven at 60–80 °C, and ensure that both components are well mixed.

4. Place the sample in the mixed solution. The volume filled should be at least double the height of the sample so that after evaporation of the water (half of the solution) the sample is still covered with solution. Label the glass container and mark the liquid level so that you will be able to monitor the evaporation progress.
5. Place the open vial with the sample in the oven at 60 °C until all water is evaporated (this might take several days).
6. Take a new vial, fill it with melted PEG 2000 and quickly transfer your sample into it. Avoid hardening of the PEG while transferring.
7. Store the new vial at 60 °C for 12 h.
8. For this step you need to provide enough molten PEG to fill your form completely! Place a small drop of PEG on the embedding form and put your sample into the form such that the surface which is intended for cutting points to the bottom of the form. Let it harden to fix the position and then quickly fill the form with PEG to ensure the formation of one homogeneous block. Doing this step too slowly or stepwise means that different PEG layers form and break apart easily afterward. Doing it too quickly could induce the formation of air bubbles that will produce cracks in the PEG block during hardening.
9. Let PEG harden for at least 12 h in an undisturbed place.
10. Carefully dismount your sections. If the temperature in your lab is too high, PEG softens. In this case place the cubes in the fridge for storage.
11. Trim your sample in order to reduce the amount of PEG surrounding the sample at the cutting face.

3.1.2 Cutting Thin Sections

For cutting sections at the micrometer scale or getting plane surfaces, microtomes are used.

The sample is carefully clamped into the sample holder and moved over a fixed blade (rotary microtome) or the knife is moved vertically (sledge microtome) (*see* **Note 3**). Avoid damaging the sample by overtighting the clamp. For thicker sections of soft tissues also vibratomes can be used, in which the blade is vibrating during cutting. To achieve a thickness below 1 μm, ultramicrotomes with glass or diamond blades are used. There exist also cryo-variations to allow the cutting of frozen samples, useful if the material structure is too fragile (e.g., soft tissues like cambium). For more information about the cryosectioning protocols *see* [13].

The sample mount is important because the sample should not be squeezed, nor should it sit too loose. Consider that the sample generates resistance against the knife, so it should be mounted in a way that it does not stick out very far from the clamps, otherwise the sample itself will move or break off from the clamps and uneven cuts are produced.

The force interaction of the blade with the sample is the crucial factor for a successful cut. If the material is very soft (parenchyma cells, cambium, cuticle) concave blades with a low blade angle are used. Hard materials like wood or bone require a higher blade angle and a wedge shape (*see* **Note 4**). Beside the blade the cutting angle is an important factor. If the cutting angle is too low, the blade glides more over the sample than it cuts, if it is too high, the blade may start vibrating, which yields cuts of alternating thickness or it even gets stuck in the material. In addition, since plant materials are anisotropic, the cutting direction with respect to the plant organ axis is important for the success of the sectioning procedure (for more information *see* **Note 5**).

Cutting thin sections of plant samples with a rotary microtome:

1. Mount your sample in the microtome (*see* **Notes 6** and **7**). The blade should be very close but unable to reach the sample. The cutting area should be adjusted vertically 90° and horizontally parallel to the blade.
2. Turn the hand wheel at moderate speed until the blade starts to cut sections. In the beginning the sections must be thin in order to prevent the sample or PEG block breaking during the approximation of the sample to the knife. The orientation of the sample will never be exactly parallel; therefore, we need to "flatten" the sample now, till the blade and sample are exactly parallel.
3. Use a brush to remove the sections.
4. As soon as the blade and sample are parallel and complete cuts with the desired thickness (*see* **Note 8**) are produced, transfer them with a brush to an objective slide with a drop of water on it (*see* **Note 9**).
5. Put the slide under a microscope and check whether the section is even, unrolled and shows the area of interest (*see* **Note 10**). Samples with varying thickness should be discarded because they will cause problems with focus adjustment during measurement (*see* Fig. 4).
6. If the section is of good quality, wash it with fresh Millipore water to remove the PEG (at least three times) (*see* **Note 11**).

3.1.3 Sealing Thin Sections

The process of Raman imaging can take up to several hours, so it is necessary to ensure that the sample maintains its condition. Most biological samples have a considerable water content and would fall dry, unless they are sealed. Conventional nail polish has proved to be a very good sealant (*see* **Note 12**).

1. Place your sample on the objective slide (preferably in the center) (*see* **Note 13**) and add water.

2. Carefully put a coverslip on it. This works best by setting it first on one edge and then lowering it with tweezers. The micro section should be in the center of the coverslip to avoid interference with the nail polish used later on and should not contain air bubbles (*see* **Note 14**).
3. Put a weight on the coverslip (*see* **Note 15**).
4. Remove the excess water by tipping with a paper towel at the edge of the coverslip.
5. Now seal the edge with nail polish (*see* **Note 16**).
6. Let the nail polish dry for at least 45 min.

3.1.4 Microtome Cut Surface of a Sample Block

If sectioning is not successful (not homogeneous, rupturing) or if picking up the sections is very difficult, it is recommended to measure the fresh cut surface of the block material directly. Sectioning follows the same principles as when cutting thin sections, but the block is kept and the sections are discarded.

3.2 Raman Experiment

3.2.1 Theoretical Background

Spectroscopy is the science of interaction between electromagnetic waves and matter [14]. Heat, light, X-rays, and radio waves are all forms of electromagnetic radiation, varying only in frequency. Reflection, diffraction and transmission are all phenomena which can tell about the properties of the material interacting with radiation. Thus, typical spectroscopic measurements involve a radiation source, a sample, and a detector.

Molecules are, when above 0 K, always in motion, thus exhibiting translational, rotational, and vibrational movement. These movements can be studied by spectroscopy, which in the case of vibrations, is called vibrational spectroscopy. Two important methods are infrared (IR) and Raman spectroscopy. Most vibrational modes can be excited by radiation in the mid-infrared range (400–4000 cm^{-1}), which therefore can be probed by IR spectroscopy to gain insights into molecular structure [15].

Raman spectroscopy also probes the vibrational states in a nearly identical range, from 17 to 4000 cm^{-1} [16], but in a different way. Unlike IR spectroscopy, which is an absorption technique, Raman spectroscopy is a light scattering technique, which means that a light source is shown onto the sample and the inelastic scattering is measured. Most of the light will pass through or gets absorbed. A small portion (10^{-6}) will be scattered elastically; this is called Rayleigh scattering and means that the frequencies of the incident wave and the emitted wave are the same. An even smaller amount of light (10^{-10}) is emitted at an altered frequency (inelastic scattering) with respect to the incident radiation. Photons scattered this way either have higher or lower energy. If the scattered photon has lower energy, the process is termed "Stokes-scattering". The energy difference excited a vibrational transition. Conversely, if the scattered photon is of higher energy ("Anti-Stokes scattering"), a concomitant decrease of molecular energy occurred [17]. This was experimentally proven for the first time in 1928 by C.V. Raman [18] and is therefore named Raman scattering. Its intensity depends on the power and wavelength of the used source. High powers and short wavelengths increase the intensity, but may damage the sample or cause fluorescence which swamps the Raman signal. For biological samples, common used wavelengths are 1064, 785 and 532 nm. The controlling molecular property for Raman scattering is the polarizability of the molecule under a certain mode of vibration, that is how easy the electron cloud can be distorted by the incident light. In infrared (IR) spectroscopy, the leading property is the change of the dipole moment during a vibrational mode [17] In practice, this leads to both methods probing different vibrations of a molecule. Vibrations of polar bonds (O-H, C=O) are generally well observed in the IR, whereas apolar bonds (C-C, C=C) tend to be strong in the Raman spectrum [19]. An important consequence is that Raman spectra have only little contribution of water which is often encountered in biological samples. This makes Raman spectroscopy a very suitable tool for studying biological systems.

Molecules can also take up energy of radiation not matching any molecular absorption band. In such cases, they are said to be excited to "virtual" states. Although this is be excited by radiation to higher energy states and emit radiation when falling back to the initial level—this is called fluorescence. If the amount of energy is not enough to lift the molecule to a higher energy level, energy can still be taken up and raises the molecule to a so-called virtual level [20]. If it now falls back, it can happen that it does not fall back instantly to the starting level, instead this amount of energy is used for translation, rotation, and vibration, and is eventually dissipated as heat [14]. To the observer, this energy gap is seen in a frequency shift of the incident light source and is called Stokes shift. On the other hand, it can happen that a molecule being not in the ground state is excited, then relaxes and falls to the ground state, therefore ending up with lower energy than it initially had and thus radiating light with higher frequency than it received—named anti-Stokes shift.

In Fig. 2 these gaps are depicted as peaks in the Raman spectrum. The peak at 0 cm^{-1} refers to the Rayleigh scattering with the same frequency as that of the light source. Thus, no energy is transferred to the molecule. To the right of the Rayleigh peak, the frequency and intensity of Stokes scattering are plotted. This radiation has interacted with the molecules, reached the detector with a lower frequency (wavenumber) and thus gives valuable structural and chemical information. The shown spectrum was acquired with a band filter completely cutting the anti-Stokes side off, because it does not contain additional information on molecular structure.

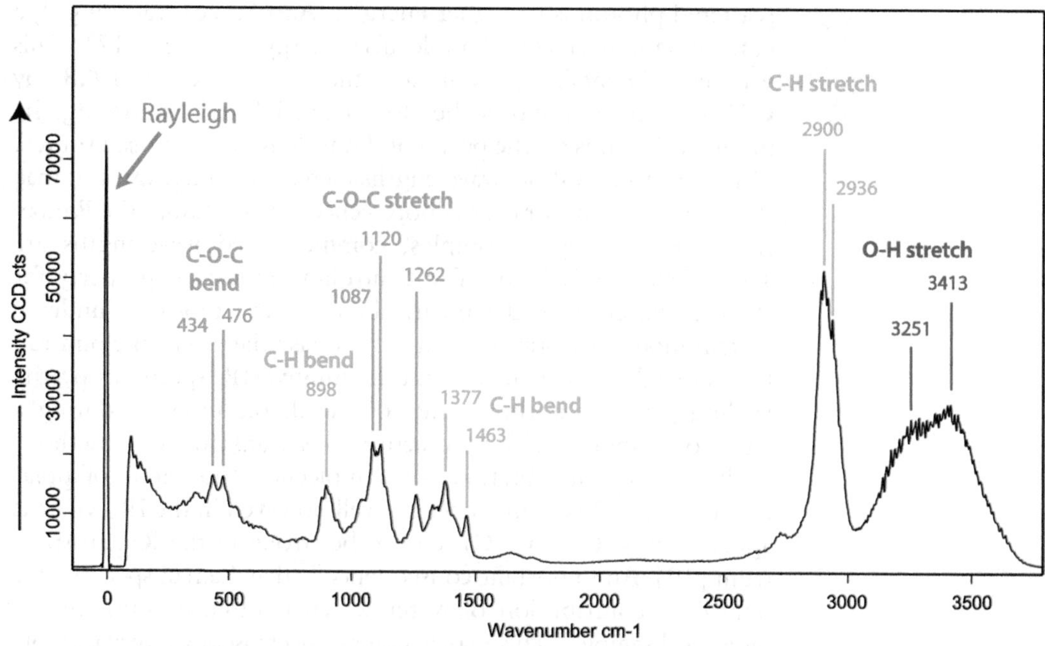

Fig. 2 Group frequency assignment of the Raman spectrum of glucomannan. The peak at 0 cm^{-1} is the exciting radiation, to the left are the intensity and frequency of Raman–Stokes scattering. Major peaks are indicated by their respective wavenumber and assignment

The stokes and anti-Stokes side of the Raman spectrum look identical, with the exception that the anti-Stokes lines are usually much weaker since much fewer molecules are in an excited state compared to the ground state [21]. The ratio of molecules being in the ground state and already excited is a matter of temperature; therefore, information on temperature changes of the sample can be monitored by comparing Stokes and anti-stokes lines [22].

The wavenumbers denote the relative shift of frequency regarding to the exciting light source and are therefore correctly written as rel. cm^{-1}, although they are often only denoted cm^{-1}. The Raman spectrum can generally be divided at 1500 cm^{-1} into two regions:

Above 1500 cm^{-1}, mainly chemical groups show up, and this region is good for a first evaluation of which chemical groups are present in the sample, because these modes appear in well-defined and -studied regions. They can often be clearly identified due to no overlap with other bands [19, 23].

The region below 1500 cm^{-1} is called fingerprint region because it is almost unique to every molecule, and peaks stem both from chemical groups and from the molecule vibrating as a whole [19]. Although not that straigtforward to interprete, this region contains rich information of the molecule and it is feasible to construct assignment tables for specific research questions, e.g. for lignin [24, 25].

With the aid of quantum mechanical calculations, it becomes more and more possible that small molecules like sucrose have their spectrum completely assigned. However, this demands knowledge about the chemical structure, which is problematic for substances of the plant cell wall like cellulose because their exact structure and interplay with other polymers in the cell wall remains poorly understood [26]. Therefore, the analysis of spectra of plant cell wall components heavily relies on the use of libraries containing reference spectra (e.g. [27, 28]) Still established, but more and more replaced by sophisticated computational approaches is the group frequency method. Here, characteristic frequencies of chemical groups, whose vibrations are largely localized within and therefore characteristic for the group, are used [19]. For example, the stretching frequencies of C-H bonds in methyl groups is always located in the range 2800–3000 cm^{-1}, regardless of where the methyl is attached to. It is therefore especially useful for rapid interpretation of a spectrum.

It is not possible to assign every peak after this method, but fortunately, this is normally not necessary. Sometimes only one peak is enough to set a substance apart from the other. In addition to the peak position, the peak shape also contains information and can tell much about a substance or structure. That is why good reference tables also contain information about shapes and intensity values. Peaks are described as "broad" if they extend over a range of wavenumbers or as "weak," when they can be hardly distinguished from the background noise.

In Fig. 2 a spectrum of glucomannan is shown, with some group frequencies indicated. Starting above 1500 cm^{-1}, the broad OH-stretch and the strong CH-stretch is noted. Only a weak, broad band (OH-bending) is found at 1600 cm^{-1}. This, together with the absence of bands at 3060 cm^{-1}, is enough to exclude the presence of an aromatic ring. In the case of a plant cell wall, (hemi)cellulose, pectin or lipids would be suitable canditates. The next step is either to compare the observed spectrum to reference spectra (e.g. by an algorithm based spectral similarity) or to subject the sample to chemical procedures to specifically target certain components (e.g. hexane extraction to remove lipids) and then record a second spectrum and check the difference. It is advisable to note down all chemical structures (at least compound classes), because this facilitates the search for reference spectra. Raman imaging is also helpful, because structural features in the image can narrow down the responsible chemical compound (e.g. layer on epidermis cells -> cuticle -> waxes). It is always useful to record spectra of chemicals used during sample preparation (ethanol, PEG, etc.) to not mistaken them for bands of the specimen. Wavenumber values have to be seen in context with instrument accuracy and calibration (*see* Sect. **3.2.4.** and **Note 1**). Whether some bump in the spectrum is a peak or just noise requires some experience. Little intensity differences may be caused by difference of focus, temperature or even by sample preparation and should not be overinterpreted (*see* **Note 1**).

3.2.2 Raman Signature of Plant Cell Wall Components

Plant cell walls are composed of cellulose microfibrils embedded in a matrix of hemicelluloses and pectin, in the case of primary cell walls, and additionally lignin, in the case of secondary cell walls. Specialized cells like the epidermis have also further components of lipidic origin (wax and cutin) helping to fulfil and reinforce its barrier function. In addition, some proteins involved in cell wall formation or maturation might also be present (e.g., extensins). Other phenolic compounds, flavonoids, stilbenes, or tannins can also be found for further customization of the cell wall. In wood, some species are characterized by the presence of extractives impregnating the cell wall which are responsible for properties such as the durability.

Each of these cell wall components has a characteristic Raman signature:

Cellulose. Cellulose is composed of units of glucose that are linked by β-1,4-glycosidic bonds forming a linear chain (Fig. 3a). A total of 18 to 36 chains [29] of several hundreds to thousands of cellulose units are arranged parallel to each other to compose a cellulose microfibril, the structural unit of the cell wall. The orientation of these fibrils in the native cell wall changes depending on the cell wall layer and mechanical stimuli or environmental conditions (e.g., reaction wood). The Raman spectrum of cellulose is composed of about 15 bands whose relative intensity changes depending on the orientation of the microfibrils (cellulose molecules) with respect to the polarization direction of the excitation laser (Fig. 3b). The spectra taken on a Ramie fiber (cellulose microfibrils aligned parallel to the fiber axis) with laser polarization at 0° and 90° with respect to the fiber axis show clearly these prominent changes. Since the orientation of cellulose microfibrils is different in the distinct cell wall layers these peaks can be used in Raman imaging to identify cell wall layers with different cellulose orientation as well as to determine the cellulose microfibril angle [30]. It is therefore important to be aware of the influence on fiber orientation by sample preparation as well as by the measurement setup.

Changes in crystallinity of the microfibrils are difficult to discriminate in the spectrum from changes in fibril orientation but normally amorphous cellulose results in peak broadening and decrease in the peak height [31]. Ratios between specific bands have also been used to determine cellulose I crystallinity [32].

Cellulose has major bands at 380 (OH-torsion of CH_2OH), 1096 (asymmetric stretching of glycosidic COC), 1380 (CH2 deformation), and 2897 cm^{-1} (CH stretching) [33–36].

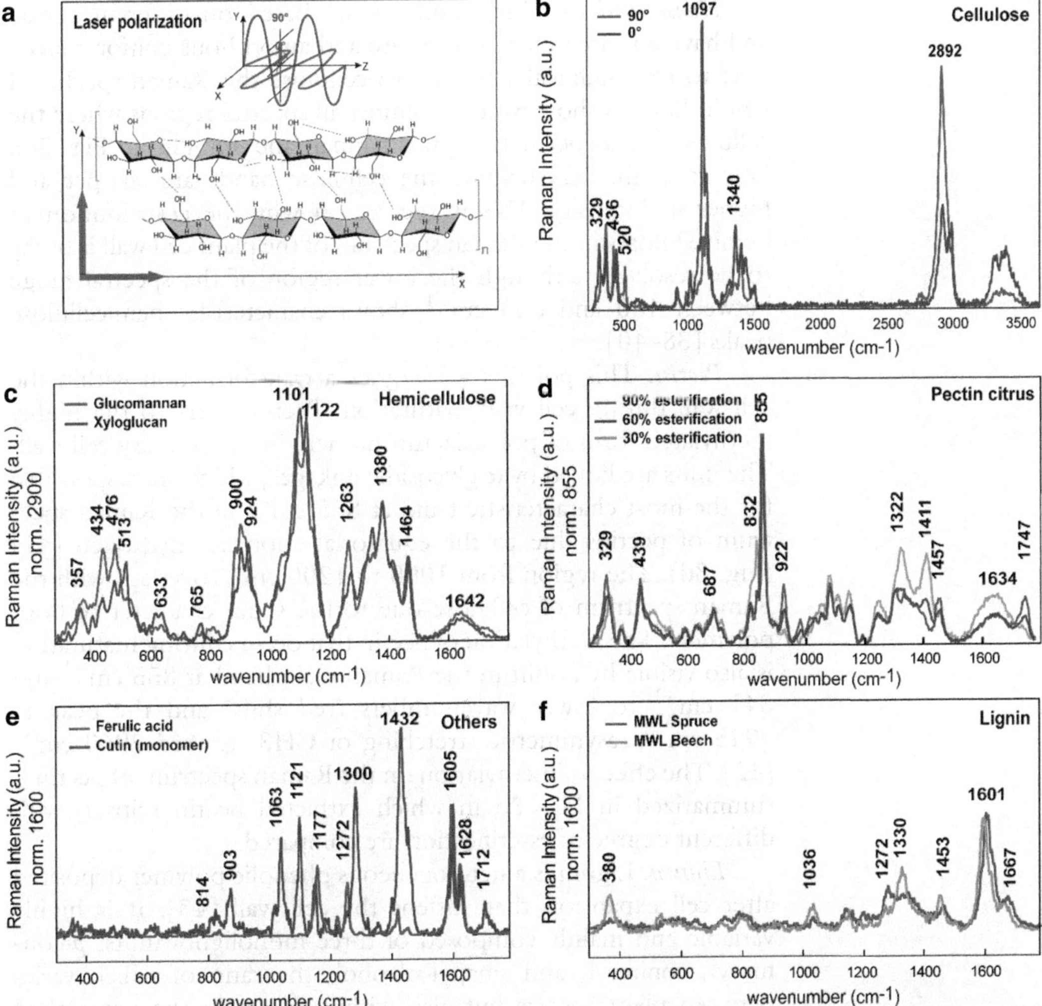

Fig. 3 Molecular structure and corresponding Raman spectra of plant cell wall polymers. (**a**) Cellulose is made of β-1-4-linked glucose molecules. Cellulose forms long chains that interact with each other through hydrogen bonds to form nano- and microfibrils. (**b**) Raman spectra acquired from a Ramie fiber (>95% cellulose, microfibrils aligned parallel to the fiber axis) depend on the angle between cellulose fibrils and laser polarization direction: in the blue spectrum (0° parallel) highest intensity is found at 1097 cm^{-1}, whereas in the red spectrum (90°) at 2892 cm^{-1}. (**c**) Raman spectra of glucomannan (Konjac, Megazyme, powder, purity >95%, Glc:Man = 37:60, acetylated) and xyloglucan (Tamarind, Megazyme, powder, purity >95%, Ara:Gal:Xyl:Glc = 3:18:34:45). (**d**) Raman spectra of citrus pectin (Sigma-Aldrich, Austria) with different grades of esterification (30%, 60%, and 90%). (**e**) Raman spectra of milled wood lignin (MWL) (extraction with a dioxane–water mixture) of spruce and beech. (**f**) Raman spectra of ferulic acid and cutin monomer (for details *see* [37]. Data acquisition: $\lambda_{ex} = 532$ nm, LPO = 45 mW, integration time = 1 s, 20 accumulations, average over three replicates, baseline corrected [rubber band, five iterations])

Hemicelluloses. Hemicelluloses are based on different sugars and have a more random structure and amorphous conformation. Due to the chemical similarity to cellulose the Raman spectra of hemicelluloses show bands in almost all spectral regions where the cellulose bands occur (*see* glucomannan and xyloglucan Fig. 3c). Due to higher crystallinity the cellulose bands are sharper and higher in intensity. This is the reason why the contribution of hemicelluloses in the Raman spectrum of the plant cell wall is tricky to be resolved, although the lower region of the spectral range between 400 and 600 cm^{-1} shows characteristic hemicellulose peaks [38–40].

Pectin. This polymer undergoes a transformation within the cell wall during cell wall maturation. Pectin refers to the highly methylated form of polygalacturonic acid in the primary cell wall. The units are linked by α-glycosidic linkages, which are responsible for the most characteristic band at 855 cm^{-1} in the Raman spectrum of pectins due to the equatorial anomeric hydrogen [41] (Fig. 3d). The region from 1000 to 1200 cm^{-1} overlaps with the Raman spectrum of cellulose due to the sugar character of both polymers. The methylation of pectin that occurs during maturation is also visible by a shift in the Raman main band at 855 cm^{-1} and 341 cm^{-1} to lower wavenumbers (red shift) and the peak at 2945 cm^{-1} (asymmetric stretching of CH3) toward 2960 cm^{-1} [42]. The effect of methylation on the Raman spectrum of pectin is summarized in Fig. 3d in which extracted pectin (citrus) with different degree of esterification are compared.

Lignin. Lignin is a heterogeneous phenolic polymer deposited after cell expansion that stiffens the cell wall [43]. It is highly variable and mainly composed of three monolignol units, *p*-coumaryl, coniferyl, and sinapyl alcohol, the ratio of which varies between plant species but also within tissues in the same plant [44]. Lignin is mostly restrained to the xylem but can also be found in wound healing or in the extern layer of the epidermis, the cuticle, and in the Casparian strip in the endodermis of roots [45]. The most characteristic bands (*see* Fig. 3e) of lignin are 1660 (C=C stretch of cinnamyl alcohols/C=O stretch of cinnamaldehydes), 1600 (aromatic ring stretching), 1330/1270 (ring substituent stretching) and 1140 cm^{-1} (ring C-H bend/aldehyde C-C stretch) [25, 46–48, 49]. The relative amount of lignin precursors can also be determined by Raman spectroscopy [50]: syringyl-(S)-to-guaiacyl (G) lignin ratio is lower in gymnosperms and vessels of angiosperms. Specific bands of these lignin types are found at 1271 (guaiacyl ring with C=O group) and 1334 cm^{-1} (syringyl lignin). *p*-hydroxyphenyl lignin has useful bands at 1250 (ring substituent stretching), 1175 (ring C-H bending) and 843 cm^{-1} (ring breathing) [51]. In Fig. 3e, the band typical for guaiacyl units at 1272 cm^{-1} is seen more pronounced in milled wood lignin

(MWL) of spruce, whereas the typical syringyl unit bands at 370–380, 1036 and 1330 and 1453 cm^{-1} are only clearly seen in MWL of beech.

Proteins. The plant cell wall is synthetized by an orchestrated process which involves the activation of several genes, the expression of transcription factors in the protoplast and the transport and function of cellulose synthases and proteins of the phenylpropanoid pathway among others [44, 52]. This is why it is not surprising to find aggregates of proteic and lipidic character (probably plasmalemma) attached to the cell wall after sectioning of plant material. Furthermore, (glycol-) proteins with structural or morphogenetic function are present in the cell wall itself [53]. Proteins main Raman bands are due to the amide bonds at 1670 cm^{-1} and the C-H bending between 1400 and 1495 cm^{-1} [54, 55].

Cutin and waxes. These hydrophobic polymers are found in the outermost layer of the epidermis of leafs, flowers and stems of plants due to its protective and regulatory purpose. They are composed of long aliphatic chains linked by ester bonds. The cuticle is formed by a crystalline layer of waxes (epicuticular wax) and a composite of waxes (intracuticular wax) embedded in a matrix of pectins and cutin [56, 57].While wax can change its crystallinity, cutin remains an amorphous polymer matrix [58]. Most of the bands are derived from the lipidic origin of these compounds: 1061 cm^{-1} (C-C stretching) [59–61] and 1121 cm^{-1} of cutin, and 1720 cm^{-1} (C=O stretching) [60] of cuticular wax. Bands at 1170–1178 (in plane deformation of ring CH and C-O-C stretching of ester bond [59, 62] or C-O stretching [63] are present in both. The main lipid bands at 1300 cm^{-1} [60] and 1434–1441 cm^{-1} [64] are also present in both components.

Other phenolic compounds. Members of the phenylpropanoid pathway (derivatives of the amino acid phenylalanine) can also be found in the cuticle and in lignin rich areas. An example is the occurrence of ferulic or cinnamic acid in the cuticle [65] which spectrum is shown in Fig. 3f (in blue). Most characteristic peaks of ferulic acid are located at 1171, 1266, and the doublet at 1606 and 1632 cm^{-1} [63].

3.2.3 Instrumentation

The basic setup of a Raman spectrometer is a laser source, a sample mount, a spectrometer and a computer. For Raman imaging, Confocal Raman microscopes are equipped with high numerical aperture (NA) objectives and an accurate *x*, *y* scan table. For the imaging approach, quick acquisition and high signal to noise (S/N) ratio are desirable. In order to achieve this, the throughput of light has to be optimized at every step.

Laser and Microscope Objective: Raman Intensity and Spatial Resolution

The Raman scattering intensity is inversely proportional to the excitation wavelength, this means, that in principal, the shorter the wavelength (typically of a laser, the more intense the spectrum [66]. The laser choice depends on the nature of the samples: Dark and colored samples are usually prone to fluorescence, whereas light samples can sustain higher energy. This means that for plant materials, nonlignified tissues like cuticle or primary cell walls usually show no problems, whereas lignified cells and bark extractives can be problematic. Especially those samples (compounds) which contain conjugated double bonds or rings can exhibit strong fluorescence even with long excitation wavelengths lasers. This not only can burn holes in the sample rendering the measured area useless for further analysis but also completely masks the recorded spectrum, such that the individual peaks are hidden under a very strong signal over the complete measurement range. To handle the problem, the excitation wavelength can be increased (e.g. 785 nm). A broad range of laser wavelengths is available on the market (Thorlabs for example offers diodes from 375 to 2000 nm) and many instruments can be equipped with more than one laser line (e.g., 532, 633 and 785 nm). Laser in the near infrared (1064 nm and higher) require different optics and detectors [15] and cannot be combined into one instrument equipped with lasers in the visible range. The drawback when using higher wavelength, is that the intensity decreases, that is, the spectra get noisier and the diffraction limited spatial (optical) resolution is inferior.

The spatial resolution is given by $r = 0.61\lambda/\mathrm{NA}$—thus together with the laser excitation wavelength (λ), the NA of the objective determines the lateral resolution (x, y) of the Raman imaging. The NA is defined by $\mathrm{NA} = n \sin\theta$ being θ the half-angle of the maximum cone of light that goes through the objective and n the refractive index of the medium (air, water, or immersion oil). The resolution of adjacent objects is good if they are separated by a minimum distance of $2r$ (Rayleigh criterion) [11] which has to be taken into account in the imaging setup. For maximal resolution (low r) immersion oil (highest n) objectives with high NA have to be used in combination with low λ. For 100× oil immersion objective (NA 1.4) and $\lambda = 532$ nm, the theoretical resolution will be about 250 nm, for fluorescent materials measured with excitation $\lambda = 785$ nm (or more) a lower lateral resolution of $r = 350$ nm will be achieved.

Depth resolution (z) depends beside laser wavelength and NA also on the confocal design of the Raman microscope and is about two times of the lateral resolution.

Spectrometer (Spectral Resolution), Detectors and Filters

All parts where the scattered light is passing through and finally detected have to be optimized for the used laser wavelength, if fast imaging with high Raman intensity is desired. The diffraction grating, which splits the beam into separate wavelengths, can be blazed according the wavelength for optimal efficiency. Usually gratings with different grooves (line spacing) are provided within one spectrometer and can be chosen according the desired spectral resolution. The more lines (l mm^{-1}), the more resolved is the spectrum in terms of wavenumbers, but the wavenumber range is reduced.

Finally, the detector (CCD camera) strongly affects the performance of the instrument. To increase the detection efficiency back-illuminated devices are available and recently electron-multiplying CCD cameras were suggested to improve S/N ratio for very small signals dominated by the readout noise [66].

Also all filters have to be adapted accordingly. There are band- and notch filters, which cut different parts of the spectrum before it reaches the spectrometer. Band filters normally cut both the anti-Stokes and the Rayleigh line, leaving only the Stokes side of the spectrum. If one is interested in also measuring the anti-Stokes side (i.e., for temperature measurements), then a notch filter is required.

Raman active vibrations can show a polarization dependent behavior; moreover, the structure can also be elucidated by varying the polarization plane. Thus, a filter to change the incident laser polarization on the sample allows to reveal differences in orientation (e.g., of cellulose fibrils/polymers) within the sample. On the other hand, polarization filters before the spectrometer allow for a separation of different vibrational modes of the sample.

3.2.4 Step by Step to High-Quality Raman Images

Checking Instrument Performance

Before starting a measurement the instrument performance should be checked by measuring a reference spectrum on a silicon wafer. Silicon shows a distinct peak at 520 cm^{-1}, whose position as well as intensity is used to evaluate instrument performance.

1. Put the silicon wafer under the microscope.
2. Use an objective without coverslip correction.
3. Focus on the aperture and/or scratches on the silicon wafer.
4. Turn the laser on its maximum power.
5. Start the oscilloscope mode.
6. Adjust the focus (*z*-direction) to maximum intensity.
7. If the intensity is lower than the usual performance the laser output intensity and pathways have to be checked and optimized (*see* **Note 17**). If the wavenumber position is not within the given error of the used grating, a spectrometer calibration has to be performed (*see* **Note 17**).
8. Record a single spectrum with always the same integration time for documentation.

Preparing for Imaging

Ideally you have checked your samples during preparation, but it is always good to get an overview and check the quality again in the light microscope mode—maybe storage had an effect, for example, drying out or bacterial/fungal growth. It is also important to orient the samples in a defined way (e.g., fibers/cell walls/layers mostly horizontally or vertically with respect to the laser polarization direction) to be able to differentiate Raman intensity changes due to changes in component (e.g., cellulose fibril) orientation from changes due to component amount.

Glass slides are fixed with a clamp and also blocks are fixed preferentially mechanically since glue does not perform well in the presence of water and/or may changes its behavior with temperature. For block measurements the objective used is not an immersion oil objective but a dry or better water immersion objective. For the latter a suitable holder (waterproof, e.g., liquid cell) has to be used to avoid damage of the equipment.

Normally the software combined with a sample positioning (stitching table) allows the adjustment of the coordinate system used to reference all the measurements on the sample. A point which can be found easily (e.g., a very distinct cell corner or the edge of the sample) should be chosen as the origin of the coordinate grid. This is particularly important if the sample might move (due to shrinking, swelling, laser energy, . . .) to match your Raman images afterward.

Then a quick measurement should be performed on a part of the sample which is not so important, because it might get damaged. This measurement will show if there are any problems with fluorescence background and/or sample degradation and the quality of the spectrum (signal-to-noise ratio) in order to adjust laser power and integration time accordingly.

Raman Imaging Setup: Point by Point Scanning

A Raman image is basically a grid of individual measurement points (individual spectra)—the denser and smaller the dots, the higher the pixel resolution. The number of points should be chosen with respect to the theoretical spatial resolution given by the used laser excitation wavelength and the numerical aperture of the chosen objective (*see* Subheading 3.2.3). The image is composed of thousands of spectra spatially ordered and carrying the chemical information. Unfortunately high resolution has two drawbacks: (1) It can take a long time to record the whole image and even worse (2) it increases the time the sample is exposed to the laser, which can cause sample damage.

The second factor is the integration time (*see* **Note 18**). This is the time the laser spot remains at one position while collecting the signal. The longer the integration time, the better the S/N ratio: the peaks become clearer in shape, the overall intensity increases and the noise decreases.

Experience and testing is required to get a feeling how much laser exposure a plant sample or special tissue can sustain in order to choose appropriate grid density and integration time.

If the scan direction can be chosen it is recommended to set it in a way that spectra are first recorded where no fluorescence is expected and at the end the parts which are very likely to absorb energy and prone to fluorescence and sample degradation (e.g. lignin or extractive rich cells). Once the sample starts to fluoresce, also those parts are affected which normally do not and the remainder of the measurement becomes useless. This can be clearly seen in Fig. 4a. The integration over the CH region at 2900 gives a beautiful image until the laser encounters the xylem in Arabidopsis. The sample then starts to absorb too much energy and burns. The spectral features (Red Cross, red spectrum) show the two prominent D and G bands at 1350 and 1595 cm^{-1}, respectively (Fig. 4b).

Figure 4c shows the problem of being out of focus. The integration over the CH stretching band shows a blurred and smeared image which is always a sign of improper focus. The spectrum also shows the typical bands of glass (coverslip or glass slide) (green

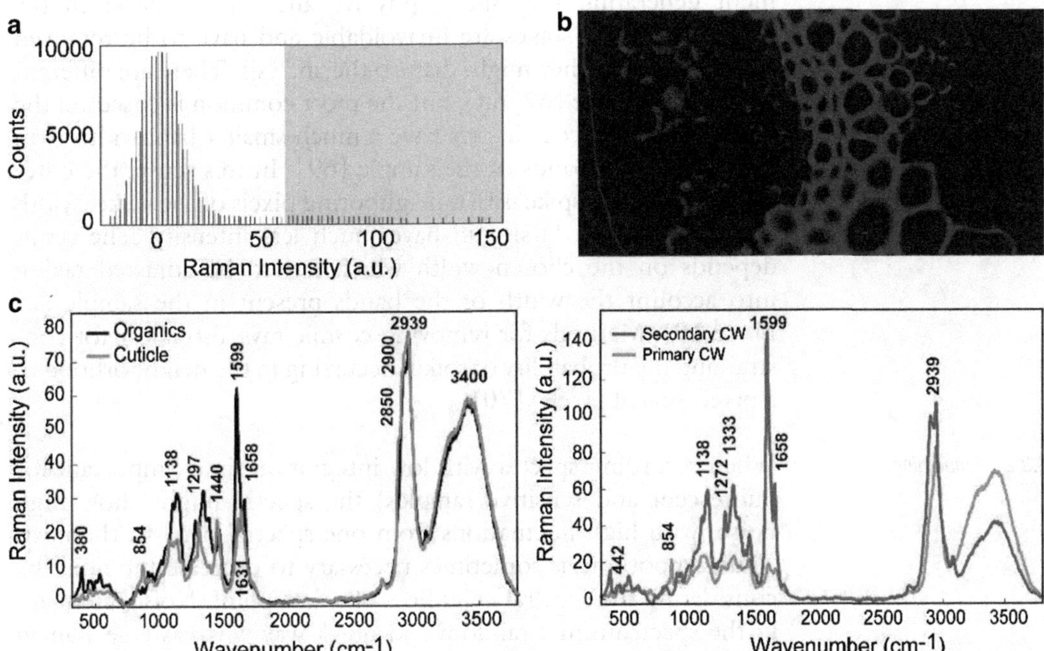

Fig. 4 Fluorescence (burning) and out-of-focus artifacts generated during the Raman imaging approach. Burning of the sample can lead to complete sample destruction (**a**). In the spectra a high background becomes visible and the two peaks around 1350 and 1595 cm^{-1} indicate the burnt carbon (**b**). If the focal plane changes (no plane surface, movement of sample, table, . . .), a change in intensity might be seen or only week signal of the cell wall bands (**c**). Instead, spectra can show high contribution of the substrate (e.g., peak at 520 and 1100 cm^{-1} of glass)

cross, green spectrum) (Fig. 4d). Therefore samples have to be flat and the focus adjusted properly prior the measurement to avoid difficulties and mistakes in data interpretations.

It is also recommended to not measure several times the same area since this could lead to sample modifications and wrong interpretation of the data. Especially lignin is prone to laser-induced changes and an area exposed to the laser has to be considered as *non-native* anymore (*see* **Note 18**). For a summary of troubleshooting options, *see* Table 1.

3.3 Data Analysis: Spectral Pretreatment

The higher the spatial resolution and the bigger the region of interest (measurement area), the bigger is the amount of data to handle and work with (storage, loading, analyze). It is important to have the research question in mind to avoid "oversampling" and too big data sets. After obtaining the raw data it might be necessary to eliminate effects not originating from the sample itself (cosmic rays, noise, ...) or not of interest (e.g., background, embedding media).

3.3.1 Cosmic Ray or Spike Removal

Cosmic rays are highly energetic particles (normally neutrons) from outer space that interact with the CCD camera during the measurement generating very sharp, positive and intense peaks in the spectrum. These spikes are unavoidable and have to be removed afterwards since they might disturb the analysis. There are different methods available [67, 68], but the most common is based on the assumption that cosmic rays have a much smaller bandwidth than the component bands of the sample [69]. In this sense, the criterion to remove a spike is that neighboring pixels of the spike (width typically 1–3 cm^{-1}) should have much less intensity. The result depends on the chosen width which has to be adjusted taking into account the width of the bands present in the sample (*see* **Note 19**). Methods for removing cosmic rays also allow for constraining the probability of spikes occurring in the neighborhood of a pixel (spatial pixels) [70].

3.3.2 Smoothing

When recording spectra with low integration time (important for fluorescent and sensitive samples) the spectra might show high noise, with high fluctuations from one spectral pixel to the next. Thus, smoothing is sometimes necessary to decrease the noise by considering the spectral neighbors of a data point. Noise fluctuates in the spectrum in a random and quick way whereas true Raman bands change gradually [69]. When a spectral position in comparison to the previous and next position changes abruptly the value is then replaced by a calculated value from those neighbors [71]. The value can be calculated in different ways: averaging all the neighbors (with or without weighting them following a Gaussian distribution for example), taking the median, using a wavelet transformation [72], and so on. One of the most applied filters is the Savitzky–

Golay algorithm [73]: instead of using an averaged value this technique fits a polynomial through the neighbors and replaces the actual value with the fitted one. The major advantage of this approach is the preservation of the spectral shape and features.

In the case of spatially resolved hyperspectral data, it makes also sense to include the spatial coordinates for the smoothing. In this case the spatial neighbors are used for calculating the value necessary for the replacement in a similar way as in the spatial spike removal tool.

However, attention is needed in this step as many similar and thus overlapping bands are observed in the Raman spectra of biological samples. A summary of the Raman spectra of pure substances of biological origin is found in [27] and for plant cell wall components in Fig. 3, where the similarity of the bands within and between different components groups can be observed. It is not surprising that mixing those substances generates a very complex Raman spectrum where small shoulders or small peaks (in comparison to others in the same spectrum) can be shadowed after the smoothing process [74]. A clear example of the effect of smoothing on the spectra is seen in Fig. 5. The single Raman spectrum taken in the cell wall (S2) of the stem of Arabidopsis is presented prior to smoothing treatment (in green) compared to smoothing based on the Savitzky–Golay algorithm with increasing window size (9 in blue and 17 in light blue) (windows size = neighbors pixels in the spectrum taken for the fitting of the curve). When considering more than five pixels (not shown) the intensity of the majority of the bands is reduced. With nine neighbors some peaks change their shape as is the case of the 1096 cm^{-1} band

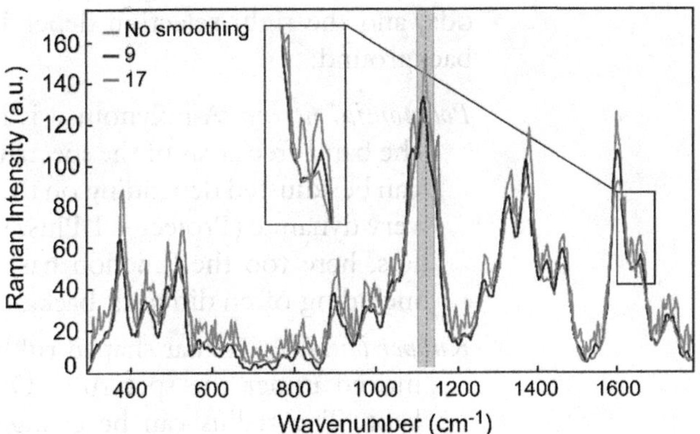

Fig. 5 Effect of the Savitzky–Golay smoothing on the Raman spectra of plant cell walls. Spectra (λ_{ex} = 532 nm, LPO = 35 mW, integration time = 0.1 s, baseline corrected (rubber band, five iterations) are shown without (green) and after nine (dark blue) and 17 points smoothing (light blue)

of cellulose or the 1660 cm^{-1} band of lignin normally attributed to the presence of coniferyl aldehyde and alcohol, precursors of lignin [47, 48]. Applying a smoothing with 17 neighbors, some important bands disappear: 1623 cm^{-1} of coniferyl aldehyde (*see* zoom in insert in Fig. 5), 500 and 1121 cm^{-1} of cellulose and 1270 cm^{-1} of guaiacyl lignin (G lignin). This is why smoothing has to be applied carefully and only if necessary, for example, for multivariate data analysis (*see* **Note 20**).

3.3.3 Baseline Correction and Background Subtraction

The small signal generated by the Raman scattering is guided to a CCD camera which adds a constant voltage in order to make the signal a positive value [69]. This indeed creates a constant background which can be easily removed by subtracting a constant value (from hundred to few thousands of CCD counts). It can be that the focal plane was not chosen properly and the substrate (typically borosilicate glass) is present all along the recorded Raman spectra. If the sample is even and the glass contribution therefore constant, the sample spectrum can be recovered by substracting the glass spectrum. In this case the right procedure is to calculate the average spectrum of the glass (e.g., in the lumen of the cells) for the subtraction. However, since laser excitation wavelengths used in Raman spectroscopy often match the excitation range of naturally fluorescent molecules in the sample (e.g. lignin), background profiles can be of different intensity and become more complicated than a simple line. This has to be taken into account when validating and analyzing the data. Often it is essential to get rid of differences in the background intensity to be able to reveal differences due to changes in peak intensities and shapes by multivariate data analysis approaches.

"Complicated" backgrounds can be tackled by different methods, and the right selection depends mainly on the shape of the background:

Polynomial fitting: A polynomial is fitted on the basis (mostly) of the band-free areas of the spectrum [69].The polynomial grade can be adjusted depending on the background shape and is thus very dynamic (Project 4.1 Plus, WiTEC, Germany). Nevertheless, here too the function has to fit for all acquired spectra, including often different background shapes (*see* **Note 21**).

Rubber band: A circular shaped rubber band with a fixed radius is moved under the spectrum. Only areas above the circle are kept. The radius can be changed as well depending on the broadness of the bands of interest. This approach can be applied using different software programs (e.g., Project 4.1 Plus [WiTEC, Germany], OPUS [Bruker, Germany]).

Asymmetric Least Squares: This method is used when in the spectrum narrow bands and a broad background (e.g., fluorescence originated) are expected [75]. The method was originally developed as smoothing algorithm proposed by Whittaker [76] in 1923 and relaunched by Eilers 2003 [77] and recently used by Felten [78]. It basically consists in drawing a line under the spectrum and assign a very small weight to the points above this line and a high weight for the points below (hence called asymmetric). Iteratively the baseline adapts to the boundaries determined by the weight factors. This type of background correction can be applied to low S/N spectra without losing any spectral feature. Two of the parameters have to be optimized (λ, p) in order to keep the spectral shapes. Normally "p" values should be minimized since no negative bands are expected (the p values determine the weighting between values under and above the baseline). The λ value determines the fitting of the baseline to the actual spectrum. Higher values would adjust to the baseline in a better way. However, if the values are too big, it can result in an over fitted baseline with removal of significant spectral features. This pretreatment can, for example, be carried out in MatLab with the application GUI 2.

Lieber algorithm: This iterative method uses a polynomial interpolation of all spectral pixels. All points above the current fitting are set as points of the polynomial. The number of iterations can be chosen until the algorithm converges to a polynomial baseline of an order previously defined [79]. This kind of algorithm is implemented in Image Lab (Epina, Austria).

It is rather complicated to find a unique algorithm that is applicable to all cases and samples and all positions in the Raman map. The optimal algorithm has to be postverified especially in the case of overlapping and multicomponent bands (as is the case of the Raman fingerprint of plant cell walls). For example areas where the presence of extractives and/or lignin is higher might need a prominent round shaped background subtraction, whereas for pure cellulose a constant extraction or polynomial fitting of low order might be enough.

In Fig. 6 a comparison of three baseline corrections is performed for the average spectra of different anatomic parts of Arabidopsis: epidermis, phloem and cambium, interfascicular fibers, and xylem. The first graphic shows the raw data after cosmic ray removal. The higher wavenumber part of all the spectra is characterized by a high-broad background whereas the fingerprint region is rather flat. Before baseline correction the spectra should be trimmed (up to 300 cm^{-1}). The dotted lines under the spectra correspond to the polynomial of order 6 drawn and extracted for each spectrum during polynomial baseline correction, whose

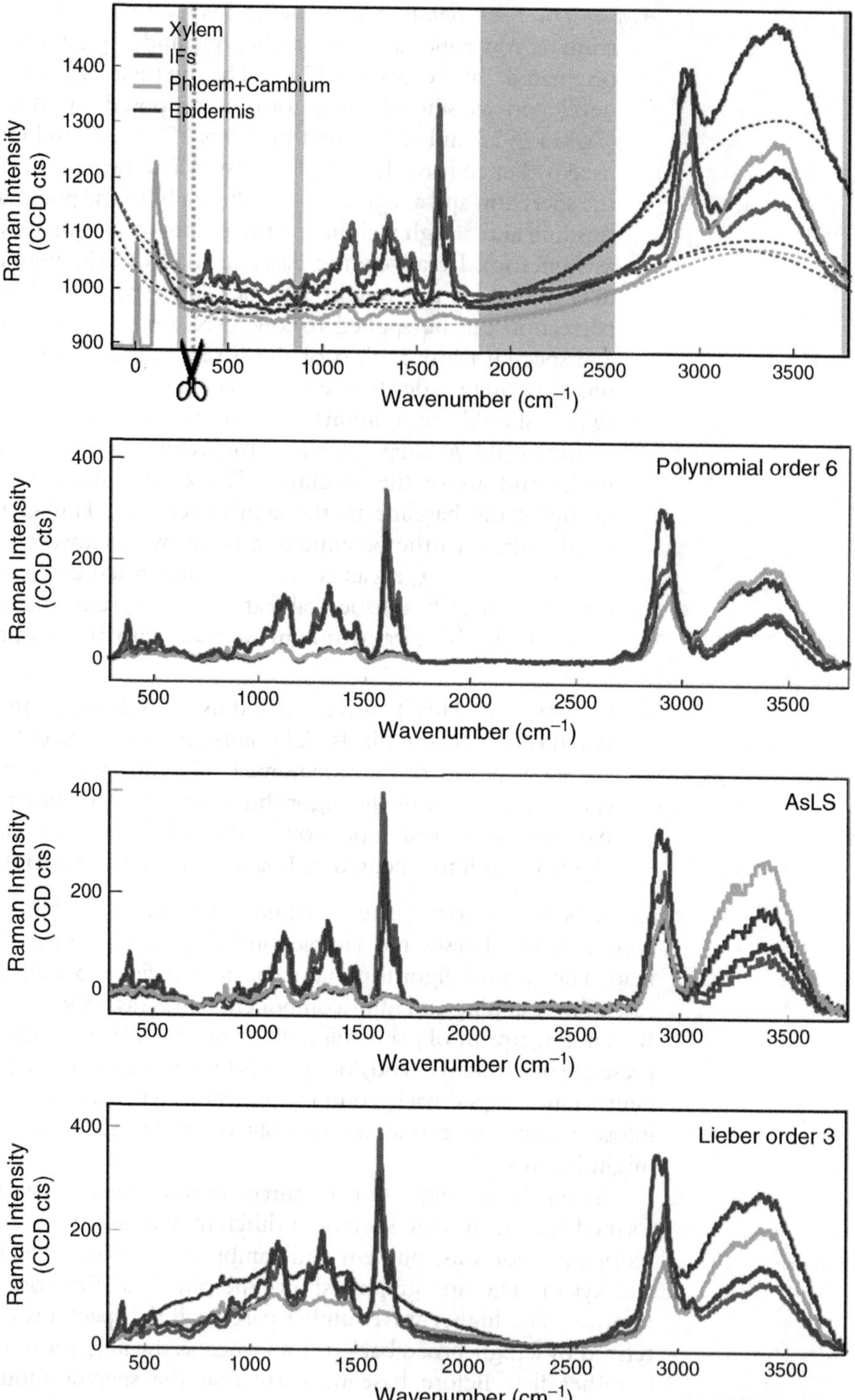

Fig. 6 Comparison of different baseline correction algorithms. (**a**) Raman average spectra of the stem of Arabidopsis extracted selectively from xylem, interfascicular fibers (IFs), phloem and cambium, and

output is displayed in the next graph. The spectral background is now regular and brought to zero. Tissues with primary cell wall (epidermis (dark green) and phloem and cambium (light green)) show less intense spectra. The iterative baseline correction by "asymmetric least squares" (AsLS) is exemplified in Fig. 6c. The trend in the intensities along the spectral region is similar and differs mainly in the OH-stretching, where the highest background occurs. This shows the higher and more different the background becomes the more difficult it is to find and decide what is the right background correction. In contrast, after applying the "Lieber algorithm" correction of order 3 with 100 iterations, the background changes its shape: the lower spectral region adopts a curved shape, whereas the CH region and water (2500–3700 cm^{-1}) becomes flatter than in the untreated spectra (Fig. 6d). This algorithm does not allow to select specific areas but is also taking into account a continuous spectral region.

This is just a demonstration for the need and challenges to find the optimal background correction mode for spectral data sets. Indeed, when the subtraction is applied to spectra comprising mainly the fingerprint region, it is usually easier and all approaches give more similar results.

3.3.4 Derivatization and Normalization

Derivation: Very tiny changes and overlapping bands can be sometimes very struggling to distinguish. The derivation of spectra offers the enhancement of these subtle changes in order to make them more visible [80–82] as has been demonstrated in FTIR and Raman spectra of wood [83, 84]. First derivative spectra perform in a way a baseline removal and is the rate of change of the intensity with respect to the wavelength. Second derivative spectra will increase the sharpness (the derivative bandwidth decreases with increasing order) of the bands independently of the signal and the spectral peaks become a negative minimum. They both can be useful to unmask overlapping bands (with similar wavenumber values) and remove baseline differences. Therefore, this kind of pretreatment can be very useful before applying multivariate data analysis methods.

Fig. 6 (continued) epidermis—having all manifold backgrounds. The areas in blue were selected to fit the polynomial background subtraction in (**b**). Further tested algorithms were asymmetric least squares (AsLS) (**c**) following the protocol in [70] (parameters: epidermis ($\lambda = 90946.083$, $p = 0.00107$), IFs ($\lambda = 220371.79$, $p = 0.001$), phloem and cambium ($\lambda = 175262.09$, $p = 0.001$), and xylem ($\lambda = 122,283,25$, $p = 0.001$)) and the Lieber algorithm of order 3 (100 iterations) (**d**). Spectra parameters: $\lambda_{ex} = 532$ nm, LPO = 35 mW, integration time = 0.1 s

Normalization: The normalization of spectra is often helpful to compare the composition within a sample (qualitative changes) without regarding the quantitative role of the intensity [71]. Normalization eliminates the influences of concentration or measuring parameters but has to be adjusted always according the research question.

3.4 Hyperspectral Image Data Analysis

The hyperspectral data cube. Before going any further it is necessary to define the multivariate nature of the spectral maps. Considering a cross section of the stem of Arabidopsis where the epidermis and cuticle is of interest to be imaged by Raman microscopy. A small measurement ($20 \times 20\ \mu m^2$) with high lateral resolution, that is, around 0.3 μm (with the 532 nm excitation wavelength laser and a NA of 1.4), consists of a total number of 3600 spectra, that is, 3600 pixels that are spatially ordered within the scanned area. Until here we are not out of the 2D space. Since we are acquiring hyperspectral data, it means that not one value is recorded, but all intensities along the spectral range. The spectral resolution depends on the grating used; for example, for 3800 cm^{-1} around 1024 positions are recorded with a 600 g/mm grating. Then to our 2D data matrix a third dimension has to be added in order to obtain the hyperspectral data cube of $60 \times 60 \times 1024$ dimensions in which each point corresponds to an intensity for a determined spatial and spectral position (*see* Fig. 7a).

This overwhelming amount of data needs to be analyzed. Based on the underlying spectra images can be generated using univariate and/or multivariate approaches. The univariate method produces abundance maps (heat maps) of single components or spectral regions without establishing any relations between them whereas

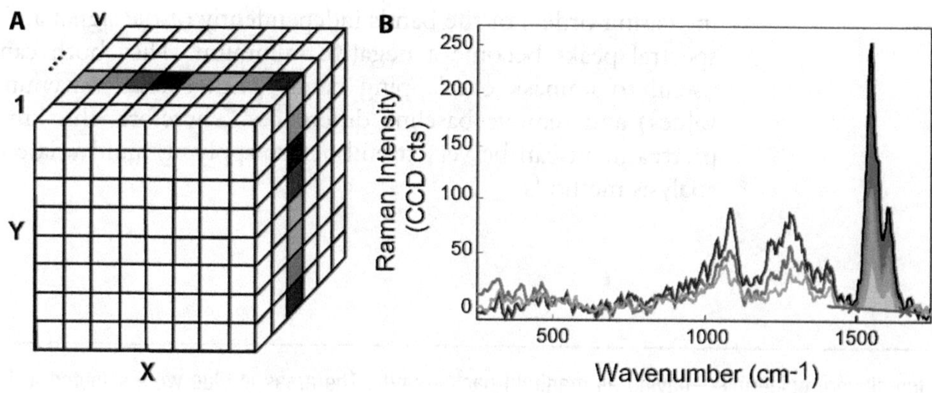

Fig. 7 Hyperspectral data cube analysis. (**a**) Hyperspectral data cube showing the spatial and spectral coordinates. Each small square represents the intensity at a specific wavenumber (*ν*) and location (*x,y* positions) in the sample. (**b**) Spectra from three different spatial positions with different intensities of the same band (spectral coordinate). For image generation, a specific wavenumber/descriptor is selected from the hyperspectral data cube e.g. lignin band at 1600 cm^{-1}. The color palette indicates the different intensities of this band in the spectra taken along the whole image

the general multivariate approach obtains distribution maps of components and the related average or basis spectra (normally weighted) taking into account all (or a chosen part) of the hyperspectral data available. Multivariate methods try to unravel the hidden relation between the spectral data based on the shape, intensity, and position along (and between) the spectra.

3.4.1 Univariate Approaches

The univariate analysis is named so because only one specific Raman band (i.e., chemical component or functional group) is "imaged", that is used for generating the heat map. This approach is helpful to get an overview of the components that are present in a sample if some knowledge about the nature of the sample is present (i.e., the user knows which bands are significant for the study case) and the component band is not overlapping with others. There are different ways to obtain images of single components: the easiest and most straightforward approach is to translate the intensity of the band maximum to a color palette in which the tonality is analogous to the intensity of the band and thus to the amount of the component along the scan (given that the intensity is not altered by uneven focus). The filter is applied to all pixels (= spectra) in the scan in a way that a false color image is generated and corresponds to the abundance of the specific band (and component) along the sample analyzed. Another strategy used very often is the integration of the whole band by selecting the boundaries at the side of the band of interest (red cross in Fig. 7b) to draw a baseline (red line). In Fig. 7a the colored pixels denote the univariate approach where only one band is considered at a time. The color tonality of the pixel is proportional to the intensity of the specific band at the different spatial positions (x, y).

The total area (including background) can be integrated but normally a line is drawn from the boundaries of the band (the line is fitted with one or more neighbors of the boundary) in order to remove the intensity from the background (*see* **Note 22**). This is the procedure to follow when no prior background subtraction has been done or the spectral background shape is very heterogeneous (as example *see* Fig. 6). Other filters are based on spectral characteristics such as band width, standard deviation, or shift of the peak position [10].

In Fig. 8, an example of Raman images based on band integration of spectra acquired from a cross section of the stem of Arabidopsis is shown. Images from **a** to **e** correspond to the integration over different spectral regions (*see* figure legend for more information). Integration over the CH stretching region visualizes all organic compounds in the stem, that is, all tissues and cells (Fig. 8a). When restricting this region to the asymmetric CH_2 stretching region (higher in lipids due to the long aliphatic chains), the cuticle in the epidermal cells of Arabidopsis is highlighted (Fig. 8b). The integration over the main pectin band generates an

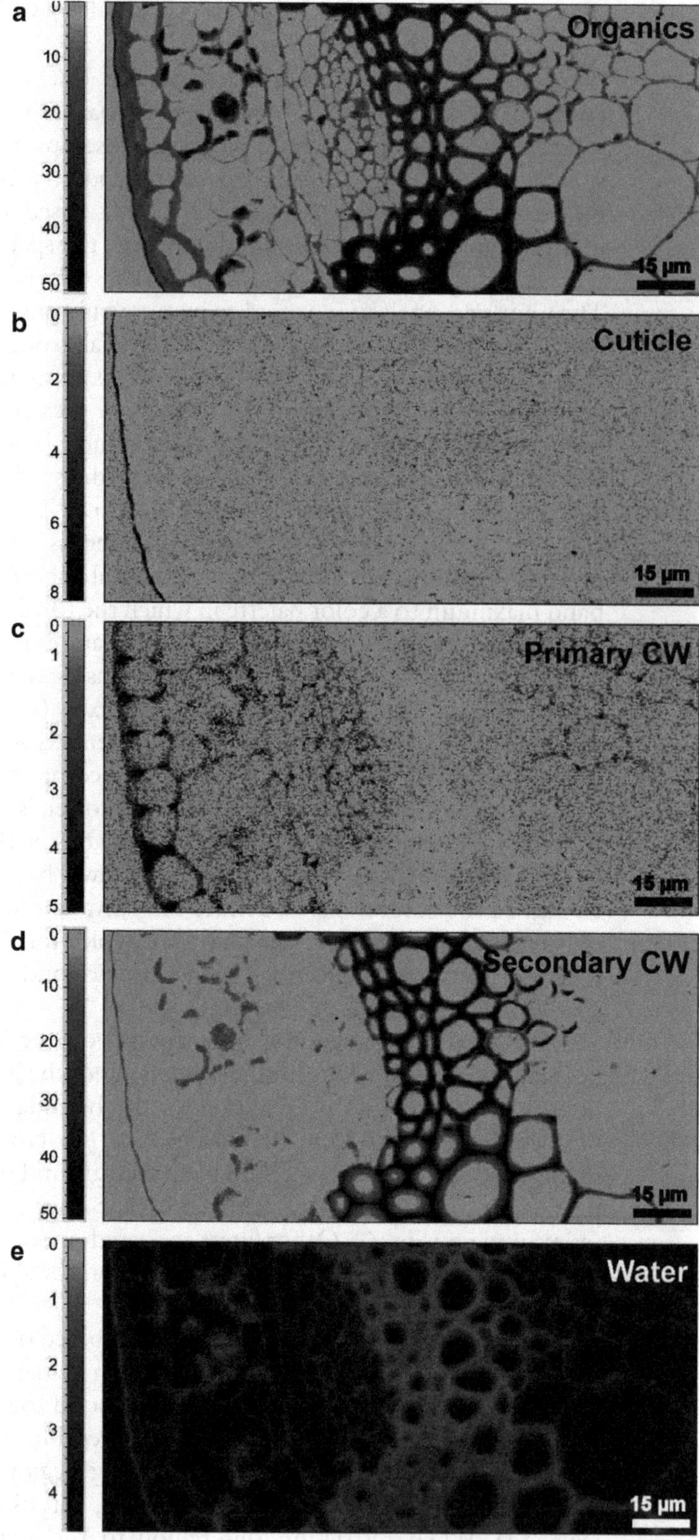

Fig. 8 Raman images based on band integration. False colored images of a cross section of Arabidopsis stem (sampled 20 cm above ground) generated by integrating over (**a**) 2760–3044 cm^{-1} (CH vibrational modes, organics), (**b**) 2825–2869 cm^{-1} (symmetric CH_2 stretch of lipids, cuticle), (**c**) 841–872 cm^{-1} (α-glycosidic bond of pectin, primary cell wall), (**d**) 1518–1714 cm^{-1} (ring deformation mode of lignin, secondary cell wall) and (**e**) 3116–3722 cm^{-1} (OH stretch of water, lumen). Spectra parameters: $\lambda_{ex} = 532$ nm, LPO = 35 mW, integration time = 0.1 s

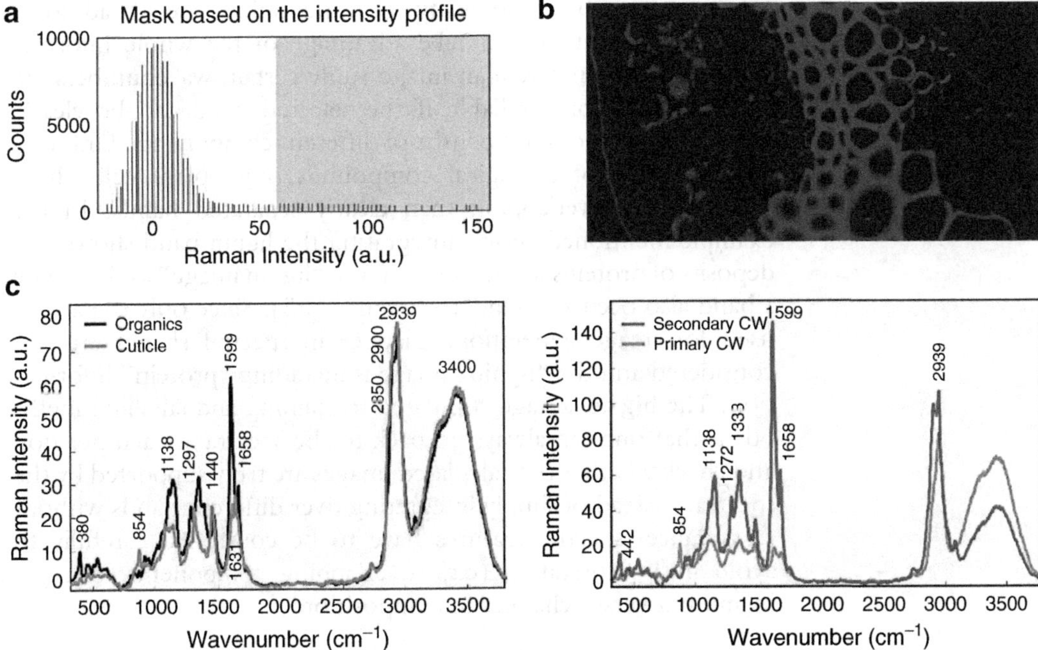

Fig. 9 Average spectra of specific areas based on intensity thresholds of band integration. Based on the integration images and their intensity profiles, pixels with defined intensities (e.g., lignin >70) can be selected (masks filters, (**a**)). By this, for example, the xylem (locations rich in lignin, blue areas, (**b**)) can be selected and average spectra calculated (**c**, **d**). The mask filter over the integration image of the organic compounds (Fig. 8a) gives an average spectrum of all tissues present in the image (**c**, in black). In the same manner, cuticle (**c**, in blue), secondary cell wall (**d**, in red), and primary cell wall (**d**, in green) average spectra were calculated from the integration images of Fig. 8b–d, respectively

overview of the primary cell wall (Fig. 8c) while considering the lignin main band at 1600 cm^{-1} shows mainly the xylem (Fig. 8d). Integrating the broad OH-stretching shows which tissues and cells contain water (Fig. 8e).

The images can be used to select or discard regions of interest in the image based on the Raman intensity of selected bands (*see* Fig. 9a, b). A mask generated in such a way can then be used to calculate average spectra of the selected areas and determine their chemical profile. In this example, average spectra of all the organics (all the stem), the cuticle, secondary and primary cell walls have been extracted (Fig. 9c) based on the distribution maps shown in Fig. 8. To compare Raman images and/or spectra taken with different parameters (e.g., different integration time or laser power) spectra have to be normalized or images calculated as the ratio of a defined band integral in relation to the integral of the band of interest (*see* **Note 23**). As shown the chemical distribution of a component might coincide with a cell type, tissue, or organ. Thus, the integration approach can be appropriate to distinguish tissue types with more specificity through the chemical components images.

However, band intensity or integration are "traditional approaches" that do not take advantage of the whole information contained in a Raman image (only certain wavenumbers are used) and are only reliable if the selected band can be clearly delineated and does not consist of different components. Unfortunately spectra of biological compounds (e.g., plant cells) have usually more overlapping than clearly separated bands. In the example mentioned above, integrating the lignin band shows also deposits of proteins in the lumen in the "lignin image" as the amide I band also occurs around 1670 cm^{-1} [55]. Since only the area is used for image generation, changes in spectral shapes are not considered and the "lignin" image is including "protein" information. The big advantage compared to staining and labelling methods is that one can always go back to the spectra at each position and check whether the calculated images are truly supported by the spectra. Instead of simply integrating over different bands without reassurance, spectral features have to be controlled carefully to avoid misinterpretation (e.g., overlapping components, changes in amount versus changes in composition).

3.4.2 Multivariate Data Analysis

Multivariate data analysis has become a very common (and necessary) approach when analyzing biological samples with Raman microscopy. Plant cells are not only composed of a few easily isolated components but are of multicomponent nature. Furthermore, components may be present in different modifications and might interact in different ways. All these varieties of components and their mixtures and interactions are represented by changes in the Raman spectra, representing a unique molecular fingerprint. As biological samples are composed mainly of the same elements (and functional groups) the Raman features are also to a certain grade similar and might overlap along the spectrum or fingerprint region. It is the task of the analyzer, with help of multivariate approaches, to find those relationships or groups hidden in the hyperspectral data cube (*see* definition below) generated after scanning a sample with a Raman microscope. Some multivariate data analysis methods are indeed designed as unmixing algorithms that will help to decrypt overlapping and hidden bands in the spectra by using all the information contained in the spatial and spectral coordinates.

Furthermore, multivariate methods (e.g., PCA) also perform a reduction of the dimensionality to a few main components and thus of the size of the data. A different way to reduce the amount of data is also working with spectral descriptors instead of the spectral raw data. They consist of band integrations, band ratios, band heights, or band widths, among others. The data reduction avoids the curse of dimensionality (small amount of relevant information in a big space) when dealing with huge amounts of data and improves the data structure [85]. The data are then represented by parts of the spectra with relevant information. The software ImageLab (Epina, Austria) allows this reduction before applying further multivariate analysis, reducing significantly the time and resources necessary.

There are plenty of algorithms and approaches. A definition of all of them is beyond the scope of this chapter, but examples are described in the following.

Cluster analysis aims to find groups in the data. The pixels are grouped to a certain cluster based on the distance or similarity calculated between the spectrum in this pixel and the average spectrum of the existent clusters previously computed [69]. There are several variations [86], hierarchical clustering (tree clustering) being the most known. However, it is not always the case that components, although related, follow a hierarchy. For such cases, (fuzzy) *k*-means clustering would be the better choice. The latter is especially useful when the defined spot size of the measurement is big enough to contain mixtures of components because it allows the assignment of one same pixel to several clusters [87]. However a clear disadvantage of *k*-means clustering is that the number of clusters has to be chosen *a priori*. A way to define the optimal number of clusters can be done by a trial-and-error process [88]. Approximations to the correct number can be done by plotting the percentage of variance explained (eigenvalues) versus the number of clusters or by other assumptions [89]. A compromise between the data reduction and the accuracy is necessary and experience or at least an idea about what to expect is advantageous.

The similarity between clusters is calculated based on the distance to the centroids (pro- or metaclusters), which can be defined in different manners. The choice of the distance will also influence the output in the analysis, as well as the number of iterations and the starting point (*see* **Note 24**). Depending on what constraints the user needs, the distance type should be chosen: in the case of the intensity of the spectra not being relevant and only the shape is to be taken into account, one can choose to preprocess the spectra with a Euclidean normalization and use the same distance for the clustering (*see* **Note 25**). Cluster analysis can be performed using many software (e.g., Witec, ImageLab, Cytospec).

Vertex Component analysis (VCA): Strives to find the most pure components (endmembers) in a sample [90]. It is an iterative method that defines a space within the projection of the most pure pixels (orthogonal subspace projection). Each spectrum is defined as an *N* dimensional space where *N* is the number of spectral pixels. Mixtures of the components are enclosed in this space whereas components with the same composition but different concentration will have parallel projections to the endmembers. The hyperspectral data cube is split after the VCA in pure spectra called endmembers (i.e., pure lignin or a constant mixture of two components) and their corresponding abundance maps, which reduces the dimensionality of the data. Each pixel is described by a mixture of pure components in a ratio defined by the abundance maps (called matrix of coefficients). The number of endmembers needs to be defined by the user and knowledge of the variability in the sample has to be taken into account. This approach is a very fast algorithm since almost not constraints are applied in the calculation.

VCA analysis is shown in Fig. 10 on a selected region (zoom) of Raman data obtained from the stem of Arabidopsis (shown in Figs. 8 and 9). Based on previous knowledge of the sample and after a trial-and-error procedure, a final number of seven endmembers (EM) was selected (*see* **Note 26**). The images on the top (Fig. 10a) indicate the abundance maps of the different endmember spectra plotted in the bottom part (Fig. 10b). For better comparison the spectra are also shown after baseline correction and normalization against the lignin main band around 1600 cm^{-1} (Fig. 10c, d). EM 1 represents the cell corners of the lignified cells (interfascicular fibers and xylem) which show the highest content of lignin in the spectrum (EM 1, in black). Note that the spectrum is very similar to the spectrum of pure lignin (1140, 1271, 1334, 1453, 1596, and 1660 cm^{-1}) (Fig. 3e). EM 2 displays the lumen-sided inner layer of the interfascicular fibers and vessels but with more intensity in the younger vessels (right-top part of the image) and the corresponding endmember shows a weak lignin spectrum with higher background. The lumen or free-tissue part of the scan is given by EM 3, with no clear bands visible (grey spectrum). Deposits in the lumen (EM 4) attached to the inner cell walls are found mostly in the phloem region and are carrying compounds of proteic and lipidic character as shown by the red spectrum in Fig. 10b, d (possibly the amino acid phenylalanine (1004 cm^{-1}), 1453, 1600, and 1663 cm^{-1} (amide I band). EM 5 highlights the cell wall of interfascicular fibers with a higher S/G ratio (in purple). After normalization against the lignin main band at 1600 cm^{-1}, the band at 1659 cm^{-1} is also raised in comparison to the EM 1 and EM 6, indicating the high amount of coniferyl aldehyde and coniferyl alcohol in interfascicular fibers. The cellulose content is also higher than in EM 1 and EM 6 (bands at 380, 437, 520, 1096, 1121, and 1380 cm^{-1}). The cellulose molecules (microfibrils) oriented more parallel to the laser polarization (x-directions) are depicted in EM 6 (in green) with an increase of the 1096 cm^{-1} band in comparison to the 1121 cm^{-1}. At last, EM 7 shows the primary cell wall composed mainly of pectin (443 and 856 cm^{-1}), hemicellulose, and cellulose. VCA is able to discriminate cell parts based on their chemical composition and elucidates further features not recognizable with univariate approaches. Also without prior knowledge of the sample, the algorithm is able to detect the most pure components and hidden relationships among them.

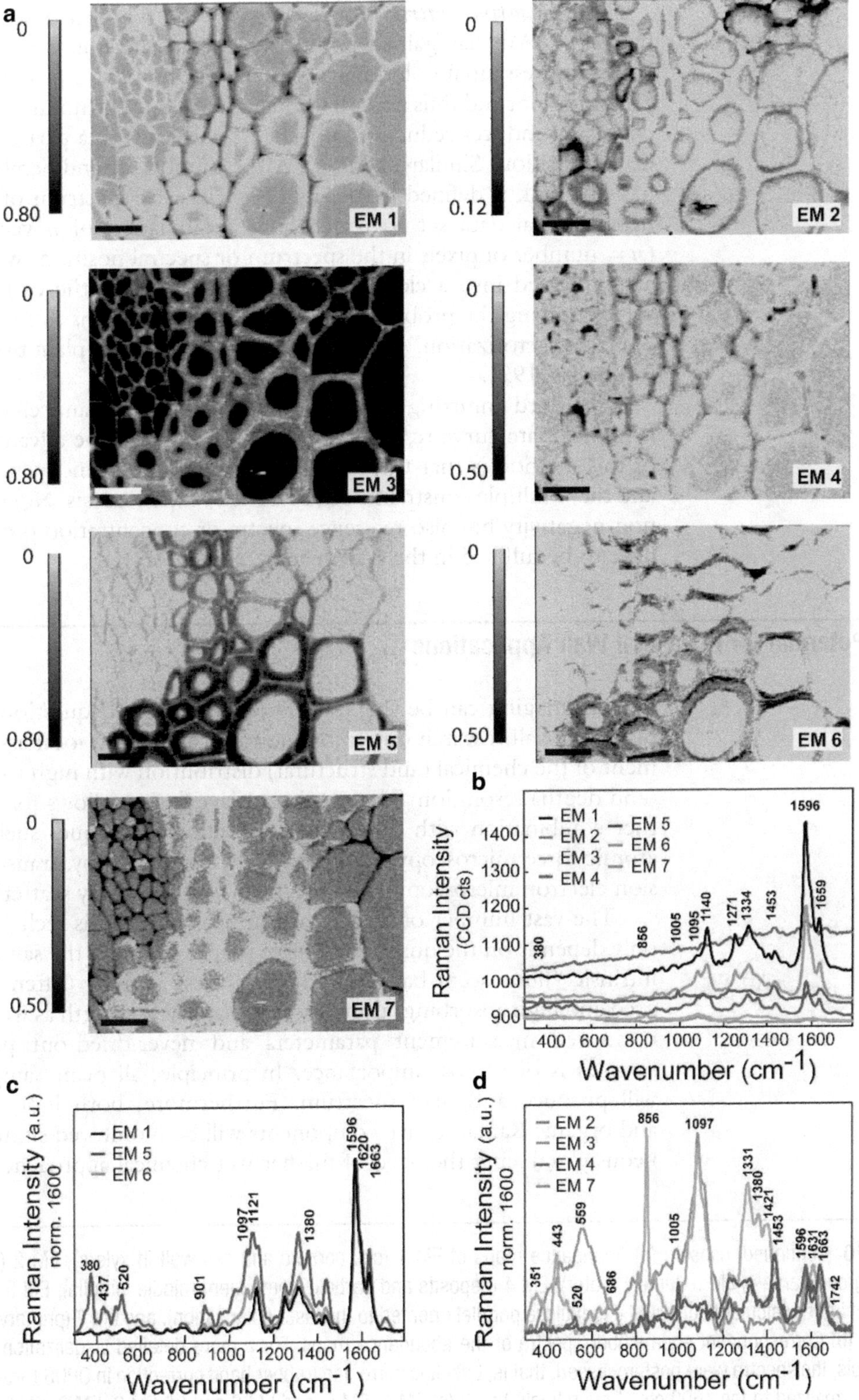

Fig. 10 Vertex component analysis (VCA) on the phloem and xylem of the Raman image of Arabidopsis. The fingerprint region of the spectra was used (300–1800 cm^{-1}) and a final number of seven endmembers was

Non-negative matrix factorization (NMF) was first described by [91]. NMF has gained popularity due to the non-negativity constraint, essential in biological systems. In biology, a product is present or not and thus never has a negative value. The nonnegative constraint induces reduction of the data and leads to a part-based decomposition. Similar like in VCA the number of endmembers, basis or rank is defined by the user. Again, each spectrum of the hyperspectral data set is treated as a combination of n vectors (n = number of pixels in the spectrum or spectral position) which are projected into a cloud of the space defined by the extreme values, solving the problem of facing multiple solutions as a result of matrix factorization. An example of NMF applied to plant tissues is found in [92].

The third unmixing approach already applied on plant cell walls is multivariate curve resolution (*MCR-ALS*) [78]. The advantage of this method is that the components must not be independent and that multiple constraints can be added to the analysis. Not only non-negativity but also reference spectra or concentration profiles have to be fulfilled in the system analyzed [78, 93].

4 Potential for Plant Cell Wall Applications

Raman imaging can be the key to many unresolved questions in plant cell wall research due to the nondestructive and local assessment of the chemical (and structural) distribution with high lateral (and depth) resolution. The nondestructive feature allows for further combination with other characterization techniques such as atomic force microscopy, scanning electron microscopy, transmission electron microscopy, mass spectrometry, and X-ray scattering.

The vast number of fields that can profit from this technique only depends on the quality of sample preparation and the samples intrinsic "fluorescent background." For these samples (often colored and high absorbing) using the right laser wavelength as well as optimized measurement parameters and never-dried-out plant samples is of utmost importance. In principle, all plant samples will produce a Raman spectrum. Furthermore, both inorganic and organic Raman-active components will be monitored simultaneously, reducing the need of further wet chemical approaches or

Fig. 10 (continued) chosen: (**a**) Abundance maps of EM 1 (cell corners and cell wall in xylem), EM 2 (high background areas), EM 3 (lumen, water), EM 4 (deposits and partially component middle lamella), EM 5 (cell wall of interfascicular fibers), EM 6 (cellulose parallel oriented to the laser polarization), and EM 7 (primary cell wall). (**b**) Correspondent endmember spectra of the abundance maps. For a more detailed visualization and analysis, the spectra were postprocessed, that is, baseline corrected (rubber band correction in OPUS (Bruker)) and normalized to the 1600 cm^{-1} main lignin band: (**c**) EM 1, EM 5, and EM 6 and (**d**) EM 2, EM 3, EM4, and EM 7. The blind unmixing given by the VCA is able to identify parts of the anatomical characteristics of the stem based on the chemical composition given by the spectral features. Scale bar: 10 μm

part-based determinations. There are multiple examples already in the literature in which Raman microscopy has been used successfully to study plant cells and their outer walls [94–102].

5 Notes

See Table 1.

1. The choice of PEG, which polymer chain length, depends on how hard the tissue is. Higher PEG number means higher melting point, whereas, for example, PEG 600 body temperature is enough to melt and therefore the handling is more difficult.
2. Sledge microtomes are mainly used for hard samples (e.g., wood) and allow to cut bigger areas. Nevertheless, the knives are more expensive, need to be treated with care, and sharpened regularly. Rotary microtomes give high-quality sections with disposable blades that ensure always a new sharp surface and are cheaper.
3. Disposable microtome blades with 35° have shown the best performance for PEG 2000 embedded wood (spruce, poplar, beech) and Arabidopsis stem samples among others. For softer tissues (not embedded), smaller angles (e.g., 22°) have been helpful.
4. In order to obtain good and flat sections, the cutting direction plays also an important role. For wood, cross sections are of better quality (flat and cell wall not ruptured) when cutting from early to late wood. For grass or smaller samples the stem needs to be intact to get high-quality sections. Broken cuticle in Arabidopsis will induce the breaking of the inner part when cutting. Cutting in radial or tangential directions (parallel to the fiber direction) is easier and rolling/folding of the sections less problematic.
5. PEG is corrosive in the presence of water. Therefore, wrap the PEG blocks in paper towels such that only the cutting side remains unwrapped. This prevents the PEG to come into contact with the instrument and makes cleaning much easier.
6. Before cutting the sample should be trimmed in a way that only the region of interest is cut. For example, PEG blocks should be trimmed to a pyramidal shape, exposing only the sample area to the knife.
7. The nature of the sample, the quality of the blade and microtome, and the human factor (experience, patience) are crucial for the section quality and thickness. Arabidopsis stems embedded in PEG 2000 show the best section quality when the

Table 1
Troubleshooting and tips to be followed during Raman imaging approach

Step	Problem	Possible reason	Solution
Sample preparation, PEG embedding	Improper sample alignment	PEG hardens too quickly, sticking on the tweezers	Heat again in the oven or use warm tweezers to avoid sticking
	The PEG breaks	Too much air inside	Be aware of the nonhumid environment in the oven during final infiltration Because air bubbles enclose the upper part, fill above the sample with an excess of PEG
Sample preparation, sectioning	A lack of good, continuous microsections	Too much air inside, poor infiltration	See advice for Step 1B(x)
		Blade is not sharp	Use an unused position on the same blade or a new one
		Wrong cutting velocity, cutting angle	Change microtome settings
		Wrong sample size	Trim the PEG or sample block. Less sample width on the part first in contact with the knife might help
	Rolling up of section	Sample guidance is needed	Guide the section during cutting with a brush on the knife or use a coverslip and let the section slide under
Sample preparation, section deposition	Air bubbles inside	Coverslip has been put on too fast	Lift the coverslip with the help of tweezers on one side until the air bubble is removed and put slowly down again
Raman measurement	Intensity is too low	Laser power is too low	Check if any filter (e.g., polarization) is in the path; check laser power (power meter)
		Dirt on the microscope objective	Clean the objective
	Low signal (S/N)	Improper focus	Careful refocusing
		Integration time is too short	Increase the integration time
	Too much fluorescence	Sample is dry	Rewet the sample
		Bleaching is necessary	Perform a measurment with short integration time to bleach fluorescent compounds. Then run a second measurement. Beware of laser-induced artefacts!
	Sample burns	Laser power too high	Decrease laser power

Adapted from [13]

thickness is 2–3 μm (blade NH35R) and the PEG block trimmed such that almost no PEG frame is present. For thin sections a coverslip is placed on the blade to make them slide under to avoid rolling up and ensure controlled transfer to the glass slide (*see* also **Note 9**). Cross sections of wood (1×2 cm) can be cut from 5 to 10 μm, whereas bigger areas usually need to be thicker (around 20 μm). Cryosectioning (sample embedded in block of water) also allows to perform cross sections of poplar stem or spruce needles with a thickness of about 2–3 μm and 1–2 μm, respectively.

8. Often cuts tend to roll up themselves. There are two possibilities: Try to prevent rolling or unroll afterward. To prevent rolling put the brush carefully on the section while it is being cut (requires experience). Another way is to take a coverslip and put it flat onto blade such that the edge of blade and glass are parallel. Press only slightly at the lower glass edge, and the section should glide between the glass and blade, and should not roll itself. For samples not embedded in PEG, putting a drop of water on the blade to allow the micro section to float on it could also work. Unfolding afterward does work, if the sections are stiff enough. Putting the section into a small petri dish of water often results in self-unfolding. Also, tweezers can be used to unfold the section manually.
9. Vary cutting speed, blade angle, and position if micro sections are not satisfying. Air humidity and temperature can affect the PEG stiffness.
10. When placing the section on the glass slide, the water droplet might cause collapsing of very fragile sections. Placing a coverslip on the top before adding carefully and slowly the water from the side will maintain the sample intact.
11. Nail polish can be too liquid for sealing. To get a thicker consistency the cap can be removed overnight. Since nail polish softens when heated up, sealing for high temperature experiments should be done with nail glue.
12. 532 nm and 633 nm lasers can penetrate common glass coverslips (the objective must be coverslip corrected to the right thickness, usually 0.17 mm). With 785 nm and 1064 nm lasers glass becomes problematic (fluorescence, penetration) and therefore working without coverslips is preferred or special accessories become necessary (CaF_2 slides, quartz coverslips, etc.).
13. Air bubbles must be avoided because they might expand when heated by the laser and dry out the sample or cause movement. Experience shows that the removal of bubbles by pressing on the coverslip does not work very well. In case of air bubbles it is

recommended to carefully lift the coverslip up again, if necessary rewet and carefully cover again with the help of tweezers.

14. Cylindrical weights work very well but also clamps might be used.

15. Ensure that both the slide and the coverslip are coated to guarantee the sealing.

16. The intensity and position of a peak are the primary source of information in Raman spectroscopy. However, it is important to consider the following: (a) Differences in peak intensity might not only arise from changes in the amount of a certain component in the sample but also from changes in composition/interaction, focus (no plane surface, drift of the scan table or *z*-axis), temperature changes (room temperature, instrumentation), and laser irradiation (heating up or even modification of the sample); (b) peak positions are very important characteristics, but values are not absolute. The error of the used grating has to be taken into account and correct calibration of the spectrometer has to be ensured. Furthermore, peaks can shift their position because some vibrations are sensitive to their surroundings and substitutions (e.g., solvent effects). Therefore, tabulated values have to be taken with a grain of salt and should always be checked for consistence with other reference literature; (c) if samples (spectra) need to be compared, the preparation and measurement procedure as well as the storage conditions have to be the same; (d) record spectra of all the materials your sample is embedded in or in contact with, for example, PEG, microscopic glass slides, plastic vials, or immersion oil, to be able to distinguish their contribution.

17. The alignment of the optical fibers entering the microscope at the top can be optimized by turning the thumb screws very carefully. Adjusting the laser output to the fibers is only recommended for experienced users. Ensure that the spectrometer is calibrated.

18. Integration time can have a similar effect on the sample like laser power. For fluorescent samples high integration times might cause overheating and burning of the section. Be aware that high integration time (or high laser power) can also cause chemical modification of the sample. Depending on the efficiency of your system (λ, CCD camera), decreasing the integration time to values of 0.05–0.1 s can yield good S/N ratio for a laser power of about 20 mW for fluorescent samples (e.g., xylem containing lignin). For lignin it was found that even very short integration times of 10 ms affect the ratio of the 1660/1600 cm^{-1} bands, which can be observed in subsequent measurements of the same spot [103]. Despite Raman spectroscopy is often referred to as being "non-destructive," it is clear

that the lasers employed in the experiments can direct high energies to the sample, especially if focused on a small spot. Although this is not an issue in many cases, it has to be kept in mind and samples should be always tested for laser susceptibility before performing measurements.

19. Cosmic ray removal has to be adjusted for the shape of the bands present in the spectra of the sample. A value to start is 2–4 data points (spectral pixels) as width; the noise factor should be adjusted as well.

20. Smoothing has to be done carefully. Check the effect of the smoothing algorithm not just on one spectrum of the image scan, but on several "representative" spectra along the scan, for example, in cell corners, cell wall, and cambium. Usually the Savitzky–Golay smoothing with 5–9 neighbors is a good compromise to not lose spectral details.

21. For the polynomial background subtraction the Rayleigh region (below the bandpass filter ~250 cm^{-1}) has to be removed before setting first anchor points at the extremes of the remaining spectrum (e.g., 300 and 3750 cm^{-1}). Then around 2000 cm^{-1} (normally no information is encountered around this frequencies) and so on until the best fitting is achieved. Control the baseline correction at different positions along the scan even at "free of sample" areas. Polynomial order of 3 usually works well as a starting point. An example of anchor areas for the fitting is shown in Fig. 7 (top) as light blue areas.

22. Integration of bands which position and/or intensity change along the scan will need a broad enough border selection without including neighboring bands. When the background fluctuates it is recommended to average a higher number of neighbor pixels for setting the boundary points. A higher number of averaged pixels often compensates the change of the base line at different points. If available in the software, set baseline points and area borders separately.

23. When comparing measurements taken at different conditions the most "common" approach is normalization against the intensity of a specific band. When trying to compare the lignin structure between species, it is helpful to normalize the 1600 cm^{-1} band, since the aryl stretching of the phenolic ring is common to all. When comparing methylation degree in pectin it will be more suitable to normalize over the main pectin band, since we consider the amount of pectin (α-glycosidic bond) the same for all samples. However, if the amount of pectin is to be assessed, the right choice will be to rather normalize against the cellulose band at 380 cm^{-1} where no contribution of pectin is expected. The C-H peak is very

sensitive to changes in the focal plane, which is the case if samples are not absolutely plane and therefore often not suitable as a reference band.

24. During cluster analysis, the algorithm starts randomly choosing some pixels to calculate the first iteration of the centroids average. This influences the output if the number of iterations is not optimized and the algorithm does not converge. Make sure that the analysis is right by performing several calculations for each parameter setup.

25. The Manhattan distance is the distance between two points (e.g., the intensity at 1600 cm^{-1} between two spectra) calculated as the sum of the differences of their corresponding components (intensity and spectral position or pixel) and is optimal when the small contributions are to be taken into account. The Euclidean distance is the shortest way between two points and is to be used when the influence of small values should be suppressed. Normalization (vector normalization or min/max on a specific band) is useful to reveal changes in composition (band shape) but no longer gives relevant changes in amounts (band intensity).

26. The result of VCA is affected by the number of endmembers and the spectral region selected for the analysis. The analysis needs supervision of an experienced user. Depending on the scientific question, the fingerprint region can be enough and give better results than taking into account the whole wavenumber range. If some conclusions about the crystallinity of the epicuticular layer or the length of the lipidic chains have to be done, the C-H stretching region might be important as well. Starting with a known number of components (e.g., 4 EM: compound middle lamella, cell wall, cellulose orientation, primary cell wall) will help to build up a trustful analysis. If, for example, a higher methyl esterification of pectin in some areas is expected, then an extra endmember will be added and so on. It is important that the EMs can be explained and make sense. Repeat the analysis for different samples (replicates) in order to obtain definitive conclusions.

Acknowledgments

The authors thank Martin Felhofer for support in the graphic design, Ivan Sumerskii in the group of Antje Potthast for providing the milled wood lignin reference samples and Marie-Theres Hauser for providing the *Arabidopsis thaliana* specimens. All people mentioned belong to the University of Natural Resources and Life Sciences, BOKU, Vienna. The work was funded by the Austrian Science Fund (FWF): START Project [Y-728-B16] and from the

European Research Council (ERC) under the European Union's Horizon 2020 research and innovation programme grant agreement No. 681885.

References

1. Graham LE, Cook ME, Busse JS (2000) The origin of plants: body plan changes contributing to a major evolutionary radiation. Proc Natl Acad Sci 97(9):4535–4540. https://doi.org/10.1073/pnas.97.9.4535
2. Agarwal UP, Atalla RH (1986) In situ Raman microprobe studies of plant cell walls – macromolecular organization and compositional variability in the secondary wall of *Picea mariana* (Mill) Bsp. Planta 169(3):325–332
3. Jones L, Ennos AR, Turner SR (2001) Cloning and characterization of irregular xylem4 (irx4): a severely lignin-deficient mutant of Arabidopsis. Plant J 26(2):205–216. https://doi.org/10.1046/j.1365-313x.2001.01021.x
4. Jarvis MC, McCann MC (2000) Macromolecular biophysics of the plant cell wall: concepts and methodology. Plant Physiol Biochem 38(1/2):1–13
5. Laschimke R (1989) Investigation of the wetting behavior of natural lignin — a contribution to the cohesion theory of water transport in plants. Thermochim Acta 151:35–56. https://doi.org/10.1016/0040-6031(89)85335-3
6. Sarkanen KV, Ludwig CH (1971) Lignins: occurrence, foramtion, structure, and reactions. In. Wiley-Intersci, New York, NY, p 916
7. Gindl W, Gupta HS, Schoberl T, Lichtenegger HC, Fratzl P (2004) Mechanical properties of spruce wood cell walls by nanoindentation. Appl Phys A Mater 79(8):2069–2073. https://doi.org/10.1007/s00339-004-2864-y
8. Salzer R, Steiner G, Mantsch HH, Mansfield J, Lewis EN (2000) Infrared and Raman imaging of biological and biomimetic samples. Fresenius J Anal Chem 366:712–726
9. Burrell M, Earnshaw C, Clench M (2007) Imaging matrix assisted laser desorption ionization mass spectrometry: a technique to map plant metabolites within tissues at high spatial resolution. J Exp Bot 58(4):757–763. https://doi.org/10.1093/jxb/erl139
10. Schmidt U, Ibach W, Muller J, Weishaupt K, Hollricher O (2006) Raman spectral imaging – a nondestructive, high resolution analysis technique for local stress measurements in silicon. Vib Spectrosc 42(1):93–97. https://doi.org/10.1016/j.vibspec.2006.01.005
11. Griffith PR (2009) Infrared and Raman instrumentation for mapping and imaging. In: Salzer R, Siesler HW (eds) Infrared and Raman spectroscopic imaging. Wiley-VCH Verlag GmbH & Co. KGaA, Weinheim, pp 3–64
12. Edwards HG (2005) In: Smith E, Dent G (eds) Modern Raman spectroscopy—a practical approach. John Wiley and Sons Ltd, Chichester, p 210
13. Gierlinger N, Keplinger T, Harrington M (2012) Imaging of plant cell walls by confocal Raman microscopy. Nat Protoc 7(9):1694–1708. https://doi.org/10.1038/nprot.2012.092
14. Harris DC, Bertolucci MD (1989) Symmetry and spectroscopy. Dover Publications, Inc., New York
15. Griffiths PR, de Haseth JA (2007) Introduction to vibrational spectroscopy. In: Fourier transform infrared spectrometry. John Wiley & Sons, Inc., Hoboken, NJ
16. Pelletier MJ, Pelletier CC (2010) Spectroscopic theory for chemical imaging. In: Šašić S, Ozaki Y (eds) Raman, infrared, and near-infrared chemical imaging. John Wiley & Sons, Inc., Hoboken, NJ
17. Colthup NB, Daly LH, Wiberley SE (1990) Introduction to Infrared and Raman Spectroscopy. Academic Press Inc.
18. Raman CV, Krishnan KS (1928) A new type of secondary radiation. Nature 121:501–502. https://doi.org/10.1038/121501c0
19. Miller FA, Mayo DW, Hannah RW (2003) Course notes on the interpretation of infrared and Raman spectra. John Wiley & Sons, Hoboken, NJ
20. Smith WE, Dent G (2005) Introduction, basic theory and principles. In: Modern Raman spectroscopy—a practical aproach. John Wiley & Sons, Chichester
21. Parson WW (2009) Modern optical spectroscopy. Springer, Dordrecht
22. Kip BJ, Meier RJ (1990) Determination of the local temperature at a sample during

Raman experiments using stokes and anti-stokes raman bands. Appl Spectrosc 44 (4):707–711. https://doi.org/10.1366/0003702904087325

23. Reichenbächer M, Popp J (2012) Vibrational spectroscopy. In: Challenges in molecular structure determination. Springer, Berlin. https://doi.org/10.1007/978-3-642-24390-5_2
24. Agarwal UP, Reiner RS, (2009) Near-IR surface-enhanced Raman spectrum of lignin. J Raman Spectrosc 40(11):1527–1534
25. Bock P, Gierlinger N (2019) Infrared and Raman spectra of lignin substructures: Coniferyl alcohol, abietin, and coniferyl aldehyde. J Raman Spectrosc
26. Agarwal UP, Ralph SA, Reiner RS, Baez C (2016) Probing crystallinity of never-dried wood cellulose with Raman spectroscopy. Cellulose 23(1):125–144. https://doi.org/10.1007/s10570-015-0788-7
27. De Gelder J, De Gussem K, Vandenabeele P, Moens L (2007) Reference database of Raman spectra of biological molecules. J Raman Spectrosc 38(9):1133–1147. https://doi.org/10.1002/jrs.1734
28. Czamara K, Majzner K, Pacia MZ, Kochan K, Kaczor A, Baranska M (2015) Raman spectroscopy of lipids: a review. J Raman Spectrosc 46(1):4–20
29. Thomas LH, Forsyth VT, Sturcova A, Kennedy CJ, May RP, Altaner CM, Apperley DC, Wess TJ, Jarvis MC (2013) Structure of cellulose microfibrils in primary cell walls from collenchyma. Plant Physiol 161(1):465–476. https://doi.org/10.1104/pp.112.206359
30. Gierlinger N, Luss S, Konig C, Konnerth J, Eder M, Fratzl P (2010) Cellulose microfibril orientation of *Picea abies* and its variability at the micron-level determined by Raman imaging. J Exp Bot 61(2):587–595
31. Agarwal UP, Reiner RS, Ralph SA (2010) Cellulose I crystallinity determination using FT-Raman spectroscopy: univariate and multivariate methods. Cellulose 17(4):721–733. https://doi.org/10.1007/s10570-010-9420-z
32. Schenzel K, Fischer S, Brendler E (2003) New method for determining cellulose I crystallinity by means of FT raman spectroscopy. Abstr Pap Am Chem S225:U279–U279
33. Wiley JH, Atalla RH (1987) Band assignments in the Raman spectra of celluloses. Carbohydr Res 160:113–129
34. Schenzel K, Fischer S (2001) NIR FT Raman spectroscopy – a rapid analytical tool for detecting the transformation of cellulose polymorphs. Cellulose 8(1):49–57. https://doi.org/10.1023/A:1016616920539
35. Denise T. B. De Salvi, Hernane da S. Barud, Oswaldo Treu-Filho, Agnieszka Pawlicka, Ritamara I. Mattos, Ellen Raphael, Sidney J. L. Ribeiro, (2014) Preparation, thermal characterization, and DFT study of the bacterial cellulose. J Therm Anal Calorim 118 (1):205–215
36. Barsberg S (2010) Prediction of Vibrational Spectra of Polysaccharides—Simulated IR Spectrum of Cellulose Based on Density Functional Theory (DFT). J Phys Chem B 114(36):11703–11708
37. Prats Mateu B, Hauser M-T, Heredia A, Gierlinger N (2016) Waterproofing in Arabidopsis: following phenolics and lipids in situ by confocal Raman microscopy. Front Chem 4. https://doi.org/10.3389/fchem.2016.00010
38. Himmelsbach DS, Akin DE (1998) Near-infrared Fourier-transform Raman spectroscopy of flax (*Linum usitatissimum* L.) stems. J Agr Food Chem 46(3):991–998. https://doi.org/10.1021/Jf970656k
39. Chu LQ, Masyuko R, Sweedler JV, Bohn PW (2010) Base-induced delignification of miscanthus x giganteus studied by three-dimensional confocal raman imaging. Bioresour Technol 101(13):4919–4925. https://doi.org/10.1016/j.biortech.2009.10.096
40. Kacuráková M, Wellner N, Ebringerova A, Hromádková Z, Wilson RH, Belton PS (1999) Characterisation of xylan-type polysaccharides and associated cell wall components by FT-IR and FT-Raman spectroscopies. Food Hydrocoll 13:35–41
41. Mathlouthi M, Koenig JL (1986) Vibrational spectra of carbohydrates. Adv Carbohydr Chem Biochem 44:7–89
42. Synytsya A, Copikova J, Matejka P, Machovic V (2003) Fourier transform Raman and infrared spectroscopy of pectins. Carbohydr Polym 54(1):97–106
43. Donaldson LA (2001) Lignification and lignin topochemistry – an ultrastructural view. Phytochemistry 57(6):859–873
44. Boerjan W, Ralph J, Baucher M (2003) Lignin biosynthesis. Annu Rev Plant Biol 54:519–546. https://doi.org/10.1146/annurev.arplant.54.031902.134938
45. Naseer S, Lee Y, Lapierre C, Franke R, Nawrath C, Geldner N (2012) Casparian strip diffusion barrier in Arabidopsis is made of a lignin polymer without suberin. Proc Natl

Acad Sci U S A 109(25):10101–10106. https://doi.org/10.1073/pnas.1205726109

46. Agarwal UP (1999) An overview of Raman spectroscopy as applied to lignocellulosic materials. In: Advances in lignocellulosics characterization. TAPPI, Atlanta, GA, pp 209–225
47. Agarwal UP, McSweeny JD, Ralph SA (2011) FT-Raman investigation of milled-wood lignins: softwood, hardwood, and chemically modified black spruce lignins. J Wood Chem Technol 31(4):324–344. https://doi.org/10.1080/02773813.2011.562338
48. Agarwal UP, Ralph SA (2008) Determination of ethylenic residues in wood and TMP of spruce by FT-Raman spectroscopy. Holzforschung 62(6):667–675. https://doi.org/10.1515/Hf.2008.112
49. Sun L, Varanasi P, Yang F, Loque D, Simmons BA, Singh S (2011) Rapid determination of syringyl: guaiacyl ratios using FT-Raman spectroscopy. Biotechnol Bioeng 109 (3):647–656. https://doi.org/10.1002/bit.24348
50. Larsen KL, Barsberg S (2010) Theoretical and Raman spectroscopic studies of phenolic lignin model monomers. J Phys Chem B 114 (23):8009–8021. https://doi.org/10.1021/jp1028239
51. Keegstra K (2010) Plant cell walls. Plant Physiol 154(2):483–486. https://doi.org/10.1104/pp.110.161240
52. Cassab GI (1998) Plant cell wall proteins. Annu Rev Plant Physiol Plant Mol Biol 49:281–309
53. Tuma R (2005) Raman spectroscopy of proteins: from peptides to large assemblies. J Raman Spectrosc 36(4):307–319. https://doi.org/10.1002/Jrs.1323
54. Zhu GY, Zhu X, Fan Q, Wan XL (2011) Raman spectra of amino acids and their aqueous solutions. Spectrochim Acta A 78 (3):1187–1195. https://doi.org/10.1016/j.saa.2010.12.079
55. Heredia A (2003) Biophysical and biochemical characteristics of cutin, a plant barrier biopolymer. BBA-Gen Subjects 1620(1–3):1–7. https://doi.org/10.1016/S0304-4165(02)00510-X
56. Bock P, Nousiainen P, Elder T, Blaukopf M, Amer H, Zirbs R, Potthast A, Gierlinger N (2020) Infrared and Raman spectra of lignin substructures: Dibenzodioxocin. J Raman Spectrosc
57. Pollard M, Beisson F, Li YH, Ohlrogge JB (2008) Building lipid barriers: biosynthesis of cutin and suberin. Trends Plant Sci 13 (5):236–246. https://doi.org/10.1016/j.tplants.2008.03.003
58. Littlejohn GR, Mansfield JC, Parker D, Lind R, Perfect S, Seymour M, Smirnoff N, Love J, Moger J (2015) In vivo chemical and structural analysis of plant cuticular waxes using stimulated raman scattering (srs) microscopy. Plant Physiol 168(1):18–28. https://doi.org/10.1104/pp.15.00119
59. Prinsloo LC, du Plooy W, van der Merwe C (2004) Raman spectroscopic study of the epicuticular wax layer of mature mango (Mangifera indica) fruit. J Raman Spectrosc 35 (7):561–567. https://doi.org/10.1002/Jrs.1185
60. Trebolazabala J, Maguregui M, Morillas H, de Diego A, Madariaga JM (2013) Use of portable devices and confocal Raman spectrometers at different wavelength to obtain the spectral information of the main organic components in tomato (*Solanum lycopersicum*) fruits. Spectrochim Acta A 105:391–399. https://doi.org/10.1016/j.saa.2012.12.047
61. Yu MML, Konorov SO, Schulze HG, Blades MW, Turner RFB, Jetter R (2008) In situ analysis by microspectroscopy reveals triterpenoid compositional patterns within leaf cuticles of *Prunus laurocerasus*. Planta 227 (4):823–834. https://doi.org/10.1007/s00425-007-0659-z
62. Heredia-Guerrero JA, Benitez JJ, Dominguez E, Bayer IS, Cingolani R, Athanassiou A, Heredia A (2014) Infrared and Raman spectroscopic features of plant cuticles: a review. Front Plant Sci 5. https://doi.org/10.3389/fpls.2014.00305
63. Ram MS, Dowell FE, Seitz LM (2003) FT-Raman spectra of unsoaked and NaOH-soaked wheat kernels, bran, and ferulic acid. Cereal Chem 80(2):188–192. https://doi.org/10.1094/Cchem.2003.80.2.188
64. Wu HW, Volponi JV, Oliver AE, Parikh AN, Simmons BA, Singh S (2011) In vivo lipidomics using single-cell Raman spectroscopy. Proc Natl Acad Sci U S A 108 (9):3809–3814. https://doi.org/10.1073/pnas.1009043108
65. Hunt GM, Baker EA (1980) Phenolic constituents of tomato fruit cuticles. Phytochemistry 19(7):1415–1419. https://doi.org/10.1016/0031-9422(80)80185-3
66. Hollricher O (2010) Raman Instrumentation for confocal Raman microscopy. In: Dieing T, Hollrichter O, Toporski J (eds) Confocal Raman microscopy. Springer, Heidelberg, pp 43–60

67. Li S, Dai LK (2011) An improved algorithm to remove cosmic spikes in Raman spectra for online monitoring. Appl Spectrosc 65 (11):1300–1306. https://doi.org/10.1366/10-06169
68. Katsumoto Y, Ozaki Y (2003) Practical algorithm for reducing convex spike noises on a spectrum. Appl Spectrosc 57(3):317–322. https://doi.org/10.1366/000370203321558236
69. Dieing T, Ibach W (2010) Software requirements and data analysis in confocal Raman microscopy. In: Dieing T, Hollrichter O, Toporski J (eds) Confocal Raman microscopy. Springer, Heidelberg, pp 61–89
70. Cappel UB, Bell IM, Pickard LK (2010) Removing cosmic ray features from Raman map data by a refined nearest neighbor comparison method as a precursor for chemometric analysis. Appl Spectrosc 64 (2):195–200
71. de Juan A, Maeder M, Hancewicz T, Duponchel L, Tauler R (2009) Chemometric tools for image analysis. In: Salzer R, Siesler HW (eds) Infrared and Raman spectroscopic imaging. WILEY-VCH Verlag GmbH & Co. KGaA, Weinheim, pp 65–108
72. Ramos PM, Ruisanchez I (2005) Noise and background removal in Raman spectra of ancient pigments using wavelet transform. J Raman Spectrosc 36(9):848–856. https://doi.org/10.1002/jrs.1370
73. Savitzky A, Golay MJE (1964) Smoothing and differentiation of data by simplified least squares procedures. Anal Chem 36:1627–1639
74. Gierlinger N, Reisecker C, Hild S, Gamsjaeger S (2013) Raman microscopy: Insights into chemistry and structure of biological materials. In: Fratzl P, Dunlop JWC, Weinkamer R (eds) Materials design inspired by nature: function through inner architecture. Royal Society of Chemistry, Cambridge
75. Peng J, Peng S, Jiang A, Wei J, Li C, Tan J (2010) Asymmetric least squares for multiple spectra baseline correction. Anal Chim Acta 683(1):63–68
76. Whittaker ET (1923) On a new method of graduation. Proc Edinb Math Soc 51:63–73
77. Eilers PHC (2003) A perfect smoother. Anal Chem 75(14):3631–3636. https://doi.org/10.1021/ac034173t
78. Felten J, Hall H, Jaumot J, Tauler R, de Juan A, Gorzsas A (2015) Vibrational spectroscopic image analysis of biological material using multivariate curve resolution-alternating least squares (MCR-ALS). Nat Protoc 10(2):217–240. https://doi.org/10.1038/nprot.2015.008
79. Lieber CA, Mahadevan-Jansen A (2003) Automated method for subtraction of fluorescence from biological Raman spectra. Appl Spectrosc 57(11):1363–1367. https://doi.org/10.1366/000370203322554518
80. Ohaver TC (1973) Wave-length modulation — applications in analytical spectrometry. Abstr Pap Am Chem S 1973:33
81. Ohaver TC, Green GL (1976) Numerical error analysis of derivative spectrometry for quantitative-analysis of mixtures. Anal Chem 48(2):312–318. https://doi.org/10.1021/Ac60366a016
82. Windig W, Stephenson DA (1992) Self-modeling mixture analysis of 2nd-derivative near-infrared spectral data using the simplisma approach. Anal Chem 64(22):2735–2742. https://doi.org/10.1021/Ac00046a015
83. Gierlinger N, Schwanninger M, Reinecke A, Burgert I (2006) Molecular changes during tensile deformation of single wood fibers followed by Raman microscopy. Biomacromolecules 7(7):2077–2081. https://doi.org/10.1021/bm060236g
84. Gierlinger N, Schwanninger M, Wimmer R (2004) Characteristics and classification of Fourier-transform near infrared spectra of the heartwood of different larch species (Larix sp.). J Near Infrared Spec 12 (2):113–119
85. Lohninger H, Ofner J (2014) Multisensor hyperspectral imaging as a versatile tool for image-based chemical structure determination. Spectrosc Eur Asia 26(5):6–10
86. Hastie T, Tibshirani R, Friedman J (2009) The elements of statistical learning. Springer, New York, NY
87. Bezdek JC, Ehrlich R, Full W (1984) Fcm – the fuzzy C-means clustering-algorithm. Comput Geosci 10(2–3):191–203. https://doi.org/10.1016/0098-3004(84)90020-7
88. Han EH, Karypis G, Kumar V (2000) Scalable parallel data mining for association rules. IEEE Trans Knowl Data Eng 12 (3):337–352. https://doi.org/10.1109/69.846289
89. Pham DT, Dimov SS, Nguyen C (2005) Selection of K in K-means clustering. Proc Inst Mech Eng C J Mech Eng Sci 219 (1):103–119
90. Nascimento JMP, Dias JMB (2005) Vertex component analysis: a fast algorithm to unmix hyperspectral data. Ieee Trans Geosci Remote 43(4):898–910. https://doi.org/10.1109/Tgrs.2005.844293

91. Lee DD, Seung HS (1999) Learning the parts of objects by non-negative matrix factorization. Nature 401(6755):788–791
92. Szymańska-Chargot M, Pieczywek PM, Chylińska M, Zdunek A (2016) Hyperspectral image analysis of Raman maps of plant cell walls for blind spectra characterization by nonnegative matrix factorization algorithm. Chemometr Intell Lab Syst 151:136–145. https://doi.org/10.1016/j.chemolab.2015.12.015
93. Tauler R, Smilde A, Kowalski B (1995) Selectivity, local rank, 3-way data-analysis and ambiguity in multivariate curve resolution. J Chemometr 9(1):31–58. https://doi.org/10.1002/cem.1180090105
94. Gierlinger N, Schwanninger M (2006) Chemical imaging of poplar wood cell walls by confocal Raman microscopy. Plant Physiol 140(4):1246–1254. https://doi.org/10.1104/pp.105.066993
95. Lehringer C, Gierlinger N, Koch G (2008) Topochemical investigation on tension wood fibres of Acer spp., *Fagus sylvatica* L. and *Quercus robur* L. Holzforschung 62 (3):255–263. https://doi.org/10.1515/Hf.2008.036
96. Zhang Z, Ma J, Ji Z, Xu F (2012) Comparison of anatomy and composition distribution between normal and compression wood of *Pinus bungeana* Zucc. revealed by microscopic imaging techniques. Microsc Microanal 18(6):1459–1466. https://doi.org/10.1017/S1431927612013451
97. Philippe S, Barron C, Robert P, Devaux MF, Saulnier L, Guillon F (2006) Characterization using Raman microspectroscopy of arabinoxylans in the walls of different cell types during the development of wheat endosperm. J Agr Food Chem 54(14):5113–5119. https://doi.org/10.1021/jf060466m
98. Hanninen T, Kontturi E, Vuorinen T (2011) Distribution of lignin and its coniferyl alcohol and coniferyl aldehyde groups in Picea abies and *Pinus sylvestris* as observed by Raman imaging. Phytochemistry 72 (14–15):1889–1895. https://doi.org/10.1016/j.phytochem.2011.05.005
99. Morikawa Y, Yoshinaga A, Kamitakahara H, Wada M, Takabe K (2010) Cellular distribution of coniferin in differentiating xylem of *Chamaecyparis obtusa* as revealed by Raman microscopy. Holzforschung 64(1):61–67. https://doi.org/10.1515/Hf.2010.015
100. Cao Y, Lu Y, Huang Y (2004) NIR FT-Raman study of biomass (*Triticum aestivum*) treated with cellulase. J Mol Struct 693 (1–3):87–93. https://doi.org/10.1016/j.molstruc.2004.02.017
101. Agarwal U (2006) Raman imaging to investigate ultrastructure and composition of plant cell walls: distribution of lignin and cellulose in black spruce wood (*Picea mariana*). Planta 224(5):1141–1153. https://doi.org/10.1007/s00425-006-0295-z
102. Richter S, Mussig J, Gierlinger N (2011) Functional plant cell wall design revealed by the Raman imaging approach. Planta 233 (4):763–772. https://doi.org/10.1007/s00425-010-1338-z
103. Prats-Mateu B, Bock P, Schroffenegger M, Toca-Herrera JL, Gierlinger N (2018) Following laser induced changes of plant phenylpropanoids by Raman microscopy. Scientific Reports 8 (1)

Chapter 16

Analysis of Plant Cell Walls by Attenuated Total Reflectance Fourier Transform Infrared Spectroscopy

Ricardo M. F. da Costa, William Barrett, José Carli, and Gordon G. Allison

Abstract

Attenuated total reflectance Fourier transform mid-infrared (ATR-FTIR) spectroscopy is widely applicable for the chemical analysis of biological materials, relatively inexpensive, requires only simple sample preparation, and is of comparatively high-throughput compared to traditional wet chemical or chromatographic methods. It is particularly well suited for the nondestructive analysis of dried and finely ground plant samples for the subsequent prediction of cell wall and other compositional or processing parameters using chemometric regression models. Furthermore, analysis of mid IR spectra by nonregression methods (e.g., principal component analysis) provides a straightforward approach for multivariate comparison of the effects of experimental, processing, and environmental treatments, and genotypic and temporal differences on chemical composition including changes in cell wall composition. There is thus great potential for using ATR-FTIR in the lignocellulosic biomass industry at a number of levels. Here we describe methods for cell wall sample preparation and generation of ATR-FTIR spectra, and suggest techniques for the statistical analysis and/or chemometric pattern recognition between the analyzed samples.

Key words Cell wall, Lignocellulose, Biomass, FTIR, PCA, PLS, Multivariate, Biofuel, Recalcitrance

1 Introduction

The full potential of lignocellulosic biomass utilization for the production of sustainable fuel and products has yet to be realized. The resilience of the plant cell wall to enzymic deconstruction, that is, cell wall recalcitrance, is a significant barrier to saccharification and biomass usability [1–3]. Scientists, breeders, and process engineers working to overcome or decrease recalcitrance in feed stocks require reliable data on stock cell wall composition if they are to develop better varieties with lower recalcitrance, optimize preprocessing and enzymic saccharification treatments, or to assess the effect of molecular interventions. Fourier transform mid-infrared (FTIR) is a nondestructive method that can provide these data at low cost and high rates of throughput. Traditionally mid-IR spectroscopy (2.5–25 μm or 4000–400 cm^{-1}) was used to determine

Zoë A. Popper (ed.), *The Plant Cell Wall: Methods and Protocols*, Methods in Molecular Biology, vol. 2149, https://doi.org/10.1007/978-1-0716-0621-6_16, © Springer Science+Business Media, LLC, part of Springer Nature 2020

the chemical structure of purified compounds. The bonds in covalent molecules are not rigid, but vibrate with specific frequencies. Bonds absorb energy in the IR when the frequency of energy matches their vibrational frequency and absorbance causes an increase in the amplitude but not the vibrational frequency of the bond. Absorption in the IR can also only occur when it results in changes in the bond's electric dipole moment, and bonds in heteronuclear molecules (e.g., between C and O, C and N, and O and H) often absorb strongly with characteristic frequencies. Most modern mid-IR instruments rely on Fourier transform principles to improve signal-to-noise ratio and decrease analysis time, see Allison [4] for further details. The reader is directed to a suitable textbook [5, 6], or review of mid-IR spectroscopy [7] for details of characteristic absorbance peaks for the major functional groups in biomolecules.

When applied to chemically complex heterogeneous samples (e.g., powdered biomass), FTIR analysis provides only averaged structural information. However, multivariate analysis of the mid-IR spectra using chemometric approaches can reveal important relationships between samples, between variables, and gives information on the effects of experimental, processing, and environmental treatments on cell wall composition. Furthermore, robust predictive regression models can be developed that allow for prediction of compositional parameters, saving much of the time and expense associated with traditional analytical methods [4, 8–11].

Until the development of attenuated total reflectance (ATR) the preparation of samples for FTIR analysis required the consuming preparation of salt discs containing ground sample [4]. ATR makes possible the collection of spectra from cell wall biomass samples without extensive preparation. The IR beam is directed by mirrors into a crystal of some IR transparent material with high refractive index (e.g., diamond, ZnSe, or germanium) mounted in the top of the ATR. The angle of the beam is such that it is totally reflected and the reflection forms an evanescent wave that extends into and interacts with the sample. The beam exiting the ATR device is then converted to a frequency-based spectrum by Fourier transform [6]. In resulting spectra, peak positions correlate with molecular structures and this information may be used to decipher the chemical makeup of samples. ATR accessories can be fitted into most if not all modern FTIR spectrophotometers.

Finally, one last consideration; plant biomass is highly heterogeneous, non–cell wall components will contribute to spectra. It may be necessary to reduce or eradicate interference from non-cell wall components for exploratory analysis, for example, by principal component analysis. In this case cell wall preparations should be obtained from the whole plant samples (see below). However, for the purpose of prediction of cell wall components, it may be possible and certainly desirable to instead use data from ground non-extracted plant biomass.

2 Materials

All solutions should be prepared using deionized water at room temperature. Some reagents used are potentially hazardous and appropriate protection equipment should be used (e.g., gloves, fume hood) particularly during sodium azide and volatile reagent handling. All reagents should be reagent grade and stored in appropriate storage areas. Proper disposal of waste chemicals should be done according to waste disposal regulations of your institute.

2.1 Biomass

The lignocellulosic biomass selected for analysis will obviously depend on the aims of the researcher and samples from different species, organs or tissues may be used. The biomass should be freeze-dried (we use an Edwards Modulyo) from frozen until dry, which generally takes between 12 and 48 h. Dryness can be accessed by the samples reaching constant weight. For sample sets with large amounts of biomass or if there is insufficient space available at −20 °C, for example, from field harvest, samples may initially stabilized by oven drying at 60 °C for 48 h. Once dry the biomass for cell wall analysis can be stored at room temperature in sealed bags of tubs, preferably in the dark. It must then be milled to be within a particle size range of 0.18–0.85 mm. The objective is to ensure the ground samples are representative of the biomass as a whole and it is likely that the amount of milled biomass will greatly exceed that required for the collection of spectra to meet this stipulation. We use a selection of mills including a Christy & Norris lab mill (Christy Turner Ltd., UK) fitted with a 1.0 mm filter mesh for large samples, an IKA A11 Handheld Analytical Mill; sieves with mesh sizes of 80 and 20 for small samples and a Spex 6870 Freezer Mill for samples that resist all normal methods for grinding. The ground samples may be sieved to ensure consistency of particle size, for example, using IKA A11 Handheld Analytical Mill; sieves with mesh sizes of 80 and 20 [12]. Given that particle size is known to influence cell wall composition [13–15] this may be an important step to ensure consistency, especially when the samples will also be analyzed using analytical procedures [16], however for the purpose of analysis by FTIR and perhaps prediction of compositional parameters it may be decided such sieving may distort results and it is best to analyze without sieving [11]. Such decisions must lie with the experimenter and be addressed on a case by case basis.

2.2 Biomass Preparation

2.2.1 General Equipment

(a) Plastic centrifuge tubes (50 mL, with screw cap) (Greiner Bio-One GmbH).

(b) Vortex mixer.

(c) Shaking incubator.

(d) Centrifuge.

(e) Fume hood.

(f) Block heater.

2.2.2 Organic Solvent Wash

(a) 70% (v/v) aqueous ethanol.

(b) Chloroform–methanol (1:1 v/v).

(c) Acetone 100%.

2.2.3 Starch Removal

(a) Type-I porcine α-amylase (Sigma-Aldrich, catalog number: A6255; saline suspension; 29 mg protein/mL; 1714 U/mg protein).

(b) 0.1 M sodium acetate buffer (pH 5.0): Prepare 0.2 M acetic acid (A) by mixing 11.55 mL glacial acetic acid in 500 mL water and adjusting to 1 L with water. Prepare 0.2 M sodium acetate solution (B) by dissolving 27.21 g sodium acetate trihydrate in 800 mL water and adjust to 1 L with water. Mix 14.8 mL of A, 35.2 mL of B, and 50 mL of water. Confirm that the pH is 5.0 with a pH meter (if needed, the pH may be adjusted with 10 M sodium acetate or with glacial acetic acid).

(c) 0.1 M ammonium formate buffer (pH 6.0): Dissolve 6,31 mg ammonium formate in 500 mL water and adjust pH to 6.0 with formic acid. Make up to 1 L with water. Confirm pH is 6.0 using a pH meter (if needed, the pH may be adjusted with formic acid).

(d) 0.001 M sodium azide: Dissolve 10 mg sodium azide in 145 mL water.

2.3 FTIR Spectroscopy

1. FTIR spectrometer (Equinox 55, Bruker Optik).
2. Golden Gate ATR accessory (Specac, Slough, UK).
3. Spatulas and razor blade for sample positioning.
4. Acetone wash bottle for surface cleaning.

2.4 Software

1. It is not the purpose of this chapter to give a definitive guide to the software you may use for analysis of data but rather to illustrate what might be achieved.
2. Data can be analyzed using proprietary software (e.g., Bruker OPUS IR spectroscopy software; version 5.0, Bruker Optik) serves to control the FTIR spectrophotometer. It also has chemometric functions, for example, for the development of regression models.

3. Instrumentational software is often more than adequate for basic analysis but greater control and sophistication is often possible if spectra are exported to R or MatLab (Mathworks) for analysis using more extensive chemometric software packages. The authors use (but have no commercial connection with) MatLab (R2014b) and the PLS toolbox (version 8.1.1) from Eigenvector Research Inc.

3 Methods

Purified cell wall is prepared following the procedures adapted from published methods: organic solvent wash [17], starch gelatinization [17], starch removal [18, 19].

3.1 Organic Solvent Extraction of Soluble Metabolites

1. Weigh approximately 1 g of the ground plant biomass into a 50 mL plastic centrifuge tube, add 30 mL 70% (v/v) aqueous ethanol, vortex mix thoroughly and agitate in a shaking incubator at 40 °C/150 rpm for 12 h.
2. Centrifuge at 900 × *g* for 10 min and discard the supernatant by decantation or aspiration.
3. Add 30 mL 70% (v/v) aqueous ethanol to the remaining pellet, vortex mix and incubate the samples for 30 min at 40 °C/ 150 rpm.
4. Centrifuge at 900 × *g* for 10 min and discard the supernatant.
5. Repeat **steps 3–4**.
6. Resuspend the pellet in 20 mL of the chloroform–methanol (1:1 v/v) solution and agitate in a shaking incubator for 30 min at 25 °C and 150 rpm.
7. Centrifuge at 900 × *g* for 10 min and discard the supernatant.
8. Repeat **steps 6–7** twice.
9. Resuspend the pellet in 15 mL of acetone and agitate in a shaking incubator for 30 min at 25 °C and 150 rpm.
10. Centrifuge at 900 × *g* for 10 min and discard the supernatant after the acetone wash.
11. Repeat **steps 9–10** twice.
12. Let the organic-washed biomass samples dry overnight in a fume hood (alternatively they can be left in an oven set at 35 °C for approximately 16 h).

3.2 Removal of Starch

1. Resuspend the dry solvent-extracted biomass in 15 mL of 0.1 M sodium acetate buffer (pH 5.0), then heat for 20 min at 80 °C in a heating block to induce starch gelatinization.
2. Cool suspensions on ice for 15 min.

3. Centrifuge at 900 × *g* for 10 min and discard the supernatant.
4. Wash the pellet twice with 30 mL water, with centrifugation (900 × *g* for 10 min) removing and discarding the supernatant after each wash.
5. Add to the pellet the following reagents: 10 mL 0.1 M ammonium formate buffer (pH 6.0), 10 μL type-I porcine α-amylase (47 U per 100 mg cell wall), and 500 μL 0.001 M sodium azide solution.
6. Leave in a shaking incubator for 48 h at 25 °C/110 rpm.
7. Terminate the digestion by heating to 95 °C/15 min, then cool the samples on ice for 15 min.
8. Centrifuge at 900 × *g* for 10 min and discard the supernatant.
9. Wash the pellet three times with 30 mL water and twice with 20 mL acetone, with centrifugation (900 × *g* for 10 min) and supernatant removal after each wash.
10. Let the prepared cell wall material samples dry overnight in a fume hood. Alternatively they can be left in an oven set at 35 °C for approximately 16 h. The samples are stable for months at room temperature if kept in a sealed container and protected from direct light.

3.3 Collection of Spectra by ATR-FTIR

1. Place approximately 10 mg of the dry cell wall powder onto the Golden Gate ATR crystal. Press the sample into contact with the ATR crystal using the anvil of the Golden Gate ATR accessory or equivalent.
2. Collect spectra in duplicate for each biomass sample.

   Spectra are generally collected between 4000 and 600 cm^{-1} and each is the average over 32 scans at a resolution of 4 cm^{-1}.
3. Before spectra from the samples are collected it is necessary to obtain a background spectrum for the ATR and FTIR spectrophotometer. This background is automatically subtracted from the data obtained with the sample in place to give a true spectrum for the sample. The background measurement is made after cleaning of the crystal by gentle wiping with a white tissue wipe wetted with a suitable solvent. We use acetone to clean the crystal but refer to manufacturer's instructions for your device. Too much solvent can result in the crystal becoming loose in its mount so use caution. The background measurement is generally made with the crystal exposed to the air and without the holding device screwed down. Generally background measurements are repeated regularly, we repeat them between duplicate of triplicate sample measurements.

4. Higher resolution and larger scan numbers can be used but in our hands these do not improve data quality with dry powdered samples whilst file size and analysis time is increased.
5. The sample may be recovered if desired using a pair of razor blades but often the sample is removed using a small domestic vacuum cleaner, and the crystal and anvil cleaned with a laboratory tissue wipe wetted with solvent between samples.

3.4 Data Export

1. The spectral data are typically saved as individual text files in which the spectral x and y Cartesian coordinates are recorded in two separate columns. In our laboratory the Bruker OPUS IR spectroscopy software is used for this conversion using a custom macro. These files can be opened in Microsoft Excel, Notepad, or some other suitable viewer.
2. The text files are then imported using R or MatLab script into a data matrix with the rows being samples and the columns being absorbance at each wavenumber variable. Our Matlab importation script is specific for Opus format text files that contain 1764 rows of data. A version of our import script is shown later in this chapter for the purpose of illustrating how data can be imported into MatLab.

4 Data Exploration

The ancient adage of data analysis "before anything else plot your data" contains much wisdom, but in the case of complex multivariate data such as mid-IR spectra, plotting in the traditional sense often reveals very little unless there are clear differences at particular wave numbers between for example spectra from samples subjected to differing experimental treatments. A much better way to explore the data is by principal component analysis (PCA), which is now a staple in the diet of many spectroscopists. Unlike multivariate statistical approaches such as multivariate analysis of variance and canonical variance analysis, in PCA the data is merely rotated and aligned to a new set of variables (components) that better explain the data, indeed the new orthogonal components explain the data so well that only a few are needed and the data can be viewed graphically as *xy* plots in the traditional manner. Data generally needs to be scaled before PCA to avoid the larger variables from dominating the analysis [20] and some software packages may allow the goodness of fit to be estimated by cross-validation [21]. The reader is warned however that misunderstandings can lead to mistaken conclusions when using chemometric methods [22] and is pointed toward some suitable multivariate statistical primer for example Manly [23] or Otto [24], to obtain a firm grounding in these methods. What is important is that the person analysing the data understands the principles of the methods and their inherent strengths and weaknesses.

In this chapter we give an example of PCA for spectra taken from an undergraduate project study of ground samples of two genotypes of the model grass species Brachypodium distachyon [25]. A Mat file of these data is available by request from the authors for the purpose of practice data. Spectra were taken from nonextracted freeze-dried and ground leaf and stem samples of plants at 24 h (T1), 48 h (T2) and 1 week (T3) to investigate if there were differences between genotypes for resistance to powdery mildew due to infection by *Blumeria graminis.* The powdered samples were not extracted with solvent or destarched before spectral analysis and so the results of PCA of the spectral data reflect both soluble and cell wall metabolites present in the samples. The 48 spectra shown in Fig. 1a are the averaged spectra of two readings. It is important that replicate spectra, that is, "machine reps" are averaged before analysis so that the variance due to machine drift and noise is not present in the data, unless of course the experimenter is not seeking answers to biological questions but rather wishes to monitor machine drift and error. Spectra that look strange or are from nontypical samples should be removed from the data set because these will dominate the model and the purpose is to develop a model that explains typical samples. Before PCA the spectra were normalized by multiplicative scatter correction to normalize for random scatter during spectral acquisition, and mean centered. Sometimes it is common practice to centre the data by autoscaling, that is, converting data to unit variance with a standard deviation of 1. This has the effect of making all variables of equal magnitude and it is effective when changes in low magnitude variables are the most important changes to observe. For chromatographic and spectra data however the peaks are generally more important than the baseline and mean centering rather than autoscaling has less tendency to inflate the effect of baseline noise. The processed spectra are shown in Fig. 1b. For PCA a seven split venetian blind cross validation protocol was chosen and a 4-component PCA model was developed that encompassed 96.9% of the variance. The fit and cross-validation are best expressed by their root mean error (RMSE); ideally these values should be as low as possible and numerically similar. The RMSE of correlation (C), that is, model fit, was 0.001457, and the RMSE of cross-validation (CV), that is, how much of the fit is based on noise, was 0.002135. There is a general tendency to use more factors or components in PCA or regression models that are necessary (overfitting) because lower components explain variance in the data that is due to noise and inclusion of some of these components in the final model seems to give a model of better fit. Experience however tells us that more robust models, that is, ones that can be applied to new data sets,

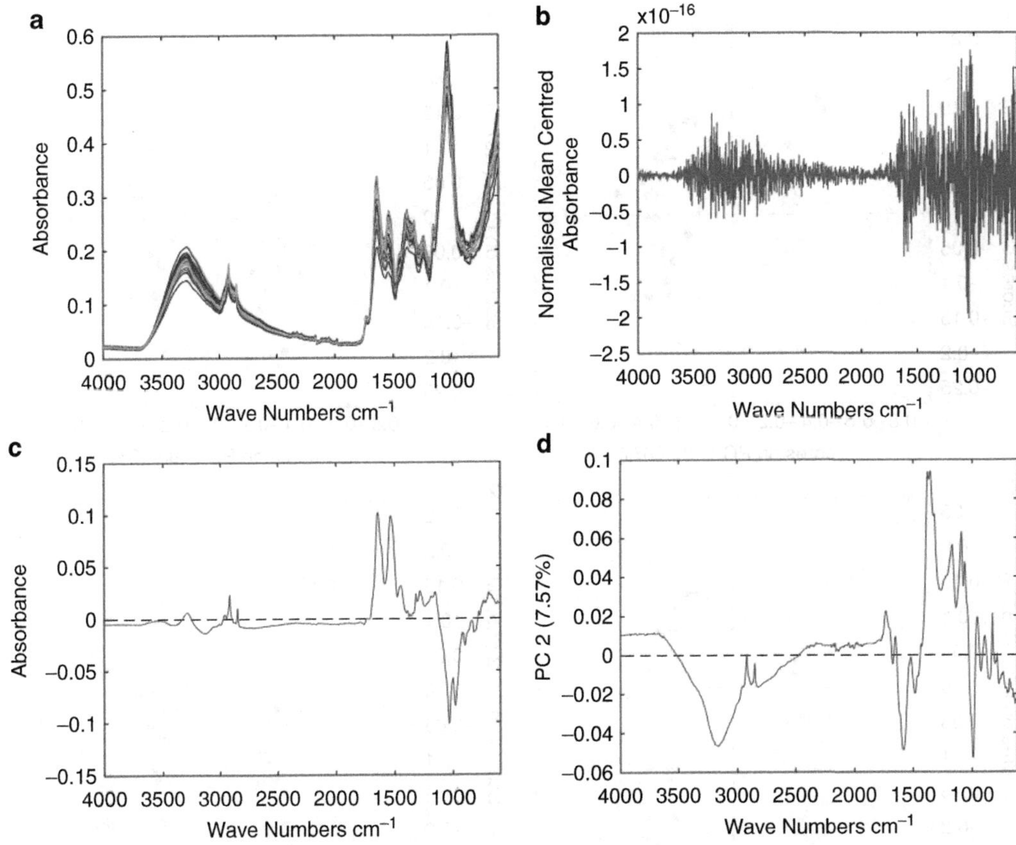

Fig. 1 (**a**) Forty-eight overlaid spectra from dried and ground leaf and stem samples taken from control Brachypodium distachyon plants and plant infected with *Blumeria graminis* at time points T1 (24 h), T2 (48 h), and T3 (1 week). (**b**) Same spectra following multiplicative scatter correction and mean centering. (**c**) PCA loadings for PC 1 (83.88% variance) and (**d**) loadings for PC2 (7.57% variance)

often contain fewer components and generally model underfitting is less of a hazard than overfitting. The cut off on which components to exclude from the model is largely for the experimenter to decide RMSECV are of great assistance in making this decision. Ideally the value of RMSECV will shadow the RMSEC curve and decrease as components are added until new components explain noise, where upon it will increase. This is a convenient guide for cut off. Sometimes it is not so straight forward but RMSECV values remain a particularly important tool. It is also important that several models are tried with different CV parameters as this has influence on RMSECV.

Plots of the scores show relationships between samples. Figure 2a–d shows score plots for PC1 (83.88% variance) vs. PC2 7.57% variance), the scores being colored according to tissue, genotype, time and infection for A–D respectively. It is plain that the largest difference between the spectra is due to tissue and this is

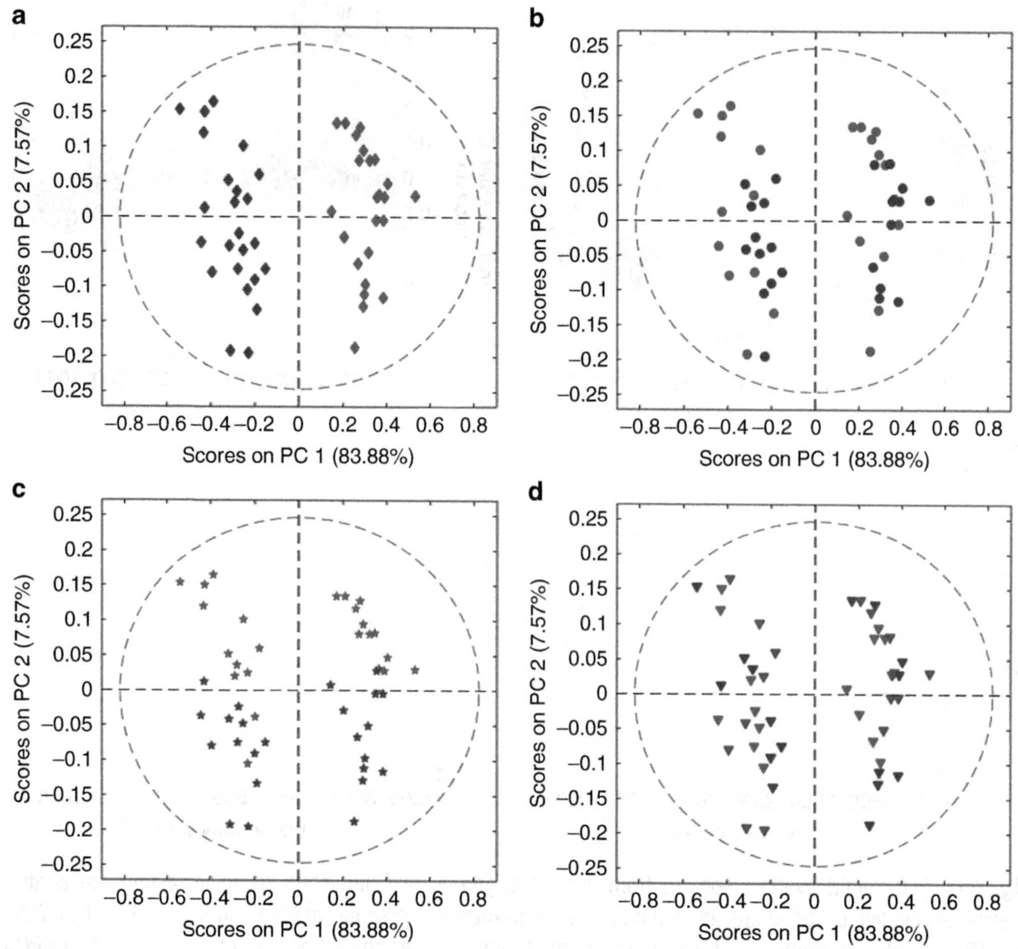

Fig. 2 PCA scores plots for PC1 (83.88% variance) and PC2 (7.57% variance) of 48 spectra from dried and ground leaf and stem samples taken from control Brachypodium distachyon plants and plant infected with *Blumeria graminis* at time points T1 (24 h), T2 (48 h) and T3 (1 week). The figure panes are coded by: (**a**) tissue (red diamonds = leaf, blue diamonds = stem); (**b**) by genotype (Genotype 4 = red circles, Genotype 5 = blue circles); (**c**) infection (red stars = infected, blue stars = control); and (**d**) Time point (T1, T2, and T3 = red, blue, and magenta inverted triangles respectively)

explained along PC axis 1 (A), with leaf samples having positive and stem negative scores along this axis. There is a lack of grouping between the samples according to plant genotype (B) or to time point (D). However, samples from infected plants are separated from noninfected along PC axis 2 (C), with samples from infected plants having positive loadings and noninfected having negative loadings along this axis. The spectral differences that contribute to sample separation by tissue or by infection can be determined by analysis of the loadings for these components (*see* Fig. 1c, d). Caution must be used when explaining positive and negative peaks in PCA loadings or spectra from complex samples as many components may contribute. However, these data show that infection results in changes in chemical composition in leaf and stem tissues even at early time points, and may support the case for a more extensive characterization of changes in soluble phenolic and carbohydrate metabolites and cell wall in infected and noninfected plants.

5 Partial Least Squares Regression Models

Often the analyst requires a robust regression model to predict something that is costly or time-consuming to measure in his or her samples. It may also be the case that the analysis wants to check the correlation between IR spectra, which are nearly always very consistent, and data from some analytical method being established. We have published previously on such models [4, 11, 26, 27] and in this chapter we consider for illustration a typical scenario in which a postgraduate student wished to develop a predictive model correlating mid-IR spectra of dry, powdered and solvent extracted samples of *Miscanthus* stem cell wall to concentrations of lignin measured in the samples by an acetyl bromide derivatization approach [28]. For a number of technical reasons the method is demanding [29, 30] but has the advantage of requiring only small amounts of sample, and the student was also interested in how well the two data sets correlated. Options for regression analysis of multivariate data include classical least squares (excellent when the user knows exactly how many components are in the mixture) and inverse least squares (e.g., multiple linear regression [MLR] and principal component regression [PCR]), but many workers routinely use partial least squares (PLS) [31, 32] as it gives a good balance between model fit and prediction and is generally robust. Regression analysis is covered in much greater detail in Otto [24]; once again the reader is warned of the danger of including outlying data in the model, whether this is in terms of uncertainty, perhaps resulting from poorly prepared samples (high Q residuals) or because the samples are atypical, perhaps with higher/lower than normal concentration of what is to be predicted (high T^2). Deciding how many components to include in the model is also very important for model prediction and comments made in the preceding section hold true here. Often great emphasis is placed on the coefficient of determination (R^2) as a measure of regression fit as this statistic explains the amount of variance captured by the model; however, this may give misleading interpretations of how good the model fits the data as a few samples with high T^2 can effectively make the regression a line connecting two distant clusters of sample scores, with the regression only poorly explaining the sample scores within each cluster. Lastly, it is vital that the experimenter develops the model on only a subset of the data and that some data is

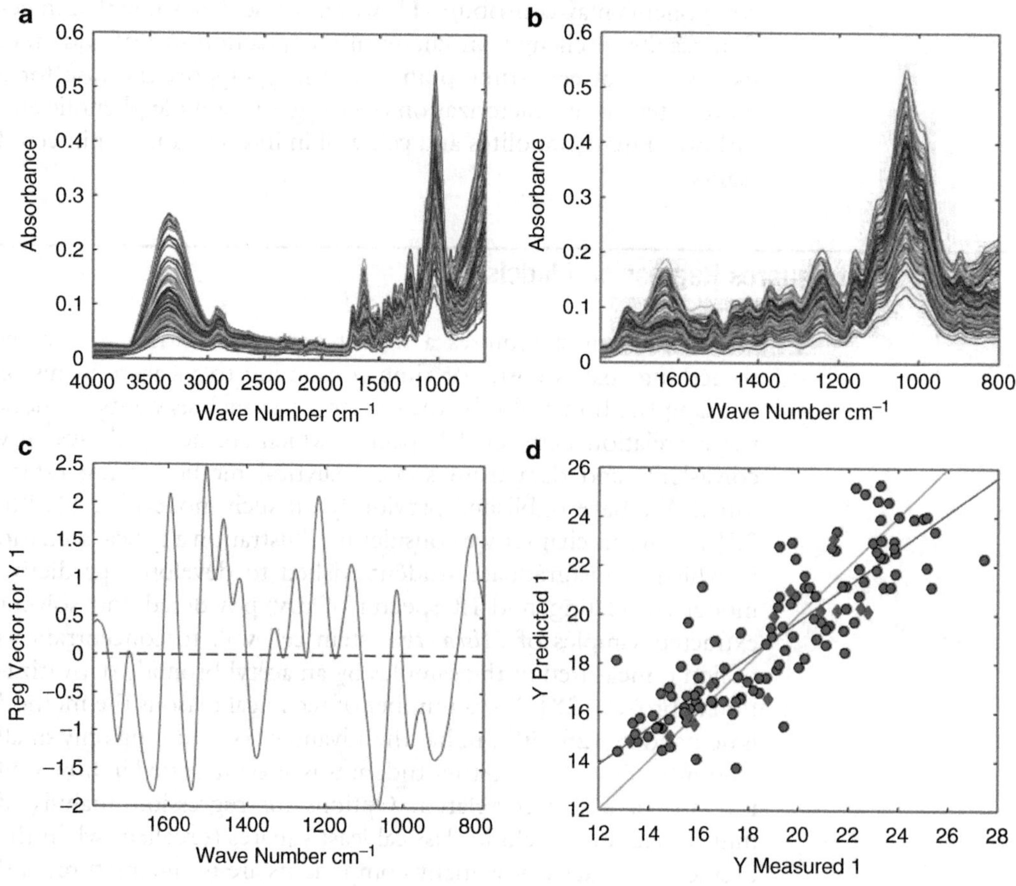

Fig. 3 PLS model for the prediction of acetyl bromide lignin in 48 samples of dry, ground *Miscanthus* cell wall preparation. (**a**) Mid-IR spectra for the 48 samples; (**b**) the fingerprint region of the spectra used for the PLS model; (**c**) the regression vector for the PLS model; and (**d**) scatter plot of predicted vs. measured concentrations of the y data (acetyl bromide lignin)

reserved for the purpose of model validation. This data set must remain virgin until the predictive accuracy of the model is determined. It is not acceptable to build the model and test the predictive accuracy in an iterative manner as this will lead to fitting on noise.

Spectra from 48 *Miscanthus* cell wall preparations were imported into MatLab (Fig. 3a) as described previously together with the acetyl bromide lignin concentrations that had been measured in the samples. These samples were obtained from a number of different *Miscanthus* genotypes collected at several time-points in the growth year from the 2TT experiment at Aberystwyth University [33]. Ten per cent of the data were reserved for model validation. The spectra were processed prior to PLS by multiplicative scatter correction and finally by being mean centered. It is good to always begin with simple processing before more complex approaches are taken, similarly, unless the experimenter has existing knowledge of wave numbers that will certainly be most important to the model, it is often good to cast a wide net at first and include all of the spectral data. This basic model was based on two components, explained 96.4% of *X* and 65.0% *y* data variance, and had RMSEC and RMSECV values of 2.09 and 2.17, respectively. Fit was not improved by additional preprocessing, for example, smoothing to a 25 variable window and baseline flattening by use of the trend algorithm. Truncating the spectra to the complex fingerprint region between 1800 and 800 cm^{-1} (Fig. 3b) a common practice used to increase the specificity of the data, resulted in a 4 component model explaining 94.1% and 73.2% of *X* and *y* variance. The fit of this fingerprint PLS model was improved compared to the basic model (RMSEC and RMSECV of 1.82 and 1.96 respectively).

Variable selection by interval PLS is a way in which the best variables for prediction can be selected from the spectra, or conversely the bad variables can be identified and removed [34, 35]. The iToolbox for MatLab can be downloaded for free from http://www.models.life.ku.dk/itoolbox. We applied three successive rounds of forward or reverse interval PLS (ten latent components with intervals of either 100 or 200 variables) but the models resulting from these explorations fit the data no better than the fingerprint model and were discarded. They did however show that no region of the sample spectra explained the lignin analytical data particularly well, suggesting the presence of random noise most likely due to inconsistencies in the acetyl bromide lignin analytical method. Figure 3c shows the regression vector of the fingerprint PLS model and it is plain that the algorithm is using many parts of the spectral data to fit the spectral *X* data to the analytical *y* data. Once the fingerprint PLS model had been finalized, it was validated against the remaining data. The root mean square error of prediction (RMSEP) for the model was 1.46, similar to the values for RMSEC and RMSECV thus indicating that the model works and is not based on noise. Figure 3d shows a plot of predicted vs. measured lignin content for the samples and it is clear that whilst the scores do not lie closely to the regression line they do regress and the prediction of the concentrations of acetyl bromide predicted in the test data points is similar in accuracy to the concentrations predicted for the correlation data set. The regression analysis of these data by PLS was useful in the development of this analytical method and as improvements were made to the analytical procedure the ability to predict acetyl bromide lignin from mid IR spectra of cell wall samples improved substantially (data not shown).

6 Conclusion

In this chapter we have attempted to give a beginners guide to making cell wall analysis by FTIR together with details on sample preparation and we have discussed at some length how to go about exploring the data and setting up a regression curve. We have deliberately chosen data that is not perfect, on those occasions when data is excellent the analysis is much easier. However, in using more challenging data we have been able to show additional approaches that may be tried and given some guidance for the experimenter new to this area to consider. The combination of chemometrics and complex data is incredibly powerful and patterns may be found in the data that surprise and delight, but there are dangers; the researcher is asked to remember that good experimental design is always essential for quality science (basing a hypothesis on what is perhaps only a very small portion of variance is risky) and to be aware of what are fairly common misconceptions [22].

7 Importing Opus Text Files into MatLab with Import_Spectra_Free

This m-file script can be viewed as an example of how data might be imported. The authors accept no liability what so ever for its use but can vouch it works with their data (Opus generated text files, see below) in MatLab R2014b. Should readers wish to use this script to import spectral data from text files into Matlab, they may need to modify the number of data rows (see comments). For this illustration we have assumed that 1764 measurements have been made, that the text file has no headers, that the wavenumbers are in column A and the absorbance data in column B. Other software may generate slightly different text file formats. To use, open a new script window in MatLab, copy in the script and save as "import_spectra_free" to a location in the MatLab path. Call the function by entering "`[X, f, w] = import_spectra_free;`".

```
function [X, f, w] = import_spectra_free

% This m.file imports OPUS tab separated text files generated
using the
% save2ascii macro in Opus
% It generates a data matrix X, a list of wave numbers w and file
names f.
% It was compiled by Gordon Allison (IBERS) 13/04/2015
% Make sure the folder with the data files is current and in the
MatLab path
% Ensure only text files are in the folder.
% IT ASSUMES 1764 rows per file, please adjust accordingly to your
data!!

diskpath = uigetdir;
NumR = input('how many rows are there in the files?: ');
filenames = dir(fullfile(diskpath, '*.txt'));

% Builds a structure with fields for name, date, bytes and isdir
for each file

f = char(filenames(:).name); % lists files as a string
f = str2cell(f); % Converts to a cell array
len = length(filenames);
X = zeros(NumR,len); % Makes an array to put in the data for each
file
w = dlmread(filenames(1).name,'\t','A1..A1764');
% Extracts all data to obtain scan numbers from the first file

for i = 1:len;
    tempdata = dlmread(filenames(i).name,'\t', 'B1..B1764');
% Change B1764 to BX, where X is the number of rows in your data
files
    X(:,i)=tempdata;
    clear tempdata;
end

X = X';
clear filenames diskpath i len NumR;

X = X';
clear filenames diskpath i len NumR;
```

Acknowledgments

This work was supported by European Regional Development Funding through the Welsh Government for BEACON Grant number 8056; the Biotechnology and Biological Sciences Research Council (BBSRC) Institute Strategic Programme Grant on Energy Grasses & Biorefining (BBS/E/W/10963A01); and Supergen Bioenergy (EPSRC GR/S28204). Special thanks are also due to Prof. Luis Mur (IBERS, Aberystwyth University) for assistance with the *Blumeria graminis/Brachypodium distachyon* infection experiment which is used in this chapter to illustrate PCA.

References

1. Pauly M, Keegstra K (2008) Cell-wall carbohydrates and their modification as a resource for biofuels. Plant J 54(4):559–568
2. Himmel ME et al (2007) Biomass recalcitrance: engineering plants and enzymes for biofuels production. Science 315:804–807
3. Torres AF et al (2016) Maize feedstocks with improved digestibility reduce the costs and environmental impacts of biomass pretreatment and saccharification. Biotechnol Biofuels 9(1):1–15
4. Allison GG (2011) Application of Fourier transform mid-infrared spectroscopy (FTIR) for research into biomass feed-stocks. In: Nokolic G (ed) Fourier transforms – new analytical approaches and FTIR strategies. Intech, London, pp 71–88
5. Stewart B (2005) Infrared spectroscopy: fundamentals and applications. John Wiley & Sons, Ltd, Hoboken, NJ
6. Smith BC (2011) Fundamentals of Fourier transform infrared spectroscopy, 2nd edn. CRC Press, Boca Raton, FL
7. Movasaghi Z, Rehman S, Rehman IU (2008) Fourier transform infrared (FTIR) spectroscopy of biological tissues. Appl Spectrosc Rev 43(2):134–179
8. Sills DL, Gossett JM (2012) Using FTIR to predict saccharification from enzymatic hydrolysis of alkali-pretreated biomasses. Biotechnol Bioeng 109(2):353–362
9. Zhou G, Taylor G, Polle A (2011) FTIR-ATR-based prediction and modelling of lignin and energy contents reveals independent intra-specific variation of these traits in bioenergy poplars. Plant Methods 7(1):9
10. Kumar R et al (2009) Physical and chemical characterizations of corn stover and poplar solids resulting from leading pretreatment technologies. Bioresour Technol 100:3948–3962
11. Allison GG et al (2009) Measurement of key compositional parameters in two species of energy grass by Fourier transform infrared spectroscopy. Bioresour Technol 100:2428–2433
12. da Costa RMF, Allison GG, Bosch M (2015) Cell wall biomass preparation and Fourier transform mid-infrared (FTIR) spectroscopy to study cell wall composition. Bioprotocol 5 (11):1–7
13. Agger J, Meyer AS (2012) Alteration of biomass composition in response to changing substrate particle size and the consequences for enzymatic hydrolysis of corn bran. Bioresources 7:3378–3397
14. Bridgeman TG et al (2007) Influence of particle size on the analytical and chemical properties of two energy crops. Fuel 86(1–2):60–72
15. Hames B et al (2008) Preparation of samples for compositional analysis. In: Laboratory analytical procedure (LAP). National Renewable Energy Laboratory, Golden, CO
16. Allison GG et al (2010) Measurement of lignocellulose composition as a tool to understand how feed-stocks can be matched to conversion process. In: Bioten. CPL Press, Birmingham
17. Foster CE, Martin TM, Pauly M (2010) Comprehensive compositional analysis of plant cell walls (lignocellulosic biomass) part II: carbohydrates. J Vis Exp 37. http://www.jove.com/video/1837/comprehensive-compositional-analysis-plant-cell-walls-lignocellulosic
18. Persson S et al (2007) The Arabidopsis irregular xylem8 mutant Is deficient in glucuronoxylan and homogalacturonan, which are essential for secondary cell wall integrity. Plant Cell 19(1):237–255

19. Kong Y et al (2011) Molecular analysis of a family of Arabidopsis genes related to galacturonosyltransferases. Plant Physiol 155 (4):1791–1805
20. Bro R, Smilde AK (2003) Centering and scaling in component analysis. J Chemom 17 (1):16–33
21. Bro R et al (2008) Cross-validation of component models: a critical look at current methods. Anal Bioanal Chem 390(5):1241–1251
22. Kjeldahl K, Bro R (2010) Some common misunderstandings in chemometrics. J Chemom 24(7–8):558–564
23. Manly BFJ (2005) Multivariate statistical methods a primer. Chapman and Hall, London
24. Otto M (2007) Chemometrics. Wiley-VCH, Weinheim
25. Mur LAJ et al (2011) Exploiting the Brachypodium tool box in cereal and grass research. New Phytol 191(2):334–347
26. Allison GG et al (2009) Quantification of hydroxycinnamic acids and lignin in perennial forage and energy grasses by Fourier-transform infrared spectroscopy and partial least squares regression. Bioresour Technol 100:1252–1261
27. Belanche A et al (2013) Estimation of feed crude protein concentration and rumen degradability by Fourier-transform infrared spectroscopy. J Dairy Sci 96(12):7867–7880
28. Foster CE, Martin TM, Pauly M (2010) Comprehensive compositional analysis of plant cell walls (lignocellulosic biomass) part I: lignin. J Vis Exp 37. http://www.jove.com/video/1745/comprehensive-compositional-analysis-plant-cell-walls-lignocellulosic
29. Hatfield RD, Brei K, Grabber JH (1996) Revising the acetyl bromide assay to optimise lignin determinations in forage plants, in 1996 research summaries, ARS, USDA. USDA, Washington, DC
30. Hatfield RD et al (1999) Using the acetyl bromide assay to determine lignin concentrations in herbaceous plants: some cautionary notes. J Agric Food Chem 47(2):628–632
31. Yeniay Ö, Göktaş A (2002) A comparison of partial least squares regression with other prediction methods. Hacet J Math Stat 31:99–111
32. Zhang L, Garcia-Munoz S (2009) A comparison of different methods to estimate prediction uncertainty using partial least squares (PLS): a practitioner's perspective. Chemom Intell Lab Syst 97(2):152–158
33. Allison GG et al (2011) Genotypic variation in cell wall composition in a diverse set of 244 accessions of miscanthus. Biomass Bioenergy 35(11):4740–4747
34. Norgaard L et al (2000) Interval partial least-squares regression (iPLS): a comparative chemometric study with an example from near-infrared spectroscopy. Appl Spectrosc 54 (3):413–419
35. Leardi R, Nørgaard L (2004) Sequential application of backward interval partial least squares and genetic algorithms for the selection of relevant spectral regions. J Chemom 18 (11):486–497

Chapter 17

Rapid Assessment of Cell Wall Polymer Distribution and Surface Topology of Arabidopsis Seedlings

Mary L. Tierney, Li Sun, and David S. Domozych

Abstract

The deposition and modulation of constituent polymers of plant cell walls are profoundly important events during plant development. Identification of specific polymers within assembled walls during morphogenesis and in response to stress conditions represents a major goal of plant cell biologists. *Arabidopsis thaliana* is a model organism that has become central to research focused on fundamental plant processes including those related to plant wall dynamics. Its fast life cycle and easy access to a variety of mutants and ecotypes of Arabidopsis have stimulated the need for rapid assessment tools to probe its wall organization at the cellular and subcellular levels. We describe two rapid assessment techniques that allow for elucidation of the cell wall polymers of root hairs and high-resolution analysis of surface features of various vegetative organs. Live organism immunolabeling of cell wall polymers employing light microscopy and confocal laser scanning microscopy can be effectively performed using a large microplate-based screening strategy (see Figs. 1 and 2). Rapid cryofixation and imaging of variable pressure scanning electron microscopy also allows for imaging of surface features of all portions of the plant as clearly seen in Fig. 3.

Key words Immunolabeling, Cell wall, Variable pressure scanning electron microscopy, Arabidopsis

1 Introduction

Plant cell walls are composite structures consisting of both cellulose and noncellulosic polysaccharides that include hemicelluloses, pectins, and mixed-linked glycans as major components. In addition, several families of structural proteins (extensins, proline-rich proteins, and arabinogalactan proteins) function in cell wall organization and structure. An integration of biochemical, genetic, and cell biology approaches has identified many of the genes involved in the biosynthesis of these structural carbohydrates and proteins. However, much less is known about the processes required for correct assembly of these polymers as new walls are synthesized during growth or how interactions between wall components change in response to environmental stress. Likewise, we are only just beginning to resolve how particular cell shapes are manifested as a result

Zoë A. Popper (ed.), *The Plant Cell Wall: Methods and Protocols*, Methods in Molecular Biology, vol. 2149, https://doi.org/10.1007/978-1-0716-0621-6_17, © Springer Science+Business Media, LLC, part of Springer Nature 2020

of cell wall composition and deposition that ultimately lead to tissue and organ growth and development behavior. As we focus on identifying plants that can tolerate changes in climate and local growth conditions as well as plants with walls that may provide efficient sources of bioenergy, it will be critical that we can easily identify changes in cell wall organization that alter cell shape and growth.

Root hairs represent a powerful and convenient model system for investigating the role of specific genes on wall biosynthesis and assembly. This is especially true in Arabidopsis where the availability of an annotated genome and the easy and inexpensive access to T-DNA insertion mutants provides loss-of-function alleles for genes of interest. Root hairs elongate in a polarized manner by tip growth after an initial period of bulge formation at the base of root hair bearing cells in Arabidopsis. As root hairs grow away from the surface of the root, changes in cell shape due to modifications in wall structure are easy to score using standard microscopy methods. Root hair length and/or shape in Arabidopsis have been shown to be affected by mutations in many genes involved in wall biosynthesis including cellulose synthase [1], xylosyltransferases [2, 3], xyloglucan galacturonosyltransferase1 [4], UDP-L-rhamnose synthase [5], UDP-4-keto-6-deoxy-D-glucose 3,5-epimerase 4-reductase 1 and UDP-D-glucuronic acid 4-epimerase 6 [6] involved in pectin biosynthesis, cellulose-synthase like genes [7–9], structural wall proteins [10, 11], and enzymes involved in controlling the hydroxylation of HRGPs (hydroxyproline rich glycoproteins) [11].

Mutations in wall biosynthetic genes that affect root hair growth are often viable, providing researchers with genetic resources to investigate the relationship between growth, wall organization and cell shape in this tip-growing system. Toward this end, a large number of monoclonal antibodies have been generated against cell wall components for which the epitopes are known [12, 13]. Thus, immunological and microscopic methods that allow for the rapid analysis of cell wall organization and its relationship to cell shape should be useful in deciphering the cellular mechanisms responsible for cell wall polymer secretion, assembly, and function.

In this chapter, we present methods for the immunolabeling of surface epitopes of root hair and root epidermal cell walls that have previously been useful in describing changes in the organization of xyloglucan within the walls of Arabidopsis seedlings [14]. This approach provides rapid, qualitative information describing the organization of wall components accessible to antibodies on the surface of epidermal tissues in live seedlings. In addition, we present a rapid cryofixation-based method for imaging the surface features of all vegetative portions of the plant using variable pressure scanning electron microscopy.

2 Materials

2.1 Supplies and Reagents for Growing Seedlings and Immunofluorescence Studies

1. *Seeds*: wt and mutant lines of *Arabidopsis* seeds are available from the Arabidopsis Biological Research Center (https://abrc.osu.edu/). Seeds can be stored in the dark at room temperature for up to 12–24 months.
2. *Seed sterilization supplies*: 20% bleach (v/v commercial bleach in deionized water, DH_2O), 70% ethanol (v/v in DH_2O), sterile DH_2O, sterile 1.5 mL microcentrifuge "Eppendorf" tubes, 1–5 μL micropipette and sterile pipette tips, 100–1000 μL micropipette and sterile pipette tips, microcentrifuge (e.g., Fisher Scientific accuSpin Micro 17), waste container, laminar flow transfer hood or biological safety cabinet, Vortex, microcentrifuge tube rotator, waste container.
3. *Growth medium supplies for seed germination and seedling maintenance*: Chemical reagents: Preweighed packets of 1× Murashige and Skoog salts with Gamborg's vitamins (Sigma Chemical, St. Louis, MO, USA), sucrose, agarose, MES (*2-(N-×10 Morpholino) ethanesulfonic acid hydrate*), DH_2O, 0.1 M KOH solution, sterile 100 × 15 mm plastic Petri dishes, Parafilm®, 1 L autoclave-resistant glass bottles, autoclave, hot plate, 70% (v/v) ethanol.
4. *Immunolabeling hardware supplies*: 24-welled non-coated tissue culture dishes (Fisher Scientific) (*see* Fig. 1a), 100–1000 μL micropipette and pipette tips, aspirator for suctioning off fluids (optional), fine-tip jeweler's forceps, 1.5 mL microcentrifuge tubes, waste container, laboratory rotator that holds Petri dishes for gently bathing seedlings with appropriate solutions, aluminum foil, single-well immuno-slides (EMS, Ft. Washington, PA, USA), and 22 × 22 mm glass coverslips.
5. *Immunolabeling reagents*: Monoclonal antibodies for probing cell wall epitopes may be obtained from Plant Probes (Leeds, UK; http://www.plantprobes.net/index.php) and the Complex Carbohydrate Research Center (Athens, Georgia, USA: https://www.ccrc.uga.edu/), anti-rat FITC or -TRITC, anti-mouse FITC or -TRITC, or other anti-rat/mouse fluorophore conjugates, nonfat milk powder, and DH_2O (*see* **Note 1**; Fig. 2).
6. *Light and confocal microscope hardware*: Olympus Fluo View™ 300 or 1200 confocal laser scanning microscope (or comparable microscope); IX70 Inverted microscope with fluorescence optics equipped with an Olympus DP71 digital camera (or comparable microscope) for preliminary assessment of root hairs.

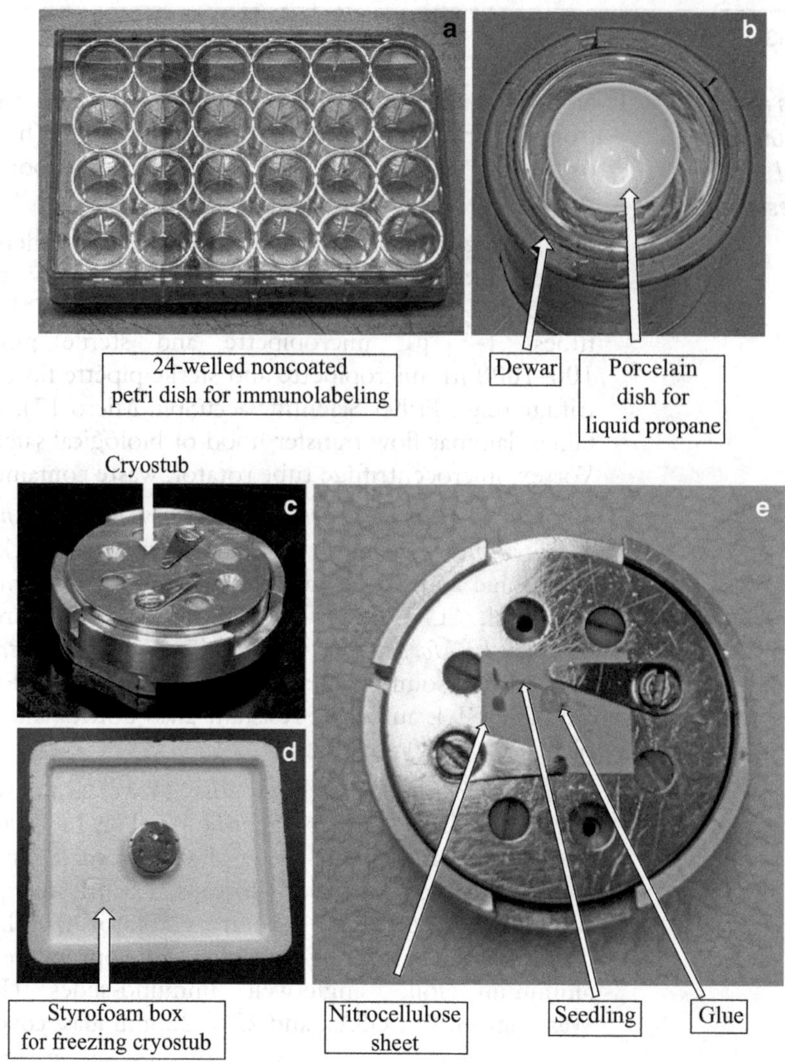

Fig. 1 Hardware for rapid assessment techniques. (**a**) Multiple antibody-based immunolabeling is conveniently performed in 24-welled Petri dishes. The small size of each well also allows for limiting the amount of antibody, that is, a money saving feature. (**b**) Liquid propane is liquefied in a porcelain dish immersed in liquid nitrogen. Propane is delivered to the dish from the propane tank with a simple tube attached to a pipette. The seedling that is attached to a nitrocellulose sheet is then plunge frozen in the propane. (**c**) The cryostub is the platform for VPSEM viewing. After it is cooled, the seedling that has been frozen on a nitrocellulose sheet is fastened to the stub and inserted into the VPSEM. (**d**) The cryostub is cooled in a shallow Styrofoam box filled with liquid nitrogen. (**e**) The seedling is glued to a nitrocellulose sheet with a drop of cryogen tolerant glue and fastened by a clip to the cryostub

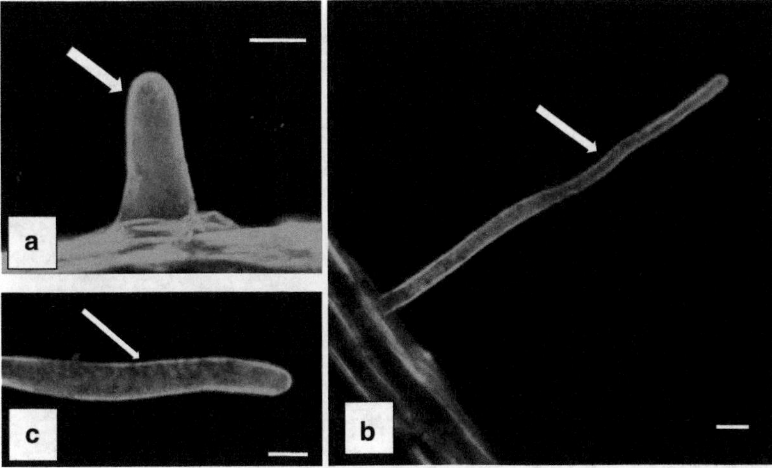

Fig. 2 LM15 immunolabeled roots hairs illustrating the distribution of xyloglucan on the surface of the wall. (**a**) A young root hair (arrow) emerges from the epidermal layer of the root. Bar = 12 μm. (**b**) Mature root hair (arrow) Bar = 13 μm. (**c**) Magnified view of mature root hair (arrow) Bar = 12 μm

2.2 Supplies for Variable Pressure Scanning Electron Microscopy (VPSEM)

1. Nitrocellulose membrane 0.45 μm pore size (Thomas Scientific) cut in 2 cm by 2 cm squares sheet.
2. Fine-tip jeweler's forceps and razor blades.
3. Low temperature glue suitable for liquid nitrogen manipulations (e.g., Tissue Tek; EMS, Ft. Washington, PA, USA).
4. Liquid nitrogen.
5. Propane tank (camping stove quality is satisfactory).
6. 10 mL porcelain dishes for freezing.
7. Liquid nitrogen-resistant thermos for holding porcelain dishes and liquefying propane.
8. Dewar for holding and transporting samples in liquid nitrogen.
9. JEOL cryostub (JEOL Corp., Peabody, MA) or comparable metal platform for holding frozen samples.
10. Shallow Styrofoam box for holding cryostub and cooling it with liquid nitrogen.
11. Fume hood.
12. Stereomicroscope for viewing and positioning specimens.
13. Variable pressure scanning electron microscope (VPSEM); our lab uses a JEOL JSM-6480 VPSEM (Fig. 3).

3 Methods

3.1 Seedling Growth, Live Cell Labeling and Analysis

Prepare all solutions using ultrapure DH_2O (18 MΩ) and analytical grade reagents. Store stock solutions in a refrigerator, unless otherwise stated. Follow all institutional and government safety and

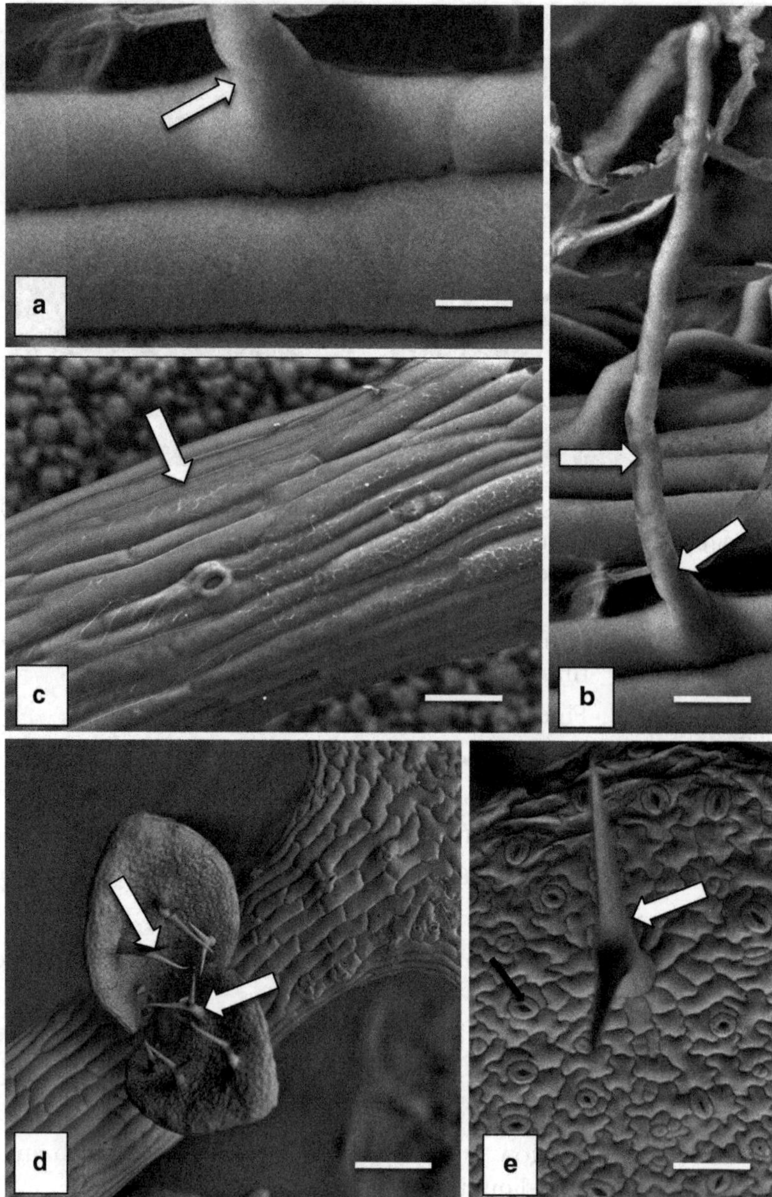

Fig. 3 Profiles of vegetative parts prepared via cryofixation and VPSEM. (**a**) The base of a root hair (arrow) emerging from the root epidermis. Bar = 9 μm. (**b**) Root hair (arrow). Bar = 25 μm. (**c**) The epidermal cells of a 7-day-old stem (arrow). Bar = 25 μm. (**d**) the surface of a 7-day-old cotyledon covered with trichomes (arrows). Bar = 75 μm. (**e**) Magnified view of the surface of a cotyledon showing the trichome (white arrow) and the numerous stomata (black arrow). Bar = 12 μm

waste disposal protocols for the laboratory. Use a laminar flow hood or biological safety cabinet that has been washed down with 20% Clorox before handling medium and performing seed sterilization procedures.

1. *Medium composition and processing*:
   (a) *MS liquid medium*: Dissolve 1 preweighed packet of 1× MS (Murashige–Skoog) salts with Gamborg's vitamins, 0.5 MES, and 10 g of sucrose in 750 mL of DH_2O. Adjust pH to 6.0 using 0.1 M KOH and then bring volume to 1 L. Dispense medium into autoclave-resistant glass bottles and autoclave at 121 °C/20 min/15 lbs/in^2.
   (b) *MS agarose medium*: Make 1 L of MS liquid medium and dispense 500 mL into each of two glass bottles. Add 6.5 g of agarose to each bottle and heat on hot plate to dissolve the agarose (*see* **Note 2**). Autoclave at 121 °C/20 min/15 lbs/in^2. Dispense 20 mL aliquots of autoclaved MS-agarose medium into sterile plastic Petri dishes and let cool under a laminar flow hood. When the agarose has solidified, the dishes may be used immediately or stored in a refrigerator until needed.
2. *Seed sterilization*:
   (a) Place approximately 100 seeds into a sterile microcentrifuge tube.
   (b) Fill the tube to 500 μL with 70% ethanol. Cap the tube, vortex for 30 s and place on microcentrifuge tube rotator/shaker for 5 min.
   (c) Centrifuge the tube with microcentrifuge at 100 × *g*.
   (d) With a 100–1000 μL micropipette and sterile tip remove and discard the ethanol and resuspend the seeds in the pellet with 500 μL of sterile DH_2O. Cap the tube and vortex.
   (e) Centrifuge the tube at 100 × *g*.
   (f) With a 100–1000 μL micropipette and sterile tip remove and discard the DH_2O and resuspend the seeds in the pellet in 500 μL of 20% Clorox. Vortex for 30 s and place on a microcentrifuge tube shaker for 30 min.
   (g) Centrifuge the microcentrifuge tube at 100 × *g*.
   (h) With a 100–1000 μL micropipette and sterile tip remove and discard the Clorox solution and resuspend the seeds in the pellet with 500 μL of sterile DH_2O.
   (i) Centrifuge the tube at 100 × *g*.
   (j) With a 100–1000 μL micropipette and sterile tip remove and discard the DH_2O. Resuspend the seeds in the pellet in 500 μL of sterile DH_2O.
   (k) Centrifuge the tube at 100 × *g*.
   (l) Repeat Subheading 3.2, **steps 10** and **11** twice more.
   (m) After removing the last DH_2O wash resuspend the seeds in 100 μL sterile DH_2O. Quickly pour the seeds into an empty sterile Petri dish.

(n) Using a 0.5–5 μL micropipette set at 3 μL, place the sterile tip near a seed and suction up a seed.

(o) Place the seed on the surface of a MS agarose Petri dish. Repeat this procedure and place approximately three rows of seeds (5–10 seeds per row) on the agar. Be sure to maintain aseptic conditions while collecting and implanting seeds on agarose.

(p) Seal the Petri dish with a thin strip of Parafilm®.

(q) Place the Petri dish on its side in a growth chamber set at 20–24 °C with diffuse light.

(r) Seeds will germinate in a 2–3 days and seedling roots will attain lengths of 1–2 cm in a week. Plants can be harvested during this period.

3.2 Root Labeling Protocol

1. *Labeling solutions:* Prior to labeling, make the following solutions:
   (a) *MS-Block*: Dissolve 1 g of nonfat instant milk in 100 mL of MS Liquid medium. Shake this solution for 5 min, dispense 1.2 mL aliquots into 1.5 mL microcentrifuge tubes and centrifuge at 8000 × *g* for 3 min. Use the supernatant as your blocking solution. *See* **Note 3**.
   (b) *Antibody dilutions*: Dilute primary antibodies 1/10 (v/v) in MS liquid medium. Dilute secondary antibodies 1/100 (v/v) in MS Liquid medium. *See* **Note 4**.
2. Dispense 200 μL of MS-Block into the wells of a 24-welled Petri dish (e.g., one seedling per well, triplicate labeling for each antibody) (*see* Fig. 1a). With fine-tip forceps gently remove a seedling from the agarose plate and submerge in the block solution. Place the plate on a Petri dish shaker and gently shake for 30 min.
3. Using a 100–1000 μL micropipette or aspirator, remove and discard the MS-Block from each well, being careful not to touch the seedling.
4. Resubmerge the seedling in 250 μL of MS Liquid medium and shake for 5 min. Remove and discard the MS with the micropipette or aspirator. Repeat this step twice more.
5. Resubmerge the seedling in 200 μL of primary antibody. Cover the dish with aluminum foil and place on a shaker for 90 min.
6. Remove and discard the antibody solution with micropipette or aspirator. Resubmerge the seedling in 250 μL of MS Liquid medium and shake for 5 min. Remove and discard the MS liquid medium. Repeat this step twice more.
7. Place the seedlings in 250 μL of MS-Block and place the dish on a shaker for 30 min.
8. Repeat Subheading 3.3, **steps 3** and **4**.

9. Place the seedlings in 200 μL of secondary antibody. Cover the dish with aluminum foil and place on a shaker for 90 min.
10. Repeat Subheading 3.3, **steps 6** and 7. Leave seedlings in 200 μL of MS Liquid medium. Seedlings can be imaged immediately or stored in a refrigerator for 24 h and then viewed.
11. Place a 100 μL drop of MS Liquid medium onto the well of an immuno-slide. Using fine-tip forceps, remove the seedling and place in the drop of MS Liquid medium. Arrange the seedling so that the root hairs are spread out. A stereomicroscope is helpful in this procedure. Place a 22 × 22 mm coverslip on the drop (*see* **Note 5**).
12. *Microscopy imaging*: Observe root hairs with a fluorescence microscope using a green filter (for TRITC) or blue filter (for FITC). For confocal laser scanning microscopy (CLSM), use a green laser with TRITC filter or blue laser with FITC filter set. Cells labeled with TRITC-coupled secondary antibody may be observed as above and the chloroplast may be viewed with the blue laser and FITC filter set. The images or image sets can then be merged. Figure 2 illustrates examples of this labeling protocol.

3.3 Variable Pressure Scanning Electron Microscopy (VPSEM)

1. Place the porcelain dish in the thermos and cool it down by adding liquid nitrogen around it. This operation should always be performed under a fume hood. Once cooled, liquefy propane in the porcelain dish (*see* Fig. 1b). Keep the dish cold by occasionally replenishing the liquid nitrogen.
2. Place a small drop of low temperature glue onto the center of a nitrocellulose square. Using fine-tip forceps, gently remove a seedling from the MS Agarose plate and adhere to the glue. Only adhere to a portion of the seedling that is not the focal point of imaging. For example, if viewing roots, adhere the stem to the glue. Rapidly plunge the nitrocellulose square sheet into the liquid propane. Let the seedling sit in the propane for 20 s.
3. Using fine-tip forceps, quickly transfer the frozen seedling/nitrocellulose sheet to a Dewar filled with liquid nitrogen (*see* **Note 6**).
4. Place the JEOL cryostub in a shallow Styrofoam box filled with liquid nitrogen and allow it to cool down (*see* Fig. 1c–e), until the cryostub chamber fills in with liquid nitrogen (after 2–3 min a clear bubbling noise indicates that the chamber is full).
5. Transfer the seedling/nitrocellulose sheet with fine-tip forceps and secure it to the top of the cryostub by tightening the screws of the clips on top of the platform.

6. Quickly place the cryostub in the VP-SEM chamber, close it and set up the microscope to low vacuum mode. Although the settings may vary, usually imaging is performed using back-scattered electrons signal (BEIW), with a pressure of 27 Pa, accelerating voltage of 10 kV, working distance of 6 mm, spot size of 60. Figure 3 provides examples of the VPSEM samples.

3.4 Concluding Remarks

In Arabidopsis, the large collection of T-DNA insertion mutants and the variety of ecotypes available to researchers provides a genetic resource to investigate the relationship between cell wall polymer organization and cell shape. This has been especially useful in studying the role of cell wall synthesis and assembly during polarized growth in root hairs. The methods described here provide two rapid, qualitative approaches to investigate the organization of wall polymers on the surface of live seedlings and the consequence of mutations affecting cell wall structure on cell and organ shape. These experimental approaches should provide additional tools for cell biologists to dissect the relationships between wall organization and cell/organ shape in plants.

4 Notes

1. A variety of fluorophore-conjugated antibodies can be used to fit the needs of the microscopist, including Alexa Fluor and CY.
2. The agarose can also be mixed in the MS liquid and dissolved during autoclaving.
3. We have had equal success with block made from either nonfat milk purchased from either a scientific supply company or supermarket.
4. The antibody dilutions reported here are those that have worked well. However, one should try out multiple dilutions of primary and secondary antibodies to determine the best labeling.
5. The single welled immunoslides have a slight depression that helps withstand compression on the seedling by the overlying coverslip. One can use deeper depression slides to prevent any compression but this will affect the choice of objective lenses that can be used for imaging.
6. One can substitute Styrofoam cups for Dewars. Be sure to wear protective gloves and work in a well-ventilated room when using liquid nitrogen.

Acknowledgments

This work was supported by NSF grant NSF-MCB-RUI-1517345 (D.S.D.) and Hatch grant VT-HO2001 (M.L.T.).

References

1. Singh SK, Fischer U, Singh M, Grebe M, Marchant A (2008) Insight into the early steps of root hair formation revealed by the procuste1 cellulose synthase mutant of *Arabidopsis thaliana*. BMC Plant Biol 8:57
2. Zabotina OA, van de Ven WT, Freshour G, Drakakaki G, Cavalier D, Mouille G, Hahn MG, Keegstra K, Raikhel NV (2008) Arabidopsis XXT5 gene encodes a putative alpha-1,6-xylosyltransferase that is involved in xyloglucan biosynthesis. Plant J 56:101–115
3. Zabotina OA, Avci U, Cavalier D, Pattathil S, Chou YH, Eberhard S, Danhof L, Keegstra K, Hahn MG (2012) Mutations in multiple XXT genes of Arabidopsis reveal the complexity of xyloglucan biosynthesis. Plant Physiol 159:1367–1384
4. Peña MJ, Kong Y, York WS, O'Neill MA (2012) A galacturonic acid-containing xyloglucan is involved in Arabidopsis root hair tip growth. Plant Cell 24:4511–4524
5. Diet A, Link B, Seifert GJ, Schellenberg B, Wagner U, Pauly M, Reiter WD, Ringli C (2006) The Arabidopsis root hair cell wall formation mutant lrx1 is suppressed by mutations in the RHM1 gene encoding a UDP-L-rhamnose synthase. Plant Cell 18:1630–1641
6. Pang CY, Wang H, Pang Y, Xu C, Jiao Y, Qin YM, Western TL, Yu SX, Zhu YX (2010) Comparative proteomics indicates that biosynthesis of pectic precursors is important for cotton fiber and Arabidopsis root hair elongation. Mol Cell Proteomics 9:2019–2033
7. Favery B, Ryan E, Foreman J, Linstead P, Boudonck K, Steer M, Shaw P, Dolan L (2001) KOJAK encodes a cellulose synthase-like protein required for root hair cell morphogenesis in Arabidopsis. Genes Dev 15 (1):79–89
8. Wang X, Cnops G, Vanderhaeghen R, De Block S, Van Montagu M, Van Lijsebettens M (2001) AtCSLD3, a cellulose synthase-like gene important for root hair growth in Arabidopsis. Plant Physiol 126:575–586
9. Bernal AJ, Yoo CM, Mutwil M, Jensen JK, Hou G, Blaukopf C, Sørensen I, Blancaflor EB, Scheller HV, Willats WG (2008) Functional analysis of the cellulose synthase-like genes CSLD1, CSLD2, and CSLD4 in tip-growing Arabidopsis cells. Plant Physiol 148:1238–1253
10. Baumberger N, Ringli C, Keller B (2001) The chimeric leucine-rich repeat/extensin cell wall protein LRX1 is required for root hair morphogenesis in *Arabidopsis thaliana*. Genes Dev 15:1128–1139
11. Velasquez SM, Ricardi MM, Dorosz JG, Fernandez PV, Nadra AD, Pol-Fachin L, Egelund J, Gille S, Harholt J, Ciancia M, Verli H, Pauly M, Bacic A, Olsen CE, Ulvskov P, Petersen BL, Somerville C, Iusem ND, Estevez JM (2011) *O*-glycosylated cell wall proteins are essential in root hair growth. Science 332:1401–1403
12. Pattathil S, Avci U, Baldwin D, Swennes AG, McGill JA, Popper Z, Bootten T, Albert A, Davis RH, Chennareddy C, Dong R, O'Shea B, Rossi R, Leoff C, Freshour G, Narra R, O'Neill M, York WS, Hahn MG (2010) A comprehensive toolkit of plant cell wall glycan-directed monoclonal antibodies. Plant Physiol 153:514–525
13. Leroux O, Sørensen I, Marcus SE, Viane RLL, Willats WGT, Knox JP (2015) Antibody-based screening of cell wall matrix glycans in ferns reveals taxon, tissue and cell-type specific distribution patterns. BMC Plant Biol 15:56
14. Larson ER, Tierney ML, Tinaz B, Domozych DS (2014) Using monoclonal antibodies to label living root hairs: a novel tool for studying cell wall microarchitecture and dynamics in Arabidopsis. Plant Methods 10:30

Chapter 18

Analysis of Plant Cell Walls Using High-Throughput Profiling Techniques with Multivariate Methods

John P. Moore, Yu Gao, Anscha J. J. Zietsman, Jonatan U. Fangel, Johan Trygg, William G. T. Willats, and Melané A. Vivier

Abstract

Plant cell walls are composed of a number of coextensive polysaccharide-rich networks (i.e., pectin, hemicellulose, protein). Polysaccharide-rich cell walls are important in a number of biological processes including fruit ripening, plant–pathogen interactions (e.g., pathogenic fungi), fermentations (e.g., wine-making), and tissue differentiation (e.g., secondary cell walls). Applying appropriate methods is necessary to assess biological roles as for example in putative plant gene functional characterization (e.g., experimental evaluation of transgenic plants). Obtaining datasets is relatively easy, using for example gas chromatography–mass spectrometry (GC-MS) methods for monosaccharide composition, Fourier transform infrared spectroscopy (FT-IR) and comprehensive microarray polymer profiling (CoMPP); however, analyzing the data requires implementing statistical tools for large-scale datasets. We have validated and implemented a range of multivariate data analysis methods on datasets from tobacco, grapevine, and wine polysaccharide studies. Here we present the workflow from processing samples to acquiring data to performing data analysis (particularly principal component analysis (PCA) and orthogonal projection to latent structure (OPLS) methods).

Key words Cell wall profiling, GC-MS, FT-IR, CoMPP, Multivariate data analysis

1 Introduction

Plant cell walls are of tremendous agricultural and industrial importance; having significant value in relation to food quality and texture aspects, fruit and vegetable ripening, biomass and bioenergy and beverage production (e.g., wine) [1, 2]. They are also important from a scientific perspective for their role in plant biology, in understanding the basis for biotic and abiotic stress; here an example is the tobacco model pathosystem where plant cell walls constitute the matrix wherein bacterial and fungal infections occur [2]. Given the advance in various systems biology based studies where it is often necessary to screen large populations of plants or in field studies (e.g., vineyards) where work occurs outside the

Zoë A. Popper (ed.), *The Plant Cell Wall: Methods and Protocols*, Methods in Molecular Biology, vol. 2149,
https://doi.org/10.1007/978-1-0716-0621-6_18,

controlled conditions present in laboratories; it has become increasingly necessary to optimize sampling and data analysis in a high-throughput manner. We routinely use three techniques in combination to evaluate large numbers of samples in parallel; and process the data using standard multivariate methods, commonly principal component analysis (PCA) and orthogonal projection to latent structures (OPLS) modelling. A common scenario is in grape sampling from a vineyard, where a single ripe grape comprises ca. 1–1.5 g fresh weight, while the majority of the material comprises water and metabolites (e.g., sugars and acids). After processing you can recover as little as five per cent of the starting weight, we can utilize the alcohol insoluble residue (AIR) at 20–40 mg (split from the same sample) and perform three analyses monosaccharide composition via gas chromatography (using 5 mg), infrared spectroscopy (using 5 mg), and comprehensive microarray polymer profiling (CoMPP) [3] (using 10 mg). The datasets are then pre-processed for analysis using multivariate data analysis software (we use SIMCA 16 from MKS (Sartorius Stedim Biotech); however, other types of analyses and software are also available). The data are then inspected using PCA, and various scenario (or treatment) models can be investigated using OPLS simulations. The protocols outlined below show detailed procedures followed in previous publications (*see* [4–6]).

2 Materials

2.1 Solvents, Reagents, and Buffers

1. Solvents: ethanol, methanol, chloroform, and acetone (analytical reagent grade from Sigma-Aldrich or Merck) and MilliQ water (Millipore, USA) for preparation of the alcohol insoluble residue (AIR).
2. HCl/Methanol 3M kit (Supelco, Bellafonte, PA, USA).
3. Analytical grade methanol (dried over molecular sieve) (Sigma-Aldrich, USA).
4. TMS derivatization (Sylon HTP) kit (contains HMDS–TMCS–pyridine in the ratio 3:1:9)) (Supelco, Bellafonte, PA, USA).
5. Analytical grade cyclohexane (Sigma-Aldrich, USA).

2.2 Solutions

1. Internal standard myoinositol (1 mg/mL) (Sigma-Aldrich, USA) in MilliQ water.
2. 4 M TFA (trifluoroacetic acid from Sigma-Aldrich, USA) solution in MilliQ water.
3. A stock solution of sugar standards (arabinose, rhamnose, fucose, xylose, mannose, galactose, glucose and galacturonic acid from Sigma-Aldrich, USA) in MilliQ (Millipore, USA) water (store aliquots frozen).

2.3 Materials

1. Liquid nitrogen (and associated storage tanks, insulated gloves and handling equipment).
2. Falcon or Greiner plastic screw-cap tubes (15 and 50 mL).
3. 10 mL borosilicate Pyrex glass culture tubes with Teflon-lined screw caps.
4. Pasteur pipettes (glass) and rubber pipetting device.
5. GC glass microvials (1.5 mL).

2.4 Equipment

1. Retsch Mixer Mill (Retsch, Haan, Germany).
2. Metal ball bearings (Retsch, Haan, Germany).
3. Centrifuge (Universal 3W, Hettich Zentrifugen, Germany) with inserts designed for 15 and 50 mL Falcon tubes and 10 mL Pyrex glass culture tubes.
4. Digital oven-incubator (Scientific Series 2000, South Africa).
5. −80 °C freezer (Snijelen Scientific, Holland)
6. Rotating wheel (e.g., Labnet II, LabRoller, USA).
7. Freeze-drier (Vacutec, Labconco, FreeZone, USA).
8. Reacti-Therm nitrogen drying unit (Thermo Scientific, MA, USA) in fume hood.
9. Gas chromatograph-coupled mass spectrometer (Hewlett Packard 5890 series II, USA) (with Chemstation software).
10. NEXUS 670 Fourier transform-infrared spectroscopy instrument (Thermo Scientific, MA, USA) containing a Golden Gate Diamond ATR (attenuated total reflectance) accessory with a type II diamond crystal (with Thermo Scientific instrument software).
11. CoMPP microarray robot (*see* CoMPP Chapter, J. Fangel, Ph. D., thesis 2012, University of Copenhagen, Denmark, unpublished).
12. PC with the following minimum system requirements for operating SIMCA 16:
    - Pentium based computer (PC) with a 1.5 GHz or faster processor.
    - 1 GB RAM or more.
    - 1 GB available hard disk space.
    - 1024 × 768 screen resolution color display.
    - Microsoft Windows 8 or 10.
    - Graphics card that has hardware 3D acceleration and supports Open GL.

13. SIMCA 16 and associated software multivariate data analysis products can be trialed and purchased from MKS Umetrics head office, Umeå, Sweden (+46 90184800; umetrics@sartorius-stedim.com) (Sartorius Stedim Biotech).
14. Microsoft Office (Microsoft Corporation, USA).

3 Methods

3.1 Plant Cell Wall Material

1. Plant material (i.e., leaves, fruit, fermentation samples) are ideally collected by flash-freezing in liquid nitrogen (to ensure cessation of metabolism) with biological replicates (at least in triplicate) [1, 2]. A number of factors including developmental state, date of collection, number of sampling points (over a season or process), can influence the data. The aim is to ensure that the sampling preserves the representativeness of the material. Sampling information should be recorded.
2. Plant material can vary with toughness and viscosity (among other properties) and so ensuring a uniform homogenization process is very important. We find that milling systems (particularly the Retsch Mixer Mill) produces highly reproducible AIR powders [4, 5]. We strongly recommend this system over pestle and mortar or other semieffective grinding systems. Samples immersed in liquid nitrogen are placed into a precooled 50 mL chamber with a single large metal ball bearing. The samples are milled in liquid nitrogen at 30 Hz for 15 s intervals (number of repeats should be kept uniform per experiment) until a fine powder is obtained which is kept at −80 °C.
3. The powders are made up to 80% ice-cold ethanol in Falcon tubes and then boiled for 15 min to deactivate endogenous plant enzymes. Additionally a destarching step should be considered with leaf samples in particular, an amylase and amyloglucosidase are conveniently available from Megazyme (Wicklow, Ireland), as starch will saturate the GC-MS trace with free glucose.
4. After cooling to room temperature the samples are centrifuged at 2500 × *g* for 10 min and the supernatant discarded.
5. To the pellets are added in a 1:10 (w/v) ratio absolute methanol. The tubes are placed on a rotating wheel for 2 h, followed by centrifugation again and replacement of the supernatant. This is repeated with methanol–chloroform (1:1) (v/v), chloroform, chloroform–acetone, and acetone solvent solutions.
6. It is important to use a series of organic solvents (particularly the combination of ethanol, methanol, chloroform, and acetone) as this opens up the AIR particularly for CoMPP analysis [3] (*see* CoMPP Chapter). It also may be necessary to repeat

washings with solvents such as chloroform (or hexane) to enhance the dewaxing of certain plant samples (e.g., tobacco) [2].

7. After discarding the final acetone supernatant, the tubes are placed with lids unscrewed in a fume hood to dry off excess of acetone, but without letting the pellet air-dry.
8. The pellet is resuspended in ice-cold MilliQ water, frozen at −80 °C and freeze-dried to form a loose fluffy powder which can be stored sealed until further analysis.

3.2 Extraction of Cell Walls for Analyses

1. At this point the AIR samples can be split for different analyses (e.g. GC-MS, FT-IR and CoMPP).
2. It is recommended that at least 10–15 mg of AIR is collected per sample for conducting all three analyses (i.e. ca. 5 mg for GC-MS, ca. 5 mg for FT-IR and ca. 10 mg for CoMPP).

GC-MS for monosaccharide composition

3. In order to obtain the monosaccharide composition via GC-MS ca. 5 mg of each AIR sample should be weighed out into a 10 mL borosilicate Pyrex glass culture tube.
4. To each sample tube should be added 500 μL of MilliQ water, 90 μL of 1 mg/mL myoinositol (as internal standard) and 500 μL of 4 M TFA.
5. A separate set of tubes should contain the standard nine sugar mixture with 90 μL of 1 mg/mL myoinositol (as internal standard) and then 500 μL of 4 M TFA.
6. All tubes should be sealed and placed in a preheated oven (set at 110 °C) for 2 h.
7. After cooling the supernatants are transferred to a set of new clean 10 mL borosilicate Pyrex glass culture tubes. The tubes are then placed in the Reacti-Therm unit and then dried at 60 °C under a nitrogen stream in a fume hood.
8. Dilute 1 volume of methanol HCl with 2 volumes of dry methanol. Add 500 μL of the acidified methanol to each tube and seal. Incubate in a preheated oven at 80 °C for 16 h.
9. Dry off the methanol at 40 °C under a nitrogen stream in a fume hood. Add 250 μL dry methanol and repeat washing (and drying) in dry methanol three times.
10. To the dried tubes add 150 μL of the silylation reagent (TMS kit from Supelco) and incubate in a preheated oven at 80 °C for 20 min.
11. After drying the tubes under a nitrogen stream at 40 °C, a further 1000 μL cyclohexane is added and mixed.
12. The redissolved cyclohexane solutions are then transferred into GC microvials and sealed for analysis.

13. Derivatives are separated and analyzed in a gas chromatograph (Hewlett Packard 5890 series II) linked to a mass spectrometer. A HPS-MS column, 30 m × 0.25 mm (i.d.), was used. The following temperature program is used; a stable oven temperature of 120 °C for 2 min, then ramped up at 10 °C intervals per minute to 160 °C, followed by further ramping at 1.5 °C/min to 220 °C, before finally an increase of 20 °C/min to 280 °C.
14. Data analysis was performed using ChemStation software (MSD Chemstation Agilent Technologies, USA). Derivatives were identified based on their retention time and mass spectra, and quantified relative to the internal standard myoinositol and the standard sugar mixtures.

ATR-FTIR spectroscopy

15. About 5 mg of the AIR powder is spread evenly across the diamond window of the Golden Gate Diamond accessory window. The powder is then clamped into position and the instrument operated in ATR mode.
16. Spectra are acquired in the region 4000–650 cm^{-1} with Geon-KBr beamsplitter and DTGS/Csl detector.
17. Each spectrum consists of at least 120 coadded scans per sample. It is also important to run sample repeats periodically to ensure instrument stability.
18. Spectral files are saved as Omnic file types (Thermo Scientific, USA) which can be imported into Microsoft Excel and SIMCA 16 (we also do this using Unscrambler X (Camo, USA) software).

CoMPP preparation

19. Samples ca. 10 mg each of AIR are placed in 96 tube and cap boxes (Greiner Bio-One GmbH, Austria).
20. Samples are processed for CoMPP (*see* CoMPP Chapter).

3.3 Processing of the Data for Analysis

1. All file types (from GC-MS, FT-IR, and CoMPP) are eventually converted into Microsoft Excel and preformatted before import into SIMCA.
2. GC-MS data is preprocessed on Agilent ChemStation software (Agilent, USA), peak assignments verified against control mixtures and integration validated manually as necessary.
3. ATR-FT-IR spectra can be converted directly from Omnic format into Microsoft Excel and/or SIMCA. It is preferable that the data is cropped to the chemically information-rich spectral region ca. 1768–770 cm^{-1} before importing.
4. CoMPP data should be provided as raw data in Microsoft Excel format for subsequent import into SIMCA for preprocessing. A heat-map is a valuable aid for initial interpretation but the values from the prenormalized heat-map should not be used for subsequent multivariate data analysis.

3.4 Multivariate Analysis Methods

1. Given that as screening tools such as GC-MS, FT-IR and CoMPP [1, 2] become more sophisticated, the number of variables detected (monosaccharides, probes, etc.) become more numerous. This can lead to greater difficulty in visually inspecting datasets for differences, that is, between samples, treatments, and so on. It is therefore imperative to use unsupervised statistical approaches to ascertain differences within sample sets.
2. We routinely choose to use the SIMCA 16 software package from MKS Umetrics, Umeå, Sweden (+46 90184 800); umetrics@sartorius-stedim.com) (Sartorius Stedim Biotech). However, other packages and software are available which do similar types of analyses. The SIMCA package conveniently contains a range of multivariate methods and also clustering tools to analyze large datasets. Data can be prepared using Microsoft Excel and the data-file preformatted to include additional nonquantitative information such as secondary IDs (background information on samples) as well as class IDs (information acquired before or after initial multivariate analysis).
3. Unsupervised methods, such as principal component analysis (PCA), allows for an unbiased overview of data structure allowing for unexpected factors in sample collection to be revealed and detection of outliers.
4. It is beyond the aim of this brief methods chapter to explain multivariate methods or the SIMCA 16 software package. Useful introductory texts and free trial versions of this particular software are available on the Umetrics website.
5. In the following unpublished example the focus is on a CoMPP cell wall CDTA (pectin-rich) dataset obtained from a fermented wine pomace sample-set where grapes were macerated (and fermented) in the absence (U = untreated) and presence of enzymes (EPG, endopolygalacturonase; EPM, endopolygalacturonase and pectin methylesterase; ARA, endopolygalacturonase, pectin methylesterase and an endoarabinanase; GAL, endopolygalacturonase, pectin methylesterase and an endogalactanase; PL, pectin lyase; Cru, a commercial pectinase mixture from an *Aspergillus niger* preparation).
6. The first unsupervised analysis is a principal component analysis (PCA) which provides an overview of the sample clustering in a score plot (Fig. 1a) in this case PL and Cru are omitted.
7. In PCA the algorithms find the maximum variance in the data assigning it to principal component 1 (PC1) and then proceed to PC2, PC3, and so on. In this case (Fig. 1a) PC1 accounts for 41.3% of the variance followed by PC2 at 23.3%.

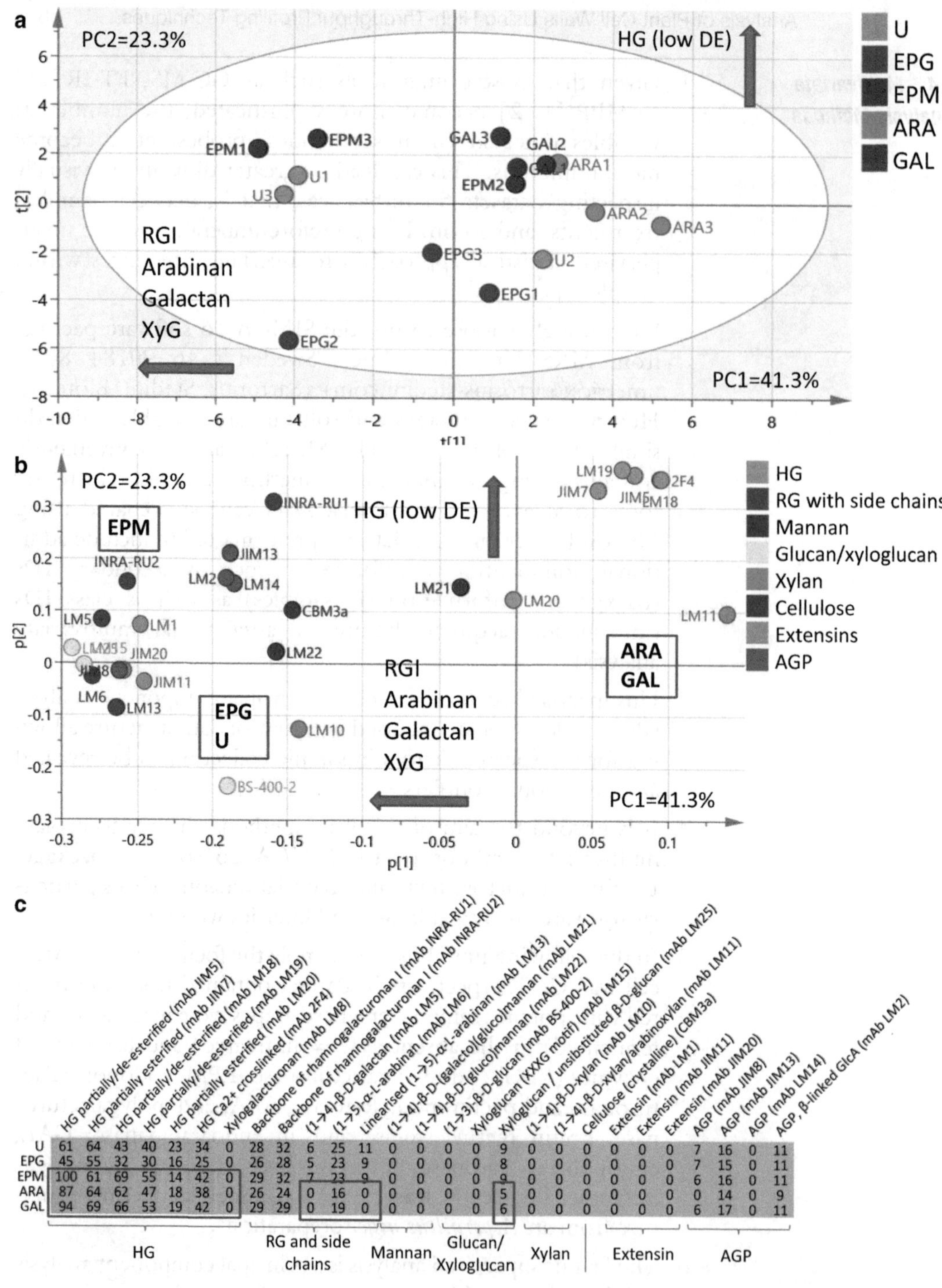

	HG partially/de-esterified (mAb JIM5)	HG partially esterified (mAb JIM7)	HG partially/de-esterified (mAb LM18)	HG partially/de-esterified (mAb LM19)	HG partially esterified (mAb LM20)	HG Ca2+ crosslinked (mAb 2F4)	Xylogalacturonan (mAb LM8)	Backbone of rhamnogalacturonan I (mAb INRA-RU1)	Backbone of rhamnogalacturonan I (mAb INRA-RU2)	(1→4)-β-D-galactan (mAb LM5)	(1→5)-α-L-arabinan (mAb LM6)	Linearised (1→5)-α-L-arabinan (mAb LM13)	(1→4)-β-D-(galacto)(gluco)mannan (mAb LM21)	(1→4)-β-D-(gluco)mannan (mAb LM22)	(1→3)-β-D-glucan (mAb BS-400-2)	Xyloglucan (XXXG motif) (mAb LM15)	Xyloglucan / unsibstituted β-D-glucan (mAb LM25)	(1→4)-β-D-xylan (mAb LM10)	(1→4)-β-D-xylan/arabinoxylan (mAb LM11)	Cellulose (crystalline) (CBM3a)	Extensin (mAb LM1)	Extensin (mAb JIM11)	Extensin (mAb JIM20)	AGP (mAb JIM8)	AGP (mAb JIM13)	AGP (mAb LM14)	AGP, β-linked GlcA (mAb LM2)
U	61	64	43	40	23	34	0	28	32	6	25	11	0	0	0	0	9	0	0	0	0	0	0	7	16	0	11
EPG	45	55	32	30	16	25	0	26	28	5	23	9	0	0	0	0	8	0	0	0	0	0	0	7	15	0	11
EPM	100	61	69	55	14	42	0	29	32	7	23	9	0	0	0	0	9	0	0	0	0	0	0	6	16	0	11
ARA	87	64	62	47	18	38	0	26	24	0	16	0	0	0	0	0	5	0	0	0	0	0	0	0	14	0	9
GAL	94	69	66	53	19	42	0	29	29	0	19	0	0	0	0	0	6	0	0	0	0	0	0	6	17	0	11

Fig. 1 CoMPP results showing a PCA score (**a**) plot and loading (**b**) plot of the CDTA extract (pectin-rich) of enzyme treated pomace. (note: *U* untreated, *EPG* endopolygalacturonase, *EPM* endopolygalacturonase and pectinmethylesterase, *ARA* endopolygalacturonase and pectinmethylesterase with endoarabinanase, *GAL* endopolygalacturonase and pectinmethylesterase with endogalactanase.) (**c**) CoMPP heatmap of abundance of antibodies on pomace CDTA extract, the values of variables are average of three biological replicates. A cutoff (<5) was applied. The colors of plots are according to the treatments in (**a**) and to the polymer category in (**b**)

8. The data do not cluster clearly in this case (Fig. 1a) showing the importance of sample replicates. It would seem that U2 and EPM2 might be anomalous samples. This indicates variability within the data structure.
9. Overall in Fig. 1a it would appear that EPM; EPG with U; and ARA with GAL may constitute separate clusters in the score plot.
10. The loading plot (Fig. 1b) shows the variables driving the separation. These loading variables correlate with the probes (mAbs and CBMs used in the CoMPP analysis). The variables appear to show that samples with higher abundance of HG epitopes (associated with mAbs LM19, LM18, LM20, JIM5, JIM7, 2F4, and LM8) are present in the upper part of the loading plot.
11. Inspection of the heatmap (Fig. 1c) supports this; samples U and EPG have a lower signal abundance for the HG epitopes than samples EPM, ARA and GAL suggesting the enzymes unraveled the cell wall structure and exposed the epitopes, leading to increased signal abundance.
12. An additional method commonly used is OPLS (orthogonal projection to latent structures) which can be used alongside the PCA to create a treatment effect PCA. In this case we have included PL and Cru treatments. Each treatment (enzyme EPG, EPM, etc.) is processed as a separate OPLS with U and the treatment (EPG, EPM, etc.).
13. From each of the individual OPLS models of U versus each treatment a p(corr) value is obtained. These p(corr) values are then used to create a treatment PCA (Fig. 2a). In Fig. 2a the centre of the plot represents the untreated state and the various points (EPG, EPM, etc.) represent the treatment directions/effects.
14. It is also then possible to provide a loading plot where the variables responsible for the various treatment directions are mapped (*see* Fig. 2b).
15. A biplot showing samples (treatments) and variables (Fig. 2c) provides the most convenient means of evaluating a treatment. Here the PL and Cru treatments both appear to have a similar effect on the pectin-rich (CDTA) extracts showing a significant deficiency in multiple pectin epitopes (mAb and CBM probes) versus controls thereby suggesting these samples were strongly depectinated.
16. Other methods particularly useful include O2PLS, OPLS-DA and various clustering methods to assess sample variability and the underlying variables strongly correlated with the discrimination.

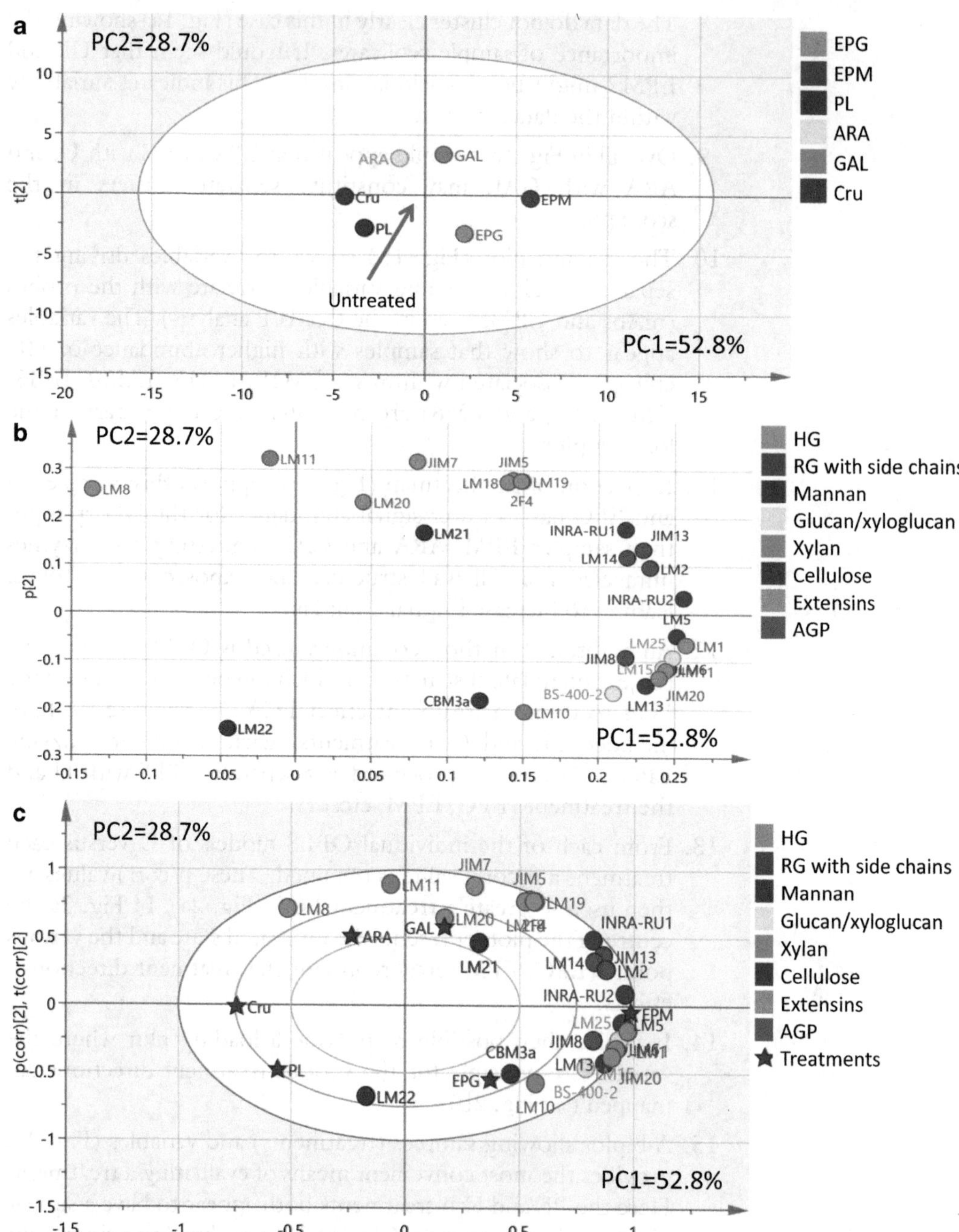

Fig. 2 CoMPP results showing an OPLS model (one model for each treatment versus untreated—center of axes) score (**a**) plot, loading (**b**) plot and biplot (**c**) of the CDTA extract (pectin-rich) of enzyme treated pomace. (note: *EPG* endopolygalacturonase, *EPM* endopolygalacturonase and pectinmethylesterase, *ARA* endopolygalacturonase and pectinmethylesterase with endoarabinanase, *GAL* endopolygalacturonase and pectinmethylesterase with endogalactanase, *PL* pectin lyase, *Cru* commercial pectinase mixture). The colors of plots are according to the treatments in (**a**) and to the polymer category in (**b**), treatments are represented as stars in the biplot (**c**)

17. Examples of combining cell wall fractionation methods and CoMPP is provided in Ref. 4, modeling using PCA and OPLS with cell wall enzyme treatments on grape skins is provided in Ref. 5, and spectral data with clustering methods and OPLS-DA in Ref. 6.

References

1. Moore JP, Fangel JU, Willats WGT, Vivier MA (2014) Pectic-(1,4)-galactan, extensin and arabinogalactan-protein epitopes differentiate ripening stages in wine and table grape cell walls. Ann Bot 114(6):1279–1294
2. Nguema-Ona E, Moore JP, Fagerström A, Fangel JU, Willats WG, Hugo A, Vivier MA (2012) Profiling the main cell wall polysaccharides of tobacco leaves using high-throughput and fractionation techniques. Carbohydr Polym 88 (3):939–949
3. Moller I, Sørensen I, Bernal AJ, Blaukopf C, Lee K, Øbro J, Pettolino F, Roberts A, Mikkelsen JD, Knox JP, Bacic A, Willats WGT (2007) High-throughput mapping of cell-wall polymers within and between plants using novel microarrays. Plant J 50:1118–1128
4. Gao Y, Fangel JU, Willats WGT, Vivier MA, Moore JP (2015) Dissecting the polysaccharide-rich grape cell wall matrix during the red winemaking process, using high-throughput and fractionation methods. Carbohydr Polym 133:567–577
5. Zietsman AJJ, Moore JP, Fangel JU, Willats WGT, Trygg J, Vivier MA (2015) Following the compositional changes of fresh grape skin cell walls during the fermentation process in the presence and absence of maceration enzymes. J Agric Food Chem 63:2798–2810
6. Moore JP, Zhang S-L, Nieuwoudt H, Divol B, Trygg J, Bauer FF (2015) A multivariate approach using attenuated total reflectance mid-infrared spectroscopy to measure the surface mannoproteins and β-glucans of yeast cell walls during wine fermentations. J Agric Food Chem 63(45):10054–10063

Chapter 19

Vibratome Sectioning of Plant Materials for (Immuno)cytochemical Staining

Olivier Leroux

Abstract

A vibrating microtome is widely used to produce good-quality sections of plant organs or tissues. This method allows for an improved preservation of antigenicity and structure and is compatible with most (immuno)cytochemical staining procedures.

Key words Vibration microtomy, Vibratomy, Arabidopsis, Lignin, Anatomy, Agarose embedding

1 Introduction

The production of high-quality sections is of key importance to successful (immuno)cytochemical investigations of plant tissues and cell walls. To ensure the preservation of tissue morphology and antigenicity, adequate fixation and tissue processing is crucial. Indeed, some antigens may not survive even moderate aldehyde fixation or are easily lost or destroyed during prolonged incubations in solvents or exposure to high temperature heating. Several techniques, including embedding in acrylic resin (e.g., LR White) or polyethylene glycol-based wax (Steedman's wax), have been optimized to combine good structural conservation and improved antigenicity [1].

An alternative to these methods is vibration microtomy (or vibratomy), which is a relatively simple technique enabling the production of thin sections of soft to medium hard plant organs or tissues. This method does not require dehydration or embedding in wax, paraffin, or resin and does not necessarily require fixation. In addition, tissues are not subjected to high temperatures during sample processing. As a result, the antigenicity is optimally preserved and epitope access is not hindered as tissues are not infiltrated with resin. Moreover, staining techniques that require organic solvents such as ethanol (e.g., staining for lipids) or

Zoë A. Popper (ed.), *The Plant Cell Wall: Methods and Protocols*, Methods in Molecular Biology, vol. 2149,
https://doi.org/10.1007/978-1-0716-0621-6_19,

concentrated hydrochloric acid (e.g., Wiesner, otherwise known as phloroglucinol, staining for lignins (*see* Fig. 2b), which destroy LR White sections, can be used on vibratome sections. While most of the advantages related to vibratomy also apply to Steedman's wax sectioning, the latter technique has the disadvantage that it necessitates dehydration as well as time and skills to produce good-quality sections. Vibration microtomy, on the other hand, requires no extensive experience and sections of either fresh or fixed samples can be produced within a short period of time. While vibration-assisted sectioning clearly offers many advantages, one should consider the disadvantages (listed in Table 1) before choosing the most appropriate sample processing and sectioning technique. LR White or Steedman's wax embedding may be preferable in some cases. This chapter discusses a variety of techniques for successful vibratome sectioning and subsequent (immuno)staining of plant materials.

Table 1
Advantages and disadvantages of vibratome sections

Advantages	
– Improved antigenicity	– Tissues can be kept at 4 °C throughout processing (if embedded in low melting point agarose, samples are exposed to a maximum temperature of 37 °C) – Tissues are not infiltrated with either wax, paraffin or resins – Dehydration with solvents is not required – Possibility of sectioning fresh, nonfixed tissue
– Fast	– Many sections can be produced within a time span of 10–30 min, depending on the sample – No extensive experience is required
– Compatibility with different staining solutions	– Absence of resin allows for use of ethanol or hydrochloric acid
Disadvantages	
– Section thickness	– Vibratome section thickness can range from 20 to 200 μm, depending on sample properties and imaging requirements
– Difficult samples	– Small samples are more difficult to handle – Some samples may be too hard for the vibratome blades (reducing sample size by preparing smaller segments is a possible solution) – Samples containing large intercellular spaces (e.g., aerenchyma) may be problematic
– Sections are fragile	– Some sections may be difficult to handle – Pretreatments with some enzymes and/or buffers may disintegrate the sections
– Sample storage	– Samples can only be stored temporarily in a buffer solution with 0.1% w/v sodium azide. Alternative techniques (LR White and PEG-embedding allow long term storage of embedded tissues)

2 Materials

2.1 Fixing and/or Storing Plant Material

1. Aldehyde-based fixative: 4% (v/v) formaldehyde (prepared from 16% TEM-grade formaldehyde solution, Electron Microscopy Sciences, Hatfield, USA) in PEM buffer (50 mM PIPES, 5 mM EGTA, 5 mM $MgSO_4$; pH adjusted to 6.9 with NaOH). Aliquots can be stored at −20 °C for up to 3 months. Other types of fixatives can be used to preserve specific structures (e.g., cytoskeleton, chromatin); consult literature to select the most appropriate fixation solution.
2. Phosphate Buffered Saline (PBS) prepared from a 10× stock (pH 7, Carl Roth, Cat. #9143.2) for washing aldehyde-fixed samples.
3. 0.1% sodium azide in buffer solution (PBS or PEM) for storage of aldehyde-fixed samples.
4. (Optional) 70% (v/v) ethanol for sample storage.

2.2 Preparing Plant Material for Sectioning

1. PBS and demineralized water.
2. Whatman No. 1 filter paper.
3. 5–7% (w/v) agarose or low-melting-point agarose in demineralized water, melted in the microwave at the lowest intensity until all agarose is dissolved. Leave to cool until 40–50 °C and most air bubbles have disappeared. If necessary, transfer to a block heater set at 40 °C to prevent solidification of the agarose.
4. Disposable plastic molds (Electron Microscopy Sciences, Hatfield, USA, Cat. #62353-15) (Fig. 1a) or Peel-A-Way disposable embedding molds (Electron Microscopy Sciences, Hatfield, USA, Cat. #70182).
5. Single-edge razor blades to excise samples to a size suitable for gluing and sectioning (e.g., Cat. #71960, Electron Microscopy Sciences, Hatfield, USA).
6. Fine tweezers (e.g., Dumont No. 5, Electron Microscopy Sciences, Hatfield, USA) or tooth-picks to manipulate and orientate samples during embedding.
7. Loctite 406 (Henkel AG & Co) or comparable cyanoacrylate-based superglue.
8. Tray with ice to speed up solidification.

2.3 Sectioning

1. Vibratome (e.g., Thermo Scientific, model HM650V) with specimen fixing plates.
2. Demineralized water or buffer (PBS or PEM), cooled to 4 °C.

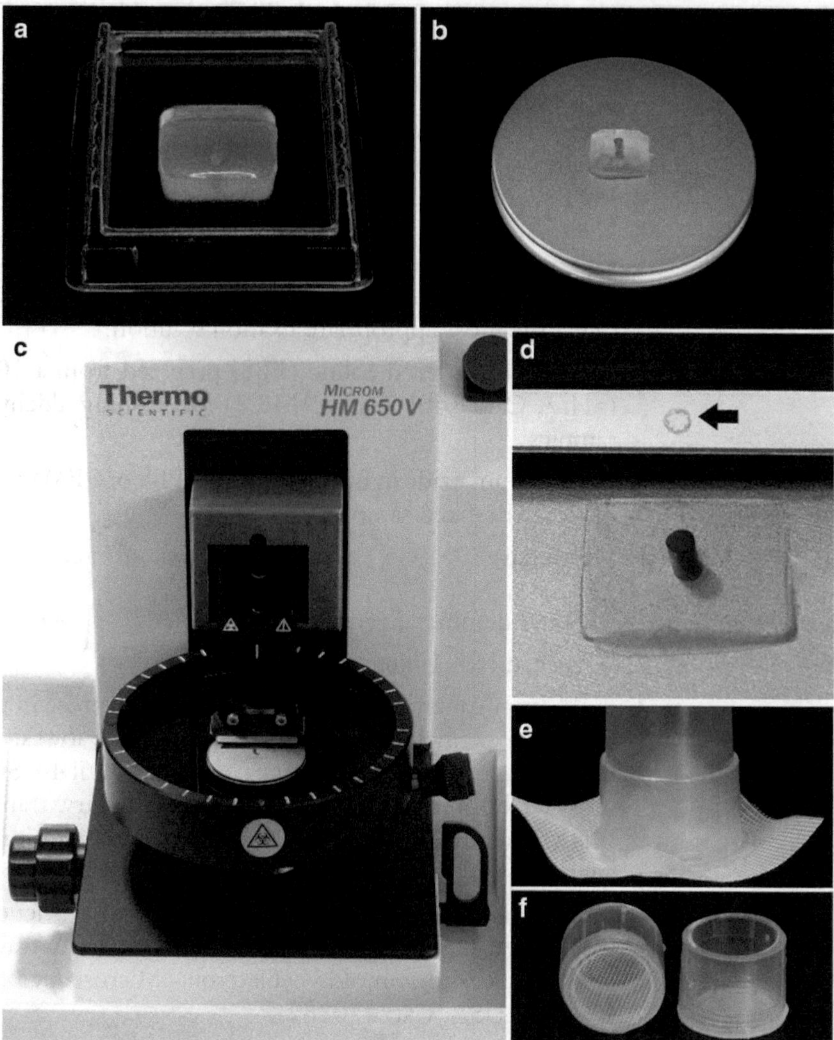

Fig. 1 Agarose embedding of an Arabidopsis inflorescence stem segment (**a–b**) and construction of staining baskets (**e**, **f**). (**a**) Arabidopsis stem segment embedded in agarose in a disposable mold. (**b**) Trimmed agarose block containing sample mounted on the vibratome sample stage with superglue. (**c**) Sample stage positioned in the water bath as seen during sectioning. (**d**) Forty micron section floating above the disposable blade edge (arrow). (**e**) Largest opening of a 1000 μL pipette tip attached to a 200 μm mesh. (**f**) Ready-to-use staining baskets after removal of excess nylon mesh and the conical part of the pipette tip

3. Good-quality double edge razor or injector blades (Cat. #72003-01 and #71990, Electron Microscopy Sciences, Hatfield, USA).
4. Fine tweezers (e.g., Dumont No. 5, Electron Microscopy Sciences, Hatfield, USA), fine hair paintbrush or a 1000 μL pipette and 1000 μL pipette tips to collect sections.

5. Recipient for sections (either glass slide or multiwell plate or weighing boats).
6. Materials for custom-made staining baskets: 1000 μL pipette tips, Bunsen burner or lighter, 120–200 μm nylon mesh, 48-well plate.

2.4 (Immuno) fluorescence Labeling and Staining

1. 3% (w/v) nonfat milk protein in 1x PBS (MP/PBS) as blocking or probe-dilution liquid (*see* **Note 1**).
2. Primary antibodies raised against specific cell wall components (e.g., rat monoclonal antibodies supplied by PlantProbes (http://www.plantprobes.net) or mouse antibodies supplied by either BioSupplies (http://www.biosupplies.com.au/) or CarboSource (https://www.ccrc.uga.edu/~carbosource/CSS_home.html).
3. Secondary antibodies that recognize the primary antibody, selected according to the class of primary antibody, the source host and the kind of label which is preferred. Use antibodies conjugated with a fluorochrome of choice (e.g., FITC, Sigma-Aldrich; Alexa Fluor, Invitrogen).
4. Calcofluor White (Fluorescent Brightener 28, Sigma-Aldrich) stock solution: 1 mg/mL in demineralized water, aliquoted and stored at −20 °C.
5. Antifade mounting medium; CitiFluor AF2 (Vector Laboratories).
6. Microscope slides and coverslips.

2.5 Brightfield Microscopy Staining and Mounting of Sections

Several methods for plant cell wall histochemical staining are extensively discussed in [2]. Here we discuss three additional staining methods.

1. *Hematoxylin staining*: fast metachromatic staining solution to check the quality of vibratome sections. Dissolve 1 g of hematoxylin in 6 mL absolute ethanol. Add this solution dropwise to 100 mL saturated ammonium aluminum sulfate. Filter this solution after a 1-week exposure to light and air. Add 25 mL glycerol and 25 mL methanol. Keep the solution at room temperature until it darkens and filter before use. Successful staining also requires a 0.2% v/v of ammonium hydroxide.
2. *ACA-staining*: triple stain to differentiate between primary (blue) and secondary cell walls (red), *see* example in Fig. 2a. Prepare three solutions: 0.5% w/v astra blue FM (Waldeck GmbH & Co; Cat. #1B-163) in 2% w/v tartaric acid; 0.5% w/v chrysoidine (Waldeck GmbH & Co; Cat. #1B-473) in 5% w/v ammonium aluminum sulfate and 0.5% v/v glacial acetic acid; and 0.5% w/v acridine red 3B (Waldeck GmbH & Co; Cat. #1B-355) in 5% w/v ammonium aluminum sulfate and

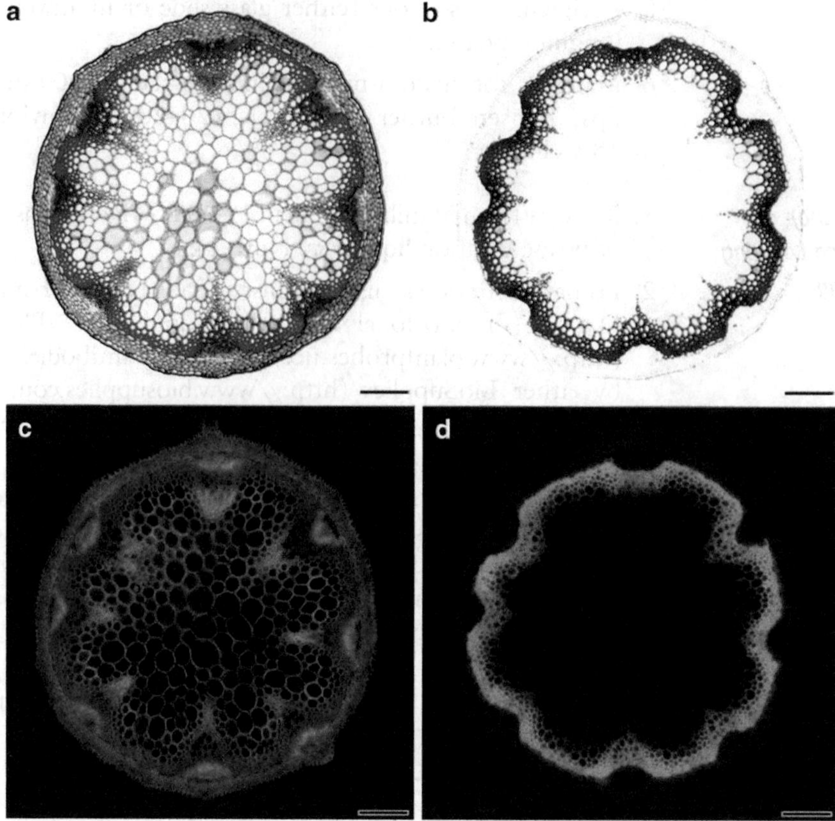

Fig. 2 (Immuno)cytochemical staining of serial transverse 40 μm sections through the inflorescence stem of Arabidopsis. (**a**) ACA triple reagent stains secondary cell walls red and primary cell walls blue. (**b**) Phloroglucinol–HCl staining following [2]; lignified cell walls appear red. (**c**) Calcofluor stains all cell walls blue. (**d**) The xylan-directed monoclonal antibody LM11 (PlantProbes) labels all secondary cell walls. Scale bars: 200 μm

0.5% v/v glacial acetic acid. Add several thymol crystals to each solution to prevent fungal growth. Prepare an ACA stock solution by mixing the astra blue, chrysoidine, and acridine red solutions in a 4:1:1 ratio.

3. *Hydroxylamine ferric chloride*: stains pectin methyl-esters red. This method requires four solutions: 14% w/v hydroxylamine, 14% w/v sodium hydroxide, 33–37% v/v hydrochloric acid, and 10% w/v ferric chloride in 0.1 N hydrochloric acid.
4. *Products and materials for mounting sections*: weighing boats or staining baskets in a 48-well plate, fine tweezers, demineralized water, 70–100% v/v glycerin, Euparal (Carl Roth, Cat. #7356.1), and isopropyl alcohol.

3 Methods

3.1 Fixing and/or Storing Samples

1. Fresh plant samples can be sectioned immediately after collection. Alternatively, they can be fixed in a cold aldehyde-based fixative for a minimum of 2 h at room temperature to overnight at 4 °C (immunocytochemistry) or stored in 70% ethanol indefinitely (anatomical investigations or histochemical staining) (*see* **Note 2**).
2. Aldehyde-fixed samples should be rinsed thoroughly with PBS or PEM buffer within 1 week to wash out the fixative as over-fixing may cause brittleness.
3. Samples can be processed for sectioning or stored in buffer (PEM or PBS) for up to 2 weeks or longer if sodium azide is added to the buffer to prevent fungal growth.

3.2 Preparation of Samples for Sectioning: No Embedding

Sample size is a crucial factor for successful vibratome sectioning. The rule of thumb is that sample height does not exceed sample width/diameter, especially for softer or pliable samples. Self-supporting samples can be glued directly onto the vibratome stage without embedding. Flat (e.g., leaves), small (e.g., roots), or soft (e.g., meristems) samples should be embedded in agarose (*see* Subheading 3.3) as the forward oscillating cutting movement of the blade will cause too much vertical deflection.

1. Rinse samples (stored in either ethanol, PBS with sodium azide or PEM-buffer) thoroughly in demineralized water (*see* **Note 3**).
2. Take the vibratome sample chuck and remove any remaining glue from the surface with a sharp single-edge razor blade. Apply and streak out a small amount of superglue to the middle of the dry sample stage. Avoid using excessive amounts of glue to avoid that samples get covered with glue.
3. Using tweezers, gently press the sample onto filter paper to remove excess water and position the dry sample surface into the glue. Gently tap on the upper surface of the sample to ensure good contact between sample and stage surface (*see* **Note 4**).

3.3 Preparation of Samples for Sectioning: Agarose Embedding

1. Fill a disposable mold with 5–7% agarose or low melting point agarose (*see* **Note 5**).
2. Place the sample in the agarose-filled mold, position in the appropriate orientation and keep it upright using tweezers or tooth picks. The tissue of interest should not be located at the bottom of the mold as this surface will be glued to the vibratome stage. Place the mold on ice to speed up solidification (Fig. 1a).

3. After the agarose has set, remove the sample block from the mold (*see* **Note 6**). Use a single-edge razor blade to cut out a small block of agarose containing the sample and trim in such a way as to yield proper final orientation of the tissue when placed on the vibratome stage. Since the agarose does not infiltrate the tissue, do not remove too much agarose from around the sample, so that the tissue will remain well-supported during sectioning.
4. Take the vibratome sample chuck and remove any remaining glue with a sharp single-edge razor blade. Apply and streak out a small amount of superglue to the middle of the dry vibratome sample stage.
5. Position the agarose block into the glue. Gently tap on the upper part of the sample to ensure good contact (Fig. 1b).

3.4 Vibratome Sectioning

1. Set up the vibratome for sectioning by inserting a double edge razor or injector blade in the blade carrier and fix into position using the clamping screws (*see* **Note 7**). Fill the tray with cold water or buffer (*see* **Note 8**).
2. Place the sample stage in the water bath of the vibratome and add demineralized water or buffer until the blade edge is submerged by 3–5 mm (Fig. 1c, d).
3. Read the vibratome instruction manual and follow all safety instructions. The following steps and settings are specific to the Thermo Scientific HM 360 V vibration microtome but are applicable to most vibratomes. Parameter settings related to successful vibratome sectioning vary with different species, organs, and tissues. It is therefore necessary to optimize these parameters for your own specimens, depending on the technical possibilities of the used vibratome. A good starting point is a frequency of 75–85 Hz, an amplitude of 0.8–1 mm and a section thickness of 50–70 μm. By trial and error it is possible to find the optimal settings to achieve good sectioning of a given tissue at a given section thickness. Consult Table 2 for more information regarding vibratome sectioning parameters.
4. Trimming the specimen is performed by serial sectioning until complete sections, with the area of interest, can be made.
5. Collect sections using tweezers, fine paint brushes or a 1000 μL pipette fitted with a pipette tip with large-hole opening (to prevent damage of the tissue). Remove agarose that surrounds the section manually on a microscope slide using preparation needles. Store sections in a microcentrifuge tube or a multiwell plate. Samples can be (immuno)stained or temporarily stored in PBS or an alternative buffer with 0.1% w/v sodium azide.
6. Replace the blade regularly, especially when sectioning hard samples.

Table 2
Optimal settings of sectioning parameters

Section thickness	– The ideal section thickness depends on the rigidity as well as the cell size and shape of the sectioned tissue – To allow good observation, sections of only one cell layer thick will be most suitable. Good quality thick sections can therefore be produced of tissues with large cell size, while samples with very small cell size require thinner sections for optimal observation – Correct orientation of the sample (i.e., main axis of the sample perpendicular to the sectioning plane) will increase the quality of the sections – Fixed tissue can be sectioned thinner (the tissue is firmer due to the fixation) – The selection of the section thickness also depends on the purpose of the study. Sections intended for observations using confocal microscopes can be thicker (100–250 μm)
Water bath temperature	– Sectioning of fresh or fixed tissues is always easier when the samples are cooled (4 °C)
Amplitude	– To get the best-quality sections, the amplitude of the blade vibration has to be maximized and the tissue has to be sectioned very slowly – The softer the tissue the lower the amplitude (to avoid large vertical deflection during sectioning)
Frequency	– The frequency can be set between 30 and 100 Hz. Higher frequencies should be used with low amplitude or speed settings
Clearance angle of the knife	– The usable range of the blade angle for most of the applications is between 15° and 25° – The harder the tissue the more shallow the blade angle (almost parallel to the cutting motion)
Cutting speed	– The cutting speed must be selected in a way that the tissue is not pushed away by the oscillating blade while cutting, but an even cut is produced

3.5 Handling Delicate Sections

Delicate sections may be damaged as a result of excessive handling. To (immuno)stain delicate sections and keep incubation volumes to a minimum (considering the cost of probes) sections can be processed in baskets made from disposable 1000 μL pipette tips and nylon mesh.

1. Hold the edge of the largest opening of a 1000 μL pipette tip into a flame for a short time to melt the plastic and push against the nylon mesh laid down on the bench surface (Fig. 1e).
2. Let stand to cool, remove excess nylon mesh with scissors and create smooth edges using a pencil sharpener. To finish the basket, cut off the conical part of the pipette tip leaving only the raised border of the pipette tip which is attached to the nylon mesh (Fig. 1f).
3. These baskets fit in most 48-well plates.

4. By pulling out baskets with sections from the multiwell plate it is possible to change incubation liquids without damaging the samples.
5. Sections can be (immuno)stained as detailed in Subheadings 3.6 and 3.7.

3.6 (Indirect) Immunofluorescence Staining with Monoclonal Antibodies

1. In a multiwell plate, incubate free-floating sections with gentle rocking in MP/PBS (1× PBS, 3% w/v nonfat milk protein) for 30 min to block nonspecific binding sites (*see* **Note 9**).
2. Remove blocking solution and add primary monoclonal antibody, diluted following the suppliers instructions in MP/PBS. Incubate at room temperature for 1.5–2 h. A negative control staining should be carried out by omitting the primary antibodies.
3. After washing in PBS three times for 5 min, incubate the sections for 1 h with a secondary antibody. This incubation and any following steps should be performed in darkness to avoid bleaching of the fluorochromes.
4. Wash with three changes of PBS (at least 5 min each).
5. Incubate with a 10- to 50-fold dilution of the Calcofluor White stock solution for 3–5 min (*see* **Note 10**).
6. Wash with three changes of PBS (at least 5 min each).
7. Mount samples in antifade reagent (e.g., CitiFluor AF2 glycerol/PBS-based antifade) and examine with an epifluorescence microscope fitted with appropriate filter cubes and optics.

3.7 Histochemical Staining

3.7.1 Hematoxylin

1. Incubate sections for 10 s in the staining solution.
2. Rinse with the ammonium hydroxide solution (*see* **Note 11**).
3. Mount in demineralized water or glycerin. Alternatively, dehydrate sections in isopropyl alcohol for 1–5 min and mount in Euparal. Small lead weights can be placed on top of the coverslips to flatten the sections. Dry at room temperature or in an oven at 40 °C.

3.7.2 Hydroxylamine–Ferric Chloride Reaction

1. Stain sections for 5 min in freshly prepared hydroxylamine reagent consisting of equal volumes of the sodium hydroxide and hydroxylamine stock solutions (*see* **Note 12**).
2. Acidify the reagent with an equal volume of concentrated hydrochloric acid.
3. Place the section on a slide and mount in the ferric chloride solution.
4. A distinct red color develops when the dye reacts with methyl esters of pectins.

3.7.3 ACA-staining

1. Stain sections for 3–10 min in a five-fold dilution of the ACA stock solution in demineralized water.
2. Rinse in demineralized water to remove excess staining solution.
3. Incubate in isopropyl alcohol for 1–5 min.
4. Mount as described in Subheading 3.7.1.
5. Lignified cell walls appear red while parenchymatous cell walls stain blue (*see* Fig. 2a).

4 Notes

1. If penetration of the probes into the samples is hindered or slow, add 0.01% v/v Triton X-100 to the blocking and incubation solutions.
2. Samples stored in 70% v/v ethanol for prolonged periods may become more rigid which, in some cases, allows production of thinner vibratome sections.
3. Insufficient removal of ethanol may hinder good attachment of the samples to the vibratome stage.
4. Removing water from the sample will allow for better attachment to the vibratome stage as the glue partly penetrates the sample.
5. As heat generally has a negative impact on antigenicity, use low melting point agarose to embed plant materials that will be sectioned for immunocytochemical experiments. For general anatomical observations both types of agarose may be used.
6. It is possible to temporarily store agarose blocks at 4 °C in a closed container with a small amount of buffer or demineralized water.
7. Caution must be exercised in the handling of the blades to not damage them. Caution must also be exercised to prevent operator injury from accidental cuts.
8. Some buffer salts can be corrosive and damage the screw threads of the blade carrier. Make sure to thoroughly rinse the blade holder with demineralized water after use of the vibratome.
9. Epitope-unmasking may be carried out by pretreatment with pectate lyase [3]. Other enzymes may be used at this stage to remove specific polymers or components.
10. Calcofluor White stains β-glucans such as cellulose and fluoresces under UV excitation. It generally stains all cell walls and can be used to show tissue- or cell type specific context of

antibody labeling (*see* Fig. 2c, d). Dilution of the stock solution depends on the sample as well as the sensitivity of the camera sensor.

11. A mild alkaline solution is necessary for differentiation of the dye. Tap water may be used instead.
12. Perform a control staining by deesterifying pectins through incubation in 0.1 M sodium carbonate, pH 10.

References

1. Hervé C, Marcus SE, Knox JP (2011) Monoclonal antibodies, carbohydrate binding modules, and the detection of polysaccharides in plant cell walls. In: Popper ZA (ed) The plant cell wall: methods and protocols, methods in molecular biology, vol 715. https://doi.org/10.1007/978-1-61779-008-9_7
2. Soukup A (2014) Selected simple methods of plant cell wall histochemistry and staining for light microscopy. In: Žárský V, Cvrcková F (eds) Plant cell morphogenesis: methods and protocols, methods in molecular biology, vol 1080. https://doi.org/10.1007/978-1-62703-643-6_2
3. Marcus SE, Verhertbruggen Y, Hervé C, Ordaz-Ortiz JJ, Farkas V, Pedersen HL, Willats WGT, Knox JP (2008) Pectic homogalacturonan masks abundant sets of xyloglucan epitopes in plant cell walls. BMC Plant Biol 8:60

Chapter 20

Monoclonal Antibodies, Carbohydrate-Binding Modules, and Detection of Polysaccharides in Cell Walls from Plants and Marine Algae

Delphine Duffieux, Susan E. Marcus, J. Paul Knox, and Cécile Hervé

Abstract

Plant and algal cell walls are diverse composites of complex polysaccharides. Molecular probes such as monoclonal antibodies (MABs) and carbohydrate-binding modules (CBMs) are important tools to detect and dissect cell wall structures in these materials. We provide an account of methods that can be used to detect cell wall polysaccharide structures (epitopes) in plant and marine algal materials and also describe treatments that can provide information on the masking of polysaccharides that may prevent detection. These masking phenomena may indicate potential interactions between sets of cell wall polysaccharides and methods to uncover them are an important aspect of cell wall immunocytochemistry.

Key words Carbohydrate-binding module, Cell wall immunocytochemistry, Immunofluorescence microscopy, Plant cell walls, Algal cell walls, Monoclonal antibody

1 Introduction

Land plants and algae have cells surrounded by a polysaccharide-rich cell wall. Compositional variation is evident in relation to phylogeny (with major cell wall biochemistry distinctions between plant and algal cell walls) and also in relation to cell types, developmental stages, season and habitats [1, 2]. The macromolecular polysaccharide components are major contributors to cell wall properties and functions. The majority of cell wall polysaccharides contain a range of structural variants that appear to be integral to polymer properties and cell wall functions. To fully understand how the diverse cell wall structures function in cell processes and organ mechanics and respond to environmental impacts, it is important to assess the presence not only of specific polymers but also specific configurations of polymers in relation to individual cell wall architectures, cell types, and cell status.

Zoë A. Popper (ed.), *The Plant Cell Wall: Methods and Protocols*, Methods in Molecular Biology, vol. 2149, https://doi.org/10.1007/978-1-0716-0621-6_20, © Springer Science+Business Media, LLC, part of Springer Nature 2020

One of the best ways to detect and assess the presence of polysaccharides in plant and algal materials is by the use of tagged proteins with specific recognition capacities. Currently, most proteins used for cell wall polysaccharide recognition are rodent monoclonal antibodies (MABs) and carbohydrate-binding modules (CBMs). In the case of complex carbohydrate polymers, purification of immunogens in sufficient amounts for antibody production is often a limiting step. In terrestrial and marine environments, microbial polysaccharide hydrolases are frequently appended to CBMs, which are used for carbohydrate recognition. These protein domains encompass a large collection of sequences and a wide range of ligand specificities [3]. When produced as separate recombinant His-tagged modules they can be readily adapted to antibody-style procedures including immunocytochemistry approaches [4]. In land plants, large sets of these MAB and CBM probes directed to the main classes of cell wall polysaccharides (e.g., cellulose, xyloglucan, heteroxylan, and pectic polysaccharides) are now available. In algae, the cell-wall contents are biochemically different and such tool development is still in its infancy. However in brown algae the cross reactivity of an anti-homogalacturonan probe can be used to detect alginates [5], and recent efforts have also been made to develop MAB against sulfated fucans and alginates [5, 6].

When using cell wall immunocytochemistry, one may consider that polysaccharide epitopes present in cell walls may not be directly detectable due to the presence of other polymers—perhaps indicating intimate associations. To date, this has been demonstrated in plants for pectic homogalacturonan (HG) polysaccharides and heteroxylan obscuring or masking other polysaccharides in primary cell walls [7–9]. Combining the use of molecular probes with specific enzymatic treatments provides a more nuanced understanding of the occurrence of polysaccharides in cell walls.

This chapter will focus on general factors relating to the use of MAB and CBM probes to detect polysaccharides in conjunction with cell wall imaging using immunocytochemistry, with an emphasis on immunofluorescence procedures. They are described for the analysis of plant material, but additional techniques relating to marine macroalgae are also presented. In the latter case they are explained for brown algae but could theoretically be extended to other algae. Treatments of plant and algal materials to uncover masked epitopes are also described.

2 Materials

2.1 Molecular Probes

1. A large range of rodent MABs that recognize plant and algal cell wall polysaccharides and proteoglycans is now available as detailed at the online sites of Biosupplies (www.biosupplies.

com.au), Carbosource Services (www.carbosource.net), PlantProbes (www.plantprobes.net) and SeaProbes (http://www.sb-roscoff.fr/en/seaprobes). Biosupplies and Carbosource MABs are derived using mouse hybridoma systems, and thus the probes require anti-mouse secondary reagents, whereas those at PlantProbes and SeaProbes are mostly rat and require anti-rat secondary reagents (*see* **Note 1**).

2. CBMs, derived from cell wall hydrolase enzymes, are generally used as recombinant proteins with polyhistidine tags to allow for detection with secondary reagents. Alternatively, they can be used directly as fusion proteins with fluorescent proteins (*see* **Note 2**). CBMs are not yet as widely available as MABs, but some are available commercially (see online sites above).

2.2 Tissue Printing

1. Scalpel or blade to cut algal or plant material.
2. Nitrocellulose membrane (size depends on the sample).
3. Phosphate-buffered saline (PBS): Prepare 10× stock with 1.37 M NaCl, 27 mM KCl, 100 mM Na_2HPO_4, 18 mM KH_2PO_4 (pH 7.4) and autoclave before storage at room temperature. Prepare working solution by dilution of one part with nine parts water. Alternatively, use prepared 10× PBS (Severn Biotech, Kidderminster, UK).
4. Blocking/antibody dilution buffers. PBS with 3% (w/v) milk protein (e.g., Marvel Milk) (PBS/MP) or 3% (w/v) bovine serum albumin (Sigma-Aldrich) in PBS (PBS/BSA).
5. Secondary antibody coupled to alkaline phosphatase (AP), for example, anti-rat, anti-mouse, or anti-polyhistidine depending on the primary probe to be used.
6. AP substrate buffer: 100 mM NaCl; 5 mM $MgCl_2$, 100 mM Tris (pH 9.5). Stock solutions of AP substrates: 54 mM of 5-bromo-4-chloro-3-indolyl phosphate disodium salt, BCIP (Sigma-Aldrich) in water and 61 mM nitrotetrazolium blue chloride, NBT (Sigma-Aldrich) in methanol.

2.3 Preparation, Fixation, and Sectioning of Plant Materials

1. Four percent solution of formaldehyde in PEM buffer: 50 mM PIPES, 5 mM EGTA, 5 mM $MgSO_4$; pH adjusted to 7.0 with KOH. Make a 12% or 16% (w/v) stock solution of paraformaldehyde in water by heating up to 70 °C and adding 1 M NaOH dropwise until the cloudy solution turns clear. Cool to RT. A good alternative is 16% formaldehyde solution (Agar Scientific, Stansted, UK). Aliquots can be stored at −20 °C for up to 6 months (*see* **Note 3**).
2. Ethanol to prepare aqueous solutions (30–100%) for dehydration procedures.
3. Wax for embedding. Steedman's wax, an ethanol-soluble low melting point polyester wax (35–37 °C), is prepared from a mixture of polyethylene glycol 400 distearate and

1-hexadecanol (cetyl alcohol) (Sigma-Aldrich, Gillingham, UK) (*see* **Note 4**).

4. LR White resin (hard grade, containing 0.5% of the catalyst benzoin methyl ether, Agar Scientific) can be used for both light and electron microscopy.
5. Disposable base molds (15 × 15 × 5 mm, Electron Microscopy Sciences, Hatfield, USA) and embedding cassettes (Simport, Beloeil, Canada) for wax embedding.
6. Gelatin capsules (Agar Scientific) for resin embedding.
7. Polysine-coated microscope slides (VWR, Lutterworth, UK).
8. Vectabond (Vector Laboratories, Peterborough, UK) coated multitest 8-well glass slides (MP Biomedicals, Solon, USA).
9. Nickel grids (Agar Scientific) for electron microscopy.

2.4 Specific Recommendation for Marine Algae

1. A good fixative to start with is a 4% (w/v) solution of formaldehyde in 50% (v/v) seawater diluted with distilled water (*see* **Note 5**). Alternatively, 2% (w/v) formaldehyde with 0.25% (w/v) of glutaraldehyde (Sigma-Aldrich), 0.001% (w/v) of Tween 80 and 342 mM NaCl in 25 mM PIPES (pH 7.2) can be used. The preparation of the stock solution of paraformaldehyde is explained in Subheading 2.3, **item 1**.
2. 0.1% (w/v) poly-L-lysine solution (Sigma-Aldrich) to pretreat the slides.
3. As for plants, LR White resin can be used for both light and electron microscopy. Tissues from marine algae remain too soft when embedded in Steedman's wax; this technique is therefore not recommended.

2.5 Immunomicroscopies

1. Super PAP hydrophobic pen (Agar Scientific) for marking buffer incubation regions on glass slides.
2. Phosphate-buffered saline (PBS): Prepare 10× stock with 1.37 M NaCl, 27 mM KCl, 100 mM Na_2HPO_4, 18 mM KH_2PO_4 (pH 7.4) and autoclave before storage at room temperature. Prepare working solution by dilution of one part with nine parts water. Alternatively, use prepared 10X PBS (Severn Biotech, Kidderminster, UK).
3. Blocking/antibody dilution buffers. PBS with 3% (w/v) milk protein (e.g., Marvel Milk) (PBS/MP) or 3% (w/v) bovine serum albumin (Sigma-Aldrich) in PBS (PBS/BSA).
4. Secondary antibodies: anti-rat immunoglobulin (whole molecule) reagents coupled to FITC and gold; mouse anti-polyhistidine; anti-mouse immunoglobulin coupled to FITC (Sigma-Aldrich), anti-polyhistidine coupled to Alexa Fluor 488 (Serotec, Kidlington, UK); anti-rat coupled to Alexa Fluor 488 (Invitrogen).

5. Anti-fade reagents. CitiFluor PBS AF3 (Agar Scientific).
6. Microscope slide coverslips (Scientific Laboratory Supplies, Nottingham, UK).
7. 0.25% (w/v) Calcofluor White (fluorescent brightener, Sigma-Aldrich).
8. General cell-wall stains for marine algae: 1.3 mM toluidine blue (TBO) (Sigma-Aldrich) in 0.1 M phosphate buffer (pH 6.8) or 0.5 mM ruthenium red (RR) (Sigma-Aldrich) in 0.1 M ammonium acetate (pH 6.85).

2.6 Enzymatic Pretreatments

2.6.1 Pectic HG Removal for Plants

1. High pH solution for pectin deesterification. 0.1 M sodium carbonate (pH 11.4).
2. Pectate lyase (from *Cellvibrio japonicus*) (Megazyme, Bray, Ireland) (*see* **Note 6**).
3. CAPS buffer: 50 mM CAPS, 2 mM $CaCl_2$, pH 10.

2.6.2 Alginate Removal for Brown Algae

1. Sodium carbonate solution: 2.8 mM sodium carbonate in water.
2. Alginate lyase (*see* **Note 6**).
3. Tris buffer: 25 mM Tris, 200 mM NaCl, pH 7.5.

3 Methods

Here we focus on three methods which provide information on the localisation of cell wall epitopes at different levels. Tissue printing gives a rapid indication of the presence and the localisation of polysaccharides in tissues as shown in Fig. 1. Soluble antigens, which might be washed out in fixatives, can still be found using this technique. Immunofluorescence microscopy is a sensitive

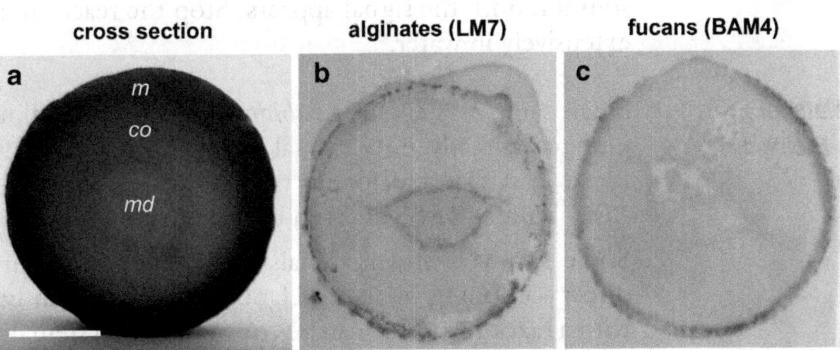

Fig. 1 Cross section of a stipe from *Laminaria hyperborea* (**a**) and corresponding tissue prints probed with the anti-homogalacturonan LM7, cross-reacting and detecting alginates in this case (**b**) and the anti-fucan BAM4 (**c**). Both epitopes are abundant within the meristoderm. LM7 binds strongly to medulla cells, while BAM4 shows a limited labeling of the area. *m* meristoderm, *co* cortex, *md* medulla. Bar = 4 mm

method that can provide an important overview of the occurrence of cell wall epitopes across an organ as well as significant detail in relation to individual cells. Electron microscopy is useful to locate cell wall polysaccharides in specific cell wall domains and other cell compartments such as the Golgi apparatus.

3.1 Tissue Printing

1. Cut and immediately firmly press the surface of the plant or alga on nitrocellulose membrane for about thirty seconds and let the membrane dry for at least 1 h.
2. Block nonspecific binding sites by incubation with PBS/MP for at least 30 min.
3. Wash with PBS for 5 min.
4. Incubate with primary monoclonal antibody diluted in PBS/MP for at least 1 h at RT or overnight at 4 °C. A five- to ten-fold dilution of a hybridoma cell culture supernatant is a good starting point for the primary antibody.
5. Wash with three changes of PBS with at least 5 min for each change.
6. Incubate with a secondary antibody diluted in the region of 1000-fold in PBS/MP for at least 1 h at RT. Anti-rat-IgG (whole molecule) linked to AP is widely used. Another good probe is anti-rat-IgG linked to HRP. HRP is less expensive, rapid, and a more stable enzyme than AP; however, AP is considered more sensitive for colorimetric detection (*see* **Note 7**).
7. Wash with three changes of water with at least 5 min for each change. Take care not to use PBS at this stage as it might interfere with the detection procedure due to the use of BCIP.
8. Substrate development of AP signal: in 10 mL of water add 82.5 μL of BCIP stock solution and 66 μL of NBT stock solution. Incubate the membrane with this freshly prepared solution until the signal appears. Stop the reaction by washing extensively in water.

3.2 Plant Material Preparation, Excision, and Fixation Procedures

1. Small samples such as *Arabidopsis* seeds or seedlings can be plunged directly into formaldehyde fixative. Maintain in fix for at least 2 h and for no more than overnight. Transfer to PEM buffer or PBS and can be stored at 4 °C until use.
2. Some relatively stiff materials such as stems are amenable to direct sectioning by hand. Hand-cut sections can be prepared with a razor blade and can be cut from a fresh stem directly into fixative solution or into water if the material is prefixed.
3. For wax- and resin-embedding procedures pieces of material (generally no thicker than 5 mm) are excised from plant organs

and placed in fixative solution. Placing material under vacuum (to expel air) can help with infiltration of the fixative.

3.3 Specific Recommendations for the Preparation of Marine Algal Material

1. For marine macroalgae, sections obtained from fresh materials without embedding give samples of reasonable quality to be used for immunofluorescence microscopy. They are obtained by hand-sectioning with a razor blade or with a microtome and directly put in the fixative solution.
2. Small filamentous algae such as *Ectocarpus* can be plunged directly into fixative. To help with subsequent labeling procedures samples can be applied/stuck to pretreated slides. To do this soak a Q-tip with 0.1% poly-L-lysine, spread the solution on the slide, and let it dry. Apply the fixed and washed sample on the slide and allow it to dry.

3.4 Wax-Embedding Protocol

The wax we use is known as Steedman's wax [10] and this is a low melting point polyester wax with good sectioning properties. It is soluble in ethanol and therefore removed prior to immunolabeling resulting in good maintenance of antigenicity.

1. Wash fixed material in PEM buffer or PBS buffer, two times 10 min.
2. Dehydrate by incubation in an ascending ethanol series (30%, 50%, 70%, 90%, and 100%) with 30 min incubation for each change at 4 °C.
3. Move samples to 37 °C for next steps.
4. Incubate in molten wax and ethanol (1:1, overnight) and then 100% wax (two times 1 h).
5. Keep wax molten using a 37 °C oven.
6. Fill base mold with molten wax and place sample in the wax. Take care to orientate the sample for optimal sectioning. Fill up with molten wax and when almost set apply embedding cassette.
7. Leave at RT overnight to solidify. Can be used 12–24 h after embedding or can be stored in a cool, dry place indefinitely.

3.5 Sectioning Wax-Embedded Material

1. These instructions are for the use of a HM 325 rotary microtome (Microm, Bicester, UK), but they will be readily adapted to other systems.
2. Sections are cut to a thickness of ~10–12 μm to produce ribbons, which are transferred to paper. Sections are selected and placed on polylysine slides over a drop of water to promote spreading.
3. Slides are allowed to dry in air overnight.

4. To dewax and rehydrate sections incubate slides with 100% ethanol 3 times 10 min, 90% ethanol–water 10 min, 50% ethanol–water 10 min, water 10 min, and water 90 min.
5. Slides are then air-dried and can be stored at RT indefinitely.

3.6 Embedding Protocol for LR White Resin

1. Wash in buffer minus fixative for three times 10 min (or overnight at 4 °C).
2. Dehydrate using an ascending ethanol series (30%, 50%, 70%, 90%, and 100%) with 30 min each change.
3. Infiltrate with resin at 4 °C by increasing from 10% resin in ethanol 1 h, 20% 1 h, 30% 1 h, 50% 1 h, 70% 1 h, 90% 1 h, 100% resin overnight, then 8 h, then overnight.
4. Transfer to gelatin capsules and ensure appropriate orientation of plant/algal material. Fill to the top with resin and seal to exclude air.
5. Allow polymerization of resin either at 37 °C for 5 days or by action of UV light at −20 °C.

3.7 Sectioning of Resin-Embedded Materials

1. These instructions relate to the use of Reichert-Jung Ultracut Ultramicrotome.
2. Prepare glass knives.
3. For light microscopy cut sections to a thickness of 1–2 μm onto water.
4. Transfer sections to a drop of water on Vectabond-coated slides and allow them to dry on to the slide in air.
5. For electron microscopy cut ultrathin sections to a thickness of ~80 nm when they are silvery gold in color.
6. Collect sections on nickel grids.

3.8 Immunolabeling of Plant and Marine Algal Cell Walls Using Monoclonal Antibodies

This procedure is for the indirect immunofluorescence labeling of sections of plant or algal material (*see* **Note 8**). Always ensure that there is a no-primary antibody control to assess the extent of cell wall autofluorescence present in the material. Here we focus on immunofluorescence procedures, but there are very effective alternatives such as immunogold with silver enhancement for light microscopy [11].

1. Use the hydrophobic pen to isolate regions around sections that will contain incubation solutions.
2. Block, wash and incubate with primary monoclonal antibody as described in Subheadings 3.1, **steps 2–4**.
3. Wash with three changes of PBS with at least 5 min for each change.

4. Incubate with a secondary antibody diluted in the region of 100-fold in PBS/MP for at least 1 h at RT. Anti-rat-IgG (whole molecule) linked to FITC is widely used. Another good photostable probe is anti-rat-IgG linked to Alexa Fluor 488.
5. Wash with three changes of PBS with at least 5 min for each change.
6. Incubate plant or algal section with a ten-fold dilution of the Calcofluor White stock solution for 5 min. (*see* **Note 9**).
7. Wash with three changes of PBS.
8. Mount samples using a small drop of antifade reagent, cover with coverslip and examine. We use CitiFluor AF3 PBS-based antifade.
9. Examine with a microscope fitted with epifluorescence optics.

3.9 Immunolabeling of Plant and Marine Algal Cell Walls Using Recombinant CBMs

1. Isolate sections on slides as appropriate and block nonspecific binding sites (as previously explained in Subheading 3.8).
2. Incubate with the CBM diluted in PBS/MP for at least 1 h at RT. The most effective working concentration should be determined by trial studies for each CBM, but most CBMs can be used effectively in the range of 5–20 μg/mL.
3. Ensure that there is a no-CBM control to assess cell wall autofluorescence in the section.
4. Wash with three changes of PBS.
5. In the case of a CBM fused with a fluorescent protein, proceed directly to **step 9**. In the case of a CBM with a polyhistidine tag, incubate with anti-polyhistidine antibody diluted in the range 1000-fold in PBS/MP for a least 1 h at RT.
6. Wash with three changes of PBS.
7. Incubate with the secondary antibody (e.g., anti-mouse coupled with FITC at 100-fold dilution) in MP/PBS for at least 1 h.
8. Wash with three changes of PBS.
9. Incubate with Calcofluor White if required as described in Subheading 3.8, **step 6**.
10. Mount slides using anti-fade reagent and examine.

3.10 Chemical Staining of Algal Cell Walls

As Calcofluor White poorly binds cell-walls in brown algae, general stains are preferentially used to indicate all cell walls in sections as shown in Fig. 2. Toluidine Blue O (TBO) binds to polysaccharides containing carboxyl or sulfated groups (alginates, fucans) and putative sulfated polyphenols. Ruthenium Red (RR) reveals carboxylated polysaccharides (alginates). The TBO staining method can be used as a postlabeling procedure where it can be very useful in quenching the autofluorescence emitted by the algal cells.

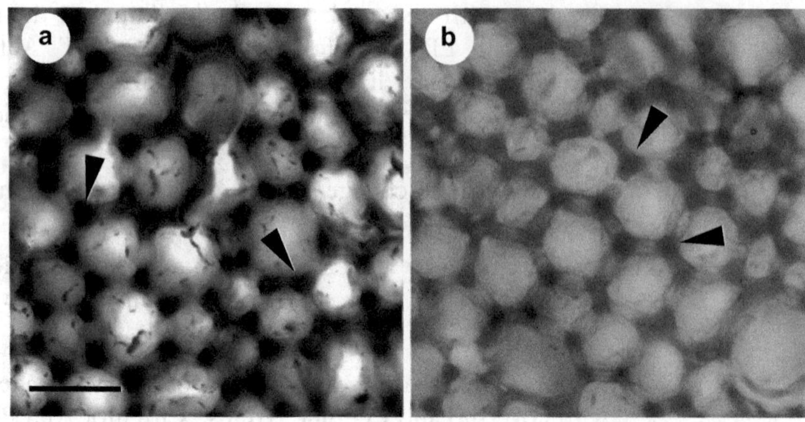

Fig. 2 Chemical staining of the outer cortex of a stipe from *Laminaria hyperborea* with Toluidine Blue O (**a**) and Ruthenium Red (**b**). Arrow heads indicate the presence of dense cell-wall material at cell junctions. Bar = 50 μm

1. For TBO staining: incubate the sample with TBO dye solution during 5 to 10 min, at room temperature. Wash five times with 0.1 M phosphate buffer (pH 6.8). Mount slide and examine in light microscopy.
2. For RR staining: incubate the sample with RR dye solution during 10 min at room temperature. Wash five times with 0.1 M ammonium acetate (pH 6.85). Mount slide and examine in light microscopy.

3.11 Immunogold Labeling for the Electron Microscope

1. Block to prevent nonspecific binding by floating the EM grid section side down on a droplet (at least 20 μL) of PBS/BSA on Parafilm for 30 min.
2. Transfer grid to a droplet of primary antibody diluted in PBS/BSA. Monoclonal antibody cell culture supernatants should be diluted between five-fold and 200-fold.
3. Wash grids by incubation in a minimum of three changes of PBS.
4. Transfer grids to secondary antibody diluted 1 in 20 with PBS/BSA. We routinely use anti-rat IgG coupled to10 nm gold.
5. Wash as in **step 3** and then extensively in distilled water.
6. Allow the grid to dry and then examine in an electron microscope.

Check Break

3.12 sSectionsPr etreatments Prior to Immunolabeling

To date in plants, most demonstrated cases of cell wall polysaccharide epitope masking are of hemicelluloses by pectic HG. For the removal of pectic HG by pectate lyase or polygalacturonase enzymes, a pretreatment of the section with a high pH is solution is required. Similar cases of epitope masking can be explored in

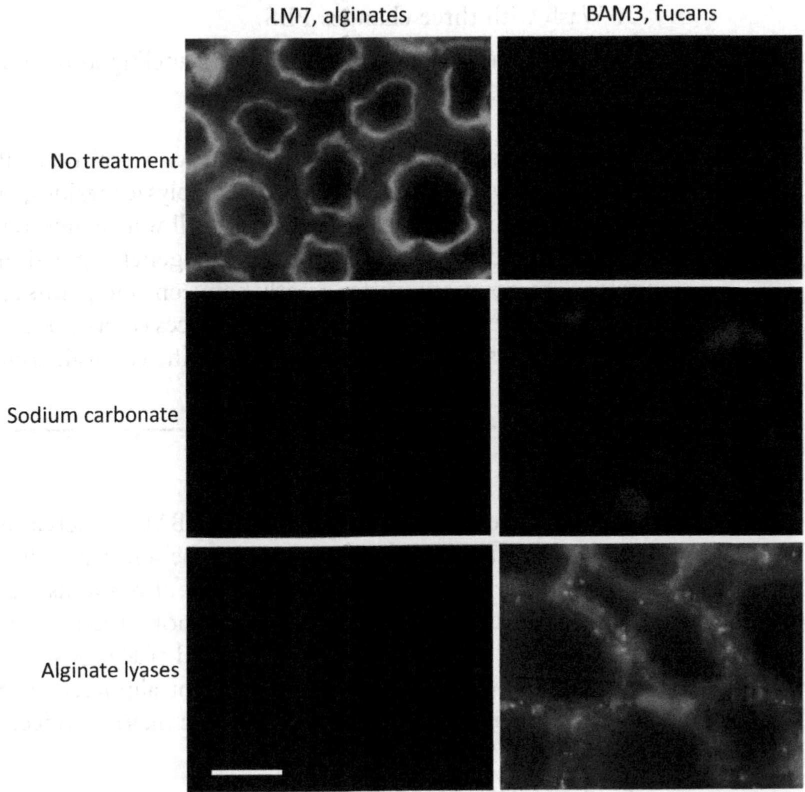

Fig. 3 Micrographs showing impacts of section pretreatments on the binding of LM7 alginates and BAM3 fucans MABs to cortical cell walls in cross sections of a stipe from *Saccharina latissima*. Immunofluorescence labeling procedures were identical in all cases and representative micrographs are shown with no pretreatment, with sodium carbonate that would solubilize most alginates and with application of alginate degrading enzymes. The BAM3 epitope is detected after the alginate lyase treatment. Bar = 20 μm

brown algae by degradation of the alginate as shown in Fig. 3. These pretreatments can be applied to all sectioned materials including wax- and resin-embedded materials (*see* **Note 6**).

3.12.1 Section Pretreatments in Plant

1. Incubate plant section with a solution of 0.1 M sodium carbonate (pH 11.4) for 2 h.
2. Wash two times 10 min with PBS.
3. Incubate with pectate lyase (10 μg/mL) in CAPS buffer for 2 h.
4. Wash with three changes PBS.
5. Sections are now ready for immunolabeling as detailed in Subheadings 3.8, 3.9 or 3.11.

3.12.2 Section Pretreatments in Brown Algae

1. To degrade alginates, incubate the brown algal section with alginate lyase (10 μg/mL) in Tris-NaCl buffer for 1 h at RT. Alternatively, most alginates can be solubilized using a 2.8 mM sodium carbonate solution for 1 h at RT.

2. Wash with three changes PBS.
3. Sections are now ready for immunolabeling as detailed in Subheadings 3.8, 3.9, or 3.11.

3.13 Conclusion

Cell wall immunochemistry, based on the use of molecular probes and the enzyme deconstructions of polysaccharides, allows for gaining real insights into the range of cell wall structures found in plant/algal cells and organs. The phylogenetic positions and/or growth environments differ greatly between land plants and macroalgae, and a comparison of these differences offers opportunities to study the physiological functions of specific cell wall components.

4 Notes

1. The range of available MABs and CBMs is increasing rapidly. Care must therefore be taken in probe selection when embarking upon an immunochemical survey of cell walls—especially if an overview is required and there is not a focus on a particular subset of cell wall polymers. A good place to start would be with probes directed to pectic HG or alginates, in plants and brown algae respectively, and also the major hemicellulose that is known for that system/taxon.
2. In the case of a GFP tag, care must be taken to assess the binding ability of the fused CBM. Indeed, depending on the recombinant target, the folding of this bulky tag may impair the recognition ability of the appended CBM by covering its binding site. In this case, the use of another tag is required.
3. Fixatives are needed to stop all cell reactions, and materials are most commonly fixed using aldehyde fixatives. For light microscopy 4% (w/v) formaldehyde is widely used. For electron microscopy 2.5% (w/v) glutaraldehyde is used - this is a good fixative but can result in sample autofluorescence and thus is not generally used for light microscopy. However, extensive glutaraldehyde-induced autofluorescence can be effectively quenched by the resin-embedding procedure and so glutaraldehyde-fixed resin-embedded material can be used for both light and electron microscopies. Aldehyde fixatives do not directly crosslink polysaccharides and some may remain soluble. This can be assessed by tissue printing. Specific fixative procedures to cross-link polysaccharides into materials have not been explored extensively.
4. Melt 900 g of polyethylene glycol 400 distearate and 100 g 1-hexadecanol in a large beaker in an incubator at 65 °C. When melted, stir wax very thoroughly using a stirring bar. Pour the wax into a tray lined with aluminum foil (or 50 mL plastic

conical tubes) and leave at RT to harden. Prepared wax can be stored at RT indefinitely. For embedding procedures, melt an appropriate amount at 37 °C and if using a water bath ensure that container is closed to keep out moisture.

5. Compared to plants, cells from marine species are hyperosmotic and specific care should be therefore taken to maintain the fixative in isoosmolarity with those cells, that is, 1100 mOsm.

6. The discovery in plants that pectic HG can mask or block the detection of hemicellulose polysaccharides requires methods for pectic HG removal from sections by the use of pectic HG-degrading enzymes. Pectate lyase or polygalacturonase can do this effectively. Both of these enzymes act on de-esterified pectic HG and therefore a high pH pretreatment to remove pectic HG methyl esters may optimize subsequent enzyme action and HG removal. Section pretreatments can also be extended for the enzymatic removal of other cell wall polysaccharides and the enzymes, buffers, and conditions required will need to be determined accordingly. By analogy, in brown algae alginate can be solubilized or enzymatically degraded by a treatment with sodium carbonate or alginate lyase, respectively. We use in-house alginate lyases [12, 13] but commercial alginate lyases (e.g., Megazyme, Bray, Ireland) should give equivalent results. Note that application of polysaccharide-degrading enzymes to material not fixed to a glass slide is likely to result in separation of cells and may cause degradation of samples.

7. In the case of a secondary antibody linked to HRP, use a stock solution of 28 mM 4-chloro-1-naphthol (Sigma-Aldrich) in absolute ethanol and hydrogen peroxide (Sigma-Aldrich). Prepare freshly the enzyme substrate solution, by adding in 5 mL of water 1.67 mL of 4-chloro-1-naphthol stock solution, 3.3 mL of absolute ethanol and 2 μL of hydrogen peroxide at 30%. Incubate the membrane with this freshly prepared solution until the signal appears. Stop the reaction by washing extensively in water.

8. Indirect procedures of immunofluorescence labeling of cell wall are widely used as these are easy and can accommodate the use of several antibodies in the same protocol and also readily allow assessments of nonspecific binding and background autofluorescence. The principles of staged incubations in the immunolabeling procedures are the same for intact materials and hand-cut sections and these materials can be incubated in tubes or plates.

9. Calcofluor White is used as a counter stain as it binds widely to β-glycans, including cellulose, and fluoresces under UV excitation and therefore can indicate all cell walls in sections and is useful for orientation and identification of immunolabeling in relation to organ and tissue anatomy.

Acknowledgments

We acknowledge funding from the UK Biotechnology & Biological Sciences Research Council and from the French National Research Agency with regard to the IDEALG program with reference ANR-10-BTBR-04.

References

1. Knox JP (2008) Revealing the structural and functional diversity of plant cell walls. Curr Opin Plant Biol 11:308–313
2. Popper ZA, Michel G, Hervé C, Domozych DS, Willats WGT, Tuohy MG, Kloareg B, Stengel DB (2011) Evolution and diversity of plant cell walls: from algae to flowering plants. Annu Rev Plant Biol 62:567–590
3. Boraston AB, Bolam DN, Gilbert HJ, Davies GJ (2004) Carbohydrate-binding modules: fine-tuning polysaccharide recognition. Biochem J 382:769–781
4. McCartney L, Gilbert HJ, Bolam DN, Boraston AB, Knox JP (2004) Glycoside hydrolase carbohydrate-binding modules as molecular probes for the analysis of plant cell wall polymers. Anal Biochem 326:49–54
5. Torode TA, Marcus SE, Jam M, Tonon T, Blackburn RS, Hervé C, Knox JP (2015) Monoclonal antibodies directed to fucoidan preparations from brown algae. PLoS One 10: e0118366
6. Torode T.A, Siméon A, Marcus S.E, Jam M, Le Moigne M.-A, Duffieux D et al. (2016). Dynamics of cell wall assembly during early embryogenesis in the brown alga Fucus. Journal of Experimental Botany 67(21), 6089–6100
7. Marcus SE, Verhertbruggen Y, Hervé C, Ordaz-Ortiz JJ, Farkas V, Pedersen HL, Willats WGT, Knox JP (2008) Pectic homogalacturonan masks abundant sets of xyloglucan epitopes in plant cell walls. BMC Plant Biol 8:60
8. Hervé C, Rogowski A, Gilbert HJ, Knox JP (2009) Enzymatic treatments reveal differential capacities for xylan recognition and degradation in primary and secondary plant cell walls. Plant J 58:413–422
9. Xue J, Bosch M, Knox JP (2013) Heterogeneity and glycan masking of cell wall microstructures in the stems of *Miscanthus* x *giganteus,* and its parents *M. sinensis* and *M. sacchariflorus*. PLoS ONE 8:e82114
10. Steedman HF (1957) A new ribboning embedding medium for histology. Nature 179:1345
11. Meloche CG, Knox JP, Vaughn KC (2007) A cortical band of gelatinous fibers causes the coiling of redvine tendrils: a model based upon cytochemical and immunocytochemical studies. Planta 225:485–498
12. Thomas F, Lundqvist LCE, Jam M, Jeudy A, Barbeyron T, Sandström C, Michel G, Czjzek M (2013) Comparative characterization of two marine alginate lyases from *Zobellia galactanivorans* reveals distinct modes of action and exquisite adaptation to their natural substrate. J Biol Chem 288:23021–23037
13. Lundqvist LCE, Jam M, Barbeyron T, Czjzek M, Sandström C (2012) Substrate specificity of the recombinant alginate lyase from the marine bacteria *Pseudomonas alginovora*. Carbohydr Res 352:44–50

Chapter 21

Electron Tomography and Immunogold Labeling as Tools to Analyze De Novo Assembly of Plant Cell Walls

Marisa S. Otegui

Abstract

High-resolution imaging of the membranous intermediates and cytoskeletal arrays involved in the assembly of a new cell wall during plant cytokinesis requires state-of-the-art electron microscopy techniques. The combination of cryofixation/freeze-substitution methods with electron tomography (ET) has revealed amazing structural details of this unique cellular process. This chapter deals with the main steps associated with these imaging techniques: selection of samples suitable for studying plant cytokinesis, sample preparation by high-pressure freezing/freeze substitution, and ET of plastic sections. In addition, immunogold approaches for identification of proteins and polysaccharides during cell wall assembly are discussed.

Key words Electron tomography, Cell walls, Cytokinesis, Cryofixation, Immunolabeling

1 Introduction

The formation of a new cell wall during cytokinesis is a highly regulated process that requires coordinated interactions between the cytoskeleton and membrane trafficking pathways. The plant cytokinetic machinery includes three distinct structures: the phragmoplast, the cell plate, and the cell plate assembly matrix (CPAM) [1–3].

The phragmoplast consists of two sets of anti-parallel microtubules (MTs) and actin filaments with their plus/barbed ends facing the division plane. The orientation of the phragmoplast MTs is established during late anaphase and is maintained by dynamic instability [4]. The Golgi apparatus plays a central role during cytokinesis, producing millions of vesicles that provide the building materials for the future cell wall [1]. These vesicles are transported along the phragmoplast MTs and they fuse with each other at the division plane, giving rise to the cell plate [5–7]. By the coordinate disassembly and reassembly of phragmoplast MTs, more vesicles are added to the growing edges of the expanding cell plate. Finally, the cell plate fuses with the parental plasma membrane and a new cell wall forms between the two daughter cells, completing

Zoë A. Popper (ed.), *The Plant Cell Wall: Methods and Protocols*, Methods in Molecular Biology, vol. 2149,
https://doi.org/10.1007/978-1-0716-0621-6_21,

cytokinesis. In addition, a filamentous matrix called CPAM has been found to enclose the cell plate growing edges, fusing vesicles, and most of the MT plus ends at the phragmoplast midline. Although the composition of this matrix is not known, it has been postulated to play an important role in both stabilization of phragmoplast MT plus ends and membrane fusion [3].

The combination of cryofixation/freeze-substitution methods with electron tomography (ET) has revealed amazing details of this complex process. By combining superb cellular preservation and three-dimensional (3D), high-resolution (4–7 nm) imaging, it has been possible to analyze the architecture of the membranous intermediates that arise during cell plate formation in different plant cell types, the changes in phragmoplast organization, and even the distribution of individual macromolecules, such as kinesin-like molecules and dynamin rings [1–3, 8]. In addition, ET has provided novel information on quantitative changes in cell plate surface area and volume that have been essential to estimate membrane dynamics during cytokinesis [1, 3]. No other method has achieved comparable 3D resolution to help us understand plant cytokinesis at the molecular level in the cellular context [9].

To obtain reliable 3D electron tomographic data it is very important to work with very well preserved biological sample. Therefore, this chapter will not only deal with the process of calculating and modeling electron tomograms but also with plant sample preparation by the best preservation method available, cryofixation and freeze substitution. Since it is also very important to be able to correlate structure with composition, protocols for immunodetection of proteins and polysaccharides are also included.

2 Materials

2.1 Plant and Cell Culture Materials

1. Sterile hood.
2. 10% (v/v) bleach.
3. 70% (v/v) ethanol.
4. Sterile water.
5. Sterile glass Pasteur pipettes or sterile 1 mL pipette tips.
6. *Arabidopsis* seeds.
7. 0.8% (w/v) agar plates containing ½ strength, that is, 2.2 g of powder per liter of Murashige and Skoog (MS) basal medium (Sigma-Aldrich, St. Louis, MO).
8. *Arabidopsis* siliques containing developing seeds (between 5 and 10 days after pollination).
9. Tobacco Bright Yellow-2 (BY-2) cultured cells.

10. BY-2 culture medium (0.43% (w/v) MS basal medium, 3 μM thiamine–HCl (B1), 0.5 mM myoinositol, 85 mM sucrose, 1 μM 2,4-dichlorophenoxyacetic acid (2,4-D), 1.3 mM KH_2PO_4, pH 5.5–5.8; autoclave and store at room temperature).
11. Aphidicolin (inhibitor of eukaryotic nuclear DNA replication) (Sigma-Aldrich).
12. Propyzamide (microtubule assembly inhibitor) (Sigma-Aldrich).

2.2 High-Pressure Freezing

1. High-pressure freezer (Leica EM HPM100, Bal-tec/ABRA Fluid AG HPM 010, or M.Wolwend HPF Compact 02).
2. Freezing brass planchettes ("hats") (e.g., type B freezing planchettes [Ted Pella, Redding, CA]).
3. Cryoprotectant: 0.1 M sucrose or 1-hexadecene.
4. 2.0 mL Cryovials (Nalge Nunc, Thermo Fisher Scientific, Rochester, NY).
5. Tweezers.
6. Tank of liquid nitrogen.

2.3 Freeze Substitution and Resin Embedding

1. Cryovials containing 1.5 mL of 2% OsO4 in anhydrous acetone (Electron Microscope Sciences, Hatfield, PA) (prepare in hood using gloves and store in liquid nitrogen).
2. Cryovials containing 1.5 mL 0.2% glutaraldehyde plus 0.2% uranyl acetate in anhydrous acetone (Electron Microscope Sciences) (prepare in hood using gloves and store in liquid nitrogen).
3. Freeze-substitution/low temperature resin embedding system (AFS, Leica, Bannockburn, IL).
4. Aluminum block with holes for fitting cryovials.
5. Eponate 12 kit (Ted Pella). Embed 812 kit from Electron Microscope Sciences can also be used.
6. Lowicryl HM20 resin kit (Polysciences, Warrington, PA, or Electron Microscope Sciences).
7. Flat embedding molds (Ted Pella).
8. Coverwell Imaging chambers (2.8 mm deep; 20 mm diameter; Electron Microscope Sciences).
9. Dry ice and Styrofoam box.
10. Plastic mounting cylinders (Ted Pella).

2.4 Preparation of Sections for Electron Tomography

1. Copper/rhodium slot grids coated with 0.7–1% (w/v) Formvar in ethylene dichloride (Electron Microscopy Sciences).
2. Ultramicrotome.
3. 2% uranyl acetate on 70% methanol.

4. Reynold's lead citrate (2.6% lead nitrate and 3.5% sodium citrate, pH 12).
5. 10- or 15-nm colloidal gold particles (Electron Microscopy Sciences) (store at 4 °C).
6. Carbon coater.

2.5 Image Acquisition and Calculation of Dual-Axis Electron Tomograms

1. Intermediate voltage (200 or 300 kV) electron microscope (e.g., FEI Tecnai G2 30 TWIN, FEI Talos, or FEI Titan Halo), equipped with high-tilt rod for tomographic image acquisition.
2. Software: SerialEM [10–12] for image acquisition and IMOD [13] package for tomogram reconstruction (can be downloaded from http://bio3d.colorado.edu/docs/software.html) (*see* **Note 1**).

2.6 Image Segmentation (Modeling)

1. IMOD package [13] (can be downloaded from http://bio3d.colorado.edu/docs/software.html) (*see* **Note 1**).

2.7 Immunolabeling

1. Nickel or gold single slot grids coated with 0.25–0.5% (w/v) Formvar in ethylene dichloride.
2. Phosphate-buffered saline (PBS): Prepare 1 L of 10× stock solution with 1.76 g of NaH_2PO4; 11.49 g of Na_2HPO4, 85 g sodium chloride, pH 6.8 (store at room temperature).
3. PBS-T 0.1%: Dilute 1 mL of 10× PBS with 9 mL of water and add 10 μL of Tween-20.
4. PBS-T 0.5%: Dilute 100 mL of 10× PBS with 900 mL of distillated water and add 0.5 mL of Tween 20.
5. Blocking buffer: 5% (w/v) nonfat milk in PBS-T 0.1%.
6. Primary antibody in blocking buffer.
7. Secondary antibody conjugated to gold particles (5, 10, or 15 nm in diameter) diluted (1:10) in blocking buffer.

3 Methods

3.1 Plant Material

3.1.1 Arabidopsis Seedlings

On average, the apical root meristem region of an 8 day-old *Arabidopsis* seedling consists of ~52 cells, each of which divides every 18 h [14]. Therefore, the chance of finding at least a few cells undergoing cytokinesis in a given root tip is relatively high. To obtain root tips from 8 to 10 day-old seedlings, it is best to germinate seeds on 0.8% agar plates containing ½ strength MS basal medium.

1. Place *Arabidopsis* seeds in plastic tube and add 10% bleach for 5 min, mixing occasionally.
2. Remove bleach using sterile glass Pasteur pipette or sterile 1 mL pipette tips (opened the pipette tip box inside the hood) and rinse three times with sterile water.
3. Add 70% ethanol, mix, and discard after 5 min.
4. Rinse three times with sterile water and place seeds on 0.8% agar plates supplemented with ½ strength MS.

3.1.2 Developing Seeds

Developing seeds are also a very good source of dividing cells. In *Arabidopsis*, the analysis of developing seeds allows for the simultaneous study of somatic cytokinesis in embryo cells and an unconventional cytokinesis that occurs during endosperm cellularization [1, 15, 16]. The endosperm in *Arabidopsis* starts to cellularize at the micropylar region, when the embryo has reached the late globular stage (approximately 5–6 days after pollination). High rate of cell divisions continues in the embryo until the torpedo/ early bent cotyledon stage (approximately 10–12 days after pollination).

3.1.3 Synchronization of BY2 Cells

The tobacco BY-2 cell line developed by Nagata and coworkers [17, 18] responds well to synchronization protocols and can provide mitotic indexes of ~39–77%.

1. Grow BY-2 cells in medium containing 3–5 μg/mL aphidicolin for 24 h.
2. Wash out the aphidicolin-containing medium and allow cells to grow in fresh medium for 3 h.
3. Add 6 μM propyzamide to the medium for 6 h [19].
4. After 90–180 min of washing out the propyzamide, most dividing cells are undergoing cytokinesis [20].

3.2 High-Pressure Freezing

1. 1 mm segments of root tips, whole developing seeds, or excised developing embryos are loaded in a type B freezing planchette containing cryoprotectan (0.1 M sucrose or 1-hexadecene).
2. Another freezing planchette is placed on top to close the chamber. It is important to completely fill the chamber with cryoprotectant, not leaving air bubbles that could collapse during high-pressure freezing.
3. If working with cultured cells grown in a medium with sucrose, a soft pellet of cultured cells can likewise be loaded directly into the freezing planchette.
4. Place freezing planchettes in the sample holder and freeze them under high pressure in a HPM 010 unit.

5. Under liquid nitrogen, split open the two freezing planchettes with the tips of a pair of forceps precooled in liquid nitrogen. The freezing planchettes containing the samples can either be stored in liquid nitrogen (*see* **Note 2**) or placed directly in cryosubstitution medium.

3.3 Freeze Substitution and Resin Embedding

The freeze-substitution medium and resin should be chosen according to the type of analysis one wants to perform. To achieve good preservation and staining of microtubules and membranes, freeze-substitution in 2% OsO_4 in acetone followed by Eponate 12 embedding is recommended. However, OsO_4 and epoxy-based resins such as Eponate 12 are not suitable for most immunolabeling approaches. Cryosubstitution in acetone without fixatives or low concentrations of glutaraldehyde (0.2%) and uranyl acetate (0.2%) [21] followed by embedding in methacrylate-based low temperature UV curing resins, such as Lowicryl HM20, are preferred for immunogold labeling applications.

It is important to keep in mind that many antibodies raised against cell wall polysaccharides, such as anti-callose (Biosupplies Australia, Victoria, Australia), anti-xyloglucan [22], and CCRC-M1 [23], and carbohydrate epitopes on arabinogalactan proteins, such as the JIM13 [24] and LM2 [25] antibodies, work well on osmicated samples embedded in Eponate 12.

3.3.1 For Structural Analysis Using Dry Ice and Styrofoam Box

1. Place freezing planchettes with frozen samples in cryovials containing 1.5 mL of 2% OsO_4 in acetone. Be sure to keep cryovials in liquid nitrogen during planchette transferring and to precool the tip of the tweezers in liquid nitrogen before touching the freezing planchettes.
2. Place the cryovials in an aluminum block precooled in dry ice (at −80 °C) inside a Styrofoam box or in a dedicated −80 °C freezer overnight (*see* **Note 3**).
3. Transfer aluminum block with cryovials to a fume hood inside a small Styrofoam box with a few dry ice pellets and keep it on a rocking shaker under gentle agitation for 5–6 h.
4. Discard substitution medium and rinse samples with fresh anhydrous acetone at least five times (every 5 min).
5. Remove freezing planchettes (freezing planchettes can be reused if they are cleaned by sonication in methanol or acetone).
6. Rinse samples one more time with fresh acetone.
7. Prepare Eponate 12 resin mix without accelerator (29 g of Eponate 12 resin, 16 g of DDSA, 14.3 g of NMA) (it can be kept at 4 °C for several days).

8. Add increasing concentration of Eponate 12 resin mix (without accelerator) in anhydrous acetone and keep the samples at least 4 h in each resin concentration: 10% Eponate 12 resin mix, 25% Eponate 12 resin mix, 50% Eponate 12 resin mix, 75% Eponate 12 resin mix.
9. Add 100% Eponate 12 resin mix (without accelerator) for at least 8 h.
10. Prepare 100% Eponate 12 resin mix with accelerator (add 2.5–3% BMPA to freshly prepared Eponate 12 resin mix) (*see* **Note 4**). It can be kept at 4 °C for several days.
11. Add 100% Eponate 12 resin mix with accelerator for at least 12 h (repeat this step twice).
12. Place samples in flat embedding molds and polymerize at 60 °C for 24 h (*see* **Note 5**).
13. Select samples in resin blocks using a dissecting microscopes and mount pieces of resin containing the samples on plastic mounting cylinders.

3.3.2 For Immunolabeling Using the Leica AFS

Freeze-substitution for immunolabeling can also be performed using "custom" freeze-substitution devices as explained in Subheading 3.3.1. However, if a low temperature UV curing resin is used, it is better to use an automatic freeze-substitution device (e.g., Leica AFS). These devices allow for a precise control of the temperature during freeze substitution and resin embedding, as well as a UV lamp that can be directly attached to the sample chamber for resin polymerization.

1. Place freezing planchettes containing samples in cryovials containing frozen cryosubstitution medium (0.2% glutaraldehyde plus 0.2% uranyl acetate in anhydrous acetone). Keep the freezing planchettes under liquid nitrogen during transfer.
2. Set the program/s in the Leica AFS according to Table 1.
3. Transfer cryovials to the Leica AFS and leave them at −90 °C for 5 days.

Table 1
Suggested steps/programs to use during freeze substitution and resin embedding using a Leica AFS

Step/program	Temperature 1	Temperature 2, °C	Ramp, °C/h	Duration, h
0	−90 °C	−90	–	120
1	−90 °C	−60	5	54
2	−60 °C	−50	5	24
3	−50 °C/h	18	5	24

4. After the AFS chamber has reached −60 °C, discard cryosubstitution medium and rinse samples and freezing planchettes with precooled anhydrous acetone at least three times.
5. Remove freezing planchettes with precooled tweezers. Samples should be free in the acetone by now, completely detached from the freezing planchettes.
6. Prepare HM20 resin mix (2.98 g Crosslinker D, 17.02 g Monomer E, 0.1 g Initiator C). Mix the three ingredients in a brown-colored glass bottle (HM20 is sensitive to light) and keep it at −20 °C.
7. Start the infiltration with HM20 by adding increasingly higher concentrations of resin in anhydrous acetone to the cryovials. Three steps can be used: 30%, 60%, and 100% HM20 resin (at least 3 h for each concentration). All the resin solutions should be precooled to −60 °C before being added to the cryovials containing the samples.
8. Add fresh, precooled 100% HM20 resin mix 4–5 times over the next 24 h.
9. Using a precooled glass Pasteur pipette remove samples from cryovials and place them in Coverwell imaging chambers with a glass coverslip on top. Fill the chamber with resin leaving no air bubbles after placing the glass coverslip.
10. Attach the UV lamp and adjust the temperature to −50 °C.
11. Polymerize at −50 °C for 24 h followed by a slow warming up to 18 °C over the following 24 h (*see* Table 1). After 48 h under UV light, the HM20 resin should be completely polymerized.
12. Remove resin blocks with samples from the imaging chambers.
13. Examine samples in resin blocks using a dissecting microscopes and mount selected pieces of resin containing the samples on plastic mounting cylinders.

3.4 Preparation of Sections for Electron Tomography

High-pressure frozen/freeze-substituted samples embedded in Eponate 12 can be sectioned in an ultramicrotome (60–70 nm thick section) and analyzed in a regular transmission electron microscope to evaluate preservation quality and to identify cells undergoing cytokinesis. Once the right specimens have been selected, semithick section for ET can be prepared.

1. Collect 250–300 nm thick sections on single slot specimen grids coated with Formvar.
2. Stain sections with a 2% uranyl acetate in 70% methanol for 10 min (*see* **Note 6**) followed by Reynold's lead citrate for 5 min.

3. Apply 10 μL of 10- or 15-nm colloidal gold solution to each side of the sections for 5 min. Remove the excess solution by touching the grid with filter paper. The gold particles are used as fiducials, aiding in the fine alignment of the tilt images during the tomographic reconstruction.
4. Carbon-coat both sides of the grids with the sections to minimize charging and drifting during imaging under the electron beam.

3.5 Image Acquisition and Calculation of Dual-Axis Electron Tomograms

The resolution of a tomographic reconstruction depends on different factors. Given that the quality of the tilt series (image focus and alignment) is optimal, the main factors affecting resolution are (a) the magnification at which the images are collected, (b) the angular interval, (c) the angular range, and (d) section thickness [26, 27]. Before collecting the images for calculating tomographic reconstructions, the variables involved in these four factors have to be carefully considered.

3.5.1 Magnification

Very often, large areas of the cells have to be imaged to analyze forming cell walls and associated phragmoplasts. However, imaging at low magnification is not recommended because of the resulting loss in resolution. If a charge-couple device (CCD) camera is used to collect the images, the magnification should be high enough that each pixel in the image is equivalent to ~1 nm in the specimen. If the structure of interest is so large that cannot be imaged in a single frame, montaged images can be used to image large areas without compromising resolution [9].

3.5.2 The Angular Interval

1° or 1.5° angular intervals are recommended.

3.5.3 Angular Range

Contrary to medical computed tomography in which images of the patient can be collected over a full 360°, the angular range allowed by the conventional tilting specimen holders used for ET is much more restricted, resulting in a wedge of missing information between the maximal tilt angle collected and 90° [10, 28]. This results in distorted tomograms with anisotropic resolution [26]. To improve the isotropy in resolution, it is recommended to collect images from two orthogonal axes and combine the two resulting tomograms into a dual-axis tomogram [10].

3.5.4 Section Thickness

If an intermediate-voltage (200–300 kV) electron microscope is used, sections thicker than 300 nm will likely result in poor resolution images, particularly at high tilt angles. If larger volumes are required for imaging cell plates and phragmoplasts, serial tomograms can be obtained from serial sections [2, 3]. A ribbon of several serial sections can be placed on the same grid and a dividing cell can then be located and imaged on each relevant section in the ribbon. However, it is important to note that this approach suffers from a 15–25 nm gap of missing information between serial tomograms [29]. For highly complex cellular structures, this gap in information may cause difficulties in the alignment of serial tomograms.

1. Place specimens in a high-tilt sample holder of an intermediate (300 kV) electron microscope and collect images at 1° angular intervals and over an angular range of ±60–70° using the free software SerialEM (*see* **Note 1**) (Fig. 1).

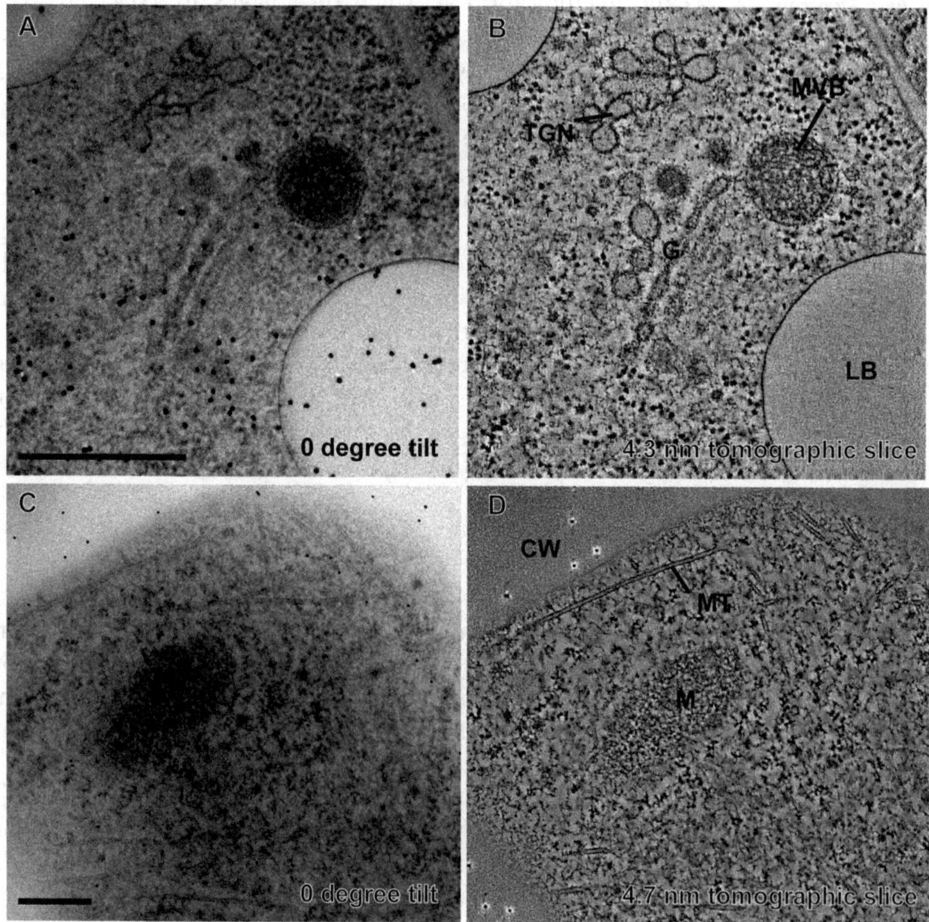

Fig. 1 Image acquisition and ET reconstruction. (**a**, **c**) Semi-thick (300 nm) sections imaged in an FEI Tecnai TF30 at 0° tilt. The small black dots are 15 nm gold particles used as fiducials for image alignment. Images were taken every 1° interval, from +60° to −60°, along two orthogonal axes. The two resulting single-axis tomograms were combined in a single dual-axis tomogram. (**b**, **d**) Single tomogramic slice show in great detail membrane profiles, microtubules, ribosomes hardly distinguishable in the original semi-thick sections (**a**) and (**c**), respectively. *CW* cell wall, *G* Golgi, *LB* lipid body, *M* mitochondrion, *MVB* multivesicular body, *TGN Trans-*Golgi Network. Scale bars = 500 nm

2. After collecting the first stack of images, rotate the grid 90° and collect images at 1° angular intervals along the second axis.
3. Process the resulting images using the *eTomo* program in the IMOD package (*see* **Note 1**) (Fig. 1).

3.6 Modeling and Quantitative Analysis of Tomographic Models

Cellular structures contained in electron tomographic reconstructions can be manually segmented (modeling) using the *3dmod* program of the IMOD package. Modeling is the creation of graphic objects that accurately represent the 3-D positions of features of interest in a tomogram. Image segmentation is the most time consuming part of the process and it is somehow a subjective task. The *3dmod* program allows the operator to draw on the image data, placing points, chosen shapes (circles, etc.), or sets of point (lines or curves that match the structure of interest) as "overlays" on the image data. Each such representation is called a contour, and these are generally drawn on a single tomographic slice extracted from the tomogram (Figs. 2 and 3). Contours can be either closed (if they represent something like the membrane that surrounds a vesicle or cisterna) or open (if they represent a fiber, like a microtubule or actin filament).

Three different types of image displays are generally used during modeling: the Zap windows that displays tomographic slices cut parallel to the surface of the physical section that was reconstructed, the Slicer window that allows the operator to display a slice at an arbitrary angle through the volume, and the Model View window that shows contours and meshed objects as they are created (Figs. 2 and 3). A general guide that provides a comprehensive description of the *3dmod* modeling program can be found at http://bio3d.colorado.edu/IMOD.

Quantitative information obtained from tomograms (see below) is based on the tomographic models that have been meshed, a process that generates 3D graphic objects with defined surfaces. However, meshes are derived from contours, so it is extremely important to draw the contours accurately.

3.6.1 Microtubules and Actin Filaments

In the simplest case, microtubules and actin filaments can be modeled as hollow tubes of a given diameter.

1. Create a new object under the "Edit" menu of *3dmod* and chose the "open contours" option (Fig. 2a).
2. Move along the stack of tomographic slices using the Zap window, identify one end of the microtubule and place the first point, as close to the center of the microtubule as possible. Place one or more points every fifth or sixth tomographic slice along the microtubule length until reaching the other end of the microtubule.

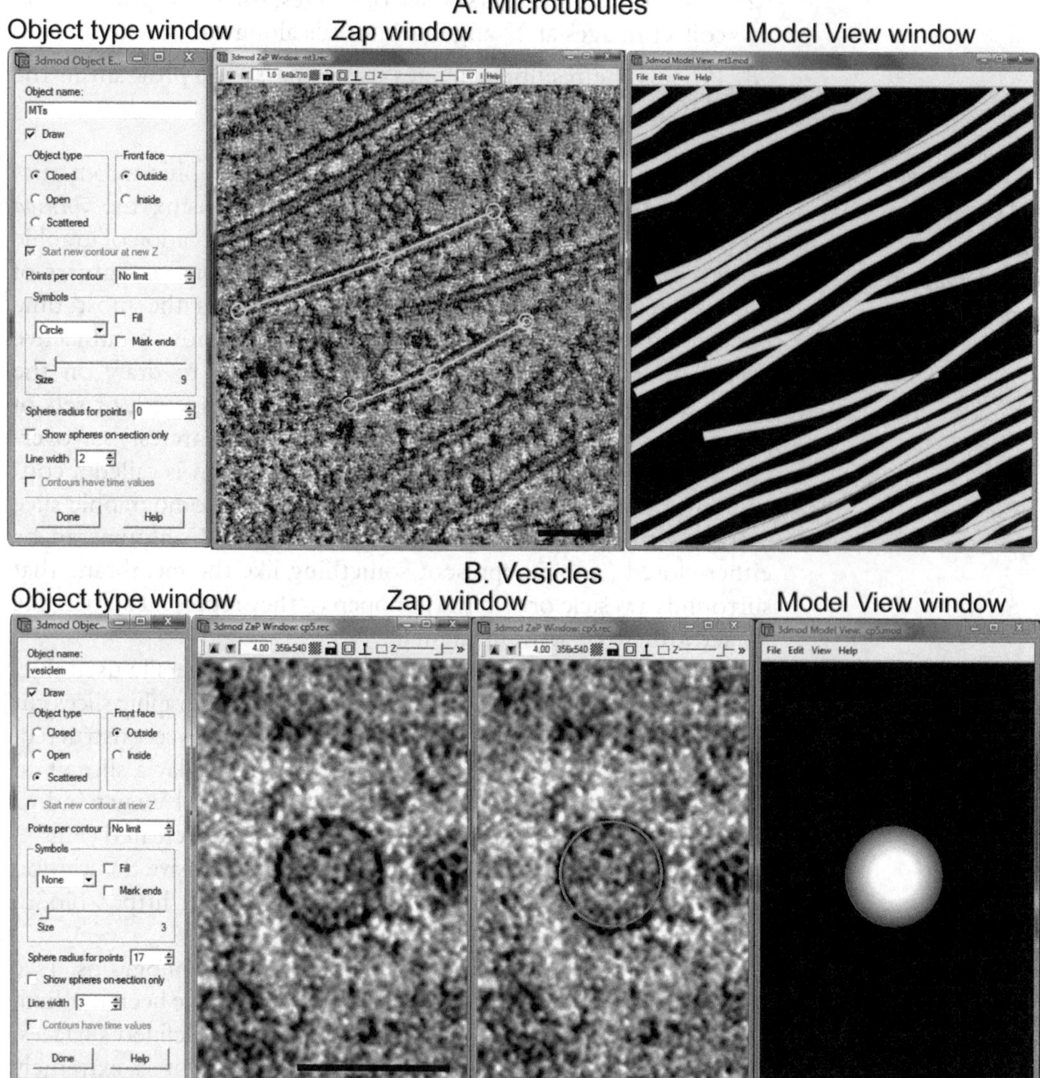

Fig. 2 Segmentation of phragmoplast microtubules (**a**) and vesicles (**b**) during cell wall assembly using *3dmod*. Microtubules are segmented as open contours (lines) in the Zap window and rendered as hollow cylinders 25 nm in diameter in the Model View window. Vesicles are rendered as spheres of variable radii. Scale bars = 100 nm

3. Alternatively, use the "slicer" window to model microtubules [28]. Adjust the X, Y, Z sliders to find a tomographic slice that contains as long a segment as possible of the microtubule axis and place points at the beginning and end of the segment. In either case, each microtubule/ filament should be considered a new contour within one object.

4. To obtain a 3D representation of the microtubule/filament, use the command *imodmesh* with the options -t, for tube, -d (the pixel diameter of the tube, 25 nm for a microtubule,

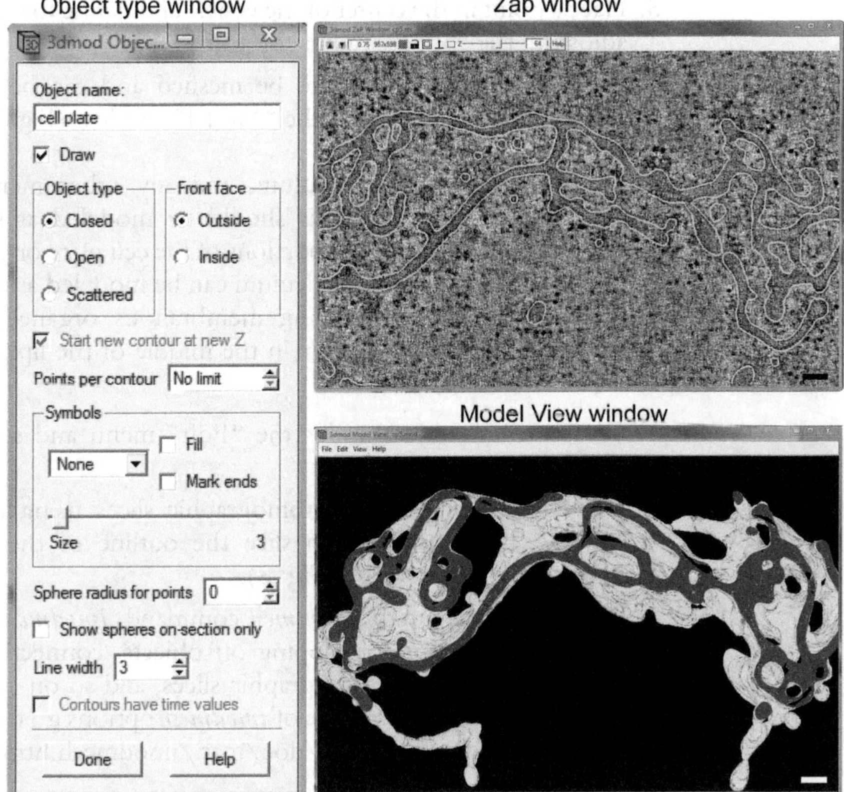

Fig. 3 Segmentation of cell plate domains using the "Closed" object type option in *3dmod*. Cell plate membranes are traced as overlay contours in the Zap window and rendered as a meshed 3-D object in the Model View window. Scale bars = 100 nm

5–9 nm for an actin filament), and -E (this will "cap" the end of the tube). Complete instructions for use of the *imodmesh* command can be found at http://bio3d.colorado.edu/imod/doc/man/imodmesh.html (*see* **Note** 7). Be sure to save the model before running *imodmesh*.

3.6.2 Vesicles

To simplify the modeling of vesicles, it can be assumed that their shape approximately correspond to a sphere. In this case, vesicles can be modeled as "scattered points," where a sphere is computed from a single point placed in the center of the vesicle, and its size can be adjusted to match the diameter of the vesicle (Fig. 2b). All similar vesicles (for example, all clathrin-coated vesicles) can be contained in a single object.

1. Create a new object under the "Edit" menu and select the "scattered" option.
2. Move along the stack of tomographic slices using the Zap window and identify the center of the vesicle.

3. Place a point in the center of the vesicle and define the radius by adjusting the "Sphere radius for points" option.
4. Spheres do not required to be meshed and can be directly displayed as 3D objects in the "Model" window (Fig. 2b).

3.6.3 Cell Plates and Membrane-Bound Organelles

Cell plates, endoplasmic reticulum, and any other membrane-bound organelle under analysis should be modeled as separate closed objects. Nonconnected portions of the cell plate or different domains of the endoplasmic reticulum can be modeled as different objects as well. When modeling membranous organelles, it is recommended to place contours in the middle of the lipid bilayer [1, 29].

1. Create a new object under the "Edit" menu and select the "closed" option (Fig. 3).
2. Move along the stack of tomographic slices using the Zap window and draw in each slice the outline of the cellular structure being modeled (Fig. 3).
3. Save model and run the *imodmesh* command. *Imodmesh* offers a number of options for capping off objects, connecting contours in nonadjacent tomographic slices, and so on (*see* **Note 7**). For a complete reference of *imodmesh* options go to http://bio3d.colorado.edu/imod/doc/man/imodmesh.html.

3.6.4 Quantitative Analysis

One of the powerful advantages of ET is the possibility of obtaining quantitative information from tomographic models. One can analyze spatial relationships between vesicles and microtubules, membrane surface area changes during cell plate assembly, frequency and distribution of microtubule plus ends in the phragmoplast [1, 3, 8]. The measurements are generally expressed in terms of the pixel size, which is described in the model header as a number in nanometers. The extent to which the section thinned in the electron beam ("thinning factor") is also described in the model header and is applied to the calculations. Therefore, it is very important to enter accurate values in the model header window (under the "Edit" menu) before performing quantitative calculations.

3.6.5 Quantifying the Volume and Surface Area of Cell Plates

1. Extract randomly located boxes of a defined size along the cell plate.
2. Mesh the portions of cell plates enclosed in these boxes and run the *imodinfo* command to calculate total surface and volume of different modeled structures, using the option -s. A complete description of all options available for the *imodinfo* command can be found at http://bio3d.colorado.edu/imod/doc/man/imodinfo.html (*see* **Note 8**).

3.6.6 Quantifying the Density of Ribosomes, Vesicles, or Any Other Structures Modeled as Spheres

1. Ensure that all the individual items (e.g., individual ribosomes) are points in the same contour of the same object.
2. Extract defined size boxes from the model and run the *imodinfo* command to obtain the number of points contained in the box volume.

3.6.7 Spatial Relationships: MTK (MicroTubule Kissing) Analysis

To measure the distances between the objects in 3D and to compute an average density of neighboring items as a function of distance between objects, one can use the *MTK* program of the IMOD package [13, 30]. This program considers structures of three kinds: open contours, like microtubules; scattered point objects, like vesicles; and meshed, closed contour objects, like cell plate domains. For microtubules, each contour is considered as a separate line. It is possible to calculate the spatial relationship between any object and a whole microtubule, or to break the microtubule into multiple segments and measure the closest distances between a chosen object and each of the microtubule segments. Distances can be measured from the central axis or surface of one object to the central axis or surface of another. A complete explanation of *MTK* and its commands can be found in http://bio3d.colorado.edu/imod/doc/man/mtk.html.

3.7 Immunolabeling

One of the main challenges in this type of imaging analysis is to be able to identify the biochemical identity/composition of the structures reconstructed in electron tomograms. One approach to identify macromolecules in tomographic reconstructions is to perform immunolabeling experiments. This can be done as a correlative approach (the immunolabeling is performed in different sections from the ones that are used for tomographic reconstruction) or a direct approach (when the same sections are used for both immunolocalization and electron tomography) [31]. As an example, correlative immunolabeling experiments with specific antibodies suggested that the electron-dense rings constricting cell plate membranes in endosperm cell plates are dynamin ADL1A/DRP1A polymers [1] (Fig. 4).

1. Float nickel or gold grids containing section on a drop of 5% milk in PBT-T 0.1% for 20 min.
2. Transfer specimen grid to a drop of primary antibody diluted in 5% milk in PBT-T 0.1% for 1 h.
3. Rinse the grid with a continuous stream of PBS-T 0.5% for 1 min.
4. Remove the excess liquid by touching the grid with filter paper.
5. Place specimen grid in secondary antibody conjugated to gold particles (5, 10, 15 nm in diameter) diluted 1:10 in 5% milk in PBT-T 0.1% for 1 h.

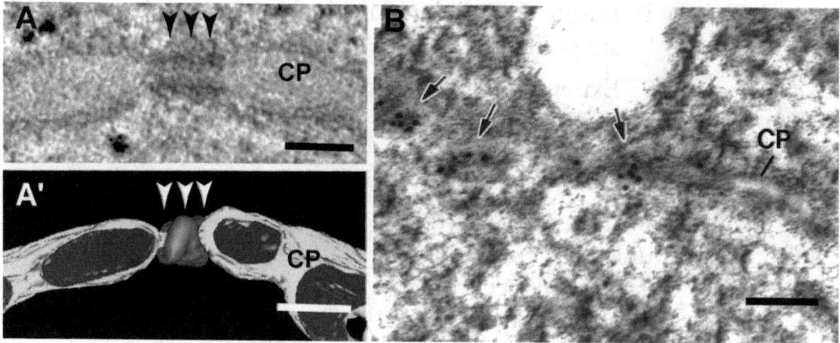

Fig. 4 Identification of protein complexes in electron tomographic reconstructions of *Arabidopsis* endosperm cell plates by correlative immunogold labeling. (**a**, **b**) Membrane-associated rings in endosperm cell plates. (**a**) Tomographic slice and (**a**′) corresponding tomographic model of a cell plate tubule. Note the presence of an electron dense collar (arrowheads) associated with a constricted tubule. The identity of the rings as dynamin ADL1A/DRP1A polymers was determined by immunolabeling with specific antibodies (**b**). Scales = 50 nm. Reproduced from Ref. 1 (Copyright © 2001 American Society of Plant Biologists)

6. Rinse the grid with a continuous stream of PBS-T 0.5% for 1 min followed by rinsing with distillated water.
7. Poststain grid with 2% uranyl acetate in 70% methanol and Reynold's lead citrate.

4 Notes

1. SerialEM is a free software for image acquisition (http:/bio3d.colorado.edu/SerialEM) that is compatible with FEI and JEOL electron microscopes. IMOD is a free software package that runs in both PC and Macintosh systems and was developed primarily by David Mastronarde, Rick Gaudette, Sue Held, and Jim Kremer at the Boulder Laboratory for 3D Electron Microscopy of Cells. It contains around 140 programs for image processing, tomogram calculation, image segmentation, display, and quantitative analysis.
2. High-pressure frozen material can be stored in liquid nitrogen for months or even years without suffering changes in cellular preservation.
3. OsO_4 in acetone is highly volatile. Even when the acetonic OsO_4 solution is kept in closed cryovials, osmication of objects around cryosubstitution vials can easily happen. If a freezer/fridge is used during cryosubstitution with OsO_4, it is advisable to have a freezer/fridge fully dedicated to this use to avoid contamination of other reagents and labware.

4. BDMA is recommended as the accelerator for the Eponate 12 resin mix instead of DMP-30 because BDMA has lower viscosity and diffuses more rapidly into tissues.
5. Unpolymerized resin waste, dirty globes, and other resin-contaminated items should be placed in the oven for 24 h for polymerization before discarding them.
6. The use of methanolic solution of uranyl acetate is recommended because it increases the contrast of membrane and cytoskeletal elements more than aqueous solutions do.
7. *imodmesh* generates triangles that connect neighboring points within a contour and nearby points on adjacent contours. The resulting triangles represent a surface in space that is an excellent approximation to all the contour information, so they are used for all subsequent quantifications of area, volume, and distance. They can also be used to generate a shaded surface that provides a good visual representation of the modeled object.
8. *imodinfo* provides information about IMOD models, such as lists of objects, contours and point data, lengths and centroids of contours; and surface areas and volumes of objects or surfaces.

Acknowledgments

This work was supported by NSF MCB1614965 grant to M.S.O.

References

1. Otegui MS, Mastronarde DN, Kang BH, Bednarek SY, Staehelin LA (2001) Three-dimensional analysis of syncytial-type cell plates during endosperm cellularization visualized by high resolution electron tomography. Plant Cell 13:2033–2051
2. Otegui MS, Staehelin LA (2004) Electron tomographic analysis of post-meiotic cytokinesis during pollen development in *Arabidopsis thaliana*. Planta 218:501–515
3. Segui-Simarro JM, Austin JR II, White EA, Staehelin LA (2004) Electron tomographic analysis of somatic cell plate formation in meristematic cells of *Arabidopsis* preserved by high-pressure freezing. Plant Cell 16:836–856
4. Smertenko AP, Piette B, Hussey PJ (2011) The origin of phragmoplast asymmetry. Curr Biol 21:1924–1930
5. Lukowitz W, Mayer U, Jürgens G (1996) Cytokinesis in the Arabidopsis embryo involves the syntaxin-related KNOLLE gene product. Cell 84:61–71
6. Steiner A, Müller L, Rybak K, Vodermaier V, Facher E, Thellmann M, Ravikumar R, Wanner G, Hauser M-T, Assaad FF (2016) The membrane-associated Sec1/Munc18 KEULE is required for phragmoplast microtubule reorganization during cytokinesis in *Arabidopsis*. Mol Plant 9:528–540
7. Boruc J, Van Damme D (2015) Endomembrane trafficking overarching cell plate formation. Curr Opin Plant Biol 28:92–98
8. Austin JR II, Segui-Simarro JM, Staehelin LA (2005) Quantitative analysis of changes in spatial distribution and plus-end geometry of microtubules involved in plant-cell cytokinesis. J Cell Sci 118:3895–3903
9. Otegui MS, Austin JR II (2007) Visualization of membrane-cytoskeletal interactions during plant cytokinesis. Methods Cell Biol 79:221–240

10. Mastronarde DN (1997) Dual-axis tomography: an approach with alignment methods that preserve resolution. J Struct Biol 120:343–352
11. Mastronarde DN (2005) Automated electron microscope tomography using robust prediction of specimen movements. J Struct Biol 152:36–51
12. Mastronarde DN (2008) Correction for non-perpendicularity of beam and tilt axis in tomographic reconstructions with the IMOD package. J Microsc 230:212–217
13. Kremer JR, Mastronarde DN, McIntosh JR (1996) Computer visualization of three-dimensional image data using IMOD. J Struct Biol 116:71–76
14. Beemster GTS, Baskin TI (1998) Analysis of cell division and elongation underlying the developmental acceleration of root growth in *Arabidopsis thaliana*. Plant Physiol 116:1515–1526
15. Otegui MS, Staehelin LA (2000) Cytokinesis in flowering plants: more than one way to divide a cell. Curr Opin Plant Biol 3:493–502
16. Otegui MS, Staehelin LA (2000) Syncytial-type cell plates: a novel kind of cell plate involved in endosperm cellularization of Arabidopsis. Plant Cell 12:933–947
17. Nagata T, Nemoto Y, Hasezawa S (1992) Tobacco BY-2 cell line as the "HeLa" cell in the cell biology of higher plants. Int Rev Cytol 132:1–30
18. Nagata T, Kumagai F (1999) Plant cell biology through the window of the highly synchronized tobacco BY-2 cell line. Methods Cell Sci 21:123–127
19. Kakimoto T, Shibaoka H (1988) Cytoskeletal ultrastructure of phragmoplast-nuclei complexes isolated from cultured tobacco cells. Protoplasma 140:151–156
20. Samuels LA, Giddings TH, Staehelin LA (1995) Cytokinesis in tobacco BY-2 and root tip cells: a new model of cell plate formation in higher plants. J Cell Biol 130:1345–1357
21. Giddings TH (2003) Freeze-substitution protocols for improved visualization of membranes in high-pressure frozen samples. J Microsc 212:53–61
22. Moore PJ, Darvill AG, Albersheim P, Staehelin LA (1986) Immunogold localization of xyloglucan and rhamnogalacturonan I in the cell walls of suspension-cultured sycamore cells. Plant Physiol 82:787–794
23. Puhlmann J, Bucheli E, Swain MJ, Dunning N, Albersheim P, Darvill AG, Hahn MG (1994) Generation of monoclonal antibodies against plant cell-wall polysaccharides. I. Characterization of a monoclonal antibody to a terminal alpha-(1→2)-linked fucosyl-containing epitope. Plant Physiol 104:699–710
24. Knox JP, Linstead PJ, Peart J, Cooper C, Roberts K (1991) Developmentally regulated epitopes of cell surface arabinogalactan proteins and their relation to root tissue pattern formation. Plant J 1:317–326
25. Smallwood M, Yates EA, Willats WGT, Martin H, Knox JP (1996) Immunochemical comparison of membrane associated and secreted arabinogalactan-proteins in rice and carrot. Planta 198:452–459
26. McEwen BF, Frank J (2001) Electron tomographic and other approaches for imaging molecular machines. Curr Opin Neurobiol 11:594–600
27. Marsh BJ (2005) Lessons from tomographic studies of the mammalian Golgi. BBA-Mol Cell Res 1744:273–292
28. O'Toole ET, Giddings JTH, Dutcher SK, McIntosh JR (2007) Understanding microtubule organizing centers by comparing mutant and wild type structures with electron tomography. In: Methods in cell biology, vol 79. Academic, Cambridge, MA, pp 125–143
29. Ladinsky MS, Mastronarde DN, McIntosh JR, Howell KE, Staehelin LA (1999) Golgi structure in three dimensions: functional insights from the normal rat kidney cell. J Cell Biol 144:1135–1149
30. Marsh BJ, Mastronarde DN, Buttle KF, Howell KE, McIntosh JR (2001) Organellar relationship in the Golgi region of the pancreatic beta cell line, HIT-T15, visualized by high-resolution electron tomography. Proc Natl Acad Sci U S A 98:2399–2406
31. Donohoe BS, Kang B-H, Staehelin LA (2007) Identification and characterization of COPIa- and COPIb-type vesicle classes associated with plant and algal Golgi. Proc Natl Acad Sci U S A 104:163–168

Chapter 22

Contributions to Arabinogalactan Protein Analysis

Romain Castilleux, Marc Ropitaux, Youssef Manasfi, Sophie Bernard, Maïté Vicré-Gibouin, and Azeddine Driouich

Abstract

Arabinogalactan proteins (AGPs) are important plant proteoglycans involved in many development processes. In roots, AGPs occur in the cell wall of root cells and root cap–derived cells as well as in the secreted mucilage. Detection, localization, and quantification techniques are therefore essential to unravel the AGP diversity of structures and functions. This chapter details root-adapted immunocytochemical methods using monoclonal antibodies, and a collection of biochemical analysis protocols using β-D-glucosyl Yariv reagent for comprehensive AGP characterization.

Key words Arabinogalactan proteins, Cell wall, β-D-Glucosyl Yariv, Electrophoresis, Immunocytochemistry, Monoclonal antibodies, Root

1 Introduction

Arabinogalactan proteins (AGPs) are widespread proteoglycans in plants, present in all organs such as leaves, stems, and roots [1]. They are specifically found in the cell wall, the plasma membrane and secretions. The binding to the plasma membrane is mediated by a glycosylphosphatidylinositol (GPI) anchor [2, 3]. AGPs are classified as hydroxyproline-rich glycoproteins (HRGPs) due to their particular structure with a protein core rich in hydroxyproline residues that are *O*-glycosylated [4]. The *O*-glycan moiety is mostly composed of D-galactose linked in β-1,3 and β-1,6, and L-arabinose linked in α-1,3, α-1,5, and β-1,3. Other monosaccharides such as L-rhamnose, L-fucose, D-xylose, or D-glucuronic acid can also be present on AGP glycans as "minor" sugars [4–6].

However, the composition and the structure of AGPs are diverse, which may explain the strong heterogeneity of their biological functions. AGPs have been shown to be important for many aspects of plant survival and development, including cell growth and morphology, pollen tube guidance, somatic

Zoë A. Popper (ed.), *The Plant Cell Wall: Methods and Protocols*, Methods in Molecular Biology, vol. 2149,
https://doi.org/10.1007/978-1-0716-0621-6_22,

embryogenesis, and root development [6, 7]. Recent studies have demonstrated that AGPs are essential in the interaction between roots and soil-borne microorganisms [8–10]. AGPs are indeed required for both root colonization by beneficial microbes such as *Rhizobium* sp. and root protection against pathogens by either repelling or trapping them [8–11]. This implies that AGPs are not only present within the cell wall but also in the root mucilage [9, 12, 13]. The mucilage is mainly secreted by root cap cells and root cap–derived cells, namely, border cells (BCs) or border-like cells (BLCs) depending on the plant species [8, 14].

The immunocytochemistry approach is a very useful technique to unravel the presence of AGP epitopes within root tissues using monoclonal antibodies. A large set of anti-AGP antibodies including LM2 [15, 16], LM14 [17], JIM4 [16, 18, 19], JIM8, JIM13, JIM14, JIM15, JIM16 [16, 20], and MAC207 [16, 21] are commercially available. The exact nature of the recognized epitopes is not always clearly identified. However, the LM2 antibody is known to recognize a β-linked glucuronic acid epitope whereas JIM4, JIM13, and MAC207 are believed to bind to a GlcAβ(1 → 3) GalAα(1 → 2)Rha motif. These antibodies are valuable and highly useful tools to investigate the diversity of AGP structure, function, and location.

It is worth noting that immunodetection is limited by the requirement of an unmasked epitope and the likelihood that some AGPs in a given sample would not be recognized by any of these antibodies. A complementary approach to immunolocalization and cell imaging is therefore needed to characterize in more details AGPs. This can be achieved by using the β-D-glucosyl Yariv reagent (β-D-GlcY) that enables purification of AGPs, as well as their biochemical analyses using electrophoresis techniques (e.g., Rocket or Cross electrophoresis) and others (e.g., radial gel diffusion) [9, 22–27]. β-D-GlcY is widely used as a probe that binds and precipitates β-1,3-galactan-containing AGPs [28–31].

In order to detect, localize, and extract AGPs, immunocytochemical and biochemical methods have been already described in the previous book edition [27, 32]. Nevertheless, significant improvements have been made since then, either for routine experiments or for specific purposes. In the present chapter, we describe new protocols for AGP immunolocalization specifically dedicated to root tissues, root border cells and root mucilage. As a matter of fact, root BCs and/or root BLCs and mucilage can be easily lost during "classical" immunolabeling procedure, and it is important to have them preserved during manipulations. Moreover, a purification step has been added to the AGP extraction and a supplemental analysis tool has been set up.

Therefore, two methods to immunolocalize AGP epitopes are presented here in this chapter. The first one is adapted for very thin roots releasing several files of root BLCs as it occurs in the model plant *Arabidopsis thaliana*. The second method is more appropriate for plants with a thicker root releasing isolated BCs and an abundant mucilage such as in the leguminous species (e.g., *Pisum sativum*). Finally, AGP extraction and purification procedures as well as analysis techniques, such as Rocket electrophoresis and radial gel diffusion, are also presented.

2 Materials

2.1 Immunolabeling of "Thin" Root Tips with BLCs (e.g., Arabidopsis thaliana)

1. Plant material.
2. Scalpel and tweezers.
3. Superfrost diagnostic slides 10 wells 8 mm (Zuzi, ref.: 30503101).
4. Micropipette.
5. 200 μL pipette tips.
6. Gloves.
7. Flow hood.
8. Stock solution paraformaldehyde (PFA) 16% (EMS, ref.: 15710). Storage at −20 °C.
9. Phosphate-buffered saline (PBS) 0.01 M: NaCl 137 mM, KCl 47 mM, KH_2PO_4 1.5 mM, and $Na_2HPO_4(H_2O)_{12}$ 7 mM (pH 7.2). Autoclave before storage at +4 °C (*see* **Note 1**).
10. Refrigerator.
11. Bovine serum albumin (BSA; AURION, ref.: 900.011) for blocking nonspecific binding sites. Storage at +4 °C.
12. Tween 20 (SIGMA, ref.: P1379).
13. Primary monoclonal antibody (PlantProbes, www.plantprobes.net/). Storage at +4 °C.
14. A wet chamber for maintaining air humidity (Fig. 1, *see* **Note 2**).
15. Secondary antibody: Anti-rat IgG coupled with FITC (fluorescein isothiocyanate; SIGMA, ref.: F6258). Storage at +4 °C.
16. Coverslip (Menzel-Glaser; 24 × 60 mm #1).
17. Nail polish (EMS, ref.: 72180).

2.2 Immunolabeling of "Thick" Root Tips with BCs and Mucilage (e.g., Pisum sativum)

1. Plant material.
2. Ultrathin tweezers.
3. Microscope slides 18 wells—Poly-L-lysine, sterile (IBIDI, ref.: 81824).
4. Micropipette.
5. 200 μL pipette tips.

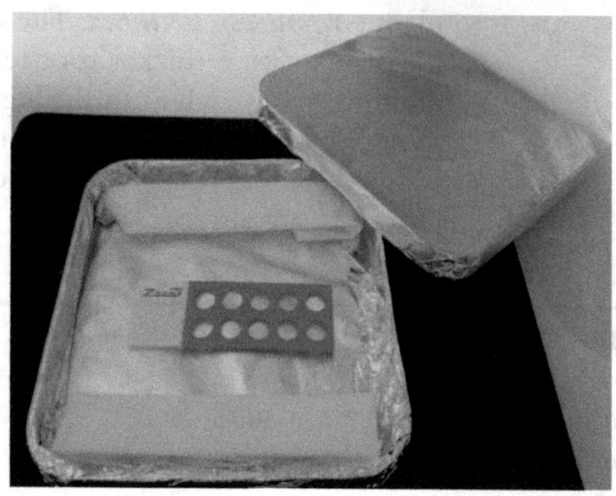

Fig. 1 Wet chamber (a square Petri dish containing wet filter paper to keep air humidity)

6. Gloves.
7. Flow hood.
8. Stock solution paraformaldehyde (PFA) 16% (EMS, ref.: 15710). Storage in −20 °C.
9. PIPES (Alfa Aesar, ref.: A16090).
10. $CaCl_2$.
11. Phosphate-buffered saline (PBS) 0.01 M: NaCl 137 mM, KCl 47 mM, KH_2PO_4 1.5 mM and $Na_2HPO_4(H_2O)_{12}$ 7 mM (pH 7.2) (*see* **Note 1**).
12. Refrigerator.
13. Bovine serum albumin (BSA; AURION, ref.: 900.011) for blocking nonspecific binding sites. Storage at +4 °C.
14. Primary monoclonal antibody (PlantProbes www.plantprobes.net/). Storage at +4 °C.
15. A wet chamber for maintaining air humidity (Fig. 1, *see* **Note 2**).
16. Secondary antibody: Anti-rat IgG coupled with TRITC (tetramethylrhodamine; SIGMA, ref.: T5778). Storage at −20 °C.
17. CitiFluor (Agar scientific, ref.: AF2 R1320).

2.3 AGP Extraction

1. Plant material.
2. Liquid nitrogen.
3. Mortar and pestle.
4. Freeze-dryer.
5. Gloves.
6. Extraction buffer: 50 mM Tris base (adjust to pH 8 with HCl), 10 mM $Na_2EDTA(H_2O)_2$, 2 mM $Na_2S_2O_5$, 1% (v/v) Triton X-100 (*see* **Notes 3** and **4**).

7. Corning tube 50 mL.
8. Refrigerator.
9. Rotating shaker.
10. Centrifuge.
11. Ethanol 96% (v/v).
12. 50 mM Tris base (adjust pH 8 with HCl).
13. Dialysis tubing 3.5 kDa.
14. Freezer −20 °C.
15. 1% (w/v) NaCl.

2.4 Purification of AGPs

1. 2 mL Eppendorf tubes.
2. Gloves.
3. 2 mg/mL β-D-glucosyl Yariv reagent (β-D-GlcY).
4. Refrigerator.
5. 1% (w/v) NaCl.
6. 100% methanol.
7. Flow hood.
8. 100% dimethylsulfoxide (DMSO).
9. Sodium hydrosulfite.
10. Vortex.
11. Ultrapure water.
12. PD-10 Desalting Columns containing 8.3 mL of Sephadex™ G-25 Medium from GE Healthcare (Fig. 6a).
13. 1.5 mL Eppendorf tubes.
14. Radial gel diffusion (protocol described in Subheading 3.4).

2.5 Radial Gel Diffusion: AGP Detection/ Semiquantification

1. Radial gel diffusion: 1% (w/v) agarose gel containing 0.15 M NaCl, 0.02% (w/v) $NaNO_3$, 10 μg/mL β-D-GlcY (*see* **Note 5**).
2. Weighing scale.
3. Square Petri dish 120 mm × 120 mm.
4. Micropipette.
5. 200 μL cut pipette tips.
6. Gum Arabic (Gum Acacia, Fisher scientific, ref.: G/1050/53).
7. 1% (w/v) NaCl.
8. Refrigerator.
9. Wet chamber as in Fig.1.
10. Shaker platform.

2.6 Rocket Electrophoresis: AGP Quantification

1. Rocket gel: 1% (w/v) agarose gel containing 25 mM Tris-base (adjust to pH 8.3 with HCl), 200 mM glycine, 20 μg/mL β-D-GlcY (*see* **Note 5**).
2. Gel tray 18 × 15 cm (*see* **Note 6**).
3. Tape (width 2 cm).
4. 10 μL pipette tips.
5. 10 μL cut pipette tips.
6. Gum arabic (Gum Acacia, Fisher scientific, ref.: G/1050/53).
7. 1% (w/v) NaCl.
8. Running buffer 1: 25 mM Tris base (adjust pH 8.3 with HCl), 200 mM glycine (*see* **Note 5**).
9. Isoelectric focusing electrophoresis system (IEF).
10. Shaker platform.

2.7 Agarose Gel Electrophoresis: Detection of AGP Subpopulations

1. Agarose gel: 1% (w/v) agarose gel containing 90 mM Tris base (adjust pH 8.3 with HCl), 90 mM boric acid, 2 mM $Na_2EDTA(H_2O)_2$ (*see* **Note 5**).
2. Gel tray 18 × 15 cm and the comb (*see* **Note 6**).
3. Tape (width 2 cm).
4. Loading buffer: 32% (v/v) glycerol, 2% (w/v) bromophenol blue.
5. Vortex.
6. Agarose gel electrophoresis system.
7. Running buffer 2: 90 mM Tris base (adjust to pH 8.3 with HCl), 90 mM boric acid, 2 mM $Na_2EDTA(H_2O)_2$ (*see* **Note 5**).
8. β-D-GlcY (*see* **Note** 7).
9. 1% (w/v) NaCl.
10. Shaker platform.

2.8 2D Electrophoresis: Characterization of AGP Subpopulations and Quantification

1. Agarose gel: 1% (w/v) agarose gel containing 90 mM Tris base (adjust to pH 8.3 with HCl), 90 mM boric acid, 2 mM $Na_2EDTA(H_2O)_2$ (*see* **Note 5**).
2. Gel tray 18 × 15 cm and the comb (*see* **Note 6**).
3. Tape (width 2 cm).
4. Loading buffer: 32% (v/v) glycerol, 2% (w/v) bromophenol blue.
5. Vortex.
6. Agarose gel electrophoresis system.

7. Running buffer 2: 90 mM Tris base (adjust to pH 8.3 with HCl), 90 mM boric acid, 2 mM $Na_2EDTA(H_2O)_2$ (*see* **Note 5**).
8. Scalpel.
9. Rocket gel: 1% (w/v) agarose gel containing 25 mM Tris base (adjust to pH 8.3 with HCl), 200 mM glycine, 20 μg/mL β-D-GlcY (*see* **Note 7**).
10. Running buffer 1: 25 mM Tris base (adjust to pH 8.3 with HCl), 200 mM glycine.
11. Isoelectric focusing electrophoresis system (IEF).
12. 1% (w/v) NaCl.
13. Shaker Platform.

3 Methods

3.1 Immunolabeling of "Thin" Root Tips with BLCs (e.g., Arabidopsis thaliana)

This protocol enables a general view of the root apex and provides optimal preservation of BLCs organization at the root tip (*see* **Note 8**). This method has been adapted from [33].

1. Cut root tips using a scalpel (Fig. 2a.), place them onto a 10-well superfrost diagnostic slide (Fig. 2b.) and fix them by adding 20 μL of 4% PFA (Paraformaldehyde) (Fig. 2C.) for 30–40 min at room temperature (RT). The liquid is removed gently by placing the pipette tip at the top of the root to limit the loss of BLCs (Fig. 2d, *see* **Note 9**).
2. Wash briefly at RT with PBS 1× (*see* **Note 10**).
3. Incubate for 30–45 min at RT with 3% (w/v) BSA diluted in PBS 1× (*see* **Note 11**).
4. Wash briefly at RT with PBS 1× and then with PBS 1× + 0.1% (v/v) Tween 20 (PBST) (*see* **Note 12**).
5. Incubate overnight (O/N) with primary antibody at +4 °C in a wet chamber. Primary antibody is diluted at 1:5 with PBST (*see* **Note 13**).
6. Wash twice briefly at RT with PBST.
7. Incubate for 2 h at RT in darkness with the secondary antibody diluted at 1:30 in PBST, in a wet chamber. The second antibody is an anti-rat IgG coupled with FITC (*see* **Notes 14** and **15**).
8. Wash briefly three times at RT with PBST.
9. Cover the slide with a coverslip and fix with nail polish.
10. Observe with a confocal laser-scanning microscope (Excitation: 488 nm; Emission: 500–535 nm). Results are shown in Fig. 3.

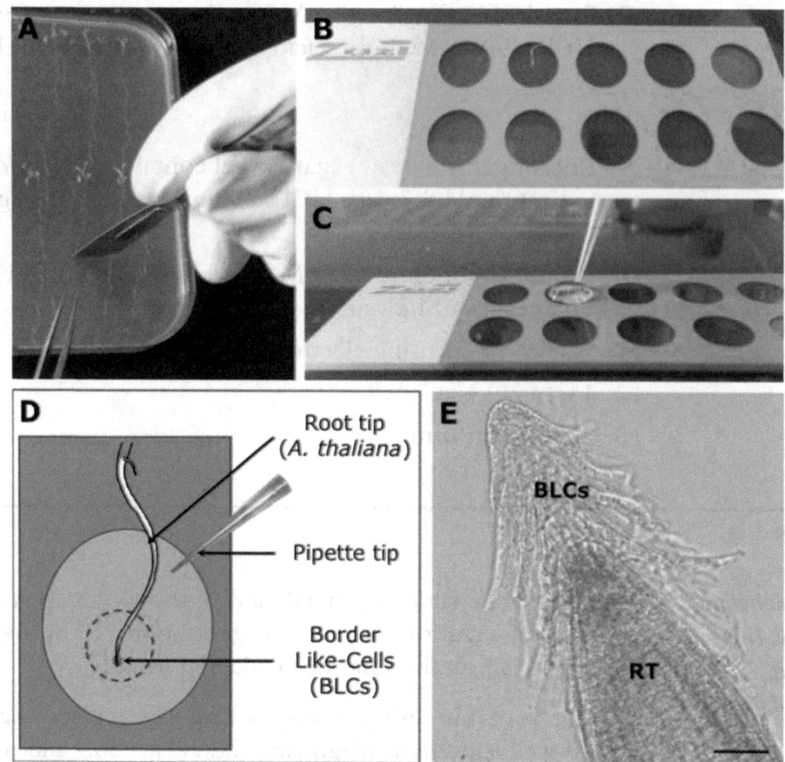

Fig. 2 Technical dissection and preparation of *A. thaliana* root tips for optimal BLCs preservation during immunofluorescence labeling. (**a**) Root tip is cut with a scalpel and collected gently with tweezers. (**b**) Root tip is placed in the center of the well and (**c**) 20 μL of PFA are gently added (30–40 min fixation). (**d**) PFA is very gently and slowly removed from the well using a pipette to avoid disruption of BLCs. (**e**) An image of a collected *A. thaliana* root tip and associated BLCs. *BLCs* Border-like cells, *RT* root tip. Scale bar in **e** = 50 μm

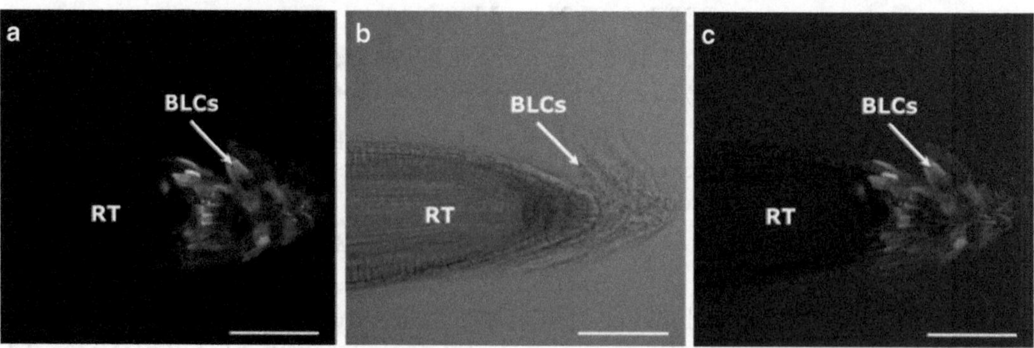

Fig. 3 Immunostaining of epitopes associated with AGPs on root tip and BLCs of *Arabidopsis thaliana* using the monoclonal antibody JIM13. (**a**–**c**) Immunofluorescence images. Observations are made with an inverted confocal laser scanning microscope Leica SP2 ($\lambda_{excitation}$: 488 nm; $\lambda_{emission}$: 500–535 nm). (**a**) Maximum intensity Z-projection of several focal planes taken in fluorescence. (**b**) Bright-field microscopy. (**c**) Overlay reconstruction. Note that labeling with JIM13 is associated with the root cap and BLCs. *BLCs* border-like cells, *RT* root tip; Scale bars = 100 μm

3.2 Immunolabeling of "Thick" Root Tips with BCs and Mucilage (e.g., *Pisum sativum*)

The immunolabeling of "thick" root tips protocol provides a good preservation of BCs and mucilage despite several washing steps (*see* **Note 16**). This method has been adapted from [9].

1. Cut root tips using ultrathin tweezers (Fig. 4a), place it onto a sterile 18-well poly-L-lysine microscope slide (Fig. 4b) and fix with 20 μL of 4% PFA diluted in 50 mM PIPES, pH 7 and 1 mM $CaCl_2$ (Fig. 4c) for 40 min at RT. The liquid is gently removed by placing the pipette tip at the top of the root to limit the loss of BCs (Fig. 4d, *see* **Note 9**).
2. Wash 10 min four times at RT with PBS 1× containing 1% (w/v) BSA (*see* **Notes 10** and **11**).
3. Incubate O/N with the primary antibody at +4 °C in a wet chamber. Primary antibody is diluted at 1:5 with PBS 1× containing 1% (w/v) BSA (*see* **Note 13**).
4. Wash for 10 min four times at RT with PBS 1× containing 1% (w/v) BSA.

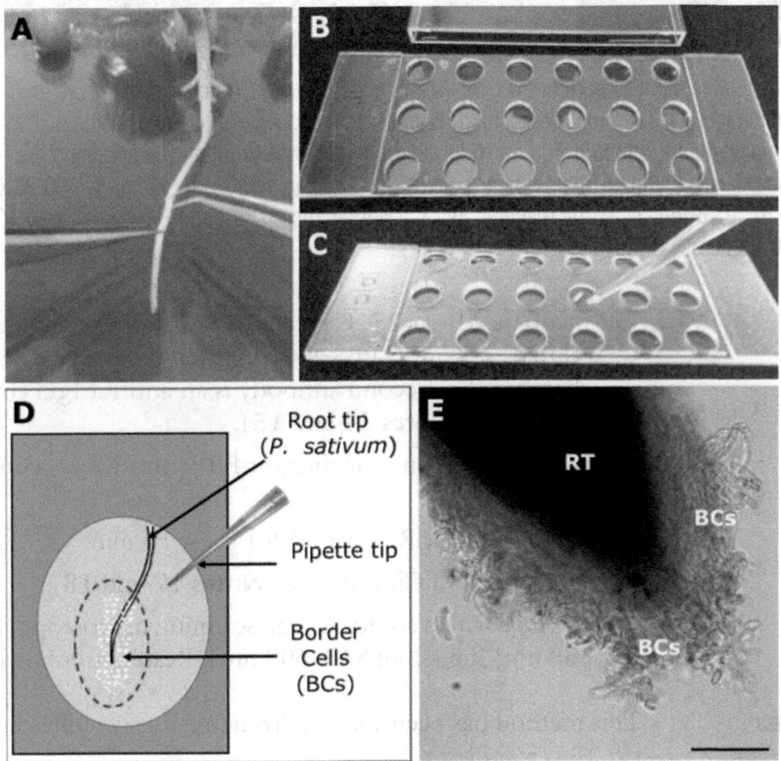

Fig. 4 Technical dissection and preparation of *Pisum sativum* root tip for optimal BC preservation during immunofluorescence labeling. (**a**) Root tip is cut and collected with two tweezers. (**b**) Root tip is placed in the center of the well and (**c**) 20 μL of PFA are gently added. (**d**) PFA is very gently and slowly removed from the well using a pipette to avoid disruption of BCs. (**e**) An image of a collected *P. sativum* root tip and associated BCs. *BCs* border cells, *RT* root tip. Scale bar in **e** = 200 μm

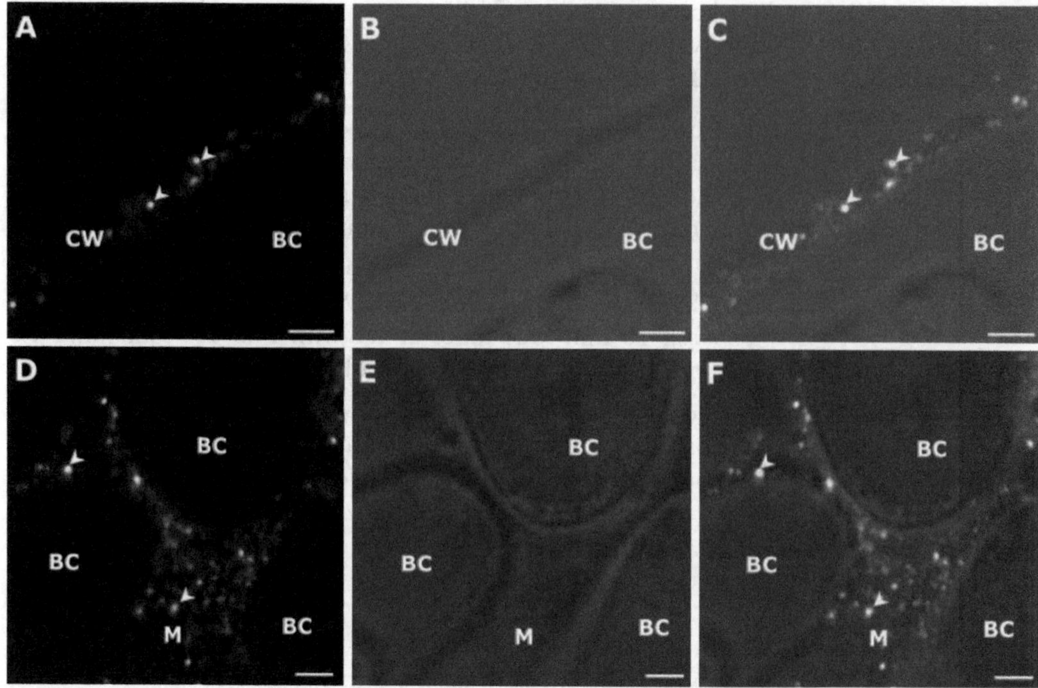

Fig. 5 Immunostaining of AGP epitopes at the surface of BCs and mucilage of *Pisum sativum* with the monoclonal antibody JIM13. A punctate staining (arrowheads) is observed (**a–c**) over the cell wall of BCs and (**d–f**) over the mucilage surrounding BCs. (**a–f**) Immunofluorescence images. Observations are made with an inverted confocal laser scanning microscope Leica SP5 ($\lambda_{excitation}$: 550 nm; $\lambda_{emission}$: 560–600 nm). (**a**, **d**) Fluorescence. (**b**, **e**) Bright-field microscopy. (**c**, **f**) Overlay reconstitution. *BC* border cell, *CW* cell wall, *M* mucilage; Scale bars = 2.5 μm

5. Incubate 2 h at +25 °C in darkness with the secondary antibody diluted at 1:50 in PBS 1× containing 1% (w/v) BSA in a wet chamber. The second antibody is an anti-rat IgG coupled with TRITC (*see* **Notes 14** and **15**).
6. Wash for 10 min four times at RT with PBS 1× containing 1% (w/v) BSA.
7. Finally wash at RT with PBS 1× for 10 min.
8. Add a drop of CitiFluor (*see* **Notes 17** and **18**).
9. Observe with a confocal laser-scanning microscope (Excitation: 550 nm; Emission: 560–600 nm). Results are shown in Fig. 5.

3.3 Extraction of AGPs

This method has been adapted from previously published protocols [9, 23, 25, 27].

1. Grind 10 g of plant material with liquid nitrogen using a mortar and a pestle.
2. Freeze the ground material then freeze-dry it (*see* **Note 19**).

3. Mix 1 g of the freeze-dried material with 40 mL of extraction buffer in 50 mL corning tube and incubate at +4 °C O/N under agitation (*see* **Note 20**).
4. Centrifuge for 10 min at +4 °C, 14,000 × *g*.
5. Recover the supernatant and add 5 volumes of 96% ethanol.
6. Incubate at +4 °C O/N to precipitate high molecular weight molecules.
7. Centrifuge for 10 min at +4 °C, 14,000 × *g*.
8. Carefully remove the supernatant and suspend the pellet in 40 mL of 50 mM Tris-base (adjust to pH 8 with HCl).
9. Centrifuge for 10 min at +4 °C, 14,000 × *g*.
10. Collect the supernatant and resuspend the pellet in 20 mL of 50 mM Tris-base previously adjusted to pH 8 with HCl.
11. Centrifuge for 10 min at +4 °C, 14,000 × *g*.
12. Collect the supernatant and pool it with the one collected **step 10**.
13. Put the pooled supernatant under dialysis at 3.5 kDa.
14. After dialysis, freeze and freeze-dry the solution.
15. Dissolve the freeze-dried sample in 1 mL of 1% (w/v) NaCl (*see* **Note 21**).

3.4 Purification of AGPs

This method has been adapted from previously published protocols [9, 24].

Because of their structural heterogeneity, binding of AGPs to β-D-GlcY can vary depending on the subpopulations present in the sample, and therefore impacts purification quality.

1. Precipitate the AGPs in 2 mL Eppendorf tube by adding an equal volume of 2 mg/mL β-D-GlcY to the solubilized AGP solution (*see* **Note 7**).
2. Incubate at +4 °C for 48 h.
3. Centrifuge for 90 min at RT, 10,000 × *g*.
4. Recover the pellet containing β-D-GlcY/AGP complex.
5. Add 1 mL of 1% (w/v) NaCl to the pellet and vortex.
6. Centrifuge for 10 min at RT, 10,000 × *g*.
7. Discard the supernatant and repeat the **steps 5** and **6** twice.
8. Add 1 mL of pure methanol to the pellet and vortex (*see* **Note 22**).
9. Centrifuge for 10 min at RT, 10,000 × *g*.
10. Discard the supernatant and repeat the **steps 8** and **9** twice.
11. Dry the pellet under a flow hood.
12. Dissolve the pellet by adding 500 μL of pure DMSO and vortex (*see* **Note 23**).

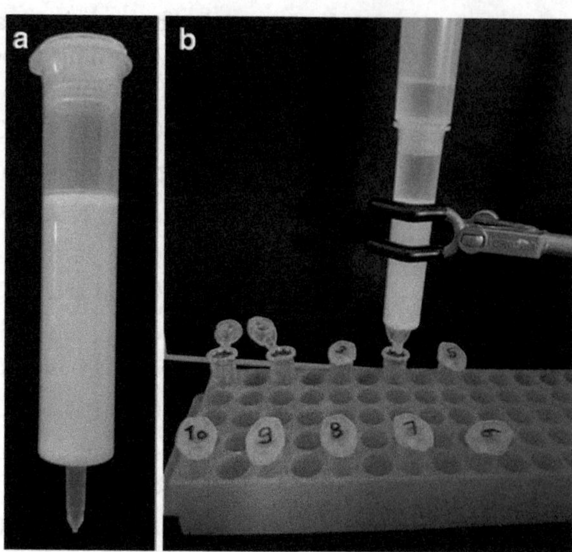

Fig. 6 (**a**) PD-10 column. (**b**) Desalting the AGP solution using a PD-10 column and recovering the eluted solution in a set of ten Eppendorf tubes 1.5 mL

13. Add 10–30% (w/v) sodium hydrosulfite and vortex. Then, add ultra-pure water until the color becomes clear yellow (*see* **Note 24**).
14. Desalt the clear yellow solution with a size exclusion chromatography (PD-10 desalting columns). Recover the drops of the eluted solution in a set of ten Eppendorf tubes of 1.5 mL (Fig. 6b).
15. Check for the presence of AGPs in each tube by radial gel diffusion (described in Subheading 3.5).
16. Collect the solutions from the tubes that contain AGPs and gather them in a 15 mL corning tube.
17. Freeze-dry the sample.
18. Dissolve in 0.5 mL 1% (w/v) NaCl (*see* **Note 25**). Store AGPs-containing solution at −20 °C until use.

3.5 Radial Gel Diffusion for Rapid AGP Detection

This method has been adapted from previous published protocols [9, 22, 23, 27].

1. Prepare 100 mL of radial gel diffusion. After heating, keep the gel under agitation and add 10 μg/mL β-D-GlcY. Then, pour it in a square Petri dish (*see* **Notes 5**, 7, and **26**).
2. Cut out wells in the gel using 200 μL cut pipette tips (Fig. 7a).
3. Load into the wells 40 μL of gum Arabic of different concentrations as standards (e.g., 0.125 mg/L, 0.25 mg/mL, 0.5 mg/mL, 1 mg/mL). As a negative control, load 1% (w/v) NaCl.

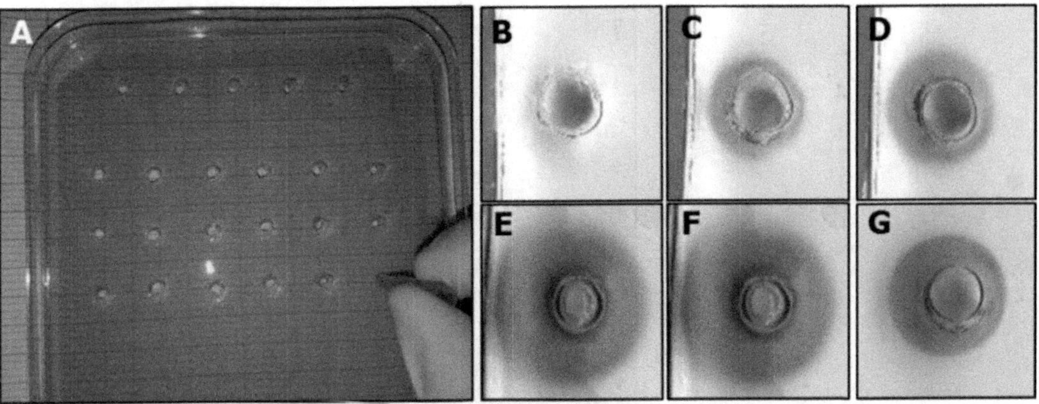

Fig. 7 Radial gel diffusion assay. (**a**) Cutting wells into the gel. (**b**) 40 μL loading of 1% (w/v) NaCl. (**c**–**f**) 40 μL loadings of 0.125, 0.25, 0.5, and 1 mg/mL of gum arabic. (**g**) 40 μL loading of AGP-containing samples

4. Load 40 μL of the sample.
5. Incubate the gel in a wet chamber at +4 °C, O/N.
6. The presence of AGPs is indicated by the appearance of a red halo around the wells due to the interaction with the β-D-GlcY. If the AGPs are not detected, concentrate the sample and load it again (get back to **step 4**).
7. Incubate the gel in 1% (w/v) NaCl under gentle agitation onto shaker platform to remove the excess of β-D-GlcY. Results are shown Fig. 7b–g (*see* **Note 27**).

3.6 Rocket Electrophoresis for AGP Quantification

This method has been adapted from previously published protocols [9, 23, 25].

1. Prepare 60 mL of rocket gel (*see* **Notes 5** and 7).
2. Pour the rocket gel into the gel tray with both ends sealed with tape and without the comb installed (*see* **Note 28**).
3. Carefully transfer the rocket gel from the gel tray to the IEF platform.
4. Load the IEF wells with the running buffer 1 and install the filter paper salt bridges (*see* **Note 29**).
5. Cut out wells in the rocket gel using 10 μL cut pipette tips.
6. Load into the wells 7 μL of gum arabic at different concentrations as standards (0.125 mg/L, 0.25 mg/mL, 0.5 mg/mL, 1 mg/mL). As a negative control, load 7 μL of 1% (w/v) NaCl.
7. Load 7 μL of AGP sample.
8. Run the IEF electrophoresis for 16 h at 200 V, 5 mA, 10 W.
9. Red precipitation triangle-shaped lines are formed due to the interaction between the β-D-GlcY and AGPs.

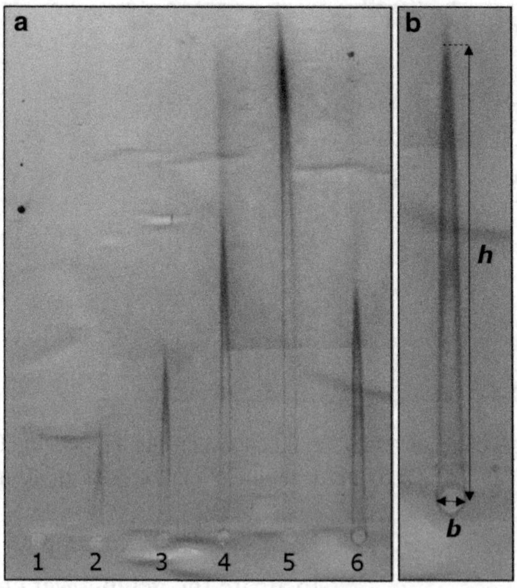

Fig. 8 (**a**) Rocket gel electrophoresis. (*1*) 7 μL loading of 1% (w/v) NaCl. (*2–5*) 7 μL loadings of 0.125, 0.25, 0.5, and 1 mg/mL of gum Arabic. (*6*) 7 μL loading of AGP-containing samples. (**b**) Peak area measurement (*h*, *b*, height and base of the peak)

10. Incubate the rocket gel in 1% (w/v) NaCl under a gentle agitation to remove the excess β-D-GlcY. Results are shown Fig. 8a (*see* **Note 27**).
11. Calculate the surface area of each peak (Formula: Area = $(b \times h)/2$; b, base of the peak and h, height of the peak), as detailed in Fig. 8b.
12. Associate the calculated surface area of each peak to the standard concentration to make a calibration range.
13. Quantify the sample using the calibration range (*see* **Note 30**).

3.7 Agarose Gel Electrophoresis: Detection of AGP Subpopulations

This method has been adapted from previously published protocols [9, 23, 25, 26].

1. Prepare 200 mL of agarose gel (*see* **Note 5**).
2. Pour the agarose gel into a gel tray with both ends sealed with tape and the comb installed (*see* **Note 28**).
3. Solubilize 10–25 μg of AGP samples in 30 μL of 1% (w/v) NaCl. Then, add 10 μL of loading buffer and vortex.
4. Take off the tape once the agarose gel is polymerized.
5. Put the gel tray with the agarose gel inside into the electrophoresis tank and add the running buffer 2.

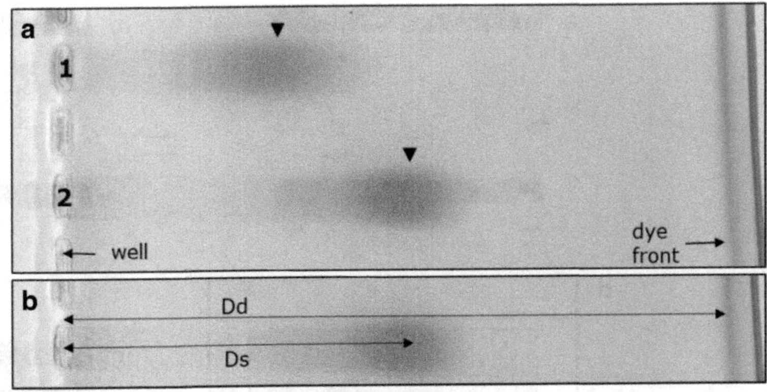

Fig. 9 (**a**) Agarose gel. Linear arablnogalactan proteins profiles revealed by agarose gel electrophoresis followed by staining with β-D-GlcY. The two samples (*1* and *2*) present different AGP subpopulations indicated by black arrows with different Rf values. (**b**) Measurement of the migration distance of the sample and the dye front to calculate the retention factor: Rf = Ds/Dd

6. Load the samples into the wells and run the agarose gel electrophoresis at 250 V, 120 mA for 50 min, or until the dye front move 8 cm. If 8 cm is not enough to separate the different AGP subpopulations, let migrating for a longer distance.
7. After electrophoresis, stain the gel in 10 μg β-D-GlcY, O/N (*see* **Note 7**).
8. A red precipitation smear is formed due to binding of AGPs by the β-D-GlcY.
9. Incubate the agarose gel in 1% (w/v) NaCl under gentle agitation to remove the excess of β-D-GlcY. Results are shown Fig. 9a (*see* **Note 27**).
10. Calculate the retention factor (Rf): Rf = Ds/Dd (Fig. 9b, *see* **Note 31**).

3.8 2D Electrophoresis: Characterization of AGP Subpopulations and Quantification

This method has been adapted from previously published protocols [9, 23, 25].

The technical steps for 2D electrophoresis are detailed in Fig. 10.

1. Prepare an agarose gel as described above (*see* Subheading 3.7, **steps 1–6**).
2. After agarose gel electrophoresis, cut the lanes of AGP migration (AGP-containing strips of gel) with a scalpel.
3. Put them in a gel tray, with both ends sealed with a tape (*see* **Note 28**).
4. Prepare a rocket gel as described above (*see* Subheading 3.6, **step 1**) and pour it into the gel tray around the cut strips.

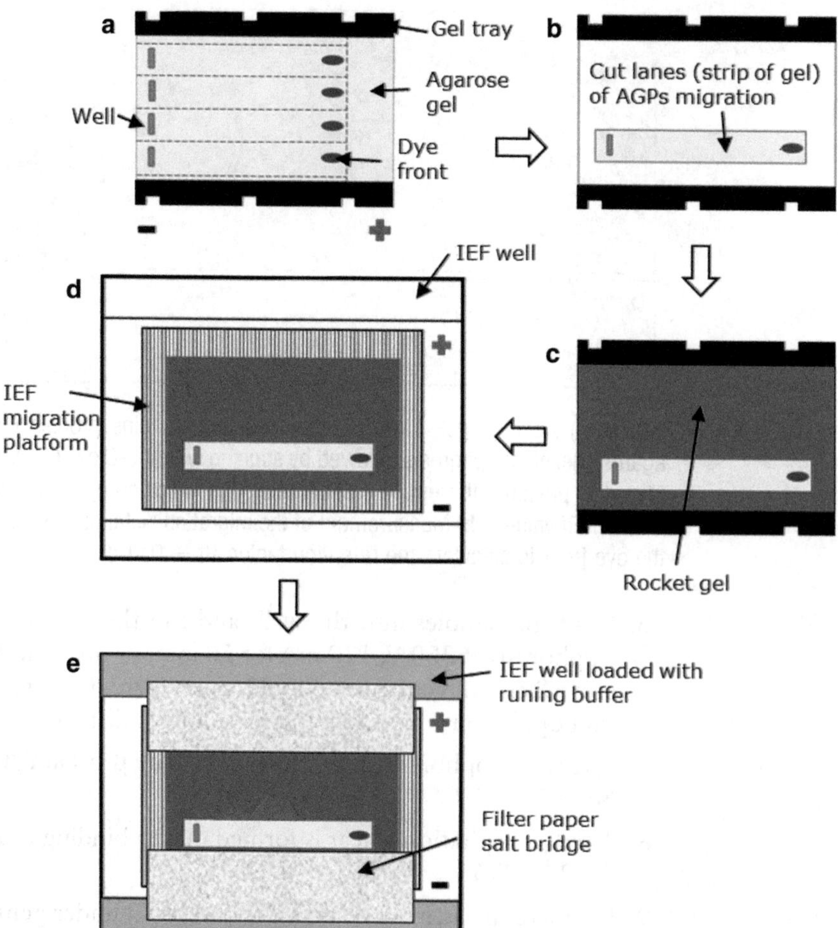

Fig. 10 Technical steps for 2D electrophoresis. (**a**) AGP samples are loaded on an agarose gel and left to migrate across the gel. (**b**) The AGP migration lane (strip of gel) is cut and placed in a gel tray (cut strips are indicated by dashed lines in **a**). (**c**) A rocket electrophoresis gel (containing β-D-GlcY) is poured into the gel tray, around the cut strip. (**d**) After polymerization, the gel is transferred into the IEF migration platform. (**e**) Load the IEF wells with running buffer 1, install the salt bridges and run the IEF electrophoresis. Red-colored precipitation areas are formed (binding of AGPs to β-D-GlcY). + and −: electrodes

5. When the rocket gel is casted around the strips, carefully transfer it from the gel tray to the IEF platform.
6. Load the IEF wells with the running buffer 1 and install the salt bridges (*see* **Note 29**).
7. Run the IEF electrophoresis for 16 h at 200 V, 5 mA, 10 W.
8. A red precipitation area is formed due to binding of AGPs to β-D-GlcY.
9. Incubate the gel in 1% (w/v) NaCl under gentle agitation onto shaker platform to remove the excess of β-D-GlcY (Fig. 11, *see* **Note 27**).

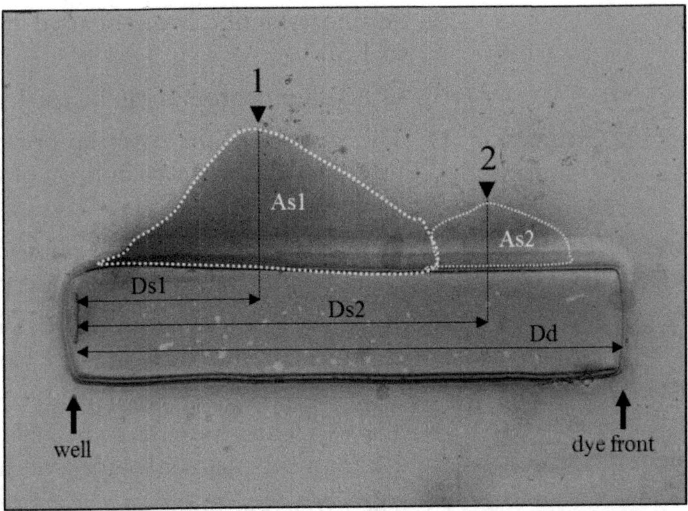

Fig. 11 Cross electrophoresis. Two dimensions profile of arabinogalactan proteins as revealed by cross electrophoresis. Two different subpopulations of AGPs (*1* and *2*) are indicated by black arrows. Ds: migration distance of the sample; Dd: migration distance of the dye front. The surface area of each subpopulation is indicated by the yellow dashed line (As1 and As2)

10. Calculate the retention factor (Rf) Rf = Ds/Dd (Fig. 11, *see* **Note 31**).
11. Calculate the relative percentage area of each population: RA = (As/At) × 100 (Fig. 11, *see* **Note 32**).

4 Notes

1. The PBS solution is autoclaved and stored at +4 °C. Before each use, this solution is filtered through a 0.2 μm filter tip.
2. The wet chamber fits one or two slides.
3. Thin roots are slightly pressed between the slide and the coverslip offering an overall view of root cap and associated BLCs.
4. The solution is prepared from a 16% stock solution PFA diluted in PBS 1×. This step has to be undertaken under a flow hood due to PFA toxicity.
5. Washes are performed by removing the liquid with a pipette and then adding 20 μL of PBS 1×.
6. BSA is used for blocking nonspecific binding sites. This solution is freshly prepared.
7. Tween 20 is used to permeabilize the membrane and help the penetration of the antibody.
8. Concentration of primary antibody should be adapted to each plant specimen.

9. Secondary antibody can be used at lower concentration down to 1:50.
10. Other fluorochromes can be used.
11. The absence of the coverslip prevents crushing of thick root tips and disturbance during observation due to instable coverslip.
12. CitiFluor limits fluorescent photobleaching and increases the resolution for microscopical observation.
13. Once CitiFluor has been added, slides can be stored in a wet chamber in the dark at +4 °C for 1 week without any labeling loss.
14. The extraction buffer is to be kept under agitation for at least 2 h for a complete dissolution of the Triton X-100.
15. The extraction buffer and the Tris base are toxic. Gloves are required during preparation.
16. The gel and the running buffer are toxic. Gloves are required.
17. Pour the gel gently into the gel tray without creating bubbles.
18. β-D-GlcY is toxic. Gloves are required.
19. The extraction can be performed on a freshly ground material.
20. The extraction buffer volume depends on the nature of the sample. Add extraction buffer until the sample is totally dissolved in the solution.
21. Residues may be present in the solution at this stage prior to purification of AGPs using β-D-GlcY.
22. Methanol is toxic and volatile. Use it under a flow hood and wear gloves.
23. The pellet should be totally dissolved. If vortexing is not sufficient, use a micro spatula to help dissolve the pellet.
24. Start with 10% (w/v) sodium hydrosulfite then add ultrapure water until the color becomes clear yellow. If the 2 mL Eppendorf tube is full and the color remains red, add more sodium hydrosulfite to a maximum of 30% (w/v). If the color is still not clear yellow, use a bigger tube and add ultra-pure water until the color becomes clear yellow.
25. If the viscosity of AGP solution is high, add more 1% (w/v) NaCl.
26. The plate containing the gel can be stored at +4 °C.
27. Change the 1% (w/v) NaCl solution every 5 h until all the excess of β-D-GlcY is removed.
28. The filter paper salt bridges (Whatman 17CHR, GE Healthcare Life Science, ref.: 3017-915) must be properly soaked in the running buffer before installation.

29. To be able to quantify the sample, the resulting peak of the sample has to be in the calibration range. If not, dilute or concentrate the sample.
30. The tape must be properly installed by pressing it hard on the edges to avoid any leak.
31. Measure the migration distance of the sample (Ds) and of the dye front (Dd). The Rf of each subpopulation can be calculated as follows: Rf = Ds/Dd.
32. Measure the area of each subpopulation using ImageJ or another appropriate software. The relative percentage area (RA) of each subpopulation can be calculated as follows: RAs = [As/(At)] × 100; As: area of one subpopulation; At: sum of all subpopulations areas.

Acknowledgments

The authors are grateful to La Région de Haute Normandie and le Grand Réseau de Recherche-Végétal, Agronomie, Sol et Innovation, l'Université de Rouen and Le Fonds Européen FEDER for financial support. Fluorescence images were obtained on PRIMACEN (http://www.primacen.fr), the Cell Imaging Platform of Normandy, Faculty of Sciences, University of Rouen, 76821 Mont Saint Aignan.

A.D. is very grateful to all former students who have contributed in "AGP Biology" work over the years, in his laboratory.

References

1. Fincher GB, Stone BA, Clarke AE (1983) Arabinogalactan-proteins: structure, biosynthesis, and function. Annu Rev Plant Physiol 34:47–70
2. Youl JJ, Bacic A, Oxley D (1998) Arabinogalactan-proteins from *Nicotiana alata* and *Pyrus communis* contain glycosylphosphatidylinositol membrane anchors. Proc Natl Acad Sci 95:7921–7926
3. Svetek J, Yadav MP, Nothnagel EA (1999) Presence of a glycosylphosphatidylinositol lipid anchor on rose Arabinogalactan proteins. J Biol Chem 274:14724–14733
4. Nguema-Ona E, Vicré-Gibouin M, Gotté M et al (2014) Cell wall *O*-glycoproteins and N-glycoproteins: aspects of biosynthesis and function. Front Plant Sci 5:499
5. Showalter AM (2001) Arabinogalactan-proteins: structure, expression and function. Cell Mol Life Sci 58:1399–1417
6. Seifert GJ, Roberts K (2007) The biology of Arabinogalactan proteins. Annu Rev Plant Biol 58:137–161
7. Ellis M, Egelund J, Schultz CJ et al (2010) Arabinogalactan-proteins: key regulators at the cell surface? Plant Physiol 153:403–419
8. Vicré M, Santaella C, Blanchet S et al (2005) Root border-like cells of Arabidopsis. Microscopical characterization and role in the interaction with rhizobacteria. Plant Physiol 138:998–1008
9. Cannesan MA, Durand C, Burel C et al (2012) Effect of Arabinogalactan proteins from the root caps of pea and *Brassica napus* on *Aphanomyces euteiches* zoospore chemotaxis and germination. Plant Physiol 159:1658–1670
10. Nguema-Ona E, Vicré-Gibouin M, Cannesan M-A et al (2013) Arabinogalactan proteins in root–microbe interactions. Trends Plant Sci 18:440–449

11. Driouich A, Follet-Gueye M-L, Vicré-Gibouin M et al (2013) Root border cells and secretions as critical elements in plant host defense. Curr Opin Plant Biol 16:489–495
12. Knee EM, Gong F-C, Gao M et al (2001) Root mucilage from pea and its utilization by rhizosphere bacteria as a sole carbon source. Mol Plant-Microbe Interact 14:775–784
13. Koroney AS, Plasson C, Pawlak B et al (2016) Root exudate of *Solanum tuberosum* is enriched in galactose-containing molecules and impacts the growth of *Pectobacterium atrosepticum*. Ann Bot 118(4):797–808
14. Durand C, Vicre-Gibouin M, Follet-Gueye ML et al (2009) The organization pattern of root border-like cells of arabidopsis is dependent on cell wall homogalacturonan. Plant Physiol 150:1411–1421
15. Smallwood M, Yates EA, Willats W et al (1996) Immunochemical comparison of membrane-associated and secreted Arabinogalactan-proteins in rice and carrot. Planta 198:452–459
16. Yates EA, Valdor J-F, Haslam SM et al (1996) Characterization of carbohydrate structural features recognized by anti-Arabinogalactan-protein monoclonal antibodies. Glycobiology 6:131–139
17. Moller I, Marcus SE, Haeger A et al (2008) High-throughput screening of monoclonal antibodies against plant cell wall glycans by hierarchical clustering of their carbohydrate microarray binding profiles. Glycoconj J 25:37–48
18. Knox JP, Day S, Roberts K (1989) A set of cell surface glycoproteins forms an early marker of cell position, but not cell type, in the root apical meristem of *Daucus carota* L. Development 106:47–56
19. Stacey NJ, Roberts K, Knox JP (1990) Patterns of expression of the JIM4 Arabinogalactan-protein epitope in cell cultures and during somatic embryogenesis in *Daucus carota* L. Planta 180:285–292
20. Knox JP, Linstead PJ, Peart J et al (1991) Developmentally-regulated epitopes of cell surface Arabinogalactan-proteins and their relation to root tissue pattern formation. Plant J 1:317–326
21. Pennell RI, Knox JP, Scofield GN et al (1989) A family of abundant plasma membrane-associated glycoproteins related to the Arabinogalactan proteins is unique to flowering plants. J Cell Biol 108:1967–1977
22. Van Holst G-J, Clarke AE (1986) Organ-specific Arabinogalactan-proteins of *Lycopersicon peruvianum* (Mill) demonstrated by crossed electrophoresis. Plant Physiol 80:786–789
23. Van Holst G-J, Clarke AE (1985) Quantification of Arabinogalactan-protein in plant extracts by single radial gel diffusion. Anal Biochem 148:446–450
24. Gane AM, Craik D, Munro SLA et al (1995) Structural analysis of the carbohydrate moiety of Arabinogalactan-proteins from stigmas and styles of *Nicotiana alata*. Carbohydr Res 277:67–85
25. Ding L, Zhu J-K (1997) A role for arabinogalactan-proteins in root epidermal cell expansion. Planta 203:289–294
26. Girault R, His I, Andeme-Onzighi C et al (2000) Identification and partial characterization of proteins and proteoglycans encrusting the secondary cell walls of flax fibres. Planta 211:256–264
27. Popper ZA (2011) Extraction and detection of Arabinogalactan Proteins. In: Popper ZA (ed) The plant cell wall: methods and protocols. Humana, Totowa, NJ, pp 245–254
28. Yariv J, Lis H, Katchalski E (1967) Precipitation of arabic acid and some seed polysaccharides by glycosylphenylazo dyes. Biochem J 105:1C
29. Willats WGT, Knox JP (1996) A role for arabinogalactan-proteins in plant cell expansion: evidence from studies on the interaction of beta-glucosyl Yariv reagent with seedlings of *Arabidopsis thaliana*. Plant J 9:919–925
30. Kitazawa K, Tryfona T, Yoshimi Y et al (2013) β-Galactosyl Yariv reagent binds to the -1,3-galactan of Arabinogalactan proteins. Plant Physiol 161:1117–1126
31. Paulsen BS, Craik DJ, Dunstan DE et al (2014) The Yariv reagent: behaviour in different solvents and interaction with a gum Arabic Arabinogalactanprotein. Carbohydr Polym 106:460–468
32. Hervé C, Marcus SE, Knox JP (2011) Monoclonal antibodies, carbohydrate-binding modules, and the detection of polysaccharides in plant cell walls. In: Popper ZA (ed) The plant cell wall: methods and protocols. Humana, Totowa, NJ, pp 103–113
33. Plancot B, Santaella C, Jaber R et al (2013) Deciphering the responses of root border-like cells of Arabidopsis and flax to pathogen-derived elicitors. Plant Physiol 163:1584–1597

Chapter 23

In Situ/Subcellular Localization of Arabinogalactan Protein Expression by Fluorescent In Situ Hybridization (FISH)

Mário Luís da Costa, María-Teresa Solís, Pilar S. Testillano, and Sílvia Coimbra

Abstract

The arabinogalactan proteins are highly glycosylated and ubiquitous in plants. They are involved in several aspects of plant development and reproduction; however, the mechanics behind their function remains for the most part unclear, as the carbohydrate moiety, covering the most part of the protein core, is poorly characterized at the individual protein level. Traditional immunolocalization using antibodies that recognize the glycosidic moiety of the protein cannot be used to elucidate individual proteins' distribution, function, or interactors. Indirect approaches are typically used to study these proteins, relying on reverse genetic analysis of null mutants or using a reporter fusion system. In the method presented here, we propose the use of RNA probes to assist in the localization of individual AGPs expression/mRNAs in tissues of Arabidopsis by fluorescent in situ hybridization, FISH. An extensive description of all aspects of this technique is provided, from RNA probe synthesis to the hybridization, trying to overcome the lack of specific antibodies for the protein core of AGPs.

Key words Whole mount, Immunolocalization, RNA probes, Arabidopsis, FISH, Arabinogalactan proteins

1 Introduction

Arabinogalactan proteins (AGPs) are cell wall glycoproteins, ubiquitous in the plant kingdom, that have been involved in several biological processes, namely, in sexual plant reproduction. The structure of these proteins has an N-terminal sequence that targets the protein to the endoplasmic reticulum; a protein core rich in Pro/Hyp, Ala, Ser, and Thr; and a C-terminal sequence for the addiction of a glycosylphosphatidyl inositol (GPI) anchor. The protein core defines the different classes of AGPs as the classical, having this typical Pro/Hyp core between the N and the C-terminal sequence, the lysine-rich AGPs that have a lysine-rich domain, the fasciclin-AGPs (FLAs) with fasciclin-like domains and

Zoë A. Popper (ed.), *The Plant Cell Wall: Methods and Protocols*, Methods in Molecular Biology, vol. 2149, https://doi.org/10.1007/978-1-0716-0621-6_23, © Springer Science+Business Media, LLC, part of Springer Nature 2020

finally the AG peptides containing small protein cores with no more than 30 amino acid residues [1].

For the past twenty years, several approaches have been developed to try to learn more about the biological way of action of this important class of proteins, but exactly due to their structure, with more than 90% of the molecule being sugars, this has been a hard task to perform. AGP glycans are polysaccharide chains O-glycosidically linked to the Hyp residues. Initial studies used the synthetic Yariv reagent that blocks AGP function and clearly demonstrated the importance of these proteins in several biological processes. The use of monoclonal antibodies available for the glycosidic part of the AGPs has shown that AGPs could be used as molecular markers for different stages of plant development and for different cell/tissues involved in sexual plant reproduction and embryogenesis [2–6].

The presence of the GPI anchor that binds AGPs to the plasma membrane, facing the cell wall, the possibility of cleavage by specific phospholipase C of the GPI anchor, in association with the complex sugar architecture that surrounds the protein core, and specific cell and tissue localization make these molecules important candidates for signaling events [7].

This chapter describes in detail one of the different procedures to study AGPs involved in sexual plant reproduction. The immunocytochemistry techniques using the available collection of antibodies that recognize the different sugar epitopes present in AGPs [8] is very useful to impart important roles for these proteins in sexual plant reproduction [2–6, 9], but lack resolving power. The use of other important molecular techniques becomes essential for unraveling the involvement of particular AGPs in a particular biological process. An inportant tool to learn about the function of a particular protein is to use reverse genetic analysis every time that null mutants are available [10, 11]. Also crucial for giving important results are the techniques of promoter analysis, using different type of reporter genes [12]. In this chapter we will go in detail over the technique for fluorescent in situ hybridization of AGP transcripts, a method that allows detection of specific mRNA sequences of AGPs with high sensitivity and accuracy, at the individual cell level.

The study of AGPs involvement in *Arabidopsis thaliana* ovule development is a challenging process, where timing and optical resolution are key to success. In the complex structure of the ovule, the gametophyte tissues are inaccessible and surrounded by a double layer of sporophyte integuments, a situation/fact that adds complexity to this study. Tools have been developed to ease this ordeal, like the use of reporter fused promoters or cellular terminator together with confocal and phase contrast microscopy. But these techniques are complex and time consuming, requiring besides the preparation of constructs, transformation and selection of transformed offspring, and yet a few more weeks for plantlets to reach maturity and start producing flowers. A swifter way to determine a gene expression pattern is by in situ hybridization [4, 13, 14].

In situ hybridization is traditionally a technique that allows for the detection and localization of genes and their transcripts [13]. Originally this technique relied on paraffin embedding and sectioning, a time-consuming technique that required optimization for the tissue or organ under study, due to variation on cell wall thickness and permeability; moreover, the structural preservation of cells and tissues after processing for paraffin embedding is very poor due to high temperature and exhaustive dehydration with potent solvents used during the procedure. Additionally the probes were labeled with harmful radioisotopes for detection [15, 16]; the later substitution of these radioactive materials by digoxigenin- [17–19] or alkaline phosphatase-labeled oligonucleotides [20] made this technique safer and more user friendly; however, the use of alkaline phosphatase or peroxidase needed optimization of the reaction time, as the overreaction can create artifacts. Additionally it was very difficult to evaluate the state of preservation of the tissues. With further development of highly stable and sensible fluorochromes that can be conjugated to a wide range of reporter molecules, as antibodies, nucleotides, biotin, or digoxigenin, cell biology localization techniques highly increased their sensitivity and field of application; this fact, together with the development of confocal laser scanning microscopy, opened the way for tracking specific protein and gene expression in tissues and organs at the single cell level, also in plant biology research [21].

In the case of plants, one of the challenges of these in situ methodologies has been to overcome the limitations associated with specific features of most plant cells such as the presence of cell wall which hinders, at least partially, penetration of antibodies and probes to subcellular targets. Regarding in situ localization methods for plant ovaries, the situation is even more complex, as stated before, due to small size and inaccessibility of the gametophyte inside tissues.

With the revised method presented here, localization of transcripts in Arabidopsis ovules is made easier, more sensitive, and less time-consuming; by making a whole-mount fixation, probe hybridization and detection, we eliminate the need for embedding and sectioning, reducing time and difficulty of the procedure. RNA probes have the advantage that RNA–RNA hybrids are very thermostable and are resistant to digestion by RNases. This allows for post-hybridization digestion with RNase to remove nonhybridized RNA and therefore reduces the possibility of background staining. Appropriate fixation and permeabilization are essential steps to optimally preserve cells and RNAs for efficient hybridization, messenger RNA detection, and high cellular preservation and visualization. Also the use of fluorescent conjugated antibodies to localize digoxigenin-labeled hybrids enables higher sensitivity, clearer images and higher resolution of labeling at a subcellular level [22, 23], its combination with DAPI staining [24] provides additional evaluation of cell structure and integrity.

2 Materials

2.1 Buffers

1. DNA Extraction Buffer: Tris–HCl pH 7.5 200 mM, NaCl 250 mM, EDTA 25 mM, SDS 0.5% (w/v), in $_{dd}H_2O$. Always prepare fresh.
2. RNA dilution buffer: 6× SSC (*see* Buffers 2.1.6), 3.2% (w/v) paraformaldehyde in RNase-free H_2O. Always prepare fresh.
3. PBS (phosphate-buffered saline): KCl 2.7 mM, NaCl 137 mM, Na_2HPO_4 10 mM, KH_2PO_4 1.8 mM. Adjust pH to 7.4 with HCl. Filter-sterilize or autoclave.
4. Hybridization buffer (*see* Table 1).
5. 50× Denhardt's: BSA 1% (w/v), Ficoll 400 1% (w/v), Polyvinylpyrrolidone 1% (w/v), in RNase-free water. Filter-sterilize with a 0.2 μM filter. Aliquot and store at −20 °C.
6. 20× SSC: trisodium citrate 300 mM, sodium chloride 3 M in RNase-free H_2O. Adjust pH to 7.0 with 1 M HCl. Aliquot and store at −20 °C.

2.2 Culture Media

1. LB/ampicillin/IPTG/X-Gal culture plates: Bacto tryptone 1% (w/v), Bacto yeast extract 0.5% (w/v), NaCl 1% (w/v), Bacto agar 1.5% (w/v), in $_{dd}H_2O$. Adjust pH to 7.5 with NaOH.

Table 1
Hybridization buffer

Stocks	Final concentration	Volume of hybridization buffer					Mixing order
		50 μL	100 μL	150 μL	200 μL	250 μL	
RNase-free H_2O	n.a.	5.5 μL	11 μL	16.5 μL	22 μL	27.5 μL	1
NaCl 5 M	300 mM	3 μL	6 μL	9 μL	12 μL	15 μL	2
Tris–EDTA solutions 3.1.1	10 mM/1 mM	5 μL	10 μL	15 μL	20 μL	25 μL	3
50× Denhardt's	1×	1 μL	2 μL	3 μL	4 μL	5 μL	4
Formamide	50%	25 μL	50 μL	75 μL	100 μL	125 μL	5
50% Dextran sulfate	10%	10 μL	20 μL	30 μL	40 μL	50 μL	6
tRNA (20 mg/mL)	200 μg/mL	0.5 μL	1 μL	1.5 μL	2 μL	2.5 μL	7

Sterilize by autoclaving, let cool to 45 °C, and add filter sterilized ampicillin to 100 μg/mL, IPTG to 0.1 mM, and X-Gal (in DMF) 30 μg/mL. Plate immediately and store at 4 °C, wrapped in aluminum foil.

2. SOC: Bacto Tryptone 2% (w/v), Bacto yeast extract 0.5% (w/v), NaCl 10 mM, KCl 2.5 mM, $MgCl_2$ 10 mM, $MgSO_4$ 10 mM, glucose 10 mM. Sterilize by autoclaving. SOC culture medium is used for the recovery of transformed bacterial cells.

2.3 Solutions

1. TE: Tris–HCl pH 7.5 10 mM, EDTA 1 mM, in $_{dd}H_2O$. This lightly saline solution helps to devolve DNA.
   (a) Tris–EDTA: Tris–HCl pH 7.5 100 mM, EDTA 10 mM, in RNase-free $_{dd}H_2O$. Aliquot and store at −20 °C. This variation of the TE solution is used in the hybridization buffer.
2. Miniprep solution 1: TRIS pH 8.0 25 mM, EDTA pH 8.0 10 mM, glucose 50 mM, in $_{dd}H_2O$. Always prepare fresh.
   After removing the growing medium in the first step of the plasmid purification protocol, the cells must be suspended in a uniform manner for the alkaline lysis to take place. This solution will allow just that.
3. Miniprep Solution 2: SDS 1% (w/v), NaOH 0.2 N, in $_{dd}H_2O$. This solution will burst bacterial cell by attacking the cell membrane with a detergent and high pH. In this way the plasmids will be released into the solution. Always prepare fresh.
4. Miniprep Solution 3: KOAc 3 M, glacial acetic acid 2 M, in $_{dd}H_2O$. Store at 4 °C, check for contaminants before use. After bursting the cells, proteins and all other cellular molecules are released to the alkaline solution. This solution will neutralize the solution and precipitate SDS, proteins, and other impurities.
5. Miniprep Solution 4: RNase (Thermo, EN0531) to 20 μg/mL in TE (*see* Solutions 2.3.1). If possible prepare fresh. RNA copurification is common with this miniprep protocol and can be a problem to the restriction of the purified plasmids. A simple and easy way to get rid of this contamination is a simple treatment with RNase.
6. BSA 5%: bovine serum albumin 5% (w/v) in PBS (*see* Buffers 2.1.3). Aliquot and store at −20 °C. Dilute at a ratio of 1:5 in PBS to prepare 1% BSA. Bovine serum albumin is used as a blocking agent in the immunodetection reactions.
7. Fixative Solution: paraformaldehyde 4% (w/v), Tween 20 0.001% (v/v), in PBS (*see* Buffers 2.1.3). Always prepare fresh. To prepare 1% paraformaldehyde in PBS make a 1:4

dilution of this solution in PBS. This fixative solution is used to prevent degradation of cellular content, prior to the dehydration step. It also helps with mRNA unfolding, easing RNA probe ligation.

8. Methanol series in PBS: Make methanol dilutions of 70% (v/v), 50% (v/v), and 30% (v/v), in PBS (*see* Buffers 2.1.3). Store at −20 °C. During the dehydration–rehydration process a methanol series is used to dehydrate and rehydrate the dissected pistils. This step serves as a first permeabilization to help further accessibility of probes to intracellular targets. The final dehydration prior to hybridization eliminates excess water in tissues that would hinder hybridization reaction; it also helps eliminating any remaining active enzyme used during permeabilization.
9. Macerozyme R-10: Macerozyme R-10∗ (cat.no. 28302, SERVA Germany) 2% (w/v), in RNase-free H_2O. Filter-sterilize with a 0.45 μM Filter, Aliquot and store at −20 °C. Macerozyme R-10 has various enzymatic activities to partially degrade cell wall components. Cell walls are a principal barrier for probe penetration; treating the tissue with a pectinase and cellulase cocktail makes cell walls become more permeable without loosening their integrity.

   ∗Activity: pectinase 0.5 U/mg, hemicellulase 0.25 U/mg, cellulase 0.1 U/mg.
10. DAPI staining solution: To make a 5 mg/mL DAPI stock solution (14.3 mM for the dihydrochloride DAPI or 10.9 mM for the dilactate DAPI), dissolve 10 mg in 2 mL of deionized water (dH_2O) or dimethylformamide (DMF). You may have to sonicate the dihydrochloride DAPI as it is harder to dissolve in water than the dilactate DAPI. Dilute the DAPI stock solution to 300 nM in PBS, and use about 20 μL directly over the pistil-containing wells. DAPI (4′, 6-diamidino-2-phenylindole) is a very useful stain for the nucleus. In this version of fluorescent in situ localization it serves a dual purpose, first it will help to locate the cell nucleus. The labeling of DAPI and the probe should not overlap since in the conditions used the RNA probe should have no affinity to the nucleic DNA. Observation of the nucleus integrityby DAPI staining, permits proper evaluation of the hybridization signal in the cytoplasm and good cell structural preservation.
11. Mowiol mounting medium: Weigh 6 g of analytical-grade glycerol in a 50 mL disposable plastic conical centrifuge tube. Add 2.4 g of Mowiol 4-88 (sigma, 81381) and stir thoroughly to mix the Mowiol with the glycerol. Add 6 mL of distilled water and let the solution rest at room temperature for 2 h. Mix in 12 mL of 0.2 M Tris buffer (2.42 g Tris/100 mL water, adjust pH to 8.5 with HCl) and incubate the solution in a

water bath at 50 °C for 10 min with occasional stirring to dissolve the Mowiol. Clarify the mixture by centrifugation at 5000 × *g* for 15 min. Aliquot and store at −20 °C.

The solution is stable, and the pH is retained for at least 12 months at −20 °C, and for about a month at room temperature. Mowiol is a water-soluble hydrocolloid mucoadhesive based on poly(vinyl alcohol) that carries both the advantage of antifade properties and creating a solid seal of the final preparation. Stained slides mounted with Mowiol retain fluorescence when stored at 4 °C in the dark.

2.4 Equipment

1. 0.2 and 1.5 μL centrifuge tubes.
2. 15 and 50 mL conical tubes.
3. Centrifuge.
4. Fine-tipped forceps.
5. G26 (or higher gauge) hypodermic needles.
6. Humidity chamber (*see* Fig. 1).

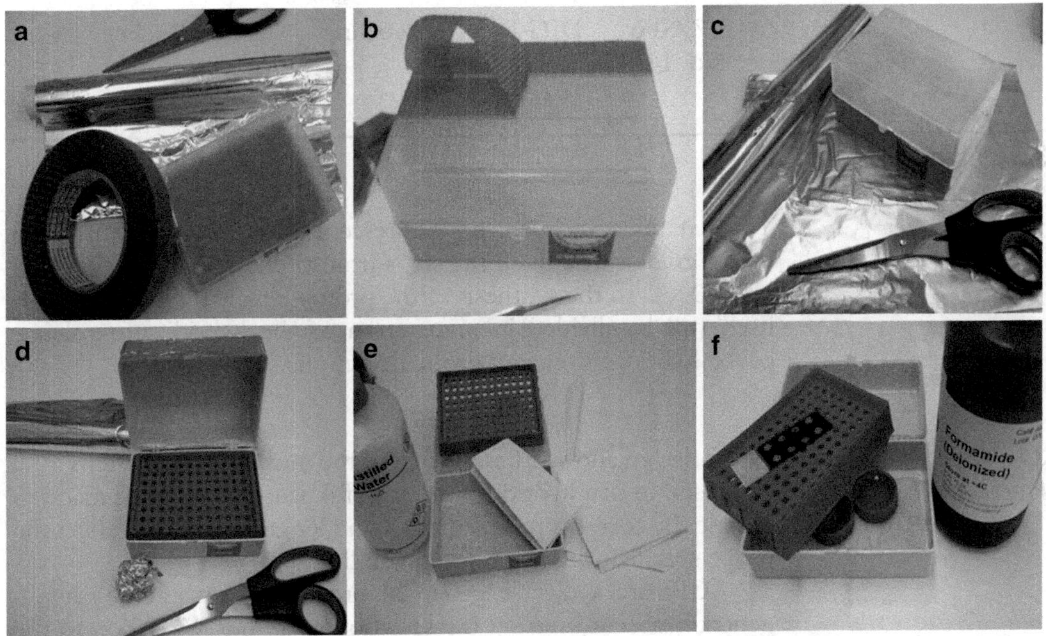

Fig. 1 Setting a hybridization box. A humidity box is a fundamental tool in this procedure; it will avoid the evaporation of solutions during the incubation periods. To make one use a tip box, tin foil, and some double-faced duct tape (**a**). Cover the top of the box with a layer of double-faced duct tape (**b**) then cover it with aluminum foil (**c**). Remove the excess foil (**d**) and remove the tip holding rack. In the bottom of the box place some damp paper towels (**e**). To make it suitable for the hybridization incubation place some 50 mL conical tubes caps field of formamide on the bottom of the box (**f**) This will saturate the air in the sealed chamber and reduce the evaporation of the hybridization solution, as after the hot incubation the leads can be easily removed to minimize the exposure to formamide vapors. Put back the tip holding rack; it will serve as a nice support for the slides.

7. Hybridization oven.
8. Vacuum filtering flask.
9. Micropipettes.
10. NanoDrop®.
11. Parafilm®.
12. Teflon-coated reaction slides.
13. Thermocycler.
14. Vacuum pump.
15. Vacuum chamber.
16. Vortex.

2.5 Enzymes

1. Taq-based PCR system.
2. PstI and NcoI.

2.6 Commercial Kits

1. NBT-BCIP® solution (sigma, 72091).
2. pGEM®-T Easy cloning kit (Promega, A1360).
3. JM109 Competent Cells (Promega, L2005).
4. T7/SP6 DIG-RNA labeling kit (Roche, Cat. No. 11175025910).

3 Method

This extended protocol for fluorescent in situ hybridization is a broad guide through all the steps from the extraction of the template DNA, to the synthesis of the probe. Finally the hybridization itself will be divided in preparation of the hybridization slides, probe hybridization, and probe detection.

3.1 Probe Preparation

Probe design and synthesis is one of the most important steps for a successful in situ hybridization. Careful planning must be made to choose the correct probe sequence and length. Has it will affect the probe specificity and penetration ability.

AGPs are a diverse group of proteins [1] where splicing is generally absent, making the selection of an mRNA region to use has target for a RNA probe fairly simple. The design process must include, however, a careful verification of the selected sequence specificity, the most preserved sequences being the N-terminal and GPI anchor signals. AG-Peptides due to their very short sequence are most challenging for this task, but short specific probe sequences can be obtained. The optimal length for a RNA probe is between 200 and 1500 bases.

Another aspect of the use of RNA probes is their sensitivity to degradation; an extra effort must be made to keep all materials and solutions RNase free. A minute amount of RNase may destroy a complete batch of precious RNA probe.

3.2 DNA Extraction

Genomic DNA will serve has a template to create the probe sequence that will be inserted in a plasmid, that in turn will be used to synthesize the RNA probe.

The first step will be to extract DNA. In the following method we propose to use a rapid and simple method to isolate genomic DNA suitable for PCR amplification, based on the method of Edwards et al. [25].

All steps are performed at room temperature and centrifugations are at 13,000 × *g*.

1. Prepare the DNA Extraction Buffer (*see* Buffers 2.1.1).
2. Use the lid of a 1.5 mL centrifuge tube to punch out a disc from a young leaf into the tube (*see* **Note 6.1a-b**), and grind it thoroughly in a small mortar and pestle. Immediately add 400 μL of extraction buffer, and vortex for 5 s (*see* **Note 6.1.c**).
3. Spin at max RPM for 2 min to pellet the tissues debris, and transfer 400 μL of the supernatant into a fresh 1.5 mL centrifuge tube (avoid taking debris from the pellet). Add 350 μL isopropanol, mix by gently inverting the tubes a few times and leave at room temperature for 2 min. Then spin for 5 min at full speed to pellet the DNA.
4. Remove all the supernatant and let air dry for a few minutes. (Do not let the pellet get too dry, otherwise it will be very difficult to suspend the genomic DNA.)
5. Dissolve the pellet in 50 μL TE (solution 2.3.1), or PCR grade H_2O by gentle shaking. DO NOT VORTEX. Check the integrity of the gDNA by agarose gel electrophoresis (*see* **Note 6.1. d**).

3.3 pGEM Cloning

After extracting the gDNA can be used as a template to amplify the target gene sequences by PCR using an enzyme with 3′ adenylation activity (*see* notes pGEM cloning 6.2).

Perform a PCR cleanup and quantify your PCR product.

3.3.1 Cloning PCR Products with pGEM®-T Easy Vectors (Promega, A1360)

1. Set up ligation reactions, with 5 μL of 2× Rapid Ligation Buffer, 1 μL of pGEM®-T Easy Vector (50 ng), 37.5 μg of purified PCR product, 1 μL of T4 DNA Ligase, and PCR grade H_2O to a final volume of 10 μL. Vortex the 2× Rapid Ligation Buffer vigorously before use (*see* **Note 6.2.b**).
2. Mix the reactions by pipetting (*see* **Note 6.2.c**). Incubate overnight at 4 °C to maximize ligations.

3.3.2 Transformation of JM109 High Efficiency Competent Cells

1. Prepare LB/ampicillin/IPTG/X-Gal plates (*see* Culture media 2.2.1).
2. Centrifuge the ligation reactions briefly. Add 2 μl of each ligation reaction to a sterile 1.5 mL tube on ice. Prepare a control tube with 0.1 ng of uncut plasmid.
3. Place the JM109 Competent Cells (Promega, L2005) in an ice bath until just thawed (~5 min). Mix cells by gently flicking the tube. Carefully transfer 50 μL of cells to the ligation reaction tubes. Use 100 μL of cells for the uncut DNA control tube. Gently flick the tubes and incubate on ice for 20 min. Then Heat-shock the cells for 45–50 s in water bath at exactly 42 °C. DO NOT SHAKE. Immediately return the tubes to the ice. After 2 min add 950 μL room temperature SOC medium (see Culture media 2.2.2) to the ligation reaction transformations and 900 μL to the uncut DNA control tube. Incubate for 1.5 h at 37 °C with shaking (~150 r.p.m.).
4. Plate 100 μL of each transformation culture onto duplicate LB/ampicillin/IPTG/ X-Gal plates. For the uncut DNA control, a 1:10 dilution with SOC is recommended. Incubate all plates overnight at 37 °C. Select white colonies.
5. Use M13 primers in conjugation with your reverse primer to confirm to orientation of your insert. This will determine which reaction T7 RNA Polymerase or SP6 RNA Polymerase will produce your antisense (or detection) probe [26, 27] (*see* **Note 6.2.d**).
6. After this initial screening select one colony and proceed with multiplication of the probe containing plasmid.

3.3.3 Miniprep Protocol [28]

The plasmid containing the probe sequence will be used as a template for SP6 and T7 RNA polymerase [26, 27]. We will now proceed with the multiplication and purification of the probe containing plasmid.

1. Inoculate 5 mL of LB/amp (75 μg/mL) (*see* **Note 6.3.a**) overnight at 37 °C with vigorous agitation.
2. Fill a 1.5 mL Eppendorf tube and centrifuge for 1 min at 10000 × *g* (*see* **Note 6.3.b**). Discard the supernatant and repeat until you have spun down 5 mL.
3. Then suspend the pellet in 100 μL of Miniprep solution 1 (*see* Solutions 2.3.2) and mix in 20 μL of 10 mg/mL of lysozyme. Let stand at room temperature for 2 min (for 1 mL: 0.01 g Lysozyme in 1 mL 0.250 mM tris pH 8.0).
4. Add 200 μL of Miniprep solution 2 (*see* Solutions 2.3.3) mix well and place on ice for 5 min, then add 150 μL of Miniprep solution 3 (material solution 3.4) vortex gently and place back on ice for 5 min.

5. Centrifuge at 4 °C for 5 min at 12000 × *g*. Transfer the supernatant to a new tube (*see* **Note 6.3.c**).
6. Add 400 μL of phenol–chloroform. Vortex and centrifuge at 4 °C, for 2 min at 12000 × *g*.
7. Transfer the aqueous upper phase to a new tube (*see* **Note 6.3.d**). Add 1 mL of 100% ethanol at room temperature.
8. Leave to stand for 2 min at room temperature.
9. Centrifuge at 12000 × *g* for 5 min, pour off the ethanol and let the pellet dry.
10. Dissolve the pellet in 50 μL of TE/RNase (*see* Solutions 2.3.1), quantify your purified Plasmid by NanoDrop and send one sample for sequencing.

3.4 RNA Probe Synthesis and Labeling

In the following steps the purified plasmid will be prepared for the synthesis and labeling of the probe with the T7/SP6 DIG-RNA labeling kit (Roche, Cat. No. 11175025910) (*see* **Note 6.4.a**).

3.4.1 Linearize the Plasmid

For the proper synthesis of the RNA probe by T7 or SP6 RNA polymerase the plasmid most first be linearized, or it will produce uneven sized and possibly unspecific sequences.

1. Carefully label two 1.5 mL tubes, with Sense (Tube 1) and Antisense (Tube 2) and add the volume equivalent to 2.5 μg purified plasmid to each.
2. Digest tube 1 with PstI and tube 2 with NcoI (*see* **Note 6.4.b**).
3. Spin down and incubate both tubes at 37 °C for 15 min then stop the reaction by heating the tubes at 95 °C for 5 min.
4. Clean up the linearized plasmids (*see* **Note 6.4.c**). Quantify both the purified linearized plasmids by NanoDrop.

3.4.2 Digoxigenin-Labeled RNA Probe Synthesis

In the next step, the Roche Dig Labeling Kit will be used to synthesize and label the RNA probe in one single step. In the described procedure both control and detection probe will be synthesized. It is of outmost importance to provide high-quality template for the reaction to occur properly.

1. Add the equivalent of 1 μg of purified linearized plasmid to a 200 μL microtube. Place the microtube on ice and add the following reagents according to the mixing order (*see* Table 2).
2. Mix gently and incubate at 37 °C for 2 h. To stop the reaction add 2 μL of 0.2 M EDTA (pH 8.0) (*see* **Note 6.4.d**). Probes may be stored at −20 °C for up to one year (*see* **Note 6.4.e**).

Table 2
Probe synthesis reaction mixture order

Vial	Reagent	Volume, μL	Mixing order
n.a.[a]	RNase-free PCR grade H_2O	To 20	1
8	10× Transcription buffer	2	2
10	Protector RNase inhibitor	1	3
7	10× NTP labeling mixture	2	4
For *sense* probe[b]			
11	RNA polymerase T7	2	5
For *antisense* probe[b]			
12	RNA polymerase SP6	2	5

[a]Sigma, W1754
[b]According to pGEM insert orientation

3.4.3 Probe Labeling Quantification (Please See **Note 6.4.f.**)

Now determine the labeling efficiency, as it may be inconsistent. This step will ensure the control of the reaction and allows for normalizing the probe for the future hybridization essays. The simplest way to determine the labeling efficiency is to compare the probe that we synthesized to standard solution, a standard solution is provided as a control in the previously used labeling kit. An efficiency test is proposed in the kit instructions, but we find it to be overcomplicated and more expensive. The following approach consists of simply comparing the diluted standard to a dilution of our probe to find out with some degree of precision the concentration of our probe.

1. Make three tenfold dilutions of the probe in RNA dilution buffer, 4% Paraformaldehyde 6× SSC in H_2O (*see* Buffers 2.1.2), starting with 1 μL of probe. And make 4 tenfold dilutions with the labeled RNA control standard, supplied with the kit (vial 5). Pipet 1 μL of the original solutions followed by 1 μL of all dilutions on a nylon membrane (*see* **Note 6. 4.g**).
2. Fix the nucleic acid to the membrane by cross-linking with UV-light or baking for 30 min at 120 °C.
3. Place the membrane strip in a 50 mL conical tube. Block for 5 min with 10 mL 5% BSA (Materials 3.6), in an orbital shaker with mild agitation. Discard the blocking solution and incubate for 90 min at room temperature with a solution of 1/1000 of anti-digoxigenin antibody conjugated to alkaline phosphatase (Sigma, A1054), in 1% (w/v) BSA.
4. Wash three times for 10 min with PBS (Materials 1.3), in an orbital shaker 80 r.p.m. Drain thoroughly and transfer to a Petri dish. Add 5 mL of NBT-BCIP® solution (sigma,

72091), cover with aluminum foil and place in an orbital shaker at 80 r.p.m. Let react until the last dilution of the control starts to appear. Stop the reaction by washing out the NBT-BCIP® solution with $_{dd}H_2O$.

5. Compare the labeling of the probe with the scale made using the RNA-labeled control (*see* **Note 6.4.h**).

4 In Situ Hybridization

The pistils will be prepared to undergo hybridization with the probe previously set, in two steps: First the pistils will be fixated and then the cells permeabilized to permit the probe penetration, and finally the hybridization and detection reaction will be performed.

4.1 Fixation

Dissecting the ovaries prior to the fixation will help with the fixation, probe penetration and visualization of the final results. The formaldehyde fixation will allow to preserve the tissues, reduce the activity of proteins and help to unfold the target mRNA.

1. Select flowers and extract pistils. Place the pistil on a piece of double-faced duct tape on a microscopy slide. Under a stereoscope remove the valves of the pistil using fine gauge needles (G26 or higher) and fine point tweezers, by cutting along the valve suture ridges. Be careful not to sever the ovules from the septum and maintain the transmitting tissue intact (Fig. 2).

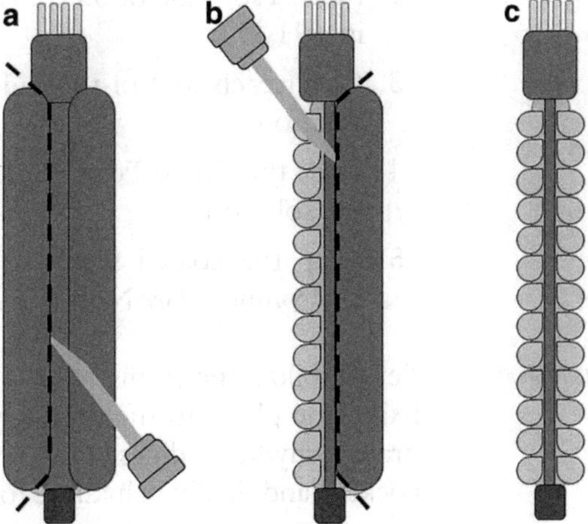

Fig. 2 Pistil dissection. To expose the ovules of the Arabidopsis pistil, place the pistil on a piece of double-faced duct tape so that the valve sutures on one of the sides face the top (**a**). With a G26 gauge hypodermic needle make a shallow incision following the suture line exposing the ovules. Proceed in the same manner on the other side (**b**). Finally using a pair of fine-tip tweezers pull the semidetached valves and finish cutting the opposite side of valve suture detaching it from the pistil (**c**)

2. Immediately immerse the samples in ice-cold fixative solution, 4% Paraformaldehyde 0.001% Tween 20 in PBS (*see* Solutions 2.3.7). Apply sustained vacuum of −70 KPa for 15 min to help the fixative to penetrate into tissues. Maintain in the fixative solution for 18 h or overnight at 4 °C.
3. Wash the fixed pistils by immersing them for 10 min in cold PBS, with mild agitation. Repeat three times. If necessary, preserve the dissected fixed pistils at 4 °C in 0.1% Paraformaldehyde in PBS until use, to avoid reversion of paraformaldehyde fixation.

4.2 Slide Preparation

To address the preparation of the slides for the hybridization, some protocols call for the use of extremely expensive slides or hybridization chambers. We propose the use of simple and affordable multi-well Teflon coated diagnostic slides, which restrict reagents to tissue samples, with some good cleaning and polylysine coating.

4.2.1 Slides Wash (See **Note 6.5.a**)

1. Place the slides in a staining rack and cover with a cleaning solution, 70% Ethanol 0.1% Triton X100, with mild agitation for 20 min.
2. Wash the slides by dipping the staining rack in $_{dd}H_2O$ with mild agitation for 10 min. Repeat this process four times. Carefully dry the rack before dipping briefly in 100% ethanol and let the slides dry out in a dust free environment.

4.2.2 Polylysine Coating

1. Place the clean slides in a clean 12 × 12 cm square Petri dish.
2. Make a solution of 0.001% (w/v) poly-L-lysine (Sigma P4707) in $_{dd}H_2O$.
3. Cover each well of the slides with 50 μL of the poly-L-lysine solution.
4. Place the closed Petri dishes in a 40 °C oven overnight to dry the solution.
5. Keep the coated slides at room temperature in a dust-free environment (*see* **Note 6.5.b**).

4.2.3 Permeabilization

Permeabilize the pistils to ease penetration of the probe; by first fixing the pistils to the poly-L-lysine-coated slides and then dehydrate, rehydrate, digest briefly with a cell wall-degrading enzyme cocktail and finally dehydrate for storage or immediate processing.

4.2.4 Pistil Adhesion to the Coated Slides

1. Transfer up to two fixed dissected pistils to each polylysine coated well (*see* **Note 6.5.c**).
2. Let air-dry for 20 min at room temperature and immediately proceed to the dehydration step.

4.3 Dehydration-Rehydration and Mild Cell Wall Enzymatic Digestion

1. Dehydrate by incubating for 5 min with an increasing concentration series of methanol in PBS, 30%, 50%, 70% 100%. Then rehydrate by incubating for 5 min with a decreasing concentration series of methanol in PBS, 70%, 50%, 30%. And finally for 5 min in PBS (*see* **Note 6.6.a**).
2. Transfer to a humidity box and immediately cover the pistils with 30 μL 2% Macerozyme (cat.no. 28302, SERVA Germany) in PBS. Seal the humidity box with Parafilm and incubate for 60 min at 45 °C.
3. Remove the Macerozyme solution and wash with PBS three times, 5 min each. Then, incubate each well with 30 μL proteinase K (20 μg/mL) for 20 min at 37 °C. Then wash three times with PBS for 5 min.
4. Wash with $_{dd}H_2O$ for 5 min (*see* **Note 6.6.b**).
5. Dehydrate again in an increasing series of methanol in PBS. Make an additional incubation with 100% methanol, and leave to air-dry.
6. Slides can be immediately used or stored in a dry sealed box at −20 °C for up to 2 years (*see* **Note 6.6.c**).

4.4 Hybridization

To proceed to the final stage of the protocol, hybridization and detection, first the previously dehydrated tissues are incubated with the labeled probe, and finally the hybridized probe is detected by immunolocalization of digoxigenin.

1. Prepare a hybridization box, by placing a few conical tube caps with formamide at the bottom of a humidity box (*see* **Note 6.7.a**).
2. After selecting the correct amount of probe to use (*see* **Note 6.7.b**).
3. Calculate the volume of hybridization solution that you will require, about 25 μL per well and prepare your hybridization buffer (300 mM NaCl, 10 mM Tris/EDTA, 1× Denhart's solution, 50% formamide, 10% dextran sulphate, 200 μg/mL tRNA) accordingly (*see* Buffers 1.4).
4. Place all aliquots required on ice and allow thawing.
5. Place the dextran sulfate aliquot in a 75 °C water bath for 15 min, as it is highly viscose, it will also help to pipette with a clipped tip. Place the unfrozen tRNA stock aliquot in boiling water for 15 s before adding to the hybridization buffer.
6. Mix all volumes from the stocks according to the mixing order in Table 2 (*see*Buffers 1.4).
7. Mix the hybridization buffer with the appropriate probe volume.
8. Add 25 μL of the hybridization solution to each well; take care to not put different probes in adjacent wells.

9. Cover the wells with a small piece of Parafilm®, place the slide in the modified humidity chamber and seal with Parafilm®. Incubate at 50 °C overnight. The Parafilm® pieces will ensure appropriate contact between tissue samples and probe, making a small hybridization chamber over tissues and avoiding the reagents evaporation during the incubation.

4.4.1 Posthybridization Washing

Extract the excess probe which has not hybridized with samples (*see* Fig. 3d).

1. Remove the conical tube caps containing formamide from the humidity chamber and carefully remove the Parafilm® covering the wells with fine tips forceps (*see* **Note 6.7.c**).
2. Wash four times for 2 min with 4× SSC at room temperature.
3. Wash four times for 2 min with 2× SSC at room temperature.

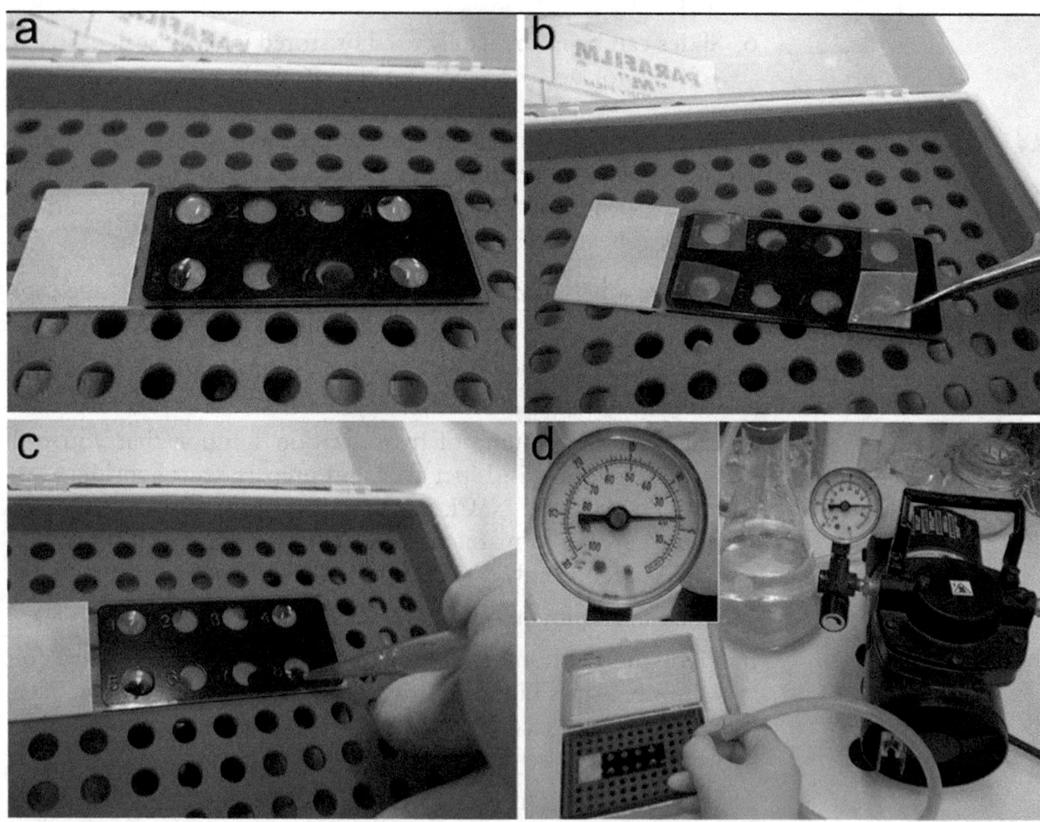

Fig. 3 Hybridization tips. Place the slides in a humidity chamber to avoid evaporation of solutions; additionally different probes can be used on the same slide if some space is left between wells; for example antisense probe is applied to wells 1 and 5 and control probe can be applied to wells 4 and 8 (**a**). Small squares of Parafilm® can be used to cover the wells to avoid excessive evaporation of hybridization solution (**b**). A pipette tip attached to a vacuum pump, under low negative pressure (−20 KPa), can be used to swiftly remove solutions from the wells (**d**), speeding the wash steps

4. Wash two times for 15 min with 0.1× SSC at 50 °C. These stringency conditions are standard, but they can be adjusted depending on the probe sequence to detect; lower stringency with lower temperature and higher SSC concentration allows hybridization between sequences that are less similar.
5. Finally wash with PBS for 5 min.

4.5 Immunolocalization of the Hybridized Probe

The final step is the immunolocalization.

1. Place your slides in a humidity box.
2. Incubate samples in wells for 5 min with 5% (w/v) BSA at room temperature to block nonspecific reactive groups of tissues.
3. Incubate for 90 min at room temperature with the primary antibody, mouse anti-digoxigenin (Sigma, 11333062910), in a solution of 1/5000 in 1% (w/v) BSA.
4. Wash three times for 5 min with 1% (w/v) BSA.
5. Incubate for 45 min at room temperature, with the secondary antibody, anti-mouse conjugated with FITC (Sigma, F0257), in a solution of 1/25 in PBS (*see* **Note 6.7.d**).
6. Wash six times for 3 min with PBS.
7. Stain for 5 min with about 20 μL of a 0.3 mM DAPI solution in PBS (*see* **Note 6.7.e**).
8. Then wash six times for 3 min each time with PBS and mount the slides with Mowiol [29, 30].
9. Observe under fluorescence or confocal microscope (*see* **Note 6.7.f**).

5 Controls

To all assays the following tests must be considered to verify and validate the specificity of the hybridization probe and also of the antibodies used.

1. In situ hybridization using the sense probe. This control evaluates the unspecific binding of the probe. The results should be negative, with no detectable hybridization.
2. Immunolocalization replacing the primary anti-DIG antibody by buffer. This control evaluates the unspecific binding of the secondary antibody. No signal should be detected.
3. Immunolocalization replacing the secondary fluorescent-labeled antibody by buffer. This control evaluates if there is any source of unspecific autofluorescence, which may contaminate the results. No autofluorescence results should be detected in the range of the fluorescent-labeled antibody.

6 Notes

6.1 DNA Extraction Notes

This crude DNA extraction method is a cheap and fast way to purify DNA.

It is not as efficient as other commercially available solutions. However, this protocol employs readily available reagents and provides DNA suitable for PCR.

(a) It is important to avoid using too much plant material.

(b) If too much material is used, it will result in an increase of gDNA degradation by interfering with the buffer denaturing capacity. It will also make it difficult to separate the DNA containing supernatant from the plant material debris.

In contrast this same method can be used to extract PCR suitable DNA from a smaller sample, like a single cotyledon, in which case all volumes should be reduced by 50% and DNA precipitation (Step 3) should be performed at −20 °C with final centrifugation at 4 °C.

(c) At this stage, the samples can be kept at room temperature for up to 1 h with no major damage to the genomic DNA, while you finish other samples that you may be preparing.

(d) You may find that some RNA have been copurified, it should not be a problem for normal PCR. However, if necessary treat by adding 1 μL of RNase solution (miniprep solution 4, see Solutions 2.3.5) and incubate at 37 °C for 20 min.

6.2 pGEM Cloning Notes

The pGEM cloning system lets you easily clone regular PCR products without any additional complicated steps.

(a) Most common PCR enzymes will leave a 3′ adenine overhang, which is necessary for pGEM cloning. Error rate for a small amplicon are low. However, if you prefer to use a proofreading enzyme verify that they have 3′ adenylation activity or perform this as an additional step.

(b) Assuming a probe size of 750 bp and a 3:1 ratio molar of insert to linearized plasmid.

The pGEM cloning system should perform perfectly with a wide range of insert-to-linearized plasmid ratios. However, if you experience difficulties with the ligation, try to increase the amount of insert.

(c) Use 0.5 mL tubes with low DNA binding capacity. In alternative regular PCR grade 200 μL may also be used.

(d) The pGEM vector features a polylinker flanked by M13 sites. You may use this feature to select plasmids that were recombined with the insert in a desirable manner. To do so, select plasmids by PCR using primers M13 forward and the reverse primer used to generate your insert.

Always select T7 to SP6 RNA polymerase ligation site insert oriented plasmids, this avoids confusion during synthesis of the RNA probe. Because of our selection T7 RNA polymerase always synthesizes Sense (or control) probe and SP6 RNA Polymerase always produce the antisense (detection) probe.

6.3 Miniprep Protocol

(a) Use 50 mL tubes as they provide better oxygenation, which will promote growth. This process can be scaled up. Most importantly, avoid letting your culture growth for too long.

(b) Two tubes may be used simultaneously to pellet down the bacterial cells, reducing the need for a counterweight in centrifugations and speeding the process, the two pellets can be joined at the resuspension step.

(c) It is best to leave some supernatant than to risk transferring of any the white precipitate. If it is too difficult to do so centrifuge again.

(d) It is of primordial importance to not disturb the phases and only transfer the aqueous upper phase, it is best to lose some than to contaminate your plasmid extract. If however you unfortunately transferred some of the other phases simply repeat **step 6** and 7.

6.4 RNA Probe Synthesis and Labeling

(a) Assume that you selected a plasmid that features an insertion with the orientation T7 to SP6. In this way SP6 polymerase will produce an antisense probe and T7 will produce the sense probe. Therefore, henceforth, to avoid confusion, we will use the term sense for the probe synthesized with T7 RNA polymerase and antisense to the probe synthesized with SP6 RNA polymerase. If you choose to do otherwise, it is of little importance; just adjust accordingly.

(b) The use of fast digest enzymes is of course optional. But it saves a lot of time to this already long procedure. To further speed and simplify the process a thermocycler may use to act as an incubator, simply perform the restriction on 200 μL tubes.

(c) Use a simple PCR purification kit (Like, Thermo K0701 "GeneJET PCR Purification Kit"), to remove enzymes and salts. Make sure to make the final elution with 20 μL of RNase-free H_2O. Run 1 μL in an agarose gel to check for nicked or supercoil plasmid, as both these forms can cause disruptions on the probe synthesis process.

(d) RNA synthesis quality should be verified by running 1 μL of the reaction next to the labeled control (Vial 5) in an agarose gel electrophoresis with ethidium bromide staining. A good RNA synthesis results in a simple band. If a smear is observed, the reaction was probably contaminated with RNase and/or the RNase Inhibitor was not added to the reaction mix. If multiple bands are visible, then most probably the plasmid linearization and purification step failed, please repeat steps 2-4 and check for nicked or coiled plasmid.

(e) RNA probes are easily degraded. To preserve the probe avoid thawing and freezing cycles; for a longer preservation consider making aliquots.

(f) It is very important to determine the labeling efficiency, as it may be inconsistent. Several times we observed different probe yields in technical replicates; furthermore, it is useful to determine the probe concentration. So that the correct amount is used in the hybridization solution.

(g) Spot immediately to avoid degradation of the probe; space the spots so that they do not overlap. It is best to have the scale and the test probe side by side for easy comparison. Figure 4 is an example of what the spot distribution should look like.

(h) The labeled control standard dilution will allow you to determine the relative concentration of your probe (*see* Fig. 4).

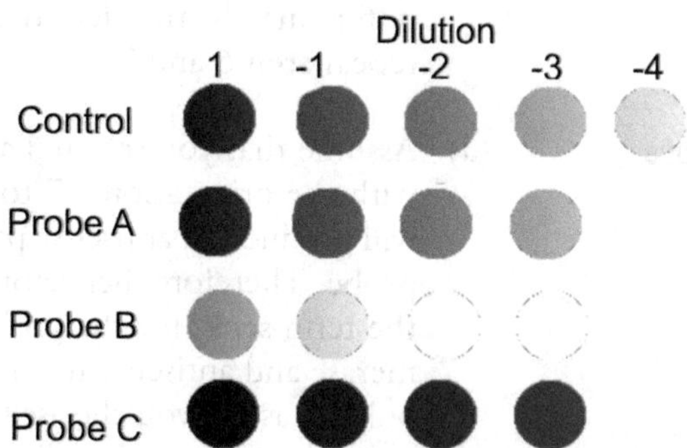

Fig. 4 Dot blot distribution example. This is an example of a dot blot comparing the five dilutions of the control probe with three test probes. The 5 tenfold dilutions of the control probe correspond to 1 μg/μL–0.1 μg/μL–0.01 μg/μL–0.001 μg/μL–0.0001 μg/μL, from left to right. This scale can be used to determine the approximate concentration of your labeled probe. If the fourth spot of your probe matches the color of the fourth spot of the control probe then your probe has a concentration of approximately 1 μg/μL, like for Probe A, but if by the time the higher dilution of the control starts to appear the last spot of your probe is still not visible, use the last spot for your probe and multiply by a factor of −10 fold for each missing spot. For Probe B if the first spot is similar to the fourth spot of the control then it has the relative concentration of 0.001 μg/μL. However, if like in the case of Probe C, the last spot is similar to a lower dilution of the control, a multiplication of tenfold must be applied for each spot to the left, in this case 1 mg/μL

6.5 Slide Preparation

(a) This step is especially important when reusing slides or if they seem to be dirty.

High-quality coating on the slides is of outmost importance, as it is what will insure that solutions do not spread and mix between wells. A good slide quality is one of the key issues for this technique. Slides can be cleaned and reused; just let them soak in a solution of ethanol 30% with some detergent for a couple of hours and gently brush them with a soft tooth brush, just prior to using this wash method. There are a lot of solution changes in this protocol. Although it is time-consuming to prepare the polylysine-coated slides, it will help significantly to keep the pistils in place [31].

(b) Polylysine coated slides can be prepared in advance, and if stored in a sealed rack at −20 °C they can be preserved for up to 1 year without losing their properties. Just let them reach room temperature before opening to avoid the accumulation of condensation. Always inspect the slides under the microscope, to check the coating and cleaning, before use.

(c) Some humidity must be present at the surface of the dissected pistils to promote adhesion to the polylysine-coated slides, but avoid transferring too much fixative solution as it will take a long time to dry. A good way to get rid of the excess moisture is to lightly touch with a lint-free tissue, before placing the pistils on the slide.

6.6 Dehydration--rehydration and mild enzymatic digestion of cell wall

(a) Going up the methanol dehydration series it is unfortunately very common for some material to loosen from the polylysine coating, please proceed with special care. If necessary do not hesitate to reposition the pistils in the reaction wells. When using higher concentrations of methanol take care not to let the solution evaporate. If necessary add more solution during incubation.

(b) Plant cell walls are the principal barrier for the penetration of the Probe. Partial digestion of them will not only facilitate the penetration of the probe but also will help with the removal off unlinked probe thus reducing nonspecific labeling.

(c) Make sure to store slides with pretreated/dried samples in small batches in clean air tight containers and also let them reach room temperature before opening the container, to avoid water condensation on samples. A cheap and easy way to prepare an airtight container is to put up to 4 slides in a 12 × 12 cm square Petri dish and seal it with a double layer of Parafilm.

6.7 Hybridization

(a) A humidity box is a versatile tool that can be purchased, or built, to maintain a saturated atmosphere around the slides avoiding the evaporation of solutions. We always build our humidity boxes by placing a few damp paper towels to bottom of a tip box. *See* Fig. 1 for how to build a simple reusable humidity chamber that also can be used for routine immunofluorescence.

(b) Use 20–100 ng of probe for each 1 mL of hybridization solution. Please note that some consideration and assays should be performed to ascertain the perfect amount of probe to use. Consider the expression level of the gene and the concentration of the probe. Assuming that your labeled probe has a concentration of near to 20 μg/μL, 1/20 or 1/50 ratio of probe to hybridization buffer is usually a good starting point to detect an averagely expressed gene. However, most probably you will need to make adjustments on the quantity of probe to use. If you do not get any labeling with this standard dilution, double the amount of probe until you do. Do the inverse if you get too much background with the antisense Probe.

(c) It is not unusual that some material may stick to the Parafilm®, make sure to check the Parafilm® before discarding. If you do find that some material did stick to the Parafilm® use your forceps to place it back in the well.

(d) This proportion is for genes with an average expression. When the signal is very weak the quantity of probe used in the hybridization buffer must be adjusted. Genes with low expression are always more difficult to detect.

(e) DAPI should only label the intact nucleus a bright blue and therefore will act as one of the controls for integrity of the tissue as well as control for the specificity of the probe, as they should not overlap. Always check your DAPI solution before staining your slides. It is not uncommon for the DAPI solution to degrade before the expected expiration date, and it is very difficult to wash out (*see* Fig. 5).

(f) Use DAPI and FITC filter setting to detect the signals. Plant cell walls have autofluorescence at almost every wavelength; for the best results use a confocal microscope if possible. Set the excitation Laser to 355 nm for DAPI and 488 nm for FITC. Emission filters between 450 and 460 nm for DAPI, and for FITC 515–525 nm. If you use an epifluorescence microscope set the filters to Excitation/Emission (nm) $^{358}/_{461}$ for DAPI and $^{485}/_{530}$ for FITC.

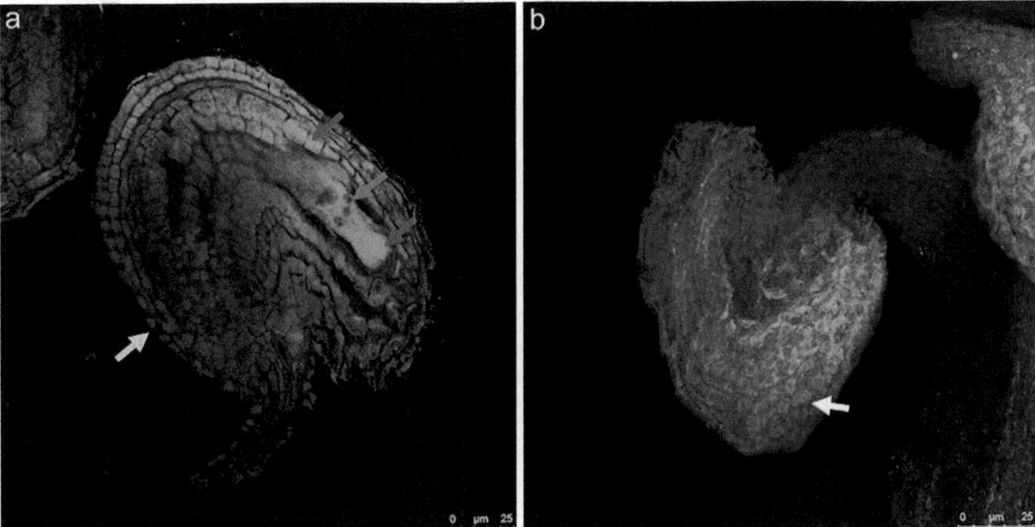

Fig. 5 Examples of results of confocal whole-mount FISH (fluorescent in situ hybridization) in Arabidopsis pistils. Examples of ovules after hybridization with the same AtAGP12 probe and counterstained with DAPI. On the left a perfectly preserved ovule was stained with a fresh solution of DAPI (**a**), on the right the ovule was not treated with paraformaldehyde and was stained with an old DAPI solution (**b**). The probe should only hybridize with the RNA present in the cell cytoplasm, giving a clear bright green image of the cells were the transcript is present (orange arrows), and should not be seen on the nucleus (pink arrow). DAPI stain is more than just a counterstain, it also serves to verify the conservation of the tissues. In perfectly preserved tissues the DAPI stain should be restricted to the cell's nucleus (yellow arrows), in poorly preserved or degenerated tissues DAPI stain spreads to the cytoplasm (white arrow) and it is very common for the probe to be detected along the cell wall (**b**). Spoiled DAPI solution will stain the cell walls (**b**)

Acknowledgments

This work was financed by FEDER through the COMPETE program, and by Portuguese National funds through FCT, Fundação para a Ciência eTecnologia (Project PTDC/AGR-GPL/115358/2009 and FCT - 02-SAICT-2017 – POCI-01-0145-FEDER-027839) and PhD grant SFRH/BD/111781/2015), and received support from Spanish–Portuguese Joint Project N° E 30/12. EU project 690946 'SexSeed' (Sexual Plant Reproduction – Seed Formation) funded by H2020-MSCA-RISE-2015.

References

1. Showalter AM (2001) Arabinogalactan-proteins: structure, expression and function. Cell Mol Life Sci 58:1399–1417
2. Pereira AM, Nobre MS, Pinto SC, Lopes AL, Costa ML, Masiero S, Coimbra S (2016) "Love is strong, and you're so sweet": JAGGER is essential for persistent synergid degeneration and polytubey block in *Arabidopsis thaliana*. Mol Plant 9:601–614
3. Costa ML, Sobral R, Ribeiro Costa MM, Amorim MI, Coimbra S (2015) Evaluation of the presence of arabinogalactan proteins and

pectins during *Quercus suber* male gametogenesis. Ann Bot 115:81–92
4. Pereira AM, Masiero S, Nobre MS, Costa ML, Solís MT, Testillano PS, Sprunck S, Coimbra S (2014) Differential expression patterns of arabinogalactan proteins in *Arabidopsis thaliana* reproductive tissues. J Exp Bot 65:5459–5471
5. El-Tantawy AA, Solís MT, Da Costa ML, Coimbra S, Risueño MC, Testillano PS (2013) Arabinogalactan protein profiles and distribution patterns during microspore embryogenesis and pollen development in *Brassica napus*. Plant Reprod 26:231–243
6. Coimbra S, Almeida J, Junqueira V, Costa ML, Pereira LG (2007) Arabinogalactan proteins as molecular markers in *Arabidopsis thaliana* sexual reproduction. J Exp Bot 58:4027–4035
7. Pilling E, Höfte H (2003) Feedback from the wall. Curr Opin Plant Biol 6:611–706
8. Knox JP (1997) The use of antibodies to study the architecture and developmental regulation of plant cell walls. Int Rev Cytol 171:79–120
9. Lopes AL, Costa ML, Sobral R, Costa MM, Amorim MI, Coimbra S (2016) Arabinogalactan proteins and pectin distribution during female gametogenesis in *Quercus suber* L. Ann Bot 117:949–961
10. Coimbra S, Costa M, Jones B, Mendes MA, Pereira LG (2009) Pollen grain development is compromised in Arabidopsis agp6 agp11 null mutants. J Exp Bot 60:3133–3142
11. Coimbra S, Costa M, Mendes MA, Pereira AM, Pinto J, Pereira LG (2011) Early germination of Arabidopsis pollen in a double null mutant for the arabinogalactan protein genes AGP6 and AGP11. Sex Plant Reprod 23:199–205
12. Xu Y, Gan ES, Ito T (2014) Misexpression approaches for the manipulation of flower development. Methods Mol Biol 1110:383–399
13. McDougall JK, Dunn AR, Jones KW (1972) In situ hybridization of adenovirus RNA and DNA. Nature 236:346–348
14. Javelle M, Marco CF, Timmermans M (2011) In situ hybridization for the precise localization of transcripts in plants. J Vis Exp 57:e3328
15. Biffo S, Tolosano E (1992) The use of radioactively labelled riboprobes for *in situ* hybridization: background and examples of application. Liver 12:230–237
16. Simmons D, Arriza J, Swanson L (1989) A complete protocol for in situ hybridization of messenger RNAs in brain and other tissues with radio-labelled single-stranded RNA probes. J Histotechnol 12:169–181
17. Komminoth P (1992) Digoxigenin as an alternative probe labeling for in situ hybridization. Diagn Mol Pathol 1:142–150
18. Höltke HJ, Ankenbauer W, Mühlegger K, Rein R, Sagner G, Seibl R, Walter T (1995) The digoxigenin (DIG) system for non-radioactive labelling and detection of nucleic acids – an overview. Cell Mol Biol 41:883–905
19. Gandrillon O, Solari F, Legrand C, Jurdic P, Samarut J (1996) A rapid and convenient method to prepare DIG-labelled RNA probes for use in non-radioactive in situ hybridization. Mol Cell Probes 10:51–55
20. Kiyama H, Emson PC (1991) An *in situ* hybridization histochemistry method for the use of alkaline phosphatase-labeled oligonucleotide probes in small intestine. J Histochem Cytochem 39:1377–1384
21. Testillano PS, Risueño MC (2009) Tracking gene and protein expression during microspore embryogenesis by confocal laser scanning microscopy. In: Touraev A, Forster BP, Mohan Jain S (eds) Advances in haploid production in higher plants. Springer, London, pp 339–347
22. Solís MT, Rodríguez-Serrano M, Meijón M, Cañal MJ, Cifuentes A, Risueño MC, Testillano PS (2012) DNA methylation dynamics and *MET1a*-like gene expression changes during stress-induced pollen reprogramming to embryogenesis. J Exp Bot 63:6431–6444
23. Solís MT, Berenguer E, Risueño MC, Testillano PS (2016) BnPME is progressively induced after microspore reprogramming to embryogenesis, correlating with pectin de-esterification and cell differentiation in *Brassica napus*. BMC Plant Biol 16:176
24. Tanious FA, Veal JM, Buczak H, Ratmeyer LS, Wilson WD (1992) DAPI (4′,6-diamidino-2-phenylindole) binds differently to DNA and RNA: minor-groove binding at AT sites and intercalation at AU sites. Biochemistry 31:3103–3112
25. Edwards K, Johnstone C, Thompson C (1991) A simple and rapid method for the preparation of plant genomic DNA for PCR analysis. Nucleic Acids Res 19:1349
26. Melton DA, Krieg PA, Rebagliati MR, Maniatis T, Zinn K, Green MR (1984) Efficient *in vitro* synthesis of biologically active RNA and RNA hybridization probes from plasmids containing a bacteriophage SP6 promoter. Nucleic Acids Res 18:7035–7056
27. Schenborn ET, Mierendorf RC Jr (1985) A novel transcription property of SP6 and T7

RNA polymerases: dependence on template structure. Nucleic Acids Res 17:6223–6236

28. Feliciello I, Chinali G (1993) A modified alkaline lysis method for the preparation of highly purified plasmid DNA from *Escherichia coli*. Anal Biochem 212:394–401
29. Longin A, Souchier C, French M, Bryon PA (1993) Comparison of anti-fading agents used in fluorescence microscopy: image analysis and laser confocal microscopy study. J Histochem Cytochem 41:1833–1840
30. Osborn M, Weber K (1982) Immunofluorescence and immunocytochemical procedures with affinity purified antibodies: tubulin-containing structures. Methods Cell Biol 24:97–132
31. Sitterley G (2008) Poly-L-lysine cell attachment protocol. BioFiles 3:8–12

Chapter 24

Localization, Extraction, and Quantification of Plant and Algal Arabinogalactan Proteins

Reina J. Veenhof and Zoë A. Popper

Abstract

Arabinogalactan proteins are a diverse group of cell wall-associated proteoglycans. While structural and molecular genetic analyses have contributed to the emerging improved understanding of the wide-range of biological processes in which AGPs are implicated; the ability to detect, localize, and quantify them is fundamentally important. This chapter describes three methods: histological staining, radial gel diffusion, and colorimetric quantification, each of which utilize the ability of Yariv reagent to bind to AGPs.

Key words Yariv reagent, Arabinogalactan proteins, Radial gel diffusion, Semiquantitative detection, Arabinogalactan protein detection

1 Introduction

One of the defining features of plants and algae is that, with some exceptions, their cells are surrounded by a cell wall consisting of structural polysaccharides and associated proteins. One class of proteins, the arabinogalactan proteins (AGPs), have been found in algal, moss, fern, and flowering plant cells walls and are strongly implicated in developmental processes [1–15]. However, they appear to have many functions. For example being involved in root–microbe interactions [16] and adaptation to abiotic stresses such as high-salt [17]. Further complexity in elucidating the function of AGPs is that microarray analysis suggests that it is likely that different AGPs have different functions [18–21].

AGPs are proteoglycans consisting of two distinct moieties, the carbohydrate and the protein domain. The carbohydrate component typically accounts for 90–98% of an AGP by weight and is rich in arabinose and galactose residues. The protein moiety, accounting for ~10% of an AGP by weight is hydroxyproline-rich [7]. However, there is a wide range of variability in the structure and composition of AGPs such that they are frequently classified as being either classical or non-classical AGPs [19–25]. Classical AGPs contain a

Zoë A. Popper (ed.), *The Plant Cell Wall: Methods and Protocols*, Methods in Molecular Biology, vol. 2149, https://doi.org/10.1007/978-1-0716-0621-6_24, © Springer Science+Business Media, LLC, part of Springer Nature 2020

hydrophobic transmembrane domain which in mature AGPs is replaced by a glycosylphosphatidylinositol (GPI) lipid anchor [20, 24, 26–28]; this domain appears to be lacking from the non-classical AGPs [20]. Non-classical AGPs also tend to be less heavily glycosylated [29]. Additionally, differences in AGP composition have been discovered that can be related to terrestrial plant taxonomy, for example bryophyte AGPs differ structurally from those of seed plants [30]. Bryophyte AGPs also differ in their composition from those found in flowering plants as they contain the sugar residue 3-*O*-methyl-L-rhamnose which appears to be absent from flowering plant AGPs [31].

AGPs are widespread in land plants [32] and more recently have been found in the extracellular matrix and cell walls of freshwater green algae [11], including *Micrasterias* [33] and *Chara corallina* [34], in green seaweeds including *Codium fragile* [35] and *Ulva* sp. (authors unpublished results), some red seaweeds (authors unpublished results), and several species of brown seaweeds [8].

There are several methods, both in situ and ex situ, which can be used to detect, quantify and localize AGPs; each method has been essential to our understanding of AGP structure, function, and localization. Each of the methods has some advantages and disadvantages, caused by the structural and compositional diversity present within both the protein and carbohydrate components of AGPs, as well as the presence of co-occurring compounds such as plant and algal pigments that can obscure staining.

Immunocytochemistry, as described by Hervé et al. [36] in Chapter 20, and specifically for AGPs, as carefully detailed by Castilleux et al. in Chapter 22 of this volume [37], is one of the most informative methods currently available for the detection and localization of AGPs within plant tissues. Furthermore, there are several monoclonal antibodies including; CCRC-M7 [28], JIM4 [29–41], JIM13-16 [1, 41], LM2 [41, 42], LM14 [43], and MAC207 [32, 41] which are known to recognize specific epitopes present in AGPs and which are commercially available (Biosupplies, Australia; CarboSource, http://www.ccrc.uga.edu/~carbosource/CSS_mabs7-07.html; PlantProbes, http://www.plantprobes.net/). This epitope specificity means that it is likely that a specific monoclonal antibody may recognize a specific AGP or a group of AGPs containing the same epitope. However, it is this specificity which currently poses some limitations because while there are several monoclonal antibodies which are known to detect and bind to epitopes present in AGPs it is likely that at present not all AGPs can currently be detected in this way. This problem will eventually be circumvented by the generation and characterization of a greater array of AGP-specific monoclonal antibodies making monoclonal antibody-based methods even more powerful for the investigation of AGP function.

Fluorescent in situ hybridization (FISH), is another elegant method that has been central to successfully determining the role (s) of individual AGPs for which the genes have been identified, and is described in detail by da Costa et al., in Chapter 23 [44] of this volume.

However, it can also be useful to apply more broad-spectrum methods of detection and several such methods for detecting the presence of AGPs have been developed. These methods nearly all employ a group of red-brown synthetic phenylglycoside dyes known as Yariv reagents, that bind to the β-1,3-galactan of AGPs [45] and are widely used in detecting and purifying AGPs. β-D-Glucosyl and β-D-galactosyl Yariv reagents bind to and precipitate AGPs whereas α D-galactosyl and α-D-mannosyl Yariv regents do not and are often used as controls [5, 46]. Yariv reagents can be used in both in situ and ex situ methods of AGP detection and quantification. They have also been a key factor enabling determination of the function of AGPs [2, 46]. A method for in situ Yariv staining of plant and algal tissues is detailed in Subheading 2 and is widely used [47–51]. However, it has the disadvantage that staining may be obscured if the plant or algal tissues are deeply pigmented or similarly colored (red-brown) to Yariv reagent. Colorimetric methods have been used to determine the concentration of AGPs present in plant tissues [52, 53]. This has the advantage that it is quantitative. However, it is not known whether some AGPs are capable of binding greater (or lesser) concentrations of AGPs than others and it is suggested that non-classical AGPs may display variable binding to β-D-glucosyl Yariv reagent [54]. What has been found is that for binding between Yariv reagent and an AGP to occur a certain length of consecutive (1,3)-β-linked galactose units is necessary [55]. Furthermore Yariv reagent is known to behave differently in different solvents, forming larger aggregates of 300 U in water compared to aggregates of 150 U in 1% (w/v) NaCl [55]. A further problem associated with in situ staining and the in situ colorimetric method is that Yariv reagents also bind to cellulose [56] giving a slightly coloured background making very low concentrations of AGPs less easy to determine and localize. Therefore, while it has the disadvantage that information regarding tissue-specific localization is lost it can be informative in some cases to extract, concentrate, and detect AGPs using Yariv in techniques. The radial gel diffusion assay developed by Van Holst and Clarke [57] and described below enables detection of low concentrations of AGPs in a cellulose free environment. The method is semi-quantitative by comparison with a dilution series of gum arabic and tissue specificity can be partially accommodated by carefully selecting the tissues prior to extraction and has been used to determine which fractions of plant extracts contain AGPs [58, 59].

2 Materials

1. 10 g plant or algal tissue of interest.
2. Extraction buffer: 50 mM Tris–HCl, pH 8, 10 mM EDTA, 0.1% v/v β-mercaptoethanol, 1% w/v Triton X-100 (*see* **Note 1**).
3. Ethanol.
4. 50 mM Tris–HCl, pH 8.
5. Bench centrifuge.
6. 1% w/v NaCl.
7. 0.15 M NaCl.
8. Freeze drier.
9. Agarose gel containing β-D-glucosyl Yariv reagent: 1% w/v agar, 0.02% w/v β-glucosyl Yariv reagent (commercially available from Biosupplies Australia, Victoria, Australia or they can be synthesized [46, 60–62] as described by Blaukopf et al. [62]), 0.15 M NaCl, 0.02% w/v sodium azide (*see* **Notes 2** and **3**).
10. Agarose gel containing α-D-galactosyl Yariv reagent: 1% w/v agar, 0.02% w/v α-galactosyl Yariv reagent (Biosupplies Australia, Victoria, Australia) (*see* **Note 4**), 0.15 M NaCl, 0.02% w/v sodium azide.
11. 2 mg/mL β-D-glucosyl Yariv reagent in 0.15 M NaCl.
12. 2 mg/mL α-D-galactosyl Yariv reagent in 0.15 M NaCl.
13. Autoclave.
14. Petri dishes.
15. 4 mg/mL gum arabic.
16. Parafilm®.
17. Aluminum foil.
18. Scanner or digital camera.
19. Automatic pipettes.
20. 50 and 50 mL screw-capped centrifuge tubes.
21. Magnetic stirrer bar and stirring plate.
22. Glass Pasteur pipette or core borer.
23. Mortar and pestle.
24. Liquid nitrogen.
25. Freezer: −20°C.
26. Microscope slides.
27. Cover slips.
28. Microscope with digital camera attachment.

29. White ceramic tile.
30. Backed razor blade.
31. Rocking platform.
32. 1.5 mL Eppendorf tubes.
33. 2 mg/mL stock solution of gum arabic in 1% w/v $CaCl_2$.
34. 1% w/v $CaCl_2$.
35. 1 mg/mL β-D-galactosyl-Yariv reagent in 2% w/v $CaCl_2$.
36. Vortex.
37. 2% w/v $CaCl_2$.
38. 20 mM NaOH.
39. 1.5 mL cuvettes.
40. Spectrophotometer.
41. Statistics software (SPSS or Excel).

3 Methods

3.1 In Situ Staining of AGPs Using Yariv Reagents [48–51]

1. Cut sections of the plant or algal tissue of interest. We find that careful hand-sectioning with a backed razor-blade on a white ceramic tile works well (*see* **Note 5**).
2. Incubate the plant materials in an Eppendorf tube containing 1 mL of 2 mg/mL β-D-glucosyl Yariv reagent in 0.15 M NaCl for 1 h, at room temperature, on a rocking platform.
3. As a control incubate a duplicate set of plant materials (e.g., sections from the same plant tissues in 1 mL of 2 mg/mL α-D-galactosyl Yariv reagent in 0.15 M NaCl under the same conditions).
4. After 1 h carefully remove the Yariv reagent taking care to damage the plant materials (*see* **Note 6**).
5. Add 1 mL of 0.15 M NaCl to each of the Eppendorf tubes and incubate, at room temperature for 5 min.
6. **Steps 4** and **5** should be repeated several times until the 0.15 M NaCl is no longer appreciably red.
7. Transfer the sections to microscope slides, examine and record using a light microscope with a digital camera attachment.

3.2 Extraction of Arabinogalactan Proteins [63]

1. Grind 10 g of plant material to a fine power in liquid nitrogen in a mortar and pestle (*see* **Note** 7).
2. Add 10 mL of extraction buffer (50 mM Tris–HCl, pH 8, 10 mM EDTA, 0.1% v/v β-mercaptoethanol, 1% w/v Triton X-100) to the plant material and incubate at 4°C for at least 3 h.
3. Centrifuge for 10 min at 4000 × *g*.

4. Carefully remove the supernatant using a Pasteur pipette and precipitate polysaccharides and glycoproteins with 5 volumes of ethanol, at 4°C for at least 16 h.
5. Centrifuge for 2 min at 2000 × *g*.
6. Carefully remove the supernatant taking care not to disturb the pellet (*see* **Note 8**).
7. Resuspend the pellet in 5 mL of 50 mM Tris–HCl, pH 8.
8. Centrifuge for 10 min at 4000 × *g*.
9. Collect the supernatant into a polypropylene tube.
10. Resuspend the remaining pellet in 5 mL 50 mM Tris–HCl, pH 8.
11. Centrifuge for 10 min at 4000 × *g*.
12. Carefully remove the supernatant and pool it with that collected in **step 9**.
13. Freeze and freeze dry the supernatant.
14. Dissolve the dried supernatant in 500 μL 1% w/v NaCl (*see* **Note 9–11**).

3.3 Detection of Arabinogalactan Proteins by Radial Gel Diffusion [57]

Extracts using methods including the one described above may be used to detect the presence of AGPs in plant materials. It is semi-quantitative if a serial dilution of a known source of AGPs (e.g., gum arabic) is included in the assay for comparison. A degree of tissue specificity is enabled by carefully selecting the plant materials from which the extracts are made. Additional samples that can be investigated for the presence and concentration of AGPs include the culture medium of plant tissue cultures and cultured algae [64]. Some AGPs are secreted into the culture medium [25, 64] which can be filtered, to remove any cell debris, and then freeze-dried prior to use in the assay.

1. Use the end of a glass Pasteur pipette or a core borer to cut out wells in the agarose gel containing β-D-glucosyl Yariv reagent.
2. Into one well load 20–50 μL of 1% w/v NaCl (*see* **Note 12**).
3. Into another well load a known amount of gum arabic. If you are investigating whether AGPs are present or not then 20–40 μL of a 4 mg/mL solution of gum arabic is suitable. However, if you are trying to quantify the concentration of AGPs in the extract or for a given quantity of plant tissue then it is advisable to include several wells on the plate which are loaded with a dilution series of gum arabic starting with 2–4 mg/mL. The minimum concentration of gum arabic that will give a positive result using this method is 0.25 mg/mL.
4. Load your extract prepared as described in Subheading 3.1 into remaining wells. It can be useful to load your extracts into two

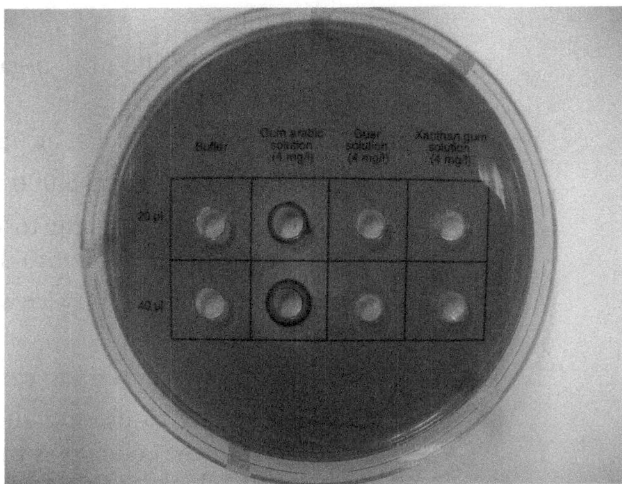

Fig. 1 Radial gel diffusion assay. A and B are 20- and 40-μl loadings of 0.15 M NaCl and C and D are 20- and 40-μL loadings of a solution of 4 mg/mL gum arabic in 0.15 M NaCl

different wells at two different loadings, that is, in one well load twice the volume that you load in another well (*see* **Note 13**).

5. Seal plates with Parafilm® to prevent them drying out.
6. Store the plates at room temperature, in darkness, for at least 48 h (*see* **Notes 14** and **15**).
7. After 48 h the results can be recorded either by scanning the plate or by taking a photograph using a digital camera (*see* **Note 16**). An example is shown in Fig. 1.

3.4 Quantification of Arabinogalactan Proteins by Colorimetric Analysis [53]

The quantity of AGPs present in culture media [25, 64], and in extracts prepared from plant and algal cells as described above (*see* Subheading 3.2) can be determined by the following colorimetric assay adapted from Lamport 2013 [53]. This method uses β-D-galactosyl-Yariv reagent to precipitate the AGPs present in sample (s), such as culture media or extracts. The absorbance of AGP-Yariv complex is measured at 440 nm and the quantity of AGPs determined via extrapolation against a standard curve of known quantities of AGPs. A serial dilution of gum arabic is typically used to prepare the standard curve. This method assumes that a given weight of Yariv reagent will precipitate a specific weight of AGPs.

Preparation of a standard curve:

1. Use a 2 mg/mL stock solution of gum arabic in 1% w/v $CaCl_2$ to prepare a dilution series of gum arabic to prepare a standard curve. Add 500 μL, 250 μL, 125 μL gum arabic stock to three 1.5 mL Eppendorf tubes and make up to 500 μL with 1% w/v $CaCl_2$ (*see* **Notes 17** and **18**).
2. Add 1.5 mL 500 μL 1% w/v $CaCl_2$ to two 1.5 mL Eppendorf tubes to use as a blank for spectrophotometry.

3. Label all tubes.
4. To each tube add 200 μL β-D-galactosyl-Yariv reagent, 1 mg/mL in 2% w/v $CaCl_2$.
5. Mix well using a vortex and incubate at 4 °C for at least 30 min.
6. Spin all tubes 15 min at 15,000 × *g* in a microfuge.
7. Use a fine tip glass pipette to carefully remove and discard supernatant, taking care not to disturb the pellet.
8. Carefully wash the pellets twice with 1 mL 2% w/v $CaCl_2$ using a fine tip glass pipette.
9. Add 1 mL 20 mM NaOH to each tube.
10. Shake vigorously to dissolve the pellet or vortex quickly for 10 seconds. If the solution is turbid, spin at 15,000 × *g* in a microfuge to clarify.
11. Carefully transfer the contents of each tube into a 1.5 mL cuvette and use a spectrophotometer to read at the absorbance at 440 nm. Use 1% $CaCl_2$ as the blank. This must be done within an hour of resuspending the pellet.
12. A standard curve can now be made plotting absorbance on the *Y* axis against the quantity of gum arabic (mg/mL) on the *X* axis (*see* **Note 19**).
13. A formula for the standard curve can be retrieved using linear regression, where *y* is the concentration of gum arabic in mg/mL in each sample and *x* is the absorbance of the sample measured at 440 nm (*see* **Note 20**).

Quantification of arabinogalactan proteins from culture media and plant/algal extracts:

1. Add 10, 50 and 100 μL of the sample for which you want to determine the quantity of AGPs to 1.5 mL Eppendorf tubes and make up to 500 μL with 1% w/v $CaCl_2$ (*see* **Note 18**).
2. Add 1.5 mL of 500 μL 1% $CaCl_2$ to two 1.5 mL Eppendorf tubes to use as a blank.
3. Next carry out **steps 3–11** in the same way as for the preparation of the standard curve detailed above.
4. The standard curve prepared using gum arabic (as described above) is used to calculate the quantity of AGPs (in mg/mL) present in the samples based on their absorbance at 440 nm.

4 Notes

1. The extraction buffer generally requires at least 2 h stirring to sufficiently dissolve the Triton X-100.
2. Caution, sodium azide is highly toxic. Gloves should be worn.

3. Inclusion of sodium azide means that sterile conditions are not required. The AGP extracts contain a high sugar content and could quickly become contaminated with microbes. However, heat sterilization could be damaging to the AGPs, and filter sterilization is not practical owing to the viscosity of many of the extracts.
4. α-D-Mannosyl Yariv reagent (Biosupplies, Australia Pty Ltd) can be used in place of α-D-galactosyl Yariv reagent.
5. Micro algae and other plant materials can also be surface-labelled using the same procedure but in this case are not sectioned prior to incubation.
6. We often find that some very thin hand-sections can be damaged if bathing solutions are removed too rapidly. One way to minimize this is by using a glass Pasteur pipette which has been heated and stretched so that the opening is narrower than normal. This also helps prevent uptake into the glass Pasteur pipette which may occur if sections or plant materials that are being stained are either very thin and/or very small.
7. In order to protect the mortar and pestle from possible damage due to sudden temperature changes it is best to cover the mortar and pestle in cling film/Saran wrap® and place in a −20°C freezer for at least 2 h prior to use.
8. One of the easiest ways to do this is by using a Pasteur pipette which has been stretched over a flame to reduce the width of the tip.
9. It is useful to weigh the freeze-dried sample so that the mg/mL of dissolve extract used in radial gel diffusion can be calculated.
10. Some of the extracts may have a high sugar content and fail to freeze at −20°C, or they once frozen they thaw rapidly. In this case the extracts may be frozen at −80 °C or alternatively can be dialyzed against distilled water, at 4°C, to remove any sugars with a low degree of polymerization (DP). To prevent microbial growth in the extract it is recommended that dialysis is against 0.05% w/v chlorobutanol (1,1,1-trichloro-2-methyl-2-propanol) which is volatile so will be removed on freeze drying [64].
11. If at this concentration the solution is extremely viscous or if there is a large amount (greater than 0.2 g) of freeze-dried extract it is advisable to dissolve the sample in a larger volume of 1% w/v NaCl.
12. If your samples are not dissolved in 1% w/v NaCl then load the buffer that they were dissolved in instead of the 1% w/v NaCl.
13. It is essential that each well on the plate is clearly labelled. We find this easiest to do by printing a grid of 1.2 × 1.2 cm squares onto an acetate sheet which can be stuck to the back of the

Petri dish. Each well is then punched into the centre of a square.

14. Darkness can be achieved by wrapping the plates in aluminum foil.
15. The plates can be stored for several weeks, even in the light, at room temperature provided that they are sealed with Parafilm® and do not dry out.
16. It takes ~48 h for the AGPs to diffuse into the gel and the Yariv reagent to bind to and precipitate them.
17. To make the standard curve more precise, the dilution series can be extended by preparing tubes with 62.5 μL of gum arabic and so on.
18. It is advisable, if possible, to prepare samples in at least triplicate, to improve accuracy.
19. When using duplicates to calculate a standard curve, use the average absorbance values for each known amount of AGP.
20. This is most easily achieved when using statistics software such as SPSS, but Excel is also able to calculate linear regressions.

Acknowledgments

Quentin Coster, a visiting MSc student from the Université Catholique de Louvain, Belgium, and Tanya Slattery, an undergraduate project student in Botany and Plant Science at NUI Galway, are thanked for their technical assistance.

References

1. Knox JP, Linstead PJ, Peart J, Cooper C, Roberts K (1991) Developmentally-regulated epitopes of cell surface arabinogalactans-proteins and their relation to root tissue pattern formation. Plant J 1:317–326
2. Lee KJD, Sakata Y, Mau S-L, Pettolino F, Bacic A, Quatrano RS, Knight CD, Knox JP (2005) Arabinogalactan proteins are required for apical cell extension in the moss *Physcomitrella patens*. Plant Cell 17:3051–3065
3. Basile DV, Kushner BK, Basile MR (1989) A new method for separating and comparing arabinogalactan proteins in the hepatica. Bryologist 90:401–404
4. Basile DV (1980) A possible mode of action for morphoregulatory hydroxyproline-proteins. Bull Torrey Bot Club 107:325–338
5. Seifert GJ, Roberts K (2007) The biology of arabinogalactan proteins. Annu Rev Plant Biol 58:137–161
6. Seifert GJ, Barber C, Wells B, Roberts K (2004) Growth regulators and the control of nucleotide sugar flux. Plant Cell 16:723–730
7. Fincher GG, Stone BA, Clarke AE (1983) Arabinogalactan-proteins: structure, biosynthesis and function. Annu Rev Plant Physiol 34:47–70
8. Hervé C, Siméon A, Jam M, Cassin A, Johnson KL, Salmeán AA, Willats WG, Doblin MS, Bacic A, Kloareg B (2016) Arabinogalactan proteins have deep roots in eukaryotes: identification of genes and epitopes in brown algae and their role in *Fucus serratus* embryo development. New Phytol 209:1428–1441
9. da Costa M, Periera AM, Pinto SC, Silva J, Pereira JG, Coimbra S (2019) In silico and expression analyses of Fasciclin-like arabinogalactan proteins reveal functional conservation during embryo and seed development. Plant Reprod 32:353–370

10. Periera AM, Lopes AL, Coimbra S (2016) Arabinogalactan proteins as interactors along the crosstalk between the pollen tube and the female tissues. Front Plant Sci 7:1895
11. Palacio-López K, Tinaz B, Holzinger A, Domozych DS (2019) Arabinoglactan proteins and the extracellular matrix of charophytes: a sticky business. Front Plant Sci 10:447
12. Lopez RA, Renzaglia KS (2014) Multiflagellated sperm cells of *Ceratopteris richardii* are bathed in arabinogalactan proteins throughout development. Am J Bot 101:2052–2061
13. Periera AM, Lopes AL, Coimbra S (2016) *JAGGER*, an AGP essential for persistent synergid degeneration and polytubey block in Arabidopsis. Plant Signal Behav 11(8):e1209616
14. Lopez RA, Mansouri K, Henry JS, Flowers ND, Vaughn KC, Renzaglia KS (2017) Immunogold localisation of molecular constituents associated with basal bodies, flagella, and extracellular matrices in male gametes of land plants. Bio-Protocol 7(21):e2599
15. Renzaglia KS, Villareal JC, Piatkowski BT, Lucas JR, Merced A (2017) Hornwort stomata: architecture and fate shared with 400-million year old fossil plants without leaves. Plant Physiol 174:788–797
16. Nguema-Ona E, Vicré-Gibouin M, Cannesan M-A, Driouich A (2013) Arabinogalactan proteins in root-microbe interactions. Trends Plant Sci 10:440–449
17. Olmos E, de la Garma J, Gomez-Jimenez MC, Fernandez-Garcia N (2017) Arabinogalactan proteins are involved in salt adaptation and vesicle trafficking in tobacco BY-2 cell cultures. Front Plant Sci 8:1092
18. Schultz CJ, Rumsewicz MP, Johnson KL, Jones BJ, Gaspar YM, Bacic A (2002) Using genomic resources to guide research directions. The arabinogalactan protein gene family as a test case. Plant Physiol 129:1448–1463
19. Gaspar Y, Johnson KL, McKenna JA, Bacic A, Schultz CJ (2001) The complex structures of arabinogalactan-proteins and the journey towards understanding function. Plant Mol Biol 47:161–176
20. Majewska-Sawka A, Nothnagel EA (2000) The multiple roles of arabinogalactan proteins in plant development. Plant Physiol 122:3–9
21. Kieliszewski MJ, Lamport DTA (1994) Extensin: repetitive motifs, functional sites, post-translational codes, and phylogeny. Plant J 5:157–172
22. Nothnagel EA, Bacic A, Clarke AE (2000) Cell and developmental biology of Arabinogalactan-proteins. Kluwer Academic, New York, NY
23. Nothnagel EA (1997) Proteoglycans and related components in plant cells. Int Rev Cytol 174:195–291
24. Du H, Simpson RJ, Clarke AE, Bacic A (1996) Molecular characterization if a stigma-specific gene encoding an arabinogalactan protein (AGP) from *Nicotiana alata*. Plant J 9:313–323
25. Mau SL, Chen CG, Pu ZY, Moritz RL, Simpson RJ, Bacic A (1995) Molecular cloning of cDNAs encoding the protein backbones for arabinogalactan proteins from the filtrate of suspension-cultured cells of *Pyrus communis* and *Nicotiana alata*. Plant J 8:269–281
26. Oxley D, Bacic A (1999) Structure of the glycosylphosphatidylinositol membrane anchor of an arabinogalactan protein from *Pyrus communis* suspension-cultured cells. Proc Natl Acad Sci U S A 6:14246–14251
27. Svetek J, Yadav MP, Nothnagel EA (1999) Presence of a glycosylphosphatidylinositol lipid anchor on rose arabinogalactan proteins. J Biol Chem 274:14724–14733
28. Youl JJ, Bacic A, Oxley D (1998) Arabinogalactan-proteins from *Nicotiana alata* and *Pyrus communis* contain glycosylphosphatidylinositol membrane anchors. Proc Natl Acad Sci U S A 95:7921–7926
29. Showalter AM (2001) Arabinogalactan proteins: structure, expression and function. Cell Mol Life Sci 58:1399–1417
30. Bartels D, Baumann A, Maeder M, Geske T, Heise EM, von Schwartzenberg K, Claussen B (2017) Evolution of plant cell walls: arabinogalactan proteins from three moss genera show structural differences compared to seed plants. Carbohydr Polym 163:227–235
31. Fu H, Yadav MP, Nothnagel EA (2007) *Physcomitrella patens* arabinogalactan proteins contain abundant terminal 3-*O*-methyl-L-rhamnosyl residues not found in Angiosperms. Planta 226:1511–1524
32. Pennell RI, Knox JP, Scofield GN, Selvendran RR, Roberts K (1989) A family of abundant plasma membrane-associated glycoproteins related to the arabinogalactan proteins is unique to flowering plants. JCB 108:1967–1977
33. Eder M, Tenhaken R, Driouich A, Lütz-Mendel U (2008) Occurrence and characterization of arabinogalactan-like proteins and hemicelluloses in *Micrasterias* (Streptophyta). J Phycol 44:1221–1234
34. Domozych DS, Sørensen I, Willats WGT (2009) The distribution of cell wall polymers during antheridium development and

spermatogenesis in the Charophycean green alga *Chara corallina*. Ann Bot 104:1045–1056

35. Estevez JM, Fernández PV, Kasulin L, Dupree P, Ciancia M (2009) Chemical and *in situ* characterization of macromolecular components of the cell walls from the green seaweed *Codium fragile*. Glycobiology 19:212–228
36. Duffieux D, Marcus SE, Knox JP, Hervé C (2011) Monoclonal antibodies, carbohydrate-binding modules, and the detection of polysaccharides in cell walls from plants and marine algae. In: Popper ZA (ed) Methods in molecular biology. The plant cell wall – methods and protocols. Humana, Totowa, NJ
37. Castilleux R, Ropitaux M, Manasfi Y, Bernard S, Vicré-Gibouin M, Driouich A (2011) Contributions to arabinogalactan proteins analysis. In: Popper ZA (ed) Methods in molecular biology. The plant cell wall – methods and protocols. Humana, Totowa, NJ
38. Steffan W, Kovác P, Albersheim P, Darvill AG, Hahn MG (1995) Characterization of a monoclonal antibody that recognizes an arabinosylated (1→6)-β-D-galactan epitope in plant complex carbohydrates. Carbohydr Res 275:295–307
39. Knox JP, Day S, Roberts K (1989) A set of cell wall surface glycoproteins forms a marker of cell position, but not cell type, in the root apical meristem of *Daucus carota* L. Development 106:47–56
40. Stacey NJ, Roberts K, Knox JP (1990) Patterns of expression of JIM4 arabinogalactan protein epitope in cell cultures and during somatic embryogenesis in *Daucus carota* L. Planta 180:285–292
41. Yates EA, Valdor JF, Haslam SM, Morris HR, Dell A, Mackie W, Knox JP (1996) Characterization of carbohydrate structural features recognised by anti-arabinogalactan-protein monoclonal antibodies. Glycobiology 6:131–139
42. Smallwood M, Yates EA, Willats WGT, Martin H, Knox JP (1996) Immunochemical comparison of membrane-associated and secreted arabinogalactan-proteins in rice and carrot. Planta 198:452–459
43. Moller I, Marcus SE, Haeger A, Verhertbruggen Y, Verhoef R, Schols H, Mikkelsen JD, Knox JP, WGT W (2008) High-throughput screening of monoclonal antibodies against plant cell wall glycans by hierarchical clustering of their carbohydrate microarray binding profiles. Glycoconj J 25:37–48
44. da Costa ML, Solís M-T, Testillano PS, Coimbra S (2011) *In situ*/subcellular localisation of arabinogalactan protein expression by fluorescent *in situ* hybridization, FISH. In: Popper ZA (ed) Methods in molecular biology. The plant cell wall – methods and protocols. Humana, Totowa, NJ
45. Kitazawa K, Tryfona T, Yoshimi Y, Hayashi Y, Kawauchi S, Antonov L, Tanaka H, Takahashi T, Kaneko S, Dupree P, Tsumurava Y, Kotake T (2013) β-Galactosyl Yariv reagent binds to the β-1,3, galactan of arabinogalactan proteins. Plant Physiol 161:1117–1126
46. Yariv M, Rapport MM, Graf L (1962) The interaction of glycosides and saccharides with antibody to corresponding phenyl glycosides. Biochem J 85:383–388
47. Willats WGT, Knox JP (1996) A role for arabinogalactan-proteins in plant cell expansion: evidence from studies on the interaction of β-D-glucosyl Yariv reagent with seedlings of *Arabidopsis thaliana*. Plant J 9:919–925
48. Anderson RL, Clarke AE, Jermyn MA, Knox RB, Stone BA (1977) A carbohydrate binding arabinogalactan-protein from liquid suspension cultures of *Lolium multiflorum* endosperm. Aust J Plant Physiol 4:143–158
49. Clarke AE, Anderson AL, Stone BA (1979) Form and function of arabinogalactans and arabinogalactan proteins. Phytochemistry 18:521–540
50. Gleeson PA, Clarke AE (1979) Structural studies on the major component of Gladiolus style mucilage, an arabinogalactan protein. Biochem J 181:607–621
51. Jermyn MA, Yeow M (1975) A class of lectins present in the tissues of seed plants. Aust J Plant Physiol 2:501–531
52. Lamport DTA, Kieliszewski MJ, Showalter AM (2005) Salt stress upregulates periplasmic arabinogalactan proteins: using salt stress to analyse AGP function. New Phytol 169:479–492
53. Lamport DT (2013) Preparation of arabinogalactan glycoproteins from plant tissue. Bioprotocol 3:1–5
54. Sommer-Knudsed J, Careke AE, Bacic A (1997) Proline- and hydroxyproline-rich gene products in sexual tissues of flowers. Sex Plant Reprod 10:253–260
55. Paulsen BS, Craik DJ, Dunstan DE, Stone BA, Bacic A (2014) The Yariv reagent: behaviour in different solvents and interaction with gum arabic arabinogalactan protein. Carbohydr Polym 106:460–468
56. Triplett BA, Timpa JD (1997) β-Glucosyl and α-galactosyl Yariv reagents bind to cellulose and other glucans. J Agric Food Chem 45:4650–4654

57. Van Holst G-J, Clarke AE (1985) Quantification of arabinogalactans-protein in plant extracts by single radial gel diffusion. Anal Biochem 148:446–450
58. Osman M, Menzies AR, Albo Martin B, Williams PA, Phillips GO, Baldwin TC (1995) Characterization of gum arabic fractions obtained by anion-exchange chromatography. Phytochemistry 38:409–417
59. Parveen S, Gupta AD, Prasad R (2006) Arabinogalactan protein from *Arachis hypogaea*: role as a carrier in drug formulations. Int J Pharm 333:78–86
60. Yariv JH, Lis E, Katchalshi E (1967) Precipitation of arabic acid and some seed polysaccharides by glycosyl phenylazo dyes. Biochem J 195:1c–2c
61. Basile DV, Ganjian I (2004) β-D-Glucosyl and α-D-galactosyl Yariv reagents: syntheses from *p*-nitrophenyl-D-glycosides by transfer reduction using ammonium formate. J Agric Food Chem 52:7453–7456
62. Blaukopf C, Krol MZ, Seifert GJ (2011) New insights into the control of cell growth. In: Popper ZA (ed) Methods in molecular biology. The plant cell wall – methods and protocols, vol 715. Humana, Totowa, NJ, pp 221–244
63. Schultz CJ, Johnson KL, Currie G, Bacic A (2000) The classical arabinogalactan protein gene family of Arabidopsis. Plant Cell 12:1751–1768
64. Fry SC (2000) The growing plant cell wall: chemical and metabolic analysis, Reprint edn. The Blackburn Press, Caldwell, NJ. [ISBN 1-930665-08-3]

Chapter 25

Plant Cell Wall Proteomes: Bioinformatics and Cell Biology Tools to Assess the *Bona Fide* Cell Wall Localization of Proteins

David Roujol, Laurent Hoffmann, Hélène San Clemente, Corinne Schmitt-Keichinger, Christophe Ritzenthaler, Vincent Burlat, and Elisabeth Jamet

Abstract

The purification of plant cell walls is challenging because they constitute an open compartment which is not limited by a membrane like the cell organelles. Different strategies have been established to limit the contamination by proteins of other compartments in cell wall proteomics studies. Non-destructive methods rely on washing intact cells with various types of solutions without disrupting the plasma membrane in order to elute cell wall proteins. In contrast, destructive protocols involve the purification of cell walls prior to the extraction of proteins with salt solutions. In both cases, proteins known to be intracellular have been identified by mass spectrometry in cell wall proteomes. The aim of this chapter is to provide tools to assess the subcellular localization of the proteins identified in cell wall proteomics studies, including: (1) bioinformatic predictions, (2) immunocytolocalization of proteins of interest on tissue sections and (3) in muro observation of proteins of interest fused to reporter fluorescent proteins by confocal microscopy. Finally, a qualitative assessment of the work can be performed and the strategy used to prepare the samples can be optimized if necessary.

Key words Bioinformatics, Cell wall protein, Fluorescent reporter protein, Immunocytolocalization, Microscopy, Plant, Proteomics, Subcellular localization

1 Introduction

The isolation of plant cell wall proteomes is a challenging task. Indeed, cell wall proteins (CWPs) are trapped within an extracellular matrix which is composed of intricate and interwoven networks of polysaccharides and structural proteins [1]. Besides, the plant cell wall is an open compartment not delimited by membranes as is the case for organelles such as chloroplasts or mitochondria. Therefore, the methods to isolate CWPs must be designed to limit the contamination of the extracellular matrix by intracellular proteins

Zoë A. Popper (ed.), *The Plant Cell Wall: Methods and Protocols*, Methods in Molecular Biology, vol. 2149, https://doi.org/10.1007/978-1-0716-0621-6_25, © Springer Science+Business Media, LLC, part of Springer Nature 2020

[2, 3]. This can be achieved either by preventing the leakage of the plasma membrane as in the so-called non-destructive methods [4], or by a careful purification of cell walls prior to protein extraction using different salt solutions as in destructive methods [5].

Discrimination between *bona fide* CWPs and intracellular contaminants is a major difficulty encountered in plant cell wall proteomics. Different methods have been employed to assess the quality of the protein extracts obtained by nondestructive or destructive methods that include measurement of intracellular enzymatic activities [6], and bioinformatic prediction of subcellular localization [7]. However, since the sensitivity of mass spectrometry techniques is much higher than those of biochemical and immunological tests, probably by 100- to 1000-fold, even minor intracellular contaminants can be revealed. Besides, although the vast majority of CWPs follow the secretory pathway starting with N-terminus signal peptide leader sequence-mediated endoplasmic reticulum (ER) targeting, the existence of alternative export mechanisms for leaderless proteins cannot be excluded. The latter noncanonical export pathway has been clearly demonstrated in animal cells [8], but so far only applies to a single plant protein, a sunflower jacalin [9]. This protein contains no leader sequence and no post-translational modification specific for the ER or the Golgi apparatus and its secretion remains unaffected by brefeldin A, a drug known to interfere with the ER-to-Golgi transport. Finally, some moonlighting proteins were described to be targeted to two different cellular compartments, like the rice α-amylase found in both the chloroplasts and the extracellular matrix [10]. These examples illustrate the difficulty to distinguish between *bona fide* CWPs and intracellular contaminants in protein extracts enriched in CWPs.

Since the beginning of plant cell wall proteomic studies, many method improvements have enabled the number of identified proteins to be increased while reducing the proportion of intracellular contaminants. Some cell wall proteomic studies are listed in Table 1 and grouped according to the method used for protein isolation, that is, destructive versus non-destructive. Based on bioinformatic predictions, the percentage of proteins predicted to be secreted varies between 31% and 93% in the case of non-destructive methods, and between 13% and 79% in the case of destructive methods (Table 1). This variability is difficult to correlate with the protein isolation procedure, the origin of plant material or the plant species. Although the non-destructive methods should theoretically allow a more efficient recovery of proteins predicted to be secreted, this is not necessarily the case, since several studies have led to the identification of at most 50% of such proteins (Table 1).

All these difficulties to clearly distinguish between *bona fide* CWPs and intracellular contaminants prompted us to check the subcellular localization of identified proteins by cell biology in

Table 1
Proteins predicted to be secreted versus proteins predicted to be intracellular in some plant cell wall proteomes

Plant species	Plant material	Total number of identified proteins	Percentage of proteins predicted to be secreted, %	Reference
Nondestructive method				
Arabidopsis thaliana	Cell suspension cultures	95	52	[39]
Oryza sativa	Roots	35	50	[40]
Arabidopsis thaliana	Seedlings	20	70	[41]
Saccharum officinarum	Stems	103	67	[42]
Populus sp.	Stems	129	38	[43]
Oryza sativa	Leaves	125	31	[44]
Arabidopsis thaliana	Leaves	93	93	[4]
Populus sp.	Leaves	133	78	[43]
Destructive method				
Arabidopsis thaliana	Cell suspension cultures (3-day-old)	792	13	[45]
Arabidopsis thaliana	Etiolated hypocotyls	173	79	[46]
Arabidopsis thaliana	Roots	516	52	[47]
Brachypodium distachyon	Stems (apical internodes)	304	61	[34]
Brachypodium distachyon	Stems (basal internodes)	225	75	[34]
Saccharum officinarum	Stems	70	56	[42]
Medicago sativa	Stems	272	73	[48]
Medicago sativa	Stems	345	63	[49]
Linum usitatissimum	Stems	151	69	[50]
Brachypodium distachyon	Young leaves	373	46	[34]

(continued)

Table 1 (continued)

Plant species	Plant material	Total number of identified proteins	Percentage of proteins predicted to be secreted, %	Reference
Brachypodium distachyon	Mature leaves	292	57	[34]
Solanum tuberosum	Leaves	364	37	[51]

It should be noted that all the proteins identified in the different cell wall proteomics studies have been reannotated using the *ProtAnnDB* tool [23]. A protein is considered to be secreted if the criteria listed in Subheading 3.1.1 **step 6** are fulfilled

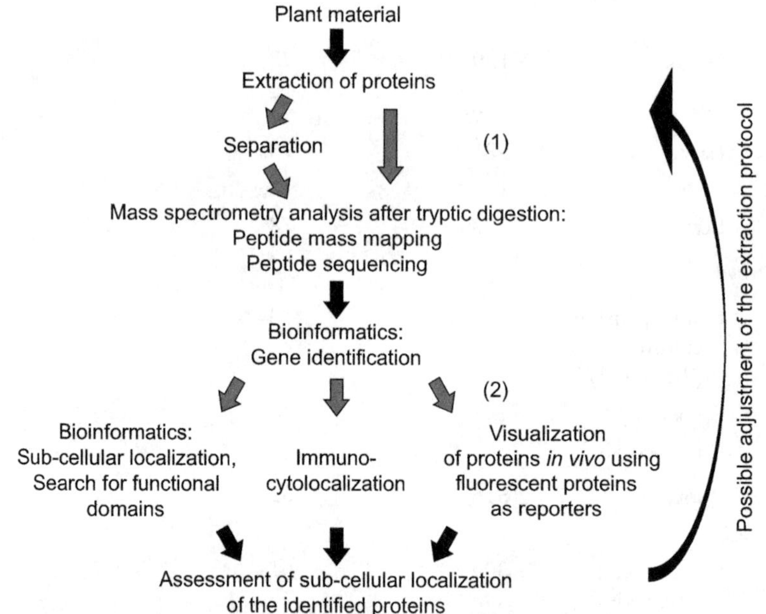

Fig. 1 Plant cell wall proteomics flowchart leading to the assessment of the subcellular localization of the identified proteins. Grey arrows correspond to optional steps: (1) separation of proteins prior to tryptic digestion and mass spectrometry analysis is not mandatory; (2) bioinformatics can be combined to either immunocytolocalization and/or in vivo visualization of proteins using fluorescent proteins as reporters

addition to the bioinformatic predictions (Fig. 1). Briefly, the flowchart of a typical cell wall proteomic experiment comprises several steps. The extraction of proteins from plant material can be done in different ways (destructive or non-destructive methods). Proteins can be separated by chromatography and/or electrophoresis. A tryptic digestion is performed prior to mass spectrometry analysis by peptide mass mapping or peptide sequencing. Finally, gene identification is done by bioinformatics software combining information provided by mass spectrometry analyses and genomic/expressed sequenced tag (EST) databases. The following three methods may be used to verify the subcellular localization of the identified proteins: (1) prediction of subcellular localization using bioinformatics including available organelle-specific proteomics databases, (2) in situ immunocytolocalization when antibodies are available, and (3) in planta visualization of candidates fused to fluorescent reporter proteins in homologous or heterologous plant cells. The calculation of the percentage of proteins predicted or experimentally demonstrated to be extracellular allows evaluation of the quality of the overall strategy and improvement of the protocol for protein extraction if required.

2 Materials

2.1 In Silico Analysis of Protein Sequences

1. NCBI Protein database: http://www.ncbi.nlm.nih.gov/protein/.
2. Phytozome: https://phytozome.jgi.doe.gov/pz/portal.html.
3. TargetP: http://www.cbs.dtu.dk/services/TargetP/ [11].
4. SignalP: http://www.cbs.dtu.dk/services/SignalP/ [12].
5. Predotar: https://urgi.versailles.inra.fr/predotar/ [13].
6. PSORT (old version; for bacterial/plant sequences): http://psort.hgc.jp/form.html [14].
7. TMHMM: http://www.cbs.dtu.dk/services/TMHMM-2.0/ [15].
8. TMpred: http://www.ch.embnet.org/software/TMPRED_form.html [16].
9. PROSITE: http://prosite.expasy.org/ [17].
10. big-PI: http://mendel.imp.ac.at/sat/gpi/gpi_server.html [18].
11. Pred-GPI: http://gpcr.biocomp.unibo.it/predgpi/info.htm [19].
12. GPI-SOM: http://gpi.unibe.ch/ [20].
13. SecretomeP: http://www.cbs.dtu.dk/services/SecretomeP/ [21].
14. Aramemnon: http://aramemnon.botanik.uni-koeln.de/ [22].
15. *ProtAnnDB*: http://www.polebio.lrsv.ups-tlse.fr/ProtAnnDB/ [23].
16. SUBA: http://suba.plantenergy.uwa.edu.au/ [24] (*see* **Note 1**).
17. *WallProtDB*: http://www.polebio.lrsv.ups-tlse.fr/WallProtDB/ [25].

2.2 Immuno-cytolocalization

2.2.1 Tissue Fixation

1. FAA: 10% Formalin (37% formaldehyde solution, Sigma-Aldrich, Saint-Quentin Fallavier, France), 50% ethyl alcohol, 5% acetic acid, 35% milli-Q water (Merck Millipore, Darmstadt, Germany).
2. 50 mL Falcon tubes.
3. Vacuum chamber.

2.2.2 Tissue Infiltration and Embedding

1. Tert-butanol.
2. Absolute ethanol.
3. Paraplast Plus® (Sigma-Aldrich).
4. Oven (60°C).

2.2.3 Tissue Embedding and Sectioning

1. Paraplast Plus®.
2. Embedding molds (Dutscher SAS, Brumath, France).
3. 60°C hot plate (slide warmer, Labscientific XH-2001, Delta Microscopies, Mauressac, France).
4. Rotary microtome (Jung AG, Heidelberg, Germany).
5. 3-Aminopropyl-triethoxysilane coated microscopy slides [26].
6. Xylene or alternative Paraplast Plus® solubilizing solvent (e.g., Histo-Clear®).

2.2.4 Immuno-cytolocalization

1. 3-Aminopropyltriethoxysilane coated microscopy slides [26].
2. TTBS: 0.01 M Tris–HCl pH 7.5, 500 mM NaCl, (0.3% w/v), Triton X100.
3. TTBS-milk: 5% (w/v) nonfat dry milk in TTBS.
4. Primary antibody specific to the CWP of interest.
5. Secondary antibody (e.g., goat anti-rabbit (GAR)-alkaline phosphatase (AP) or GAR-fluorochrome (e.g., fluorescein isothiocyanate, FITC) IgG conjugates if primary antibody has been raised in rabbit).
6. 50 mg/mL 5-bromo-4-chloro-3-indolyl phosphate *p*-toluidine salt (BCIP, Kalys, Bernin, France) stock solution in dimethyformamide (DMF) (stored at −20°C).
7. 50 mg/mL Nitro Blue Tetrazolium (NBT, Kalys) stock solution in 70% DMF (stored at −20°C).
8. AP buffer: 100 mM Tris–HCl pH 9.5, 100 mM NaCl, 10 mM $MgCl_2$.
9. Working solution for AP reaction: 150 μg/mL BCIP and 300 μg/mL NBT in AP buffer.
10. Classical immersion oil or Eukitt® quick-hardening mounting medium (Sigma-Aldrich, ref. 03989).

2.2.5 Microscopy

1. Light microscope for AP reaction observation.
2. Epifluorescence or confocal microscope for FITC detection (*see* **Note 2**).

2.3 In Planta Visualization of Proteins Using Fluorescent Proteins as Reporters

2.3.1 Biological Material

1. *Nicotiana benthamiana* seeds.
2. *Agrobacterium tumefaciens* strain GV3101 [27].

2.3.2 Constructs and Cloning

The gene of interest (GOI) with its own signal peptide (SP) sequence can be cloned in a pEAQ vector [28] upstream of a sequence encoding a fluorescent protein, for example, the TagRFP (red fluorescent protein) [29] in the pEAQ-gwTR vector [30] (Fig. 2a), using the Gateway® cloning technology [31] (*see* **Note 3**). Alternatively, the GOI deleted of its own SP sequence can be cloned downstream of the *A. thaliana* At3g12500 chitinase SP-driven TagRFP using the pEAQ-SPTRgw vector [30] (Fig. 2b). As a control of the latter vector, the pEAQ-TRgw vector should be used (Fig. 2c).

1. pEAQ binary vectors carrying fluorescent protein tags (*see* **Note 3**).

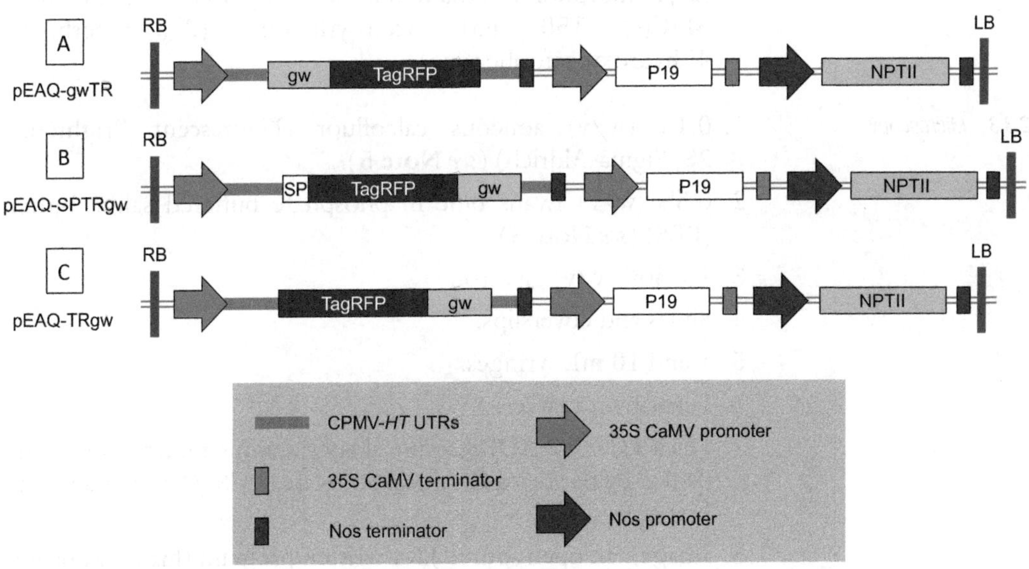

Fig. 2 Schematic representation of the T-DNA regions of major derivatives of pEAQ [52]. pEAQ-gwTR (**a**), pEAQ-SPTRgw (**b**), and pEAQ-TRgw (**c**). T-DNAs are shown with corresponding color and illustration code. The illustrations are not drawn to scale. *gw* Gateway® cloning cassette, *LB* left border of T-DNA, *NPTII* nopaline synthase gene, *P19* TBSV RNA silencing suppressor, *RB* right border of T-DNA, *SP* signal peptide from the A. thaliana At3g12500 chitinase

2. Binary vector carrying a specific green (GFP) or yellow (YFP) fluorescent reporter protein addressed to the plasma membrane (*see* **Note 4**).
3. Luria-Bertani (LB) solid medium: 10 g/L tryptone, 5 g/L yeast extract, 10 g/L NaCl, pH 7.0 (adjust using 1 N NaOH), 15 g/L agar.
4. Yeast extract beef (YEB) liquid medium: 5 g/L Bacto tryptone, 1 g/L yeast extract, 5 g/L beef extract, 5 g/L Bacto peptone, 2 mM $MgSO_4$, 50 g/L sucrose, pH 7.4 (adjust using 1 N NaOH).
5. 50 mg/mL rifampicin (stock solution stored at −20°C).
6. 20 mg/mL gentamycin (stock solution stored at −20°C).
7. 100 mg/mL kanamycin (stock solution stored at −20°C).
8. Petri dishes (8 cm diameter).
9. Oven (28°C).
10. 13 mL polypropylene tubes.
11. Shaking incubator (28°C).
12. 50 mL Falcon tubes.
13. Centrifuge.
14. Vortex-mixer.
15. MMA (MES, $MgCl_2$, acetosyringone): 10 mM MES (2-[N-morpholino]ethanesulfonic acid) pH 5.6, 10 mM $MgCl_2$, 150 μM acetosyringone (3′,5′-dimethoxy-4′-hydroxyacetophenone).

2.3.3 Microscopy

1. 0.1% (w/v) aqueous calcofluor (Fluorescent Brightener 28, Sigma-Aldrich) (*see* **Note 5**).
2. 0.5% (w/v) Evans blue in phosphate buffered saline buffer (PBS) (*see* **Note 5**).
3. 10–30% (v/v) glycerol.
4. Slides and coverslips.
5. 1 and 10 mL syringes.
6. Permanent marker.
7. Leica TCS SP2 AOBS confocal laser-scanning microscope controlled by the Leica Confocal Software (LCS) (Leica, Nanterre, France).
8. ImageJ, an open source Java-written program (https://imagej.nih.gov/ij/).

3 Methods

3.1 Bioinformatic Tools

3.1.1 In Silico Analysis of Protein Sequences

1. Collect the amino acid sequence of the protein of interest in a relevant database (e.g., the NCBI Protein database or Phytozome) (*see* **Note 6**).
2. Paste it in the query form of at least two different bioinformatic tools to predict subcellular localization: TargetP, SignalP, Predotar, and PSORT.
3. Predict the presence of transmembrane domains using TMHMM and TMpred.
4. Predict the presence of a GPI-anchor using at least two bioinformatic tools such as bigPI, PredGPI, and GPI-som.
5. Predict the presence of an ER targeting signal such as a C-terminal HDEL or KDEL sequence using PROSITE (ER_TARGET, PS00014, http://prosite.expasy.org/PDOC00014) (*see* **Note 7**).
6. Analyze the data. A protein can reasonably be considered as extracellular if: (a) extracellular localization is predicted by at least two bioinformatics tools with the highest score as compared to other possible localizations, (b) no more than two transmembrane domains are predicted (*see* **Note 8**), and (c) no ER targeting signal sequence is predicted (*see* **Note 7**).
7. Check if the experimentally demonstrated or the predicted function of the protein is compatible with its presence in the extracellular matrix by literature search.
8. If necessary, predict if proteins predicted to be intracellular, could be secreted via a nonclassical secretion mechanism, that is, not SP-dependent using SecretomeP (*see* **Note 9**).

3.1.2 Existing Databases Collecting Annotations by Different Bioinformatic Tools and/or Experimental Work

Instead of performing bioinformatics predictions of subcellular localization for each protein and using each of the different software detailed in Subheading 2.1, **item 1** separately, it can be helpful to first look in databases collecting such information.

1. Query Aramemnon. Initially devoted to membrane proteins of *Arabidopsis thaliana*, Aramemnon has been extended to all *A. thaliana* proteins, and to membrane proteins of eight other plants (*Brachypodium distachyon*, *Cucumis melo*, *Musa acuminata*, *Oryza sativa*, *Populus trichocarpa*, *Solanum lycopersicum*, *Vitis vinifera*, *Zea mays*). Predictions of subcellular localization, transmembrane domains and GPI-anchors are done with 20, 18 and 4 different bioinformatics tools respectively.
2. Query *ProtAnnDB*. This database collects predictions of subcellular localization, presence of transmembrane domains, presence of GPI anchors and of functional domains of proteins

from 20 plant species. These predictions were performed with online software and the scores of the predictions are reported. Whenever possible, the annotations of cell wall protein families by experts were included.

3. Query SUBA. This database is dedicated to *A. thaliana* proteins. It integrates bioinformatics predictions of subcellular localization using 22 different tools and experimental data including subcellular proteomics, fluorescent protein tagging and protein–protein interaction experiments. Comparisons between the different results are provided.
4. Check if the identified proteins are listed in previous cell wall proteomic studies, for example, in *WallProtDB* which collects experimental data from 13 plant species (*see* **Note 10**).

3.2 Immunocytolocalization

3.2.1 Fixation

1. Sample tissues with a razor blade and directly immerse in FAA into 50 mL tubes.
2. Vacuum infiltrate the fixative by 5 cycles of vacuum infiltration (1 min each)/vacuum release in a vacuum chamber.
3. Incubate for 6–16 h at 4 °C in FAA and rinse four times in 50% ethanol for 10 min, each (*see* **Note 11**).

3.2.2 Dehydration and Paraffin Infiltration

1. Infiltrate samples with a progressive dehydration in milliQ water/absolute ethanol/tert-butanol gradient series as indicated in Table 2 (adapted from [32]) (*see* **Note 12**).

3.2.3 Tissue Embedding and Sectioning

1. Position the sample in the desired orientation (e.g., to allow further cross- or longitudinal sectioning) in melted Paraplast Plus® inside embedding molds onto a 60°C heated-plate (*see* **Note 13**).
2. Allow the Paraplast Plus® solidification for about 15 min at room temperature and cool the embedded samples at 4°C for at least 1 h.
3. Separate the samples from the mold for direct microtomy or for storage at 4°C for years.
4. Cut serial sections (10 μm thick) of samples with a rotary microtome.
5. Carefully place the ribbons of sections on a large Kimwipes® and gently separate the sections with needles and paint brush.
6. Float individual sections on a degassed water layer on precoated microscopy slides and spread in the same orientation and position (*see* **Note 13**).
7. Carefully remove excess water with a pipette and allow slides to dry overnight at 35°C for optimal section adherence.

Table 2
Dehydration and paraffin infiltration series

H_2O, %	Absolute ethanol, %	Tert-butanol, %	Σ Alcohol, %	Paraplast, %	Timing and temperature
50	40	10	50		Day1: 1 h (RT)
30	50	20	70		Day1: 1 h (RT)
15	50	35	85		Day1: 1 h (RT)
	45	55	100[a]		Day1: 1 h (RT)
	25	75	100		Day1: 1 h (RT)
		100	100		
		100			Day2: 1 h (RT)
		50		50	Day2: 5 h (60 °C)
		50		50	Day2: O/N (60 °C)
				100	Day3: 10 h (60 °C)
				100	Day3: O/N (60 °C)
				100	Day4: 10 h (60 °C)
				100	Day4: O/N (60 °C)

[a]At this step, add 0.025% (w/v) erythrosine solution to a final concentration of 0.025% (w/v) to temporarily stain the depigmentated samples

8. Sequentially deparaffinize/rehydrate under a fume hood as follows: xylene or alternative Paraplus Plus® solubilizing solvent (2 × 15 min), 100% ethanol (2 × 5 min), 95% ethanol, 70% ethanol, 50% ethanol (5 min each), milliQ water (2 × 5 min) (*see* **Notes 14** and **15**).

3.2.4 Immunocytolocalization

For the following steps excluding primary and secondary antibody incubations, slides can be either individually covered with 1 mL of solution, or gathered in dedicated racks and immersed together in a larger volume of solution. For antibody labeling, incubation is typically performed in 50–150 μL of solution spread on the slide and covered with a coverslip. At the end of incubation, stand the slide upright in vertical position to carefully remove the coverslip. If necessary, gently shake the slide in TTBS to help removing recalcitrant coverslips, but do not manipulate the coverslip since it can result in section damage. Never allow the slides to dry out during the labeling process. An important rule to avoid unspecific background labeling is to keep the slides in a humid chamber at all steps (*see* **Note 16**), and to apply the next solution directly after having removed the previous one.

1. Block non-specific protein–protein interactions for 30 min in TTBS-milk.
2. Incubate in the primary antibody diluted in TTBS-milk (*see* **Note 17**).
3. Wash six times for 5 min with TTBS (*see* **Note 18**).
4. Incubate in secondary antibody diluted as indicated by the manufacturer (usually 1:100–1:500) in TTBS-milk, for 1–2 h at room temperature.
5. Wash in TTBS (three times 5 min) and quickly rinse in milliQ water to remove the detergent.
6. Incubate in the BCIP/NBT chromogenic substrate for 5–30 min at room temperature (*see* **Note 19**).
7. Wash in distilled water to stop the reaction.
8. Allow the slides to dry on a slide warmer before mounting under coverslip using mounting agent such as immersion oil that will remain viscous or Eukitt® that will dry.

3.2.5 Microscopy

A simple light microscope is sufficient for AP (or HRP and gold-silver enhancement, [33]) detection using bright field visualization. Care should be taken to frame the same zone on serial sections when comparing different slides (several antibodies or antibody and negative control). Epifluorescence or confocal microscope must be used when the secondary antibody is coupled to a fluorochrome (*see* **Note 20**).

This technique may provide sufficient resolution to allow the fine localization of CWPs in remote sublayers of the cell wall [34] (*see* **Note 21**).

3.3 In Vivo Visualization of Proteins Using Fluorescent Proteins as Reporters

3.3.1 Plant Material and Growth Conditions

1. Cultivate the *N. benthamiana* seedlings during 5 weeks in a growth chamber in the following conditions: 75% humidity, 16 h light at 25°C/8 h dark at 23°C.
2. Select young or just fully expanded leaves.

3.3.2 Constructions and Gateway® Cloning

1. Design oligonucleotide primers for polymerase chain reaction (PCR) according to manufacturer's instruction (Gateway®, Invitrogen™, Thermo Fischer Scientific, Illkirch, France) (*see* **Note 22**).
2. PCR amplify the GOI coding DNA sequence, including the signal peptide but without the stop codon for cloning in the pEAQ-gwTR destination vector. Alternatively, amplify the

coding sequence devoid of the encoded signal peptide but with the stop codon. In this case, cloning should be performed in the pEAQ-TRgw and pEAQ-SPTRgw destination vectors.

3. Insert the amplified fragment into a Gateway® pDONR vector, and then, into the pEAQ binary destination vector according to standard Gateway® procedures. The reactions are mediated by the Gateway® BP and LR clonases, respectively.
4. Amplify the plasmid in *E. coli*, check its sequence and transform *A. tumefaciens* strain GV3101.

3.3.3 Preparation of Agrobacterium Culture for Agroinfiltration

1. Spread the *A. tumefaciens* GV3101 strain harboring the appropriate binary plasmid on a Petri dish containing solid LB medium supplemented with 50 μg/mL rifampicin, 25 μg/mL gentamycin and 50 μg/mL kanamycin.
2. Incubate 2 days at 28°C.
3. Grow a liquid preculture of transformed *A. tumefaciens* in 3 mL of YEB medium supplemented with 25 μg/mL gentamycin and 50 μg/mL kanamycin.
4. Incubate overnight at 28°C, with vigorous shaking.
5. Grow a 20 mL culture inoculated with 200 μL of the overnight culture in the same medium conditioned in a 50 mL Falcon tube.
6. Incubate at 28°C under vigorous shaking, and allow to grow to an A_{600nm} between 0.6 and 0.8.
7. Pellet the bacteria by centrifugation at room temperature at 2500 × *g* for 10 min.
8. Resuspend the bacterial pellet by vortexing in MMA. Adjust to an A_{600nm} of 0.5 with MMA.
9. Incubate 1 h to overnight at room temperature.

3.3.4 Agroinfiltration Procedure

1. Water the plants the day before agroinfiltration for optimal stomata aperture. For the same reason, infiltrate during the light time of the photoperiod.
2. Bacterial suspensions are pressure-infiltrated into *N. benthamiana* abaxial leaf face with a 1 mL syringe (*see* **Note 23**).
3. Infiltrated areas are marked with a permanent marker.
4. Observations can be done 1–4 days after infiltration (*see* **Note 23**).

3.3.5 Preparation of Leaf Samples for Microscopic Observation

1. Collect leaf samples from the infiltrated areas with a puncher.
2. Put the samples (several leaf discs) in a syringe with water.
3. Degas tissues using the piston to replace the air by water inside the tissues for an optimal observation under the microscope.

4. Screen leaf discs from a given sample with the confocal microscope for the optimal transgene expression.
5. Select leaf discs with highest expression levels of the fluorescent protein tag for cell wall staining.

3.3.6 Staining of Plant Cell Walls

1. Stain selected leaf discs in a small container with 0.1% calcofluor for 15 min.
2. Rinse twice in water.
3. Observe discs with confocal microscope between slide and coverslip in water.

3.3.7 Plasmolysis of Plant Cells

1. Screen the samples for epidermis cells displaying simultaneous transgene expression and cell wall calcofluor staining.
2. Acquire images to illustrate the colocalization between tagged CWPs and calcofluor stained cell walls.
3. Plasmolyze cells under the microscope by sucking water by capillarity from one side of the slide, and replace with 10–30% glycerol solution from the opposite side (*see* **Note 24**).
4. Observe plasmolysis that gradually starts a few minutes after replacing water by glycerol, first at the periphery of the leaf discs, and later on in deeper tissues (after about 15 min).

3.3.8 Confocal Microscopy Imaging

1. Observe TagRFP fluorescence in the 590- to 630-nm range after excitation at 561 nm.
2. Observe YFP fluorescence in the 515- to 545-nm range after excitation at 488 nm.
3. Observe calcofluor fluorescence in the 439- to 472-nm range after excitation at 405 nm.
4. Process confocal-acquired images with ImageJ.

4 Notes

1. Apart from SUBA, other cell wall proteomics databases can be used such as: PPDB (Plant Proteome DataBase, http://ppdb.tc.cornell.edu/), AtNoPDP (Arabidopsis Protein Nucleolar DataBase, http://bioinf.scri.sari.ac.uk/cgi-bin/atnopdb/home), AT_chloro (http://at-chloro.prabi.fr/at_chloro/), Eukaryotic Subcellular Localization DataBase (eSLDB, http://gpcr2.biocomp.unibo.it/esldb/index.htm).
2. Higher throughput can be reached using a slide scanner for both methods [26, 34].
3. New pEAQ binary vectors have been developed [30]. They carry the TagRFP (Evrogen, Moscow, Russia) as the fluorescent protein tag (Fig. 2) [29]. The TagBFP (Blue Fluorescent

Protein, Evrogen) could also be used [35]. Due to their low p*K*a, 3.8 and 2.7, respectively, both TagRFP and TagBFP are very good fluorescent reporters for cell wall labeling [36].

4. For transient colocalization studies, use a specific green (GFP) or yellow (YFP) fluorescent reporter protein for the plasma membrane (e.g., binary vector reference, pm-yb CD3-1006) [36, 37].
5. Use Evans blue instead of calcofluor to label the plant cell wall when TagBFP is used as the fluorescent reporter. Be aware that Evans blue does not label exclusively the cell wall, but also the plasma membrane. Plasmolysis allows for adequate discrimination between the cell wall and plasma membrane.
6. Many of the plant protein sequences can be found at the NCBI Protein database and/or at Phytozome. However, for newly sequenced genomes or genome which sequencing is in progress, it might be necessary to look for dedicated websites.
7. Some proteins have no consensus endoplasmic reticulum targeting signal. Other domains indicating their localization in this organelle might be found, such as PF07749 (ERp29) (http://pfam.xfam.org/family/PF07749) or IPR011679 (Endoplasmic reticulum, protein ERp29, C-terminal) (https://www.ebi.ac.uk/interpro/entry/IPR011679).
8. The number of predicted transmembrane domains can be misleading. Actually the peptide signal can be predicted as a transmembrane domain because it contains hydrophobic amino acids in its central part [38]. In the same way, the C-terminal amino acid sequence of a protein carrying a GPI anchor is rich in aromatic hydrophobic residues and can be predicted as a transmembrane domain [18].
9. The SecretomeP bioinformatics software was used in several studies to predict the possible cell wall localization of proteins devoid of predicted signal peptide [7]. However, since this software has been designed for mammalian and bacterial proteins, the results of the prediction should be taken cautiously.
10. All the proteins present in *WallProtDB* have been reannotated in a homogeneous way using *ProtAnnDB* and they fit the criteria defined in Subheading 3.1, **step 6** [25]. A prerequisite to enter the data into *WallProtDB* is to rely on genomics or transcriptomics data of the plant of interest in order to ensure the precise identification of proteins by mass spectrometry analysis and bioinformatics.
11. At this step, the samples can be directly processed (*see* Subheading 3.2.2) or stored safely for weeks to months at 4°C in 50% ethanol.

12. At this stage, it is possible to directly perform tissue embedding or to store tubes with caps at 4°C. In the latter case, on restart, the tubes have to be incubated for a day at 60°C before recovering the samples.
13. To allow simultaneous sectioning of multiple samples and further observation and comparison at higher throughput of several conditions (e.g., different antibodies or antibody and negative control) for the same exact sample, numerous samples can be organized in tissue arrays [26]. The careful positioning of the serial sections of the same tissue array on different slides is important to facilitate medium-throughput during final comparative analysis of the results on the same exact tissues with different conditions (e.g., different antibodies or antibody and negative control).
14. Note that the pink erythrosine color used to temporarily stain the depigmentated samples in the paraffin (Table 2) is solubilized in the aqueous baths.
15. Slides can be directly used for immunocytolocalization or air dried and stored for weeks under clean and dust-free conditions.
16. Commercial humid chambers can be purchased but a homemade model can easily be assembled using a large Petri dish with two lines of broken Pasteur pipets taped at the bottom to allow the safe storage and handling of slides, preventing from flooding in the water placed at the bottom of the Petri dish.
17. The antibody incubation is typically performed for 2 h at room temperature but may be extended for a few hours or replaced by overnight incubation at 4°C. The specificity of the antibody has to be analyzed by classical immunological methods (ELISA, western blot) prior to use for immunocytolocalization [33]. In addition, serial dilutions of the antibody have to be performed to determine the optimal dilution resulting in the best signal–background ratio. Negative controls can be (a) comparison with preimmune serum when using polyclonal serum, (b) competitive inhibition test when the antigen is available, or (c) comparison of labeled wild-type and knocked down or knocked out mutants corresponding to the studied CWP when available. These controls are particularly important when studying CWPs belonging to multigenic families [33, 34].
18. Washing may be adapted to each antibody if necessary, using longer incubations or larger volumes and even performing the washes using the blocking buffer at higher washing temperature to increase the signal to background ratio [33].

19. Usually the reaction takes 5–30 min to reach good signal to background ratio as compared to negative control. The reaction may be monitored under the microscope. One advantage of AP is that it allows long substrate incubation without excessive background. In most cases, 5–30 min are sufficient but overnight incubation may reveal specific faint labeling details [32]. Other detection systems may also be applied, for example, secondary antibody conjugated to horseradish peroxidase (HRP), fluorochrome, or even immunogold followed by silver enhancement [33]. Other buffers such as phosphate buffer saline or other blocking agent such as bovine serum albumin may also be used.
20. Higher throughput can be reached using a slide scanner for both methods [34].
21. Higher resolution may require the use of protocols adapted to transmission electron microscopy (TEM). Even if specifically adapted to TEM, such protocols follow the same overall principle (e.g., [32, 33]).
22. pEAQ binary vectors have been designed to respect the reading frame and to generate a 15 amino acid spacer, between the CWP and the TagRFP amino acid sequences, after the LR reaction.
23. For coinfiltration of several bacterial strains, cultures are mixed at equal densities and equal volumes. If several constructs with distinct optimum observation times (e.g., 24 and 48 h post-infiltration) have to be coexpressed, it is possible to infiltrate the bacterial strain carrying the corresponding constructs with an appropriate delay (e.g., second strain delivered 24 h after the first infiltration).
24. Cell plasmolysis is induced under the microscope in order to observe the same cells before and after plasma membrane detachment from the cell wall.

Acknowledgments

The authors are grateful to *Université Paul Sabatier* (Toulouse III), *Université de Strasbourg* and CNRS for support. They also wish to thank François Berthold and Caroline Hemmer for their great contribution to the cloning of pEAQ derivatives, Alain Jauneau for providing access to cell imaging facilities (http://trigenotoul.com/), and the bioinformatics platform of GenoToul Midi-Pyrénées for providing calculation facilities (http://bioinfo.genotoul.fr/).

References

1. Carpita NC, Gibeaut DM (1993) Structural models of primary cell walls in flowering plants, consistency of molecular structure with the physical properties of the walls during growth. Plant J 3:1–30
2. Albenne C, Canut H, Jamet E (2013) Plant cell wall proteomics: the leadership of *Arabidopsis thaliana*. Front Plant Sci 4(111). https://doi.org/10.3389/fpls.2013.00111
3. Lee SJ, Saravanan RS, Damasceno CM, Yamane H, Kim BD, Rose JK (2004) Digging deeper into the plant cell wall proteome. Plant Physiol Biochem 42:979–988. https://doi.org/10.1016/j.plaphy.2004.10.014
4. Boudart G, Jamet E, Rossignol M, Lafitte C, Borderies G, Jauneau A, Esquerré-Tugayé M-T, Pont-Lezica R (2005) Cell wall proteins in apoplastic fluids of *Arabidopsis thaliana* rosettes: identification by mass spectrometry and bioinformatics. Proteomics 5:212–221
5. Feiz L, Irshad M, Pont-Lezica RF, Canut H, Jamet E (2006) Evaluation of cell wall preparations for proteomics: a new procedure for purifying cell walls from Arabidopsis hypocotyls. Plant Methods 2:10. https://doi.org/10.1186/1746-4811-2-10
6. Chivasa S, Ndimba BK, Simon WJ, Robertson D, Yu X-L, Knox JP, Bolwell P, Slabas AR (2002) Proteomic analysis of the Arabidopsis thaliana cell wall. Electrophoresis 23:1754–1765
7. Jamet E, Albenne C, Boudart G, Irshad M, Canut H, Pont-Lezica R (2008) Recent advances in plant cell wall proteomics. Proteomics 8:893–908
8. Nickel W, Seedorf M (2008) Unconventional mechanisms of protein transport to the cell surface of eukaryotic cells. Ann Rev Cell Dev Biol 24:287–308
9. Pinedo M, Regente M, Elizalde M, Quiroga I, Pagnussat LA, Jorrin-Novo J, Maldonado A, de la Canal L (2012) Extracellular sunflower proteins: evidence on non-classical secretion of a jacalin-related lectin. Protein Pept Lett 19:270–276
10. Chen MH, Huang LF, Li HM, Chen YR, Yu SM (2004) Signal peptide-dependent targeting of a rice alpha-amylase and cargo proteins to plastids and extracellular compartments of plant cells. Plant Physiol 135:1367–1377
11. Emanuelsson O, Brunak S, Von Heijne G, Nielsen H (2007) Locating proteins in the cell using TargetP, SignalP and related tools. Nat Protoc 2:953–971
12. Petersen TN, Brunak S, von Heijne G, Nielsen H (2011) SignalP 4.0: discriminating signal peptides from transmembrane regions. Nat Method 8:785–786
13. Small I, Peeters N, Legeai F, Lurin C (2004) Predotar: a tool for rapidly screening proteomes for N-terminal targeting sequences. Proteomics 4:1581–1590
14. Nakai K, Horton P (1999) PSORT: a program for detecting sorting signals in proteins and predicting their subcellular localization. Trends Biochem Sci 24:34–35
15. Sonnhammer EL, von Heijne G, Krogh A (1998) A hidden Markov model for predicting transmembrane helices in protein sequences. Proc Int Conf Intell Syst Mol Biol 6:175–182
16. Hofmann K, Stoffel W (1993) TMbase – a database of membrane spanning proteins segments. Biol Chem Hoppe Seyler 374:166
17. Sigrist CJA, de Castro E, Cerutti L, Cuche BA, Hulo N, Bridge A, Bougueleret L, Xenarios I (2013) New and continuing developments at PROSITE. Nucleic Acids Res 41(Database issue):D344–D347
18. Eisenhaber B, Wildpaner M, Schultz CJ, Borner GH, Dupree P, Eisenhaber F (2003) Glycosylphosphatidylinositol lipid anchoring of plant proteins. Sensitive prediction from sequence- and genome-wide studies for Arabidopsis and rice. Plant Physiol 133:1691–1701. https://doi.org/10.1104/pp.103.023580
19. Pierleoni A, Martelli PL, Casadio R (2008) PredGPI: a GPI-anchor predictor. BMC Bioinformatics 9:392
20. Fankhauser N, Mäser P (2005) Identification of GPI anchor attachment signals by a Kohonen self-organizing map. Bioinformatics 21:1846–1852
21. Bendtsen JD, Jensen LJ, Blom N, von Heijne G, Brunak S (2004) Feature based prediction of non-classical and leaderless protein secretion. Protein Eng Des Sel 17:349–356
22. Schwacke R, Schneider A, van der Graaff E, Fischer K, Catoni E, Desimone M, Frommer WB, Flugge UI, Kunze R (2003) ARAMEMNON, a novel database for Arabidopsis integral membrane proteins. Plant Physiol 131:16–26
23. San Clemente H, Pont-Lezica R, Jamet E (2009) Bioinformatics as a tool for assessing the quality of sub-cellular proteomic strategies and inferring functions of proteins: plant cell wall proteomics as a test case. Bioinform Biol Insights 3:15–28

24. Tanz SK, Castleden I, Hooper CM, Vacher M, Small I, Millar HA (2013) SUBA3: a database for integrating experimentation and prediction to define the SUBcellular location of proteins of Arabidopsis. Nucleic Acids Res 41(Database issue):D1185–D1191
25. San Clemente H, Jamet E (2015) *WallProtDB*, a database resource for plant cell wall proteomics. Plant Methods 11:2
26. Francoz E, Ranocha P, Pernot C, Le Ru A, Pacquit V, Dunand C, Burlat V (2016) Complementarity of medium-throughput in situ RNA hybridization and tissue-specific transcriptomics: case study of Arabidopsis seed development kinetics. Sci Rep 6:e24644
27. Shamloul M, Trusa J, Mett V, Yusibov V (2014) Optimization and utilization of *Agrobacterium*-mediated transient protein production in *Nicotiana*. J Vis Exp 86:e51204
28. Peyret H, Lomonossoff GP (2013) The pEAQ vector series: the easy and quick way to produce recombinant proteins in plants. Plant Mol Biol 83:51–58
29. Merzlyak EM, Goedhart J, Shcherbo D, Bulina ME, Shcheglov AS, Fradkov AF, Gaintzeva A, Lukyanov KA, Lukyanov S, Gadella TW, Chudakov DM (2007) Bright monomeric red fluorescent protein with an extended fluorescence lifetime. Nat Method 4:555–557
30. Berthold F, Roujol D, Hemmer C, Jamet E, Ritzenthaler C, Hoffmann L, Schmitt-Keichinger C (2019) Inside or outside? A new collection of Gateway vectors allowing plant protein subcellular localization or over-expression. Plasmids 105:102436
31. Karimi M, Inzé D, Depicker A (2002) Gateway™ vectors for *Agrobacterium*-mediated plant transformation. Trends Plant Sci 7:193–195
32. Oudin A, Mahroug S, Courdavault V, Hervouet N, Zelwer C, Rodríguez-Concepción M, St-Pierre B, Burlat V (2007) Spatial distribution and hormonal regulation of gene products from methyl erythritol phosphate and monoterpene-secoiridoid pathways in *Catharanthus roseus*. Plant Mol Biol 65:13–30
33. Burlat V, Kwon M, Davin LB, Lewis NG (2001) Dirigent proteins and dirigent sites in lignifying tissues. Phytochemistry 57:883–897
34. Douché T, San Clemente H, Burlat V, Roujol D, Valot B, Zivy M, Pont-Lezica R, Jamet E (2013) *Brachypodium distachyon* as a model plant toward improved biofuel crops: search for secreted proteins involved in biogenesis and disassembly of cell wall polymers. Proteomics 13:2438–2454
35. Kremers GJ, Goedhart J, van Den Heuvel DJ, Gerritsen HC, Gadella TW (2007) Improved green and blue fluorescent proteins for expression in bacteria and mammalian cells. Biochemistry 46:3775–3783
36. Albenne C, Canut H, Hoffmann L, Jamet E (2014) Plant cell wall proteins: a large body of data, but what about runaways? Proteomes 2:224–242
37. Nelson BK, Cai X, Nebenführ A (2007) A multicolored set of *in vivo* organelle markers for co-localization studies in Arabidopsis and other plants. Plant J 51:1126–1136
38. von Heijne G (1985) Signal sequences. The limits of variation. J Mol Biol 184:99–105
39. Borderies G, Jamet E, Lafitte C, Rossignol M, Jauneau A, Boudart G, Monsarrat B, Esquerr-é-Tugayé MT, Boudet A, Pont-Lezica R (2003) Proteomics of loosely bound cell wall proteins of *Arabidopsis thaliana* cell suspension cultures: a critical analysis. Electrophoresis 24:3421–3432
40. Zhou L, Bokhari SA, Dong CJ, Liu JY (2011) Comparative proteomics analysis of the root apoplasts of rice seedlings in response to hydrogen peroxide. PLoS One 6:e16723
41. Casasoli M, Spadoni S, Lilley K, Cervone F, De Lorenzo G, Mattei B (2008) Identification by 2-D DIGE of apoplastic proteins regulated by oligogalacturonides in *Arabidopsis thaliana*. Proteomics 8:1042–1054
42. Calderan-Rodrigues MJ, Jamet E, Douché T, Rodrigues Bonassi MB, Regiani Cataldi TR, Guimaraes Fonseca JG, San Clemente H, Pont-Lezica R, Labate CA (2016) Cell wall proteome of sugarcane stems: comparison of a destructive and a non-destructive extraction method showed differences in glycoside hydrolases and peroxidases. BMC Plant Biol 16:14
43. Pechanova O, Hsu CY, Adams JP, Pechan T, Vandervelde L, Drnevich J, Jawdy S, Adeli A, Suttle JC, Lawrence AM, Tschaplinski TJ, Séguin A, Yuceer C (2010) Apoplast proteome reveals that extracellular matrix contributes to multistress response in poplar. BMC Genomics 11:674
44. Jung YH, Jeong SH, Kim SH, Singh R, Lee JE, Cho YS, Agrawal GK, Rakwal R, Jwa NS (2008) Systematic secretome analyses of rice leaf and seed callus suspension-cultured cells: workflow development and establishment of high-density two-dimensional gel reference maps. J Proteome Res 7:5187–5210
45. Bayer EM, Bottrill AR, Walshaw J, Vigouroux M, Naldrett MJ, Thomas CL, Maule AJ (2006) Arabidopsis cell wall proteome defined using multidimensional protein

identification technology. Proteomics 6:301–311

46. Irshad M, Canut H, Borderies G, Pont-Lezica-R, Jamet E (2008) A new picture of cell wall protein dynamics in elongating cells of *Arabidopsis thaliana*: confirmed actors and newcomers. BMC Plant Biol 8:94. https://doi.org/10.1186/1471-2229-8-94
47. Nguyen-Kim H, San Clemente H, Balliau T, Zivy M, Dunand C, Albenne C, Jamet E (2016) Arabidopsis thaliana root cell wall proteomics: increasing the proteome coverage using a combinatorial peptide ligand library and description of unexpected Hyp in peroxidase amino acid sequences. Proteomics 16:491–503. https://doi.org/10.1002/pmic.201500129
48. Verdonk JC, Hatfield RD, Sullivan ML (2012) Proteomic analysis of cell walls of two developmental stages of alfalfa stems. Front Plant Sci 3:279
49. Printz B, Dos Santos Morais R, Wienkoop S, Sergeant K, Lutts S, Hausman JF, Renaut J (2015) An improved protocol to study the plant cell wall proteome. Front Plant Sci 6:237
50. Day A, Fénart S, Neutelings G, Hawkins S, Rolando C, Tokarski C (2013) Identification of cell wall proteins in the flax (*Linum usitatissimum*) stem. Proteomics 13:812–825
51. Lim S, Chisholm K, Coffin RH, Peters RD, Al-Mughrabi KI, Wang-Pruski G, Pinto DM (2012) Protein profiling in potato (*Solanum tuberosum* L.) leaf tissues by differential centrifugation. J Proteome Res 11:2594–2601
52. Sainsbury F, Thuenemann EC, Lomonossoff GP (2009) pEAQ: versatile expression vectors for easy and quick transient expression of heterologous proteins in plants. Plant Biotechnol J 7:682–693

Chapter 26

Bioinformatic Identification of Plant Hydroxyproline-Rich Glycoproteins

Xiao Liu, Savannah McKenna, Lonnie R. Welch, and Allan M. Showalter

Abstract

Hydroxyproline-rich glycoproteins (HRGPs) are a superfamily of plant cell wall proteins that function in diverse aspects of plant growth and development. This superfamily consists of three members: arabinogalactan-proteins (AGPs), extensins (EXTs), and proline-rich proteins (PRPs). Hybrid and chimeric HRGPs also exist. A bioinformatic software program, BIO OHIO 2.0, was developed to expedite the genome-wide identification and classification of AGPs, EXTs, and PRPs based on characteristic HRGP motifs and biased amino acid compositions. This chapter explains the principles of identifying HRGPs and provides a stepwise tutorial for using the BIO OHIO 2.0 program with genomic/proteomic data. Here, as an example, the genome/proteome of the common bean (*Phaseolus vulgaris*) is analyzed using the BIO OHIO 2.0 program to identify and characterize its set of HRGPs.

Key words Hydroxyproline-rich glycoproteins, Arabinogalactan-proteins, Extensins, Proline-rich proteins, Bioinformatics, Plant cell wall, *Phaseolus vulgaris*

1 Introduction

Hydroxyproline-rich glycoproteins (HRGPs) are a diverse superfamily of plant cell wall glycoproteins that are implicated to play various roles in plant growth and development [1–5]. Based on their patterns of proline hydroxylation and subsequent glycosylation, HRGPs are divided into three families: arabinogalactan-proteins (AGPs), extensins (EXTs), and proline-rich proteins (PRPs).

AGPs are rich in proline (P), alanine (A), serine (S), and threonine (T). The P residues are generally clustered as noncontiguous residues (e.g., APSPTP) and reside in certain common dipeptide sequences such as AP, PA, SP, TP, VP, and GP. These P residues are posttranslationally hydroxylated and glycosylated with arabinogalactan (AG) polysaccharides [6–8]. The AGP family can be divided into several subfamilies, including the classical AGPs, AG peptides, and chimeric AGPs including fasciclin-like AGPs (FLAs), and the plastocyanin AGPs (PAGs). EXTs typically contain repeating units

Zoë A. Popper (ed.), *The Plant Cell Wall: Methods and Protocols*, Methods in Molecular Biology, vol. 2149, https://doi.org/10.1007/978-1-0716-0621-6_26,

of SP_{3-5}, in which the contiguous P residues are hydroxylated and subsequently glycosylated with arabinose oligosaccharides [5, 9]. Many EXTs also contain YXY amino acid repeats [where X can be any amino acid] and are involved in intramolecular and intermolecular crosslinking in the cell wall [5]. EXTs can be divided into several subfamilies, including the classical EXTs, short EXTs, leucine-rich repeat extensins (LRXs), proline-rich extensin-like receptor kinases (PERKs), formin-homolog EXTs (FH EXTs), and other chimeric EXTs [5, 10]. PRPs typically contain contiguous P residues (e.g., PPVYK), which may reside in repeating amino acid sequences such as PPVX(K/T) [where X represents any amino acid] or KKPCPP. In addition to an abundance of P residues, PRPs are generally rich in valine (V), lysine (K), cysteine (C), tyrosine (Y), and threonine (T) residues. The P residues may be hydroxylated and subsequently glycosylated with arabinose oligosaccharides. PRPs can be divided into subfamilies, including the classical PRPs, PR peptides, and chimeric PRPs [11].

Most HRGPs have an N-terminal signal peptide that results in their insertion into the endomembrane system and delivery to the plasma membrane/cell wall. Certain HRGPs, particularly within the AGP family, are also modified with a C-terminal glycosylphosphatidylinositol (GPI) membrane anchor, which tethers the protein to the plasma membrane to allow the rest of the glycoprotein to extend toward the cell wall in the periplasm [4, 12].

Several bioinformatic approaches to identify HRGPs are reported [10, 11, 13–15]. All of these approaches take advantage of the characteristic amino acid compositions, repeating sequences, and sequence features present in HRGPs, but vary with respect to how each of these characteristic features is integrated into the program. Here, we present the most recent version of the HRGP bioinformatic program developed in our laboratory called BIO OHIO 2.0. This program and its predecessor allow for the effective identification and classification of HRGPs from proteomic databases by utilizing bioinformatic approaches involving biased amino acid composition searches and HRGP amino acid motif searches [11, 16].

Successful identification and classification of HRGPs will facilitate both basic and applied research on these important cell wall proteins, particularly as they relate to emulsifiers [1], adhesives [17], and biofuel [18]. Moreover, HRGP comparisons among different plant species will provide further insight to the roles these HRGPs play in plants and in evolution. This chapter provides a tutorial on how to use the BIO OHIO 2.0 bioinformatic program to identify and classify HRGPs from proteomes revealed by genomic sequencing (Fig. 1). To provide a practical example of using the BIO OHIO 2.0 program, we have chosen to analyze the proteome of the common bean (*Phaseolus vulgaris*).

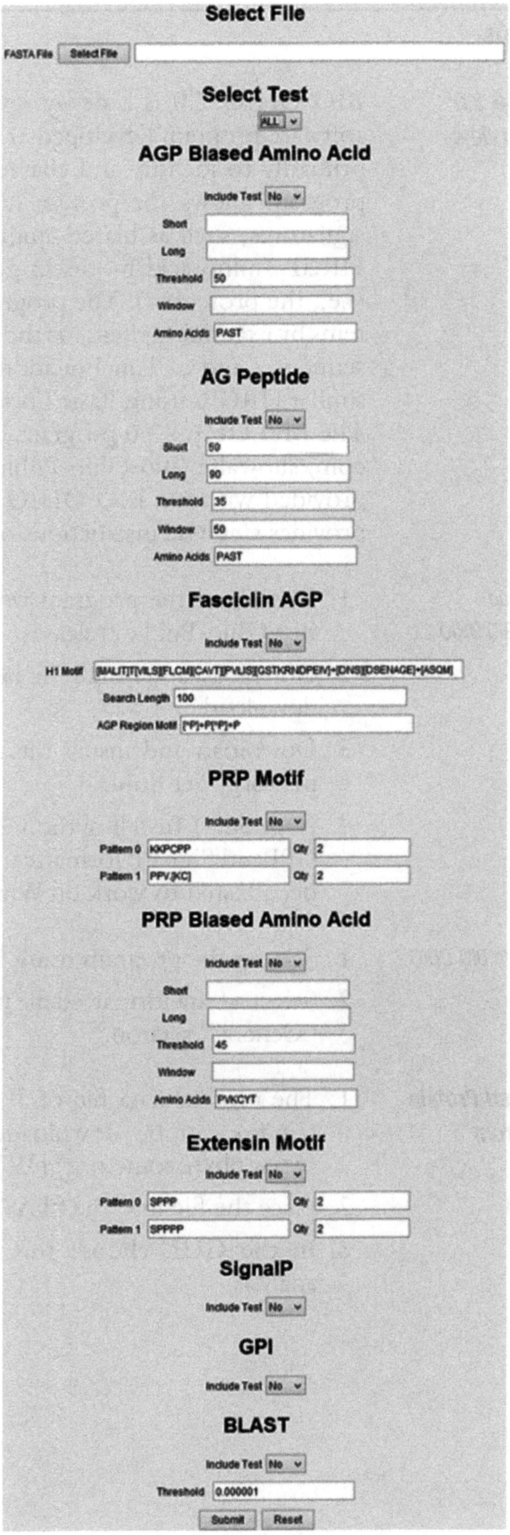

Fig. 1 Graphical user interface of the BIO OHIO 2.0 program

2 Materials

2.1 BIO OHIO 2.0 Program Overview

BIO OHIO 2.0 is a newly revised and improved bioinformatics software program developed at Ohio University that was designed primarily to identify and characterize plant HRGPs (Fig. 1). The program allows for protein identification based on amino acid signatures, such as biased amino acid compositions and common HRGP amino acid motifs in genome-encoded protein sequences (i.e., the proteome). The program further analyzes identified proteins by examining them for the presence of potential signal peptide sequences and GPI anchor addition sequences and by searching for similar HRGPs using Basic Local Alignment Search Tool (BLAST). The BIO OHIO 2.0 program is freely available at https://github.com/showalte/Bio-Ohio-Public/releases. A program guide is also provided with the BIO OHIO 2.0 program at the same URL. It provides stepwise instructions on how to use the program.

2.1.1 Download and Install the BIO OHIO 2.0 Program

1. Download the program from https://github.com/showalte/Bio-Ohio-Public/releases.
2. Download and install the Java program: https://java.com/en/download/.
3. Download and install the ActivePerl program: https://www.perl.org/get.html.
4. (Optional) Install python 2 or 3 from python.org to use the GPI and SignalP formatter. Note that BIO OHIO 2.0 has only been tested to work on Windows.

2.1.2 Open the BIO OHIO 2.0 Program

1. Startup the program using BioOhio.bat in the program folder.
2. Select a fasta formatted file using the "Select File" button in the "General" section.

2.2 Predicted Protein Sequences from the Genome of P. vulgaris

1. The protein data file of *P. vulgaris* (Pvulgaris_218_v1.0.protein.fa) can be downloaded from the Phytozome website (www.phytozome.org) [19].
2. Place the file into the "FASTAfiles" folder in BIO OHIO 2.0.
3. In the GUI, choose this file in the "General" section for analysis.

3 Methods

3.1 Methods for Identifying AGPs

3.1.1 Identification of Classical AGPs

Classical AGPs were identified by examining proteins in the proteome for biased amino acid compositions. Candidate classical AGPs were identified as having 50% or greater of the amino acids P, A, S, and T (PAST). These sequences were also examined for potential signal peptide and GPI anchor addition sequences.

BIO OHIO 2.0 Modules Used for Identifying Classical AGPs

1. "Biased AA" module.

   The "Biased AA" module was used to search for biased amino acid composition of proteins. "Include Test" denotes whether this test is to be performed (Yes) or not (No). "Short" and "Long" denotes the proteins to be included in the test based on protein lengths as defined by a lower and upper limit to the number of amino acids in a given protein. "Threshold" denotes the minimum percentage of the selected amino acids to pass the test. The default is set at 50%. "Window" denotes the length of sequence for calculation. "Amino Acids" denotes the specific biased amino acids (e.g., PAST).
2. "SignalP" module.

   The "SignalP" module was used to search for the presence of a signal peptide by running each sequence through the SignalP website (http://www.cbs.dtu.dk/services/SignalP/) [20]. This module is used for all of the HRGP tests as most HRGPs have predicted N-terminal signal peptides.
3. "GPI" module.

   The "GPI" module was used to search for the presence of a GPI anchor addition sequence by running each protein sequence through the big-PI plant predictor website (http://mendel.imp.ac.at/gpi/plant_server.html) [21]. This module is used for all of the HRGP tests given the majority of AGPs have GPI anchor addition sites, while most EXTs and PRPs do not.
4. "Blast Analysis" module.

   The "Blast Analysis" module subjects each identified sequence to a protein BLAST search against the Arabidopsis proteome on the NCBI website, and returns any hits of Arabidopsis HRGPs previously identified by Showalter et al. [11]. This module is used for all of the HRGP tests because BLAST results provide valuable information for identification at the subfamily level.

Operating BIO OHIO 2.0 for Identifying Classical AGPs

1. Choose the default values (50 for the threshold and PAST for the amino acids) in the "Biased AA" module. Click "Yes" to include the test.
2. Click "Yes" to include the "SignalP," "GPI," and "BLAST Analysis" modules in the test.

3. Click the "submit" button located at the bottom of the GUI to run the test.
4. Once the "Submit" button is clicked, a folder will be created with several subfolders at the destination "protclass\cgi-bin" with the time and date as the folder name. Meanwhile, the program GUI will freeze until the program finishes running. Once finished, two additional files will be created: an Excel file that shows the results of the test and a FASTA file that contains all protein sequences that passed the test.

Output Results for the Biased Amino Acid AGP Test

In the *P. vulgaris* proteome, 49 candidate AGPs were found that had more than 50% of PAST.

Among them, 15 sequences had a predicted signal peptide and clustered noncontiguous proline residues residing in dipeptide sequences such as AP, PA, SP, TP, VP, and GP. These 15 proteins were therefore deemed as classical AGPs. In addition, the BLAST results of one sequence (Phvul.005G073200) revealed significant similarities with PAGs found in Arabidopsis. Thus, this sequence was deemed a PAG (Fig. 2 and Table 1).

3.1.2 Identification of AG Peptides

For AG peptides, a reduced PAST threshold percentage of 35% and a protein length of 50–90 amino acids were used, since AG peptides usually contain an N-terminal signal peptide and possibly a C-terminal GPI anchor addition sequence.

Modules in BIO OHIO 2.0 Used for Identifying AG Peptides

The "AG Peptide" module was used to search for AG peptides. The length of the AG peptide was restricted from 50 (Short) to 90 (Long) amino acids, and the threshold for the biased amino acids (i.e., PAST) was set to 35%.

Operating the BIO OHIO 2.0 for Identifying AG Peptides

Include the "AG Peptide" module, the "SignalP" module, the "GPI" module, and the "BLAST Analysis" module in the test.

Output Results for the AG Peptide Test

The AG peptide test for *P. vulgaris* returned 156 candidate proteins, among which 19 proteins had predicted signal peptides and were rich in noncontiguous proline residues residing in dipeptide sequences such as AP, PA, SP, TP, VP, and GP; many of these 19 proteins had predicted GPI anchor addition sites. These 19 sequences were deemed AG peptides (Fig. 2 and Table 1).

3.1.3 Identification of FLAs

FLAs were identified by searching for one or more fasciclin motifs; a typical fasciclin motif was identified by comparing all known Arabidopsis FLAs and consisted of the following peptide sequence: [MALIT]T[VILS][FLCM][CAVT][PVLIS] [GSTKRNDPEIV] + [DNS][DSENAGE] + [ASQM] [22]. Candidate sequences were also examined for potential signal peptide and GPI anchor addition sequences.

Classical AGP

```
>Phvul.003G238200
MDRNSIFSLAFICIVIAGVGGQSPASAPSGTQPTPAASTPAAAPSTTKSPAPVASPKSSTPAASPKAVTPASSPVAS
PPTVVAPAPATKPPAASPPAAAPVSSPPAPVPVSSPPAPVPIAAPVAAPPTPVAPAPAPGKHKKSKKHGAPAPSPSL
LGPPAPPTGAPGPSEDASSPGPASSANDESGAETIMCLKKVLGGLALGWATLVLVF
```

AG Peptide

```
>Phvul.011G150500
MDMRKVTCAILIAAASVSATMAAAEVPAPAPGPSSGASAAFPLVGSLVGASVLSFFALFH
```

FLA

```
>Phvul.001G058300
MSFKSSSLLCIAFLLAFSSAVYGFDITKMLEKEPELSSFNKYITEAKLADQINSRNTITVLAVGNDAISSIAGKSPE
LIKAIISTHVILDYYDEKKLVEAQASTPQLTTLFQSSGNAVKEQGFVKVSLIGEGEIAFSSVGSSDYSELVKPIASE
PYNISILQVTKPIIPPGLDSQTAQSPQQAKASPPTSSKTAKAPAPSKTAKAPSPAKSSEAPAPSDAAAAPSPSETVA
ESPLGGSDAPATAAEGPAADDGDAASDSSSSSTVKMGLVAVMALASLFIVS
```

PAG

```
>Phvul.005G073200
MELRISVFCLLSFLFSLLSGSQAYTFNVGGKDGWLLYPSENYNHWAERMRFQVSDTLVFKYKKGSDSVLVVNKEDYE
KCNKKNPIKKFEDGASQFQFDRSGPFYFISGKDYNCEKGQKLIIVVLAVREPPPYSPPKAPYPPHQTPPPVYTPPEA
PSPPTNLPFPPKAQPPYAPIPPNTPSPTYAPSPNNHPPHAPLPPNTPSPHHQPPFVPITPSPFSQSPYVPPQPNTPS
PTSQPPYTPSPPNTPSPISQPPYIPSPPNTTPSPISQPPYISTPPSTTPSPISQPPSYISTPPSTTPSPISQPPYIP
SPPNTPSPTSQSPYPYAPTPNSNSPVTQPPSSSSPPSPPSLSPYPSTIPPSYPPSPFAPATTPSPPLSSPPSESTPA
ATSPSSSSSPGSSSNETTPSRPNGASSMSKSRFGVYSLTILVGAALSTILG
```

Fig. 2 Representative AGPs in *P. vulgaris*. Highlighted sequences indicate predicted signal peptide (green) and GPI anchor addition sequences (light blue). AP, PA, SP, TP, GP, and VP dipeptide sequences (yellow) are also indicated

Modules in BIO OHIO 2.0 Used for Identifying FLAs

The "Fasciclin" module is used to search for FLAs. The "H1 Motif" denotes a regular expression of a characteristic, known fasciclin motif in AGPs. "Search Length" denotes the maximum length of the window used in searching for this motif. "AGP Region Motif" denotes the motif to be searched in the AGP domain of a protein.

Operating BIO OHIO 2.0 for Identifying FLAs

Include the "Fasciclin" module, the "SignalP" module, the "GPI" module, and the "BLAST Analysis" module in the test.

Output Results for the FLA Test

The fasciclin test for *P. vulgaris* returned 13 candidate proteins that have a core fasciclin domain. Among the 13 sequences, eight sequences had a predicted signal peptide and had significant similarities with Arabidopsis FLAs as revealed by the BLAST Analysis results. Therefore, these eight proteins were deemed FLAs (Fig. 2 and Table 1).

3.2 Methods for Identifying EXTs

The regular expression of two or more SPPP repeats was used to search for EXTs. The resulting candidate extensin sequences were analyzed for the locations of the SP_n repeats (where $n \geq 3$) and YXY cross-linking motifs (where X can be any amino acid), the presence of signal peptide sequences, the presence of GPI anchor addition sequences, and for similar sequences using BLAST searches against

Table 1
Identification and classification of *P. vulgaris* AGPs

Protein ID	AGP subfamily	Amino acids	PAST %	SP	GPI	Top 5 Arabidopsis HRGP BLAST hits
Phvul.003G289900	Classical AGP	140	68	Y	Y	AGP5C AGP10C AGP18K PEX4 AGP2C
Phvul.003G238200	Classical AGP	210	63	Y	Y	AGP18K AGP17K PEX2
Phvul.009G033400	Classical AGP	197	74	Y	N	None
Phvul.004G059300	Classical AGP	146	61	Y	N	AGP6C PAG17 AGP18K AGP17K
Phvul.007G058600	Classical AGP	222	50	Y	Y	AGP29I
Phvul.006G057200	Classical AGP	193	64	Y	Y	AGP10C PEX2
Phvul.006G204100	Classical AGP	133	66	Y	Y	AGP1C
Phvul.002G149200	Classical AGP	144	74	Y	Y	AGP4C AGP10C AGP2C AGP7C AGP3C
Phvul.002G211000	Classical AGP	133	63	Y	Y	AGP1C AGP10C
Phvul.003G281900	Classical AGP	419	63	Y	N	AGP18K
Phvul.003G024400	Classical AGP	261	65	Y	Y	None
Phvul.006G126400	Classical AGP	283	68	Y	N	PRP10 EXT51
Phvul.002G272600	Classical AGP	681	67	Y	N	None
Phvul.002G272800	Classical AGP	470	62	Y	N	None
Phvul.002G014100	Classical AGP	297	74	Y	Y	AGP18K PRP1 AGP17K PRP2 PRP14
Phvul.003G278100	AG Peptide	80	35	Y	N	None
Phvul.009G137200	AG Peptide	63	44	Y	Y	AGP16P AGP22P AGP20P AGP41P AGP43P
Phvul.005G019400	AG Peptide	67	47	Y	Y	AGP16P AGP22P
Phvul.005G019300	AG Peptide	66	46	Y	Y	AGP16P AGP43P
Phvul.005G036200	AG Peptide	66	43	Y	Y	AGP43P AGP23P

(continued)

Table 1
(continued)

Protein ID	AGP subfamily	Amino acids	PAST %	SP	GPI	Top 5 Arabidopsis HRGP BLAST hits
Phvul.005G162900	AG Peptide	66	43	Y	N	PERK1
Phvul.011G150500	AG Peptide	60	50	Y	Y	AGP43P AGP23P AGP40P AGP14P AGP13P
Phvul.011G150600	AG Peptide	60	50	Y	Y	AGP43P AGP40P AGP23P AGP14P AGP13P
Phvul.008G250300	AG Peptide	53	39	Y	N	None
Phvul.004G002700	AG Peptide	88	35	Y	N	None
Phvul.004G080600	AG Peptide	55	41	Y	Y	AGP14P AGP12P AGP21P AGP13P AGP22P
Phvul.007G255800	AG Peptide	58	44	Y	Y	AGP43P AGP23P AGP40P AGP14P AGP15P
Phvul.001G137100	AG Peptide	58	44	Y	Y	AGP43P AGP23P AGP40P AGP14P AGP12P
Phvul.001G033400	AG Peptide	59	52	Y	N	AGP15P AGP14P AGP21P AGP13P AGP12P
Phvul.001G111400	AG Peptide	57	45	Y	Y	AGP14P AGP12P AGP13P AGP21P AGP22P
Phvul.001G079000	AG Peptide	59	40	Y	Y	AGP16P AGP20P AGP22P AGP41P AGP15P
Phvul.006G179100	AG Peptide	69	37	Y	N	None
Phvul.002G162600	AG Peptide	72	37	Y	Y	AGP16P AGP20P AGP22P AGP41P AGP15P
Phvul.002G296000	AG Peptide	74	39	Y	Y	AGP20P AGP16P AGP22P AGP41P AGP12P
Phvul.009G218400	FLA	207	40	Y	Y	FLA20 FLA6 FLA21 FLA11 FLA9
Phvul.011G100600	FLA	450	28	Y	N	FLA16 FLA17 FLA18 FLA15 FLA11
Phvul.008G288800	FLA	403	31	Y	Y	FLA2 FLA1 FLA8 FLA10 FLA14
Phvul.008G075000	FLA	419	42	Y	Y	FLA10 FLA8 FLA2 FLA1 FLA14
Phvul.008G287700	FLA	257	42	Y	Y	FLA7 FLA13 FLA11 FLA6 FLA9
Phvul.007G037300	FLA	326	32	Y	N	FLA20 FLA21 FLA19 FLA12 FLA7
Phvul.001G058300	FLA	282	44	Y	Y	FLA8 FLA3 FLA14 FLA5 FLA10
Phvul.006G057400	FLA	420	35	Y	Y	FLA1 FLA2 FLA8 FLA10 FLA4
Phvul.005G073200	PAG	436	54	Y	Y	PAG17 PAG10 PAG11 PAG14 PAG2

known Arabidopsis EXTs. A candidate EXT is deemed to be an EXT if it has the predicted signal peptide (except for PERKs which are known to lack signal peptides) and a cluster of EXT motifs (SP_{3-5}); otherwise, it is called a possible EXT. In occasional instances, however, a candidate sequence may still be deemed as an EXT if it has clustered EXT motifs but lacks the predicted signal peptide.

3.2.1 Modules in BIO OHIO 2.0 Used for Identifying EXTs

The "Extensin" module was used to identify EXTs. "Pattern 0" and "Pattern 1" denote specific extensin motifs used in the search, such as SPPP for Pattern 0 and SPPPP for Pattern 1. "Qty" denotes the minimum number of occurrences to pass the test. The default searching criteria are any protein sequences with two or more SPPP or SPPPP sequences.

3.2.2 Operating BIO OHIO 2.0 for Identifying EXTs

Include the "Extensin" module, the "SignalP" module, the "GPI" module, and the "BLAST Analysis" module in the test.

3.2.3 Output Results for the EXT Test

The EXT test for the *P. vulgaris* proteomics (protein data) file returned 79 candidate EXT sequences, each with at least two SPPP repeats. Analysis of these 79 sequences revealed: (1) eight classical EXTs that had predicted signal peptide sequences and EXT motifs ($SPPP_{3-5}$ and possibly YXY motifs) throughout the sequence; (2) 11 short EXTs (<200 AA) that had predicted signal peptide sequences and EXT motifs; (3) four LRXs that had signal peptide sequences, clusters of EXT motifs at the C terminus, and significant similarities to Arabidopsis LRXs. Two of the LRXs (Phvul.003G011500 and Phvul.009G062000) are possibly pollen-specific LRXs (PEXs) as indicated by BLAST results; (4) six PERKs that had clusters of EXT motifs near the N terminus and significant similarities to Arabidopsis PERKs; (5) one FH EXT that had a predicted signal peptide and was homologous to formins; (6) one chimeric EXT that had a predicted signal peptide, clustered EXT motifs, and a non-HRGP domain; and finally (7) one hybrid AGP-EXT (HAE) that had a predicted signal peptide, clustered EXT motifs and an AGP domain (Fig. 3 and Table 2).

3.3 Methods for Identifying PRPs

PRPs were identified using two approaches. One approach involved searching for proteins having a biased amino acid composition of greater than 45% PVKCYT [23]. The other approach involved using regular expressions in order to identify proteins which contain two or more PPVX(K/T) sequences [where X represents any amino acid] or KKPCPP sequences [23]. Candidate PRP proteins from both approaches were then examined for the location and distribution of these two amino acid sequences as well as PPV sequence repeats. These sequences were also examined for potential signal peptide and GPI anchor addition sequences.

Classical EXT

>Phvul.004G098000
MGSLMASITLTLVLAIVSLSLPSQTSADNYPYSSPPPPPKNPYYYHSPPPQEYSPPKHPYHHPSPPYKYPSSPPPIY
KNKSPPPPYKYSSPPPPPKKPYKYPSPPPPVYKYKSPPPPVYKYKSPPPPPKKPYKYPSPPPPVYKYKSPPPPVYKY
KSPPPPVYKYKSPPPPYKYPSPPPPPYKYPSPPPPVYKYKSPPPPYKYPSPPPPPYKYPSPPPPVYKYKSPPPPYKY
PSPPPPPYKYPSPPPPAYYYKSPPPPPKYPSPPPPHYVYASPPPPHHY

Short EXT

>Phvul.011G059100
MGTRQWPRLILAFAFSLMAITLAADYDKPYYSQPSTYYPHPTPPYQQQRSPPYYYKSPPPPPYVRKFPPYYYKSPP
PPSPSPPPPYVYKSPPPPSPSPPPPYLY

LRX

>Phvul.003G011500
MAHHCSNKALGLLIFLSLLSTICSAQLVPEFTHSEPVDPADAVAPAPEVDEDGAELAPAPSIEFENDEPLTPPPPAK
TNERLKKAHIAFKAWKKAIHSDPLNITGNWVGEDVCSYNGVYCAPALHDPTINVVAGIDLNNADIAGNLPEELLHLE
DIALFHINSNRFCGVIPQNLQNLTLMHEYDISNNRFVGSFPSVVLTWPNLKYLDLRFNDFEGAIPPELFHKNLDAIF
LNNNRFTSIIPDSLGNSSASVITFAYNNFKGCVPNSMGNMRNLNEIVFIGNNLGCCFPQEIGTMENLRVLDLSGNGF
VGTLPNLSGLKSVEVIDIAYNKMSGYVSNSVCQLPALKNFTFSHNYFNGEAQSCVAEGNSSVALDDSWNCLPGRKNQ
KASMKCLPVLTRPVDCVKQCGGGKENEHSHSPKPSPSPKPLTPKVVHSPPPPVHSPPPPPVHSPPPPPPPVNSP
PPPPPVNSPPPPPVFSPPPPVFSPPPPPVSSPPPPPVHSPPPPVHSPPPPVHSPPPPVSSPPPPPVHSPPPPVFS
PPPPPVSSPPPPVHSPPPPVSSPPPPVSSPPPPPVSSPPPPVHSPPPPVFSPPPPVFSPPPPPVHSPPPPVFSPPPP
PPVHSPPPPVFSPPPPVFSPPPPPLVSSPPPPVSSPPPPPVSSPPPPVNSPPPPPVSSPPPPVSSPPPPIFSPPPPV
SSPPPPVYSPPPPVYPPPPPTWDDVFLPPHFGASYKSPPPPVIVGY

PERK

>Phvul.003G024000
MSAASPSPAASTASPPSQTPSSSNTGPSPSNTTTPPPQQTPSSTSPPQQPASPPPSPPSSPPSSPPSSSPPPVSGTT
PPSVPPPSPPPSPPPAPDAPPPVTPSSPSPPPPVTPSSPSPPPPVTPSSPPPPPSAPIPSRSPPSPPPPSNPPNNT
SPPPLPPPPQPSPSAPPPNNTPPPPPRGNSPPPPATTPPPASPPRNSPPAPAAPPPSNSTRSPPPVNSPPPAAHAPP
PRSSAPPAPEPSNPPSSISPPPTPSSSPPPPSNSTPSSSPPSPPSLTPLAPPPPPSPPSPPPNATSDSTPGGDGIGT
AGVVAISVVGGFLLLGFIGVLMWCMRRQKRKIPVNGGYVMPSTLASSPESDSSFFKTHSSAPLVQSGSGSDVVYTPS
DPGGLGHSRSWFSYEELIKATNGFSSQNLLGEGGFGCVYKGQLPDGREIAVKELKIGGGQGEREFKAEVEIISRIHH
RHLVSLVGYCIEDNRRLLVYDFVSNDTLYFHLHGESQPVLEWANRVKIAAGAARGLAYLHEDCNPRIIHRDIKSSNI
LLDFNYEAKVSDFGLAKLALDANTHITTRVMGTFGYVAPEYASSGKLTEKSDVYSFGVVLLELITGRKPVDVSQPLG
DESLVEWARPLLSQAIDTEEFNTLADPRLEKNYIESELYCMIEVAAACVRHSASKRPRMGQVVRAFDSLGGSDLTNG
MQLGESEVFDSAQQSEEIRLFRRMAFGSQNYSTDFFSRASMNP

FH EXT

>Phvul.008G050300
MQISSFFFFFYLFLLCALASSQPLFFNRRILHEPFIPLTSLSPSDPPKPPPSHPSPSQPPKPPPPHPSPSASKQKP
KYPSSSTIPTTTITTTPETATTTTQSPFFPLYPSSPPPPSPITFASFPANISSLILPHSPKPSSSSNKLLPVALSA
VVAAALVISISTFVCYRRRRNAPPSPAGKVLRSETGLRPLRRNAETSVETRKLRHTSSASSEFLYLGTVVNSHMVEE
AEVGDGDRKMESPELRPLPPLARQLSVPPAPRDEAGFMTAEEDEDEFYSPRGSSLGGSGGTGSVSRRVFAADRSVTS
SSRSSSSSGSPERSITNLLPRASSSYGNTLPKSPENYNHQHVHSSSSSMCSTPDRVFAERDNDALSACAHADAAPSS
LHEGTLEKNENALSSPPPQRLSNASSSSAFSLPSSPENVTRHHTFDQSPRMSSVSDGLMLPGLSSLPLSPALLSSPE
TERGSFGAQRKHWSIPVLSMPITTPFDEIRSIPAPPPLPQRKHWEIPGPAPPPPPPLPRQRKQWGVQAPGPSTPVGQ
PVSRPPELVPPSRPFVLQNQATNVELPGSLREIEETGRPKLKPLHWDKVRTSSEREMVWDQMKSSSFKLNEKMIETL
FVVNTPNPKGKDAATNSVSHPPNQEERILDPKKSQNISILLKALNVTIEEVCEALSEGSTDALGTELLESLLRMAPS
KEEERKLREHKDESQTKLGLAEKFLKAVLDVPFAFKRIEAMLFIASFESEVEYLRTSFQTLEAACEELRHCRMFIKL
LEAVLKTGNRMNVGTNRGDAEAFKLDTLLKLADVKGADGKTTLLHFVVQEISRTEGARLSDTNQTPSSSLNEDGKCR
RLGLQVVSSLSSELSNVKKAAAMDSEVLSSDVSKLSKGIATIAEVVQLNQSSENFTESVKKFISMAEEEIPKIQAQE
SVASSLVKEITEYFHGNLAKEEAHPFRLFMVVRDFVAVLDRVCKEVGMMNERTMVSSAHKFPVPVNPMLPQPLPGSP

Fig. 3 Representative EXTs in *P. vulgaris*. Highlighted sequences indicate predicted signal peptide (green), SP_3 (blue), SP_4 (red), SP_5 (purple), and YXY (dark red) sequences

Table 2
Identification and classification of *P. vulgaris* EXTs

Protein ID	EXT subfamily	Amino acids	$SP_3/SP_4/SP_5/$ YXY counts	SP	GPI	Top 5 Arabidopsis HRGP BLAST hits
Phvul.011G059000	Classical EXT	338	7/0/7/27	Y	N	EXT22 EXT21
Phvul.008G003200	Classical EXT	884	79/0/0/47	Y	N	None
Phvul.004G098300	Classical EXT	557	0/36/3/25	Y	N	EXT22
Phvul.004G098000	Classical EXT	279	2/14/7/21	Y	N	EXT3/5 EXT22 EXT18
Phvul.007G099700	Classical EXT	233	1/8/5/7	Y	N	EXT3/5 PRP1
Phvul.007G002400	Classical EXT	870	1/54/6/48	Y	N	EXT22 EXT3/5 HAE3
Phvul.007G084600	Classical EXT	679	1/50/1/33	Y[a]	N	EXT22
Phvul.007G078600	Classical EXT	427	0/27/5/21	Y	N	EXT3/5
Phvul.003G023700	Short EXT	160	1/4/0/1	Y	Y	EXT31 EXT33
Phvul.003G110300	Short EXT	174	1/2/1/4	Y	Y	None
Phvul.009G253400	Short EXT	165	0/4/0/2	Y	Y	EXT37
Phvul.008G234100	Short EXT	160	1/0/2/2	Y	Y	EXT33 PERK6
Phvul.004G155200	Short EXT	159	2/0/0/0	Y	Y	None
Phvul.007G216700	Short EXT	175	0/3/1/0	Y	N	None
Phvul.001G231800	Short EXT	173	0/0/2/1	Y	N	None
Phvul.006G169900	Short EXT	138	0/2/1/0	Y	N	None
Phvul.002G093100	Short EXT	131	0/2/0/0	Y	Y	EXT31 EXT33 PERK3 PERK4 PERK6
Phvul.011G059100	Short EXT	105	0/4/1/4	Y	N	AGP45P EXT21
Phvul.011G166600		180	1/0/2/0	Y	N	None

(continued)

Table 2
(continued)

Protein ID	EXT subfamily	Amino acids	$SP_3/SP_4/SP_5/$ YXY counts	SP	GPI	Top 5 Arabidopsis HRGP BLAST hits
	Short EXT					
Phvul.003G011500	LRX	739	0/23/14/0	Y	N	PEX1 PEX2 PEX4 LRX5 LRX4
Phvul.009G062000	LRX	634	1/10/1/0	Y	N	PEX1 PEX3 PEX4 PEX2 LRX3
Phvul.004G161500	LRX	725	3/15/13/3	Y	N	LRX4 LRX5 LRX3 PEX4 PEX2
Phvul.002G314200	LRX	740	3/2/3/1	Y	N	LRX4 LRX5 LRX3 LRX7 PEX2
Phvul.010G149300	PERK	566	0/1/1/1	N	N	PERK1 PERK15 PERK4 PERK5 PERK3
Phvul.003G024000	PERK	736	9/5/1/0	N	N	PERK8 PERK12 PERK13 PERK1 PERK15
Phvul.008G021000	PERK	743	3/2/0/2	N	N	PERK10 PERK12 PERK1 PERK15 PERK4
Phvul.004G164000	PERK	663	5/0/0/0	N	N	PERK1 PERK5 PERK4 PERK15 PERK3
Phvul.002G070900	PERK	695	9/3/0/1	N	N	PERK12 PERK8 PERK10 PERK15 PERK5
Phvul.002G029200	PERK	631	6/0/0/1	N	N	PERK4 PERK5 PERK7 PERK6 PERK15
Phvul.008G050300	FH	1001	1/1/0/0	Y	N	None
Phvul.004G097800	Chimeric EXT	248	0/4/9/0	Y	N	None
Phvul.002G052800	HAE	213	15/0/0/0	Y	Y	AGP17K
Phvul.010G159300	HAE	384	1/5/0/1	Y	N	None

[a]Signal peptide sequence was found using the sensitive mode on the SignalP website

3.3.1 Identifying PRPs Through Biased Amino Acid Compositions

The "Biased PRP" module was used to search for PRPs with biased amino acid compositions. The default values are set as 45% PVKCYT amino acids.

Modules in BIO OHIO 2.0 Used for Identifying PRPs Based on Biased Amino Acid

Operating BIO OHIO 2.0 for Identifying PRPs with Biased Amino Acid Compositions

Include the "Biased PRP" module with the default values, as well as the "SignalP," "GPI," and "BLAST Analysis" modules in the test.

Output Results for the Biased Amino Acid Test for PRPs

The test in *P. vulgaris* returned 125 candidate PRP sequences, in which: (1) 19 PRPs were identified that had a predicted signal peptide, contiguous proline residues and were rich in PVKCYT; (2) three PR peptides (<200 AA) that had predicted signal peptides, contiguous proline residues and were rich in PVKCYT; (3) two chimeric PRPs that had a predicted signal peptide, a domain rich in contiguous prolines and VKCYT residues and a non-HRGP domain; (4) one hybrid AGP-PRP (HAP) that had a predicted signal peptide, an AGP domain and a domain typical of PRPs; and (5) one chimeric AGP-PRP that had a predicted signal peptide, an AGP domain, a PRP domain, and a non-HRGP region (Fig. 4 and Table 3).

PRP

```
>Phvul.009G211000
MASLSFLVLLFAALVLSPQGLANYDKPPVYKPPIQKPPVYKPPVEKPPVYKPPVEKPPVYKPPVEKPPVYKPPVQKP
PVYKPPYGKPPVHESQYEKAPLYKSPPLYKPPVHKPPVEKPPVYKPPVHKPPVEKPPVHKPPYGKPPVQESQYEKPP
VYKSPPVYKPPVHKPPVEKPPVYKPPVHKPPVEKPPVYKPPYGKPPHPKYPPGSN
```

PR Peptide

```
>Phvul.005G027300
MASFLSFLVLLLAALILIPQGLATYYKPIKKHPIYKPPVYKHPIYKPPVYKHKPPYYKPPYKKPPYKKHPPVEENNN
HA
```

Chimeric PRP

```
>Phvul.003G169100
MESSKIHAYLFLSMLFISSATPILGCGYCGKPPKKPNHGKKPKTPVVNPPVTHPPIVKPPVIVPPITVPPVTVPPVT
VPPIVKPPIPLPIPPVTVPPVLNPPTTPTTPGKGGNTPCPPPNSPAQATCPIDTLKLGACVDLLGGLVHVGVGDPAA
NQCCPVLQGLVELEAAACLCTTLKLKLLNLNIYVPIALQLLVACGKSPPPGYTCSL
```

Fig. 4 Representative PRPs in *P. vulgaris*. Highlighted sequences indicate predicted signal peptide (green), PPVX(K/T) (gray), and PPV (pink) sequences

Table 3
Identification and classification of *P. vulgaris* PRPs

Protein ID	PRP subfamily	Amino acids	PVKCYT %	PPVX [KT]/KKPCPP/ PPV counts	SP	GPI	Top 5 Arabidopsis HRGP BLAST hits
Phvul.009G213600	PRP	315	46	0/0/0	Y	N	PRP13
Phvul.005G018100	PRP	326	47	0/0/1	Y	N	PRP10
Phvul.005G048800	PRP	375	45	0/0/1	Y	N	PRP13 PEX4
Phvul.011G209600	PRP	338	71	0/0/0	Y	N	PEX2 PRP2
Phvul.011G209800	PRP	286	55	0/0/0	Y	N	PEX2 PEX4
Phvul.009G092000	PRP	162	47	0/0/0	Y	N	PRP15 PRP14 PRP16 PRP17 HAE4
Phvul.009G211100	PRP	137	58	7/0/10	Y	N	PRP1 PEX4
Phvul.009G211200	PRP	331	72	34/0/39	N	N	PRP1 PRP2 PRP5
Phvul.009G211000	PRP	209	72	24/0/26	Y	N	PRP1 PRP2 PRP5
Phvul.011G026700	PRP	386	55	2/0/5	Y	N	PRP2 PRP4 PEX1
Phvul.011G208200	PRP	266	55	0/0/0	Y	N	PEX2
Phvul.011G209700	PRP	232	53	0/0/0	Y	N	PEX2
Phvul.004G019900	PRP	179	58	0/0/4	Y	N	PRP18 PRP1 PEX2 PRP15
Phvul.007G127700	PRP	392	69	0/0/19	Y	N	PRP1 PRP5
Phvul.001G031400	PRP	295	54	0/0/0	Y	N	PRP10 PRP9 PEX2 PRP15
Phvul.006G003700	PRP	332	66	4/0/11	Y	N	PRP17 PRP15 PRP14 PRP16 HAE4
Phvul.002G259700	PRP	437	63	0/0/3	Y	N	PRP1 PRP5
Phvul.002G212200	PRP	176	52	0/0/8	Y	N	PRP10 PRP15 PRP9
Phvul.009G217800	PRP	186	50	0/0/1	Y	N	PRP1 PRP15 EXT20 PRP2 PEX4
Phvul.005G027300	PR Peptide	79	55	2/0/3	Y	N	PRP1 PRP2
Phvul.003G219200	PR Peptide	131	45	1/0/1	Y	N	PRP14 PRP16 PRP15 PRP17 HAE4
Phvul.003G219100	PR Peptide	131	45	1/0/1	Y	N	PRP14 PRP16 PRP15 PRP17 HAE4

(continued)

Table 3 (continued)

Protein ID	PRP subfamily	Amino acids	PVKCYT %	PPVX [KT]/KKPCPP/ PPV counts	SP	GPI	Top 5 Arabidopsis HRGP BLAST hits
Phvul.002G229200	HAP	286	53	0/0/10	Y	N	AGP30I AGP31I PRP7 PRP1 PRP11
Phvul.003G219000	Chimeric PRP	171	46	0/0/2	Y	N	PRP14 PRP15 PRP17 PRP16 HAE4
Phvul.003G169100	Chimeric PRP	210	53	0/0/6	Y	N	PRP15 PRP14 PRP16 PRP17 PRP2
Phvul.004G042800	Chimeric AGP PRP	365	46	0/0/0	Y	N	PRP11 PEX2

3.3.2 Identifying PRPs Using a Motif Search

Modules Needed for Identifying PRPs with Characteristic Motifs

The "PRP" module was used to search for proteins with two regular expressions designed to identify KKPCPP or PPV.[KT] sequences, where "." indicates any amino acid in this second regular expression. The default threshold is set to 2, such that two or more occurrences of either sequence will trigger a "hit."

Operating BIO OHIO 2.0 for Identifying PRPs with Characteristic Motifs

Include the "PRP" module with the default values, as well as the "SignalP," "GPI," and "Blast Analysis" modules in the test.

Output Results for the Test of PRPs with Characteristic Motifs

The PRP test in *P. vulgaris* returned ten candidate PRP sequences, among which one PR peptide and four PRPs were identified. These five PRPs and PR peptides, however, were already found using the biased PRP test (Fig. 4 and Table 3).

3.4 HRGPs Identified in P. vulgaris

A total of 102 HRGPs were identified and classified in *P. vulgaris* using BIO OHIO 2.0, including 43 AGPs, 31 EXTs, 25 PRPs, and three hybrid HRGPs. These *P. vulgaris* HRGPs represent virtually all subfamilies previously identified in Arabidopsis [11]. Compared with Arabidopsis, *P. vulgaris* has smaller numbers of HRGPs in most subfamilies, particularly with respect to the numbers of FLAs and PAGs in the AGP family, and the number of classical EXTs in the EXT family (Table 4).

4 Notes

1. Highlighting amino acid sequences corresponding to characteristic HRGP features, such as signal peptides (green), GPI anchor addition sequences (light blue), AP, PA, SP, TP, GP, and VP dipeptide sequences (yellow), SP_3 sequences (blue), SP_4 sequences (red), SP_5 sequences (purple), and YXY sequences (dark red), PPVX(K/T) sequences (gray), and PPV sequences (pink) is beneficial to the identification process (Figs. 2, 3, and 4).
2. When running BIO OHIO 2.0, the BLAST module, as well as the SignalP and GPI modules, often takes much more time to complete than the other modules. Depending on the number of protein sequences to be analyzed, the power of local computers, and the responsiveness of servers, it may take several hours to complete. However, if these three modules are not included, the other modules, such as biased amino acid composition analysis and motif searches, generally are completed within a few minutes.
3. Generally HRGPs have signal peptides that allow them entry to the endomembrane system and delivery to the plasma membrane/cell wall. In our analysis of *P. vulgaris*, all AGPs and EXTs (except for the PERKs) and 94% of the PRPs have predicted signal peptides.
4. Some HRGPs, especially the AGP family, are also modified with a C-terminal GPI membrane anchor that tethers the protein to the plasma membrane. In our analysis of *P. vulgaris*, 67% of AGPs, 12% of EXTs, and none of the PRPs have predicted GPI anchor addition sequences.
5. The same sequence can show up in multiple test results if it meets the criteria, which indicates the likelihood of a hybrid HRGP. Hybrid HRGPs consist of domains from more than one HRGP family. They often show up in multiple tests and may have BLAST hits in more than one HRGP family. For instance, one HAE (Phvul.002G052800) showed up in both the EXT test with 15 SPPP repeats and the biased amino acid AGP test with 73% of PAST and a GPI anchor addition sequence.
6. Chimeric HRGPs consist of domain(s) typical of an HRGP family and a non-HRGP domain. Additional chimeric HRGPs could be identified through BLAST analysis against the *P. vulgaris* proteome itself.

Table 4
Comparison of HRGPs identified in *P. vulgaris* and *A. thaliana*

HRGP family	HRGP subfamily	*P. vulgaris*	*A. thaliana*[a]
AGPs	Classical AGPs	15	25
	AG-Peptides	19	16
	(Chimeric) FLAs	8	21
	(Chimeric) PAGs	1	17
	Other Chimeric AGPs	0	6
	All AGP subfamilies	*43*	*85*
EXTs	Classical EXTs	8	20
	Short EXTs	11	12
	(Chimeric) LRXs	4	11
	(Chimeric) FHs	1	6
	(Chimeric) PERKs	6	13
	Other Chimeric EXTs	1	3
	All EXT subfamilies	*31*	*59*
PRPs	PRPs	19	11
	PR Peptides	3	1
	Chimeric PRPs	3	6
	All PRP subfamilies	*25*	*18*
Hybrid HRGPs	Hybrid AGP EXT	2	4
	Hybrid AGP PRP	1	0
	All hybrid HRGPs	*3*	*4*
Total		*102*	*172*

[a]The Arabidopsis HRGP data are from Showalter et al. [11] with the exceptions of the six chimeric FH EXTs that were added later and the one PR-peptide that was subsequently reclassified within the originally identified 12 PRPs

References

1. Showalter AM (1993) Structure and function of plant cell wall proteins. Plant Cell 5:9–23
2. Kieliszewski MJ, Lamport DTA (1994) Extensin: repetitive motifs, functional sites, post-translational codes and phylogeny. Plant J 5:157–172
3. Cassab GI (1998) Plant cell wall proteins. Annu Rev Plant Physiol Plant Mol Biol 49:281–309
4. Seifert GJ, Roberts K (2007) The biology of arabinogalactan proteins. Annu Rev Plant Biol 58:137–161
5. Kieliszewski MJ, Lamport DT, Tan L et al (2010) Hydroxyproline-rich glycoproteins: form and function. Annu Plant Rev 41:321–342
6. Tan L, Leykam JF, Kieliszewski MJ (2003) Glycosylation motifs that direct arabinogalactan addition to arabinogalactan-proteins. Plant Physiol 132:1362–1369
7. Tan L, Qiu F, Lamport DTA et al (2004) Structure of a hydroxyproline (Hyp)-arabinogalactan polysaccharide from repetitive Ala-Hyp expressed in transgenic *Nicotiana tabacum*. J Biol Chem 279:13156–13165
8. Tan L, Showalter AM, Egelund J et al (2012) Arabinogalactan-proteins and the research challenges for these enigmatic plant cell surface proteoglycans. Front Plant Sci 3:1–10
9. Shpak E, Barbar E, Leykam JF et al (2001) Contiguous hydroxyproline residues direct hydroxyproline arabinosylation in *Nicotiana tabacum*. J Biol Chem 276:11272–11278
10. Liu X, Wolfe R, Welch LR et al (2016) Bioinformatic identification and analysis of extensins in the plant kingdom. PLoS One 11:e0150177
11. Showalter AM, Keppler B, Lichtenberg J et al (2010) A bioinformatics approach to the identification, classification, and analysis of hydroxyproline-rich glycoproteins. Plant Physiol 153:485–513

12. Youl JJ, Bacic A, Oxley D (1998) Arabinogalactan-proteins from *Nicotiana alata* and *Pyrus communis* contain glycosylphosphatidylinositol membrane anchors. Proc Natl Acad Sci U S A 95:7921–7926
13. Schultz C, Rumsewicz M, Johnson K et al (2002) Using genomic resources to guide research directions. The arabinogalactan protein gene family as a test case. Plant Physiol 129:1448–1463
14. Ma H, Zhao J (2010) Genome-wide identification, classification, and expression analysis of the arabinogalactan protein gene family in rice (*Oryza sativa L.*). J Exp Bot 61:2647–2668
15. Newman AM, Cooper JB (2011) Global analysis of proline-rich tandem repeat proteins reveals broad phylogenetic diversity in plant secretomes. PLoS One 6:e23167
16. Lichtenberg J, Keppler B, Conley T et al (2012) Prot-Class: a bioinformatics tool for protein classification based on amino acid signatures. Nat Sci 4:1161–1164
17. Huang Y, Wang Y, Tan L et al (2016) Nanospherical arabinogalactan proteins are a key component of the high-strength adhesive secreted by English ivy. Proc Natl Acad Sci U S A 113(23):E3193–E3202. 1600406113v1-201600406
18. Fleming M, Decker S, Bedinger P (2016) Investigating the role of extensin proteins in poplar biomass recalcitrance. BioResources 11:4727–4744
19. Schmutz J, McClean PE, Mamidi S et al (2014) A reference genome for common bean and genome-wide analysis of dual domestications. Nat Genet 46:707–713
20. Petersen TN, Brunak S, von Heijne G et al (2011) SignalP 4.0: discriminating signal peptides from transmembrane regions. Nat Methods 8:785–786
21. Eisenhaber B, Wildpaner M, Schultz CJ et al (2003) Glycosylphosphatidylinositol lipid anchoring of plant proteins. Sensitive prediction from sequence- and genome-wide studies for Arabidopsis and rice. Plant Physiol 133:1691–1701
22. Johnson KL, Jones BJ, Bacic A et al (2003) The fasciclin-like arabinogalactan proteins of Arabidopsis: a multigene family of putative cell adhesion molecules. Plant Physiol 133:1911–1925
23. Fowler TJ, Bernhardt C, Tierney ML (1999) Characterization and expression of four proline-rich cell wall protein genes in Arabidopsis encoding two distinct subsets of multiple domain proteins. Plant Physiol 121:1081–1091

Chapter 27

Bioinformatics Analysis of Plant Cell Wall Evolution

Elisabeth Fitzek, Rhiannon Balazic, and Yanbin Yin

Abstract

In the past hundreds of millions of years, from green algae to land plants, cell walls have developed into a highly complex structure that is essential for plant growth and survival. Plant cell wall diversity and evolution can be directly investigated by chemically profiling polysaccharides and lignins in the cell walls of diverse plants and algae. With the increasingly low cost and high throughput of DNA sequencing technologies, cell wall evolution can also be studied by bioinformatics analysis of the occurrence of cell wall synthesis-related enzymes in the genomes and transcriptomes of different species. This chapter presents a bioinformatics workflow running on a Linux platform to process genomic data for such gene occurrence analysis. As a case study, cellulose synthase (CesA) and CesA-like (Csl) protein families are mined for in two newly sequenced organisms: the charophyte green alga *Klebsormidium flaccidum* (renamed as *Klebsormidium nitens)* and the fern *Lygodium japonicum*.

Key words Cellulose synthesis, Hemicellulose synthesis, CesA, Csl, GT2, Plant cell walls

1 Introduction

Celluloses, lignins, hemicelluloses, and pectins are essential building components of plant cell walls. They provide developing plant cells with their shape, structural support, as well as acting barriers against insects and pathogens. Carbohydrate-active enzymes (CAZyme) are responsible for the biosynthesis, degradation and modification of cell wall components [1]. CAZymes consist of a total of six classes: glycosyltransferases (GTs), glycoside hydrolases (GHs), polysaccharide lyases (PLs), carbohydrate esterases (CEs), carbohydrate-binding modules (CBMs), and the newest member auxiliary activities (AAs).

With the advent of the low-cost, high-throughput, and high accuracy next-generation sequencing technologies, ~100 plant genomes have been sequenced. Many of these genomes are from model organisms. For nonmodel organisms, a large number of transcriptomes are available in the public genomic databases such as the National Center for Biotechnology Information (NCBI)'s sequence read archive (SRA) database and transcriptome shotgun

Zoë A. Popper (ed.), *The Plant Cell Wall: Methods and Protocols*, Methods in Molecular Biology, vol. 2149,
https://doi.org/10.1007/978-1-0716-0621-6_27, © Springer Science+Business Media, LLC, part of Springer Nature 2020

assembly (TSA) database. Websites such as Phytozome host genomes spanning the plant kingdom from aquatic algae to early land plants to more advanced flowering plants [2].

Two approaches have been used to study the diversity and evolution of plant cell walls [3, 4]. The first one, polysaccharide profiling approach, uses chemical and biochemical techniques to probe the polysaccharide compositions in different plant and algal taxonomic groups. These techniques can directly determine the compositions and structures of polysaccharides in the walls. Nevertheless, they are not suitable for large-scale sampling of a large amount of organisms and tissues due to expensive labor, time and financial cost. The second approach, gene occurrence approach, performs data mining of genomes and transcriptomes to investigate the presence/absence of the enzymes responsible for the synthesis of certain polysaccharides. Obviously it is an indirect approach and has to rely on preexisting knowledge about what enzymes catalyze the biosynthesis of what plant cell wall biopolymers. However, it is much cheaper and faster than the polysaccharide profiling approach, with the DNA/RNA sequencing becoming increasingly less expensive. After genomic sequence data is obtained, bioinformatics data mining techniques play a key role to identify orthologous genes, as analyzing genomes and transcriptomes will need in-depth bioinformatics data analyses [5]. Overall these two approaches are complementary to each other for the study of cell wall evolution.

This chapter describes a bioinformatics protocol to analyze CAZyme gene occurrence in two newly sequenced organisms *Klebsormidium flaccidum* (charophyte green alga or CGA) and *Lygodium japonicum* (fern) [6, 7], with the focus on cellulose synthase (CesA) and hemicellulose backbone synthesis-related CesA-like (Csl) proteins. CesA and Csl proteins belong to the CAZyme family GT2, with Csl proteins further divided into nine subgroups (CslA, CslB, CslC, CslD, CslE, CslF, CslG, CslH, CslJ, and CslK) [5, 8, 9]. This protocol has been primarily used in our recent research papers [5, 9–11], and could be easily modified by replacing the query and/or database and apply to other cell wall-related CAZyme families to study their occurrence in the genomic data to infer the evolution of cell walls.

2 Materials

2.1 Computing Environment, Workflow, and Project Folder

Most bioinformatics analyses introduced here are Unix command line operations. *A terminal* is used to type in commands and to manipulate datasets. Single commands and command one-liners (commands connected with spaces and vertical bars in a single line), shown in *italicized* `Courier font` throughout this paper, are used, for example, to list files (*`ls file`*), view files (*`less file`*) or count how many Fasta sequences are in a file (`cat file | grep '^>' | wc -l`) (note that the spaces and the vertical bars must exist). For more complex data processing needs, it is advantageous to write a small script in a text editor program such as Notepad++ (Windows) or gedit (Ubuntu Linux) and implement it as a command.

Unix operating systems (OS) such as Linux and Mac have integrated terminal and valuable preinstalled programming languages (Perl, Python, etc.). If you have Windows OS, you may install Cygwin, a Linux-like environment for Windows (https://www.cygwin.com/), in order to run Unix commands and programming languages.

The bioinformatic pipeline described below is performed on a Linux system Ubuntu 12.04.5 LTS running on a computer with eight CPU processors (64-bit) and 8 Gb of RAM.

In order to reproduce the analysis described in this chapter, the reader should have some medium Unix command line skills and basic Perl or Python programming experiences.

The computational workflow (Fig. 1) will be detailed in this protocol chapter. In brief, protein sequences of species of interest serve as subject (or database) for sequence similarity searches (*blastp* and *hmmsearch*). Known CesA/Csl protein sequences or domain models will be used as query in the searches. In the end, an annotated phylogenetic tree will be generated and used to infer the evolution of CesA/Csl protein families.

In the following sections, we will describe how to download and install the needed bioinformatics tools (Subheading 2.2), the databases (Subheading 2.3), and the query datasets (Subheading 2.4).

On Ubuntu Linux computers, by default, all downloaded files from the web browser are automatically put in the Download folder. In our computer the ***absolute path*** of this folder is /home/elfitzek/Downloads/ ("elfitzek" is the user account of the first author). The "path" is a very important concept in using all Linux systems. When you run a command in a terminal environment, you must provide the correct path of the program or command or file or folder so that the computer will be able to find it. We will install all tools in the tools folder (/home/elfitzek/project/tools/), all query data in the query folder (/home/elfitzek/project/query/), and all database files in the database folder (/home/elfitzek/project/database/). We assume our readers already have the knowledge of creating, moving, and copying files/folders between different folders.

2.2 Bioinformatics Tools

As shown in Fig. 1, we will use *blastp* (a command of the BLAST package) and *hmmsearch* (a command of the HMMER package) to search the protein sequence sets of the two organisms (*K. flaccidum* and *L. japonicum*) for CesA/Csl homologs. BLAST (*blastp*) takes protein sequences as query to search against protein sequence

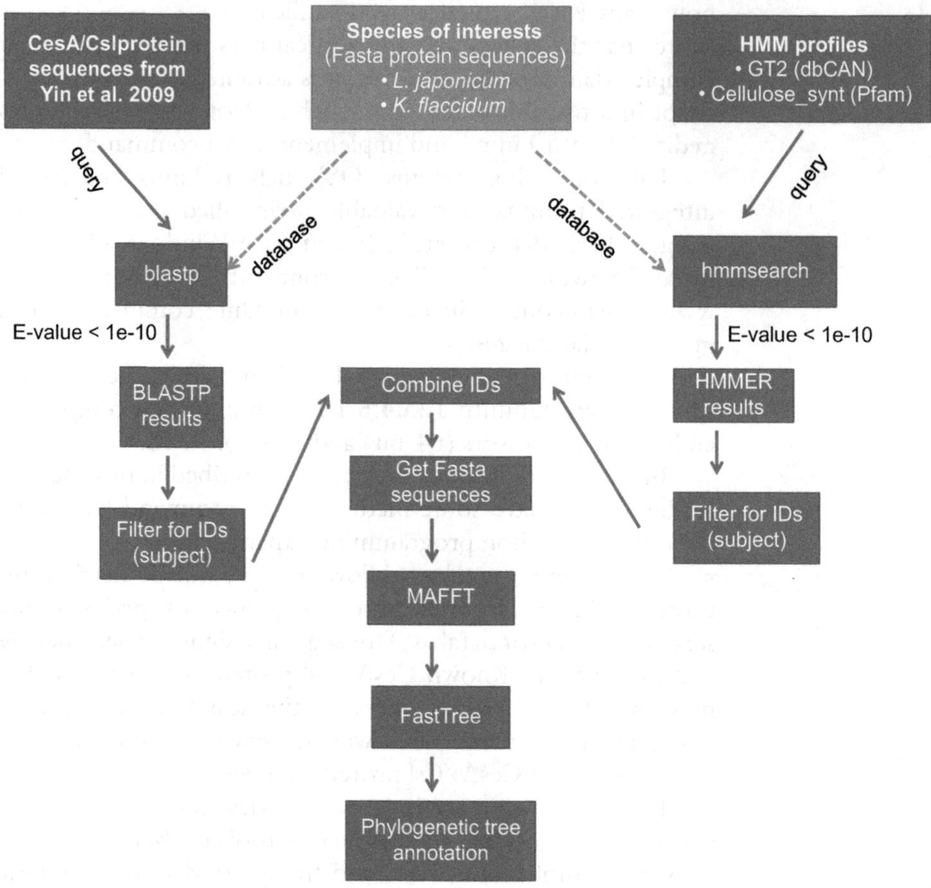

Fig. 1 Workflow of bioinformatics analysis of CesA/Csl protein families. Protein sequences (Fasta format) are retrieved from various sites (*see* Subheading 2). BLAST (the blastp command) package is used to obtain CesA and Csl homologs with *E*-value <1e−10 [19]. HMMER3 (the hmmsearch command) package is used to identify GT2 and CesA domain-containing proteins with *E*-value <1e−10 [20]. The resulting hits of both searches are combined and subjected to multiple sequence alignment with MAFFT [13] and then phylogenetic tree reconstruction using FastTree [14]. The phylogenetic tree of CesA/Csl hits is visualized and phylograms are made using iTOL [21]

database, while HMMER (*hmmsearch*) takes HMM (hidden Markov model) profiles (*see* Subheading 2.4 for details) as query to search against protein sequence database. In addition, we also need MAFFT and FastTree tools to build phylogenetic trees.

2.2.1 Download and Install HMMER

(a) Download the latest version (v3.1b2) of HMMER from http://hmmer.org/ and uncompress the .gz file [12] (*see* **Note 1**).

(b) Copy the "binaries" folder within the "hmmer-3.1b2-linux-intel-x86_64" folder and transfer it into the tools folder and rename it as "hmmer." The *hmmsearch* command will be in this hmmer folder. The absolute path of this folder in our computer is /home/elfitzek/project/tools/hmmer/.

2.2.2 Download and Install BLAST

(a) Download the latest version (version 2.3.0) of stand-alone BLAST package (ftp://ftp.ncbi.nih.gov:/blast/executables/LATEST/) and uncompress the .gz file (*see* **Note 2**).

(b) Copy the "bin" folder within the "ncbi-blast-2.3.0+" folder and transfer the "bin" folder into the tools folder and rename it as "blast." The *blastp* command will be in this blast folder (absolute path: /home/elfitzek/project/tools/blast/).

2.2.3 Download and Install the Multiple Sequence Alignment Tool MAFFT

For the multiple sequence alignment, a variety of tools are available such as Clustal Ω, MAFFT, MUSCLE, etc. In this study MAFFT is chosen for its high speed and high accuracy [13].

(a) Download the latest version of MAFFT from http://mafft.cbrc.jp/alignment/software/linuxportable.html and install it according to **Note 3** [13].

(b) Choose from "mafft-linux32" and "mafft-linux64" the correct folder that matches your computer bit system and copy it to the tools folder. In our case, we choose mafft-linux64. The *mafft.bat* command in the mafft-linux64 folder (/home/elfitzek/project/tools/mafft-linux64/) is the executable MAFFT program.

2.2.4 Download and Install FastTree [14] for Phylogenetic Analysis

(a) Download the latest version of FastTree executable program (Linux 64-bit executable (+SSE)) [14] from www.microbesonline.org/fasttree/#Install and place it in the tools folder (/home/elfitzek/project/tools/) (**Note 4**).

2.3 Genome Data of Species of Interests (Database)

Protein sequences of the two newly sequenced organisms, *K. flaccidum* and *L. japonicum*, are available at the individual sites of the research groups who generated the sequence data.

2.3.1 Download Data from K. Flaccidum Genome Website

(a) Download the protein sequences of the charophyte green algal *K. flaccidum* from http://www.plantmorphogenesis.bio.titech.ac.jp/~algae_genome_project/klebsormidium/index.html following the "Download" link. The genome of *K. flaccidum* was published in 2014 and the protein sequences were predicted from this genome [7]. Choose "Predicted Protein" (red arrow in Fig. 2) and download it (*see* **Note 5**). Note that the species name has been changed to *Klebsormidium nitens* NIES-2285 later on the website, but we still used the old name *Klebsormidium flaccidum* in this book chapter. This is because when we wrote this chapter, the website has not been updated yet.

(b) Change the downloaded file name to "Kfaccidum_protein.fa".

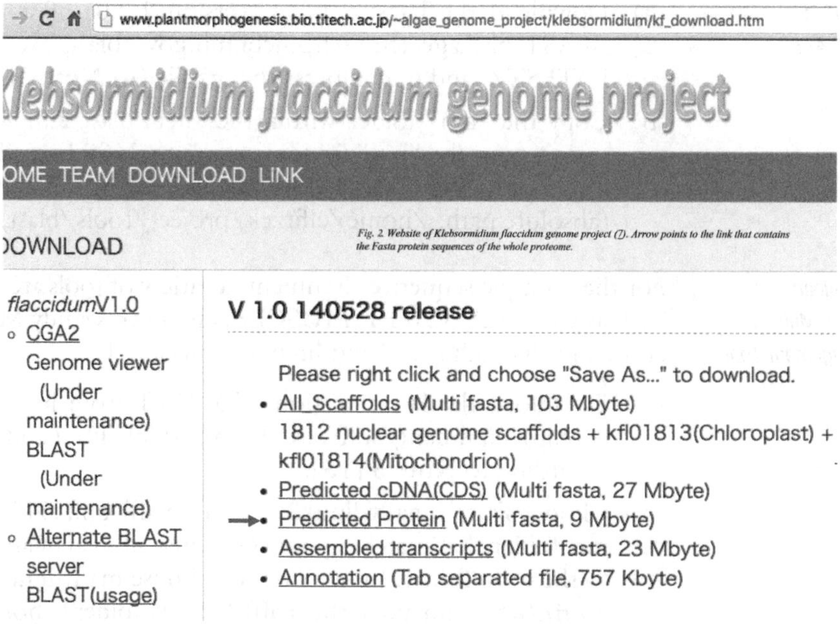

Fig. 2 Website of *Klebsormidium flaccidum* genome project [7]. Arrow points to the link that contains the Fasta protein sequences of the whole proteome

2.3.2 Download Data from L. japonicum *Genome Website*

(a) Open the website http://bioinf.mind.meiji.ac.jp/kanikusa/download.php, and download the compressed "lygodium_-predicted_protein_ver1.0RC.fasta.tar.gz" file (red arrow in Fig. 3). Different from the other organisms used in this study, these protein sequences were predicted from assembled transcriptome data instead of genome data [6].

(b) Decompress the Fasta file and copy it to the database folder.

(c) In the Fasta file, the sequence headers contain found different codes: "F," "N," "T," and "P," meaning the sequences were predicted from different bioinformatics methods. These acronyms are: F = FrameDP, N = Newbler, T = Transdecoder, and P = longest amino acids sequence predicted by in-house Perl script (personal communication with Kentaro Yano). We only select sequences containing "P" in their Fasta header for further analysis to avoid including redundancy in the analysis. This is done by using a self-developed Perl script (*see* **Note 6**). The output Fasta file is named "Lygodium_filtered.fa" and put in the database folder.

2.4 CesA/Csl Protein Sequences and Domain Models as Queries

As shown in Fig. 1, we will use both *blastp* and *hmmsearch* methods for searching CesA/Csl homologs. The reason was explained in [9] (*see* **Note 7**). The query for *blastp* search is protein Fasta sequences, while for *hmmsearch* the query is HMMs (hidden Markov models).

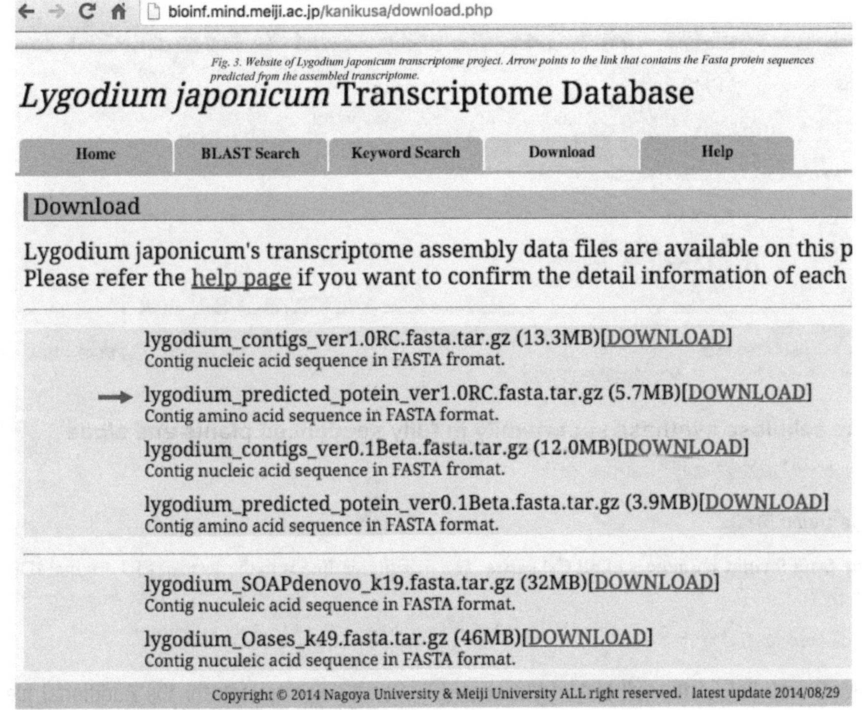

Fig. 3 Website of *Lygodium japonicum* transcriptome project. Arrow points to the link that contains the Fasta protein sequences predicted from the assembled transcriptome

2.4.1 CesA/Csl Sequences from Yin et al. [9]

CesA/Csl proteins from fully sequenced plant and algae (e.g., *Arabidopsis thaliana* and *Oryza sativa*) have been cataloged in Yin et al. [9]. We will use them as the query for the *blastp* search. Here we show how to download these sequences from PubMed Central website:

(a) Open the webpage: http://www.ncbi.nlm.nih.gov/pmc/articles/PMC3091534/, download the Additional file 3 (Fig. 4), open it using a plain text editor (gedit on our Linux computer, Notepad++ on Windows computer), and copy the *A. thaliana* and *O. sativa* sequences and the six CslA/C-like (renamed CslK later) protein sequences of chlorophyte green algae, and paste them into a new plain text editor and save as a new file: /home/elfitzek/project/query/GT2-query.fa.

2.4.2 GT2 and CesA HMM Profiles

Hidden Markov model (HMM) profiles are widely used to represent protein domains or families (*see* **Note 8**). CslA and CslC proteins contain strong GT2 domain signals (Pfam model ID: Glycos_transf_2), while other CesA/Csl proteins contain strong CesA domain signals (Pfam model ID: Cellulose_synt).

To download these two HMM profiles:

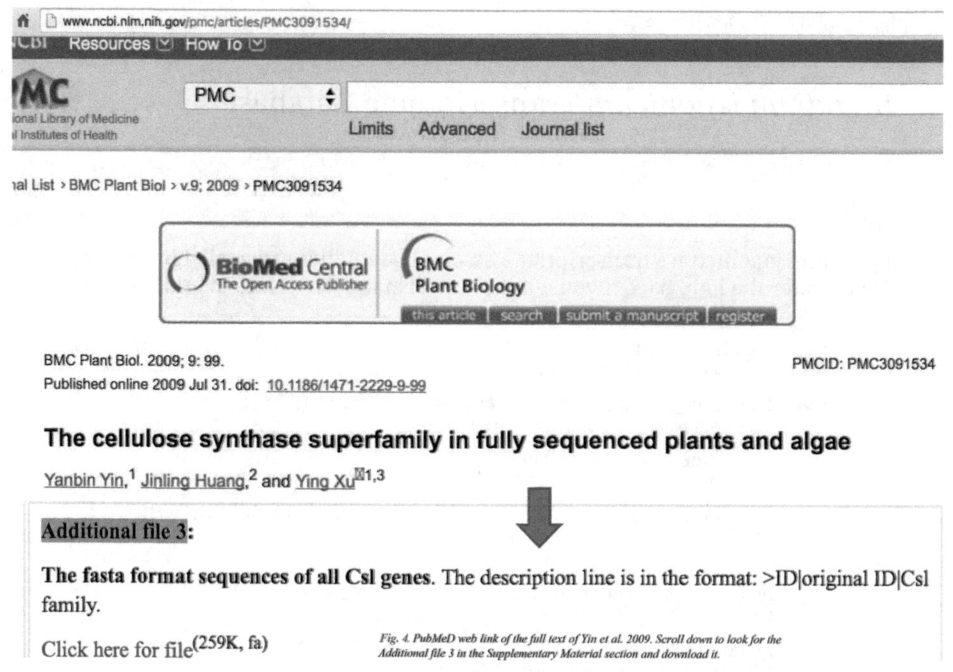

Fig. 4 PubMeD web link of the full text of Yin et al. [9]. Scroll down to look for the Additional file 3 in the Supplementary Material section and download it

(a) Go to the Pfam website and download the Glycos_transf_2.hmm file (https://pfam.xfam.org/family/Glycos_transf_2) (Fig. 5a), and rename it as GT2.hmm.

(b) Go to the Pfam website and download the Cellulose_synt.hmm file (http://pfam.xfam.org/family/Cellulose_synt) (Fig. 5b).

(c) Transfer the two HMM profiles files to the query folder (/home/elfitzek/project/query/).

3 Methods

With the tools, query, and database sets ready, we will search for CesA/Csl homologs in the two newly sequenced organisms. By now the project folder contains (folders end with "/"):

/home/elfitzek/project/tools/

- hmmer/
- blast/
- mafft-linux64/
- FastTree

/home/elfitzek/project/query/

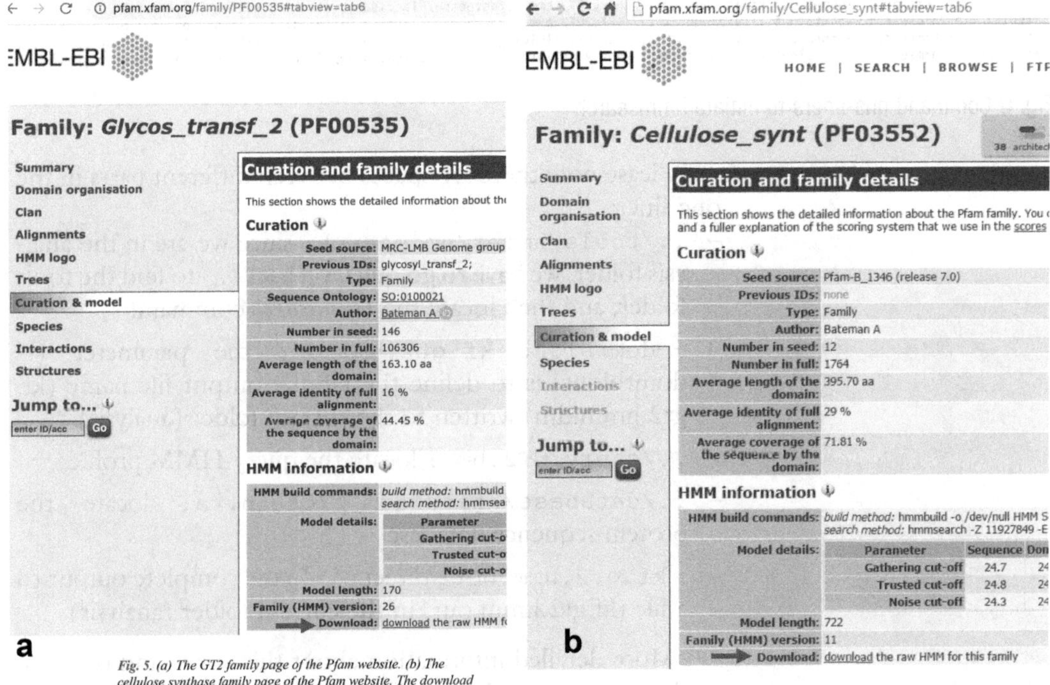

Fig. 5 (**a**) The GT2 page of the Pfam website. (**b**) The cellulose synthase family page of the Pfam website. The download links are marked with red arrows

- GT2-query.fa
- GT2.hmm
- Cellulose_synt.hmm

/home/elfitzek/project/database/

- Kfaccidum_protein.fa
- Lygodium_filtered.fa

We will create another folder called "analysis" and put it under the project folder. All analysis result files will be written to /home/elfitzek/project/analysis/.

3.1 Run hmmsearch in a Terminal

(a) Open a terminal and change directory to the analysis folder:

```
cd /home/elfitzek/project/analysis/
```

(b) Run *hmmsearch* (/home/elfitzek/project/tools/hmmer/) in a terminal on the two hmm files (/home/elfitzek/project/query/) and the two .fa files (/home/elfitzek/project/database/) (four times in total) using the command one-liners (Fig. 6):

```
../tools/hmmer/hmmsearch --domtblout kf.gt2.hmm.dm ../query/GT2.hmm ../database/Kfaccidum_protein.fa > kf.gt2.hmm.out &
../tools/hmmer/hmmsearch --domtblout lj.gt2.hmm.dm ../query/GT2.hmm ../database/Lygodium_filtered.fa > lj.gt2.hmm.out &
../tools/hmmer/hmmsearch --domtblout lj.cesa.hmm.dm ../query/Cellulose_synt.hmm ../database/Lygodium_filtered.fa > lj.cesa.hmm.out &
../tools/hmmer/hmmsearch --domtblout kf.cesa.hmm.dm ../query/Cellulose_synt.hmm ../database/Kfaccidum_protein.fa > kf.cesa.hmm.out &
```

Fig. 6 Command one-liners to initiate hmmsearch

Please note there are spaces between different parts in the one-liner:

- `../tools/hmmer/hmmsearch`: since we are in the analysis folder, we have to go one level back (..) to find the tools folder, and then locate the *hmmsearch* command
- `--domtblout kf.gt2.hmm.dm`: the parameter (--domtblout) is to define the tabular output file name (kf.gt2.hmm.dm) written to the current folder (analysis)
- `../query/GT2.hmm`: locate the query HMM profile
- `../database/Kfaccidum_protein.fa`: locate the protein sequence database
- `>kf.gt2.hmm.out`: direct (">") the complete output to a file (kf.gt2.hmm.out) in the current folder (analysis)

More detailed information about this command one-line could be found in the HMMER user guide: http://eddylab.org/software/hmmer3/3.1b2/Userguide.pdf. It is also useful to run the below command to get the option list:

```
../tools/hmmer/hmmsearch -h
```

(c) The analysis folder now should have four .hmm.dm files and four .hmm.out files. The .hmm.dm files will be further analyzed to retrieve CesA/Csl homologs.

3.2 Run blastp in a Terminal

(a) In a terminal, change directory to the analysis folder:

```
cd /home/elfitzek/project/analysis/
```

(b) In order to run *blastp*, we have to format the two .fa files in the database folder (/home/elfitzek/project/database/) using the `makeblastdb` command (/home/elfitzek/project/tools/blast/) (Fig. 7):

Running the below command will print the option list on the screen:

```
../tools/blast/makeblastdb -help
```

```
../tools/blast/makeblastdb -in ../database/Kfaccidum_protein.fa -dbtype 'prot' -parse_seqids
../tools/blast/makeblastdb -in ../database/Lygodium_filtered.fa -dbtype 'prot' -parse_seqids
```

Fig. 7 Command one-liners to make blastable databases

```
../tools/blast/blastp -query ../query/GT2-query.fa -db ../database/Kfaccidum_protein.fa -outfmt 6 -out kf.blast.out -num_threads 10
../tools/blast/blastp -query ../query/GT2-query.fa -db ../database/Lygodium_filtered.fa -outfmt 6 -out lj.blast.out -num_threads 10
```

Fig. 8 Command one-liners to run blastp search

In the database folder, for each .fa files, there will be six additional files generated with different suffixes (*see* **Note 9**).

(c) Run *blastp* (/home/elfitzek/project/tools/blast/) using the GT2-query.fa file as query (/home/elfitzek/project/query/) and the two .fa files (/home/elfitzek/project/database/) as database using the command one-liners (Fig. 8):

Running the below command will print all the options on the screen (*see* **Note 10**):

```
../tools/blast/blastp -help
```

(d) The analysis folder now should have two .blast.out files, which will be further analyzed to retrieve CesA/Csl homologs.

3.3 Extract Significant Hits

Now the analysis folder should contain the following files:

/home/elfitzek/project/analysis/

- kf.gt2.hmm.dm
- kf.cesa.hmm.dm
- kf.blast.out
- lj.gt2.hmm.dm
- lj.cesa.hmm.dm
- lj.blast.out

1. Save the IDs of the significant hits of *blastp* results (.blast.out files) using a command one-liner. An example input file (*blastp* output) and this one-liner are explained in details in Fig. 9 (*see* **Note 11**).
2. Save the IDs of the significant hits of *hmmsearch* results (.hmm. dm files) using a command one-liner. An example input file (*hmmsearch* output) and this one-liner are explained in details in Fig. 10 (*see* **Notes 11** and **12**).

1. Query ID; 2. Subject ID; 3. % Identity; 4. Alignment Length; 5. Mismatch Count; 6. Gap Open Count; 7. Query Start; 8. Query End; 9. Subject Start; 10. Subject End; 11. Evalue; 12. Bit-score

```
AT2G21770.1|cesA    kfl00459_0100  50.281  1068  453  12  5    1068  117  1110  0.0       1054
AT2G21770.1|cesA    kfl00459_0080  51.442  1075  459  17  7    1068  154  1178  0.0       1048
AT2G21770.1|cesA    kfl00889_0040  48.708  1084  462  20  6    1076  106  1108  0.0       983
AT2G21770.1|cesA    kfl00234_0060  41.006  1034  503  16  39   1065  243  1176  0.0       820
AT2G21770.1|cesA    kfl00766_0040  48.771  814   388  9   261  1068  110  900   0.0       811
AT2G21770.1|cesA    kfl00518_0010  32.540  126   78   2   528  648   539  662   3.53e-12  70.5
AT2G21770.1|cesA    kfl00518_0010  24.893  233   159  7   779  1003  667  891   3.52e-05  47.4
AT2G21770.1|cesA    kfl00518_0010  33.333  51    33   1   373  423   484  533   0.33      34.7
AT2G21770.1|cesA    kfl00007_0020  33.803  71    45   1   778  848   634  702   1.00e-05  49.3
AT2G21770.1|cesA    kfl00007_0020  27.068  133   85   4   525  648   501  630   0.008     39.7
```

```
cat kf.blast.out | awk '$11<1e-10' | cut -f2 | sort -u > kf.blast.out.1e-10.id
cat lj.blast.out | awk '$11<1e-10' | cut -f2 | sort -u > lj.blast.out.1e-10.id
```

1. Print the file content
2. Keep lines with 11th column < 1e-10
3. Keep just the 2nd column
4. Remove duplicate
5. Direct the output into a file

Fig. 9 Top: The tabular space delimited format of blastp output; Bottom: Command one-liners to parse this tabular format file to extract hit IDs with significant *E*-values $<1e{-}10$

```
#                                                                        --- full sequence --- -------------- this domain -------------  hmm coord  ali coord  env coord
# target name    accession  tlen query name           accession   qlen   E-value  score  bias   #  of  c-Evalue  i-Evalue  score  bias  from   to  from   to  from   to  acc  desc
ription of target
#--------------- ---------- ----- -------------------- ---------- -----  --------- ------ -----  --- --- --------- --------- ------ ----- ----- ----- ----- ----- ----- ----- ---- ----
----------------
kfl00148_0230    -           240 GT2.hmm              -            168   1.8e-38  132.0   0.0   1   1   5.6e-41   2.2e-38  131.7   0.0     1   168     7   178     7   178 0.93 -
kfl00004_0480    -           870 GT2.hmm              -            168   9.4e-30  103.7   0.0   1   1   4.8e-32   1.9e-29  102.7   0.0     1   123   522   646   522   678 0.94 -
kfl00248_0140    -           349 GT2.hmm              -            168   7.4e-27   94.3   0.0   1   1   2.8e-29   1.1e-26   93.7   0.0     1   167    71   274    71   275 0.84 -
kfl00518_0010    -          1371 GT2.hmm              -            168   5.5e-23   81.7   0.0   1   1   2.2e-25   8.6e-23   81.0   0.0     2   167   477   664   476   665 0.89 -
kfl00126_0150    -           674 GT2.hmm              -            168   1.2e-22   80.6   0.0   1   1   4.4e-25   1.8e-22   80.0   0.0     2   167    55   231    54   232 0.88 -
kfl00271_0180    -           720 GT2.hmm              -            168   3.7e-21   75.7   0.0   1   1   1.4e-23   5.5e-21   75.2   0.0     2   167   140   313   139   314 0.88 -
kfl00603_0080    -           871 GT2.hmm              -            168   4.6e-20   72.1   0.0   1   1   1.9e-22   7.6e-20   71.4   0.0     2   167   285   467   284   468 0.83 -
kfl00117_0160    -           340 GT2.hmm              -            168   4.3e-18   65.7   0.0   1   1   1.5e-20     6e-18   65.3   0.0    24   111    22   109     8   176 0.89 -
kfl00335_0090    -           301 GT2.hmm              -            168   1.3e-16   60.9   0.0   1   1   4.9e-19   1.9e-16   60.4   0.0     1   102    34   130    34   137 0.91 -
kfl00282_0190    -          1009 GT2.hmm              -            168   8.4e-16   58.3   0.0   1   1   3.4e-18   1.4e-15   57.6   0.0     1   162   779   941   779   947 0.73 -
```

```
cat kf.cesa.hmm.dm | grep -v '^#' | awk '$13<1e-10' | awk '{print $1}' | sort -u > kf.cesa.hmm.dm.1e-10.id
cat kf.gt2.hmm.dm | grep -v '^#' | awk '$13<1e-10' | awk '{print $1}' | sort -u > kf.gt2.hmm.dm.1e-10.id
cat lj.cesa.hmm.dm | grep -v '^#' | awk '$13<1e-10' | awk '{print $1}' | sort -u > lj.cesa.hmm.dm.1e-10.id
cat lj.gt2.hmm.dm | grep -v '^#' | awk '$13<1e-10' | awk '{print $1}' | sort -u > lj.gt2.hmm.dm.1e-10.id
```

1. Print the file content
2. Remove the header lines
3. Keep lines with 13th column < 1e-10
4. Keep just the 1st column
5. Remove duplicate
6. Direct the output into a file

Fig. 10 Top: The regular space delimited format of hmmsearch output; Bottom: Command one-liners to parse this file to extract hit IDs with significant *E*-values $<1e{-}10$

3.3.1 Combine Significant Hits from the Three Methods

Now the analysis folder should contain the following .id files:
/home/elfitzek/project/analysis/

- kf.gt2.hmm.dm.1e-10.id
- kf.cesa.hmm.dm.1e-10.id
- kf.blast.out.1e-10.id
- lj.gt2.hmm.dm.1e-10.id

```
cat kf.*.1e-10.id | sort -u > kf.all.id
cat lj.*.1e-10.id | sort -u > lj.all.id
```

Fig. 11 Command one-liners to combine significant hits and remove duplicates. Asterisk is the Unix wildcard character to represent any characters

```
../tools/blast/blastdbcmd -db ../database/Kfaccidum_protein.fa -entry_batch kf.all.id > kf.all.id.fa
../tools/blast/blastdbcmd -db ../database/Lygodium_filtered.fa -entry_batch lj.all.id > lj.all.id.fa
```

Fig. 12 Command one-liners to extract Fasta sequences

- lj.cesa.hmm.dm.1e-10.id
- lj.blast.out.1e-10.id

We run the follow command one-liner to combine the hits from the three different queries/methods and remove duplicates (Fig. 11):

3.4 Prepare Fasta Sequence File for Alignment

3.4.1 Extract Fasta Sequences of the Significant Hits in the Two Searched Organisms

In the BLAST package, there is a command called *blastdbcmd* that can take a given list of IDs and retrieve their Fasta sequences from a Fasta database file. This can only be possible if we have applied the -parse_seqids option when we run *makeblastdb* command (as we did in Fig. 7). We ran the following commands (Fig. 12) to extract the Fasta sequences of our significant hits in the two organisms.

Running the below command will print all the options on the screen:

```
../tools/blast/blastdbcmd -help
```

3.4.2 Combine Fasta Sequences from the Query Organisms and the Database Organisms

```
cat kf.all.id.fa lj.all.id.fa ../query/GT2-query.fa > all.species.fa
```

3.5 Build Alignment and Phylogeny

3.5.1 Run MAFFT to Build Multiple Sequence Alignment Using the Command One-Liner

```
../tools/mafft-linux64/mafft.bat --maxiterate 1000 --localpair all.species.fa > all.species.fa.1
```

Running the below command will print some help information on the screen:

```
../tools/mafft-linux64/mafft.bat
```

An online version of MAFFT can be found at: http://www.ebi.ac.uk/Tools/msa/mafft/, where users can upload a Fasta format file for alignment remotely.

3.5.2 Run FastTree to Build Phylogeny Using the Command One-Liner

```
../tools/FastTree all.species.fa.1 > all.species.fa.1.nwk
```

This uses the default parameters for protein phylogeny reconstruction. Running the below command will print all the options on the screen:

```
../tools/FastTree
```

3.6 Creating and Annotating Phylograms with iTOL

iTOL is web application that allows users to upload a Newick format phylogeny tree file for making publishable phylograms. Moreover, it offers very useful utilities to annotate the tree graphs such as coloring branches and leaves automatically by uploading color definition files. To make these color definition files, we will run the command one-liners as shown in Fig. 13. After all the steps, we get two definition files, one for defining branch color (all.species.br.txt) and the other for defining leaf colors (all.species.lab.txt). Then we are ready to upload these files to iTOL website for making the tree graph:

(a) Open http://itol.embl.de/upload.cgi in a web browser.

```
#1: make color definition file to specify the branch colors
cat ../query/GT2-query.fa | grep '>' | sed 's/>//' | grep '^AT' | awk '{print $1,"branch","#0000ff","normal","1"}' > at.all.id.br
cat ../query/GT2-query.fa | grep '>' | sed 's/>//' | grep '^LOC' | awk '{print $1,"branch","#ff0000","normal","1"}' > os.all.id.br
cat ../query/GT2-query.fa | grep '>' | sed 's/>//' | grep 'like' | awk '{print $1,"branch","#ffa500","normal","1"}' > alg.all.id.br
cat kf.all.id | awk '{print $1,"branch","#00ff00","normal","1"}' > kf.all.id.br
cat lj.all.id | awk '{print $1,"branch","#00ffff","normal","1"}' > lj.all.id.br

#2: make color definition file to specify the leaf colors
cat ../query/GT2-query.fa | grep '>' | sed 's/>//' | grep '^LOC' | awk '{print $1,"label","#ff0000","normal","1"}' > os.all.id.lab
cat ../query/GT2-query.fa | grep '>' | sed 's/>//' | grep '^AT' | awk '{print $1,"label","#0000ff","normal","1"}' > at.all.id.lab
cat ../query/GT2-query.fa | grep '>' | sed 's/>//' | grep 'like' | awk '{print $1,"label","#ffa500","normal","1"}' > alg.all.id.lab
cat kf.all.id | awk '{print $1,"label","#00ff00","normal","1"}' > kf.all.id.lab
cat lj.all.id | awk '{print $1,"label","#00ffff","normal","1"}' > lj.all.id.lab

#3: combine all species into one file
cat *.br > all.species.br.txt
cat *.lab > all.species.lab.txt

#4: open the two files in a plain text editor and add the three lines at the top
TREE_COLORS
SEPARATOR SPACE
DATA

# the top 10 lines of all.species.br.txt and all.species.lab.txt look like:
```

```
TREE_COLORS
SEPARATOR SPACE
DATA
kf100004_0480 branch #00ff00 normal 1
kf100007_0020 branch #00ff00 normal 1
kf100025_0050 branch #00ff00 normal 1
kf100029_0240 branch #00ff00 normal 1
kf100032_0270 branch #00ff00 normal 1
kf100053_0150 branch #00ff00 normal 1
kf100053_0170 branch #00ff00 normal 1
```

```
TREE_COLORS
SEPARATOR SPACE
DATA
kf100004_0480 label #00ff00 normal 1
kf100007_0020 label #00ff00 normal 1
kf100025_0050 label #00ff00 normal 1
kf100029_0240 label #00ff00 normal 1
kf100032_0270 label #00ff00 normal 1
kf100053_0150 label #00ff00 normal 1
kf100053_0170 label #00ff00 normal 1
```

Fig. 13 Command one-liners to make color definition files

Fig. 14 iTOL tree browser and control panel (right-top corner). Selecting different options in the control panel will change the tree graph simultaneously

(b) Choose the newick file (all.species.fa.l.nwk) from the computer and upload it or simply drag the file to the "Tree text" area to upload.

(c) The un-colored tree graph will be shown in a tree browser.

(d) Drag the two color definition file into the tree browser (*see* **Note 13**), then we will see the tree branches and leaves are colored (Fig. 14).

3.7 Interpretation of the Phylogeny

The last and probably most important step in this protocol is to manually inspect the phylogeny to make meaningful/interesting evolutionary interpretations. Figure 15 includes CesA/Csl homologs from ten organisms: two newly sequenced organisms: *K. flaccidum* (genome) and *L. japonicum* (transcriptome), two model organisms (*A. thaliana*, *O. sativa*), and six chlorophyte green algae (*Micromonas pusilla CCMP1545*, *Micromonas* strain *RCC299*, *Ostreococcus lucimarinus*, *Ostreococcus tauri*, *Chlamydomonas reinhardtii*, and *Volvox carteri f. nagariensis*). The interpretation of this phylogeny is largely based on the groupings of *K. flaccidum* and *L. japonicum* proteins into known CesA/Csl clades according to the already annotated *A. thaliana*, *O. sativa*, and chlorophyte algal proteins.

With regard to *K. flaccidum*, we can make the following interpretations from the phylogeny, most of which are in agreement with findings reported in our paper [5]: (1) no CslA orthologs are found in the CGA (charophyte green alga) *K. flaccidum*; (2) three *K. flaccidum* proteins are monophyletically clustered with CslC proteins of land plants; (3) there is a single protein of

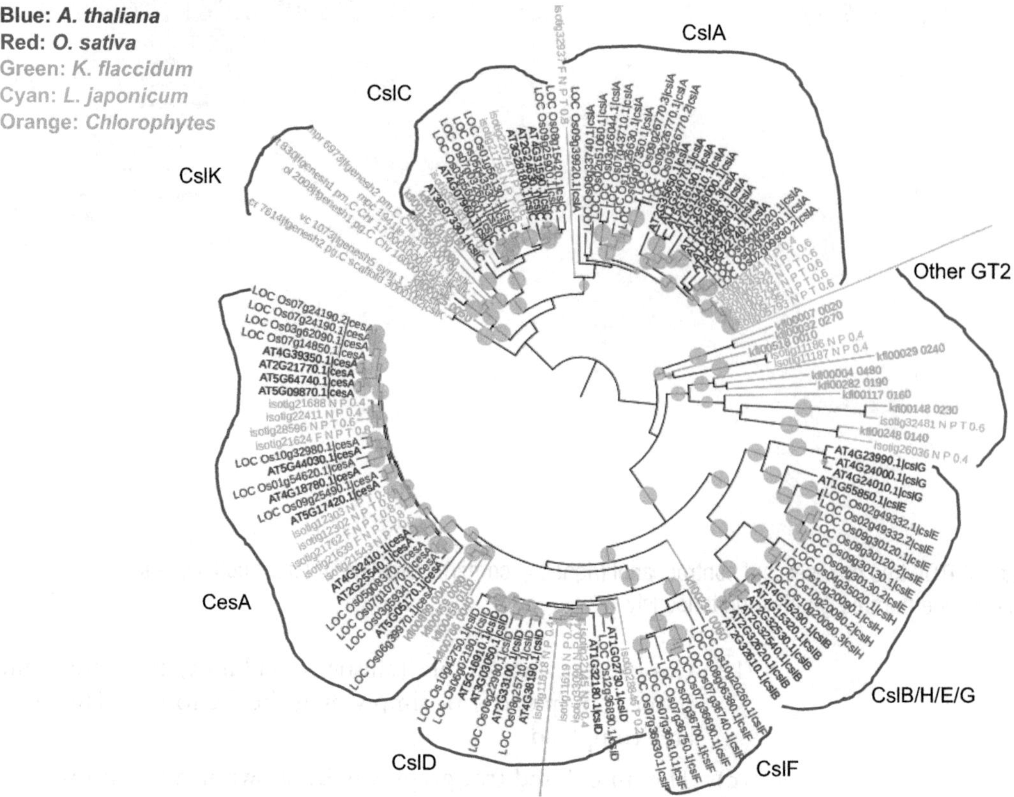

Fig. 15 Phylogeny of CesA/Csl proteins from *A. thaliana*, *O. sativa*, *K. flaccidum*, *L. japonicum*, and six chlorophyte green algae. Nodes that have >80% support values are indicated with light blue circles

K. flaccidum is clustered with chlorophyte CslK proteins, more specifically monophyletically clustered with *Micromonas* and *Ostreococcus* proteins (support value = 99%), which is very interesting because CslK was thought to be chlorophyte specific; (4) four *K. flaccidum* proteins are clustered with land plant CesAs with a support value = 95%; (5) there is one *K. flaccidum* protein clustered within CslD clade but with very long branch, suggesting this is not a reliable clustering; another *K. flaccidum* protein is basal to all CesA/CslD/CslF proteins. For the last point, we and others have found that some *Penium* and *Spirogyra* CGA have CesAs and *Coleochaete* CGA have CslD [5, 15].

For the fern *L. japonicum*, we can conclude that it has CesA, CslD, CslA, and CslC proteins. Although no *L. japonicum* proteins are found in CslB/H/E/G clades, they are likely to have not been expressed in the transcriptome data that we searched. In fact, we have previously shown that some other fern species do have expressed proteins in these clades [5].

Lastly, we see that there are some proteins from *K. flaccidum* and *L. japonicum* not clustered with any of the CesA and Csl clades in the phylogeny. They are either non-CesA/Csl GT2 proteins that are homologous to Arabidopsis dolichyl phosphate β-glucosyltransferase (AT2G39630) and dolichol phosphate mannose synthase (AT1G20575), or even bacteria linear CesA-like proteins. For the latter, it could be tested by including published bacterial, plant and algal linear CesA proteins in the phylogenetic analysis [16–18].

4 Notes

1. In Ubuntu Linux, this can be done by right-clicking the .gz file and select "Extract Here" from the menu. Or open a terminal, go to the download folder (/home/elfitzek/Downloads/), and run the command: *tar xvf hmmer-3.1b2-linux-intel-x86_64.tar.gz*

   This will create a folder called "hmmer-3.1b2-linux-intel-x86_64".
2. In Ubuntu Linux, this can be done by right-clicking the .gz file and select "Extract Here" from the menu. Or open a terminal, go to the download folder, and run the command:

```
tar xvf ncbi-blast-2.3.0+-x64-linux.tar.gz
```

   This will create a folder called "ncbi-blast-2.3.0+".
3. Open a terminal, go to the download folder, and run the command:

```
tar xvf mafft-7.273-linux.tgz
```

   This will release two folders, "mafft-linux32" and "mafft-linux64". The latter is what we need for our computer.
4. In order to run this program, we need to change the permission of the downloaded FastTree executable program with the below command in a terminal:

```
chmod 777 /home/elfitzek/project/tools/FastTree
```

5. In the webpage, right click on the link in Fig. 2 and select "Copy Link Address." Open a terminal and move to the database folder (/home/elfitzek/project/database/), type the `wget` command and paste the copied web link (there is a space after `wget`):

```
wget http://www.plantmorphogenesis.bio.titech.ac.jp/~algae_-
genome_project/klebsormidium/kf_download/131203_kfl_initial_-
genesets_v1.0_AA.fasta
```

6. The Perl script is written as a one-liner:

```
cat lygodium_predicted_potein_ver1.0RC.fasta | perl -e
'@a=&lt;&gt;;;for($i=0;$i&lt;=$#a;$i=$i+2){print $a[$i].$a[$i
+1] if $a[$i]=~/_P_/;}' &gt; Lygodium_filtered.fa
```

We do not provide the explanation of this one-liner as it is beyond the scope of this chapter.

7. Briefly, this will make sure all CesA/Csl proteins and their closely related GT2 proteins to be identified.
8. Well-known protein family/domain HMM profile databases include Pfam, PIRSF and SMART, which are very popular for functional annotation of newly sequenced proteomes.
9. These are the index files that will be used for *blastp* search. They have to be in the same folder as the Fasta file and the file name must not be changed.
10. Particularly the "-outfmt 6" parameter specifies the blast output in tabular format, which is very easy to parse.
11. *E*-value threshold can be changed and may affect the result very significantly. Here we use 1e-10 because we wanted to be very conservative in keeping significant hits. There is no universally "best" *E*-value. People usually explore different *E*-values and make decision based on their own experience. For this chapter, we have tried *E*-value <1e−2, *E*-value <1e−5, and *E*-value <1e−10. Different *E*-value thresholds do not change any of our conclusion made in the interpretation step (Subheading 3.7).
12. The ***hmmsearch*** output reports three *E*-values (Fig. 10): (a) the *E*-value refers to the full sequence; (b) the c-*E*-value is known as conditional *E*-value; and (c) the i-*E*-value is independent *E*-value. We choose to use the i-*E*-value to parse the file to extract significant hits. The detailed explanation of ***hmmsearch*** output could be found in the HMMER user guide: http://eddylab.org/software/hmmer3/3.1b2/Userguide.pdf.
13. Using the Control panel in the tree browser, we can display the tree in circular/normal/unrooted modes, show/hide bootstrap values, and export the tree graph as an image file in different formats (e.g., PDF, PNG). The more detailed user guide is available in the help page of iTOL: http://itol.embl.de/help.cgi.

Acknowledgments

E.F. is supported by the Research & Artistry Award of Northern Illinois University and partially supported by the National Institutes of Health (1R15GM114706) to Y.Y. R.B. was a University Honors Program Undergraduate Student of Northern Illinois University. We acknowledge the Department of Computer Science of NIU for providing free access to the Linux computing cluster Gaea and the Yin lab members for helpful discussions.

References

1. Lombard V, Golaconda Ramulu H, Drula E, Coutinho PM, Henrissat B (2014) The carbohydrate-active enzymes database (CAZy) in 2013. Nucleic Acids Res 42:D490–D495
2. Goodstein DM, Shu S, Howson R, Neupane R, Hayes RD, Fazo J, Mitros T, Dirks W, Hellsten U, Putnam N et al (2012) Phytozome: a comparative platform for green plant genomics. Nucleic Acids Res 40: D1178–D1186
3. Popper Z, Michel G, Herve C, Domozych DS, Willats WG, Tuohy MG, Kloareg B, Stengel DB (2011) Evolution and diversity of plant cell walls: from algae to flowering plants. Annu Rev Plant Biol 62:567–590
4. Fangel JU, Ulvskov P, Knox JP, Mikkelsen MD, Harholt J, Popper ZA, Willats WG (2012) Cell wall evolution and diversity. Front Plant Sci 3:152
5. Yin Y, Johns MA, Cao H, Rupani M (2014) A survey of plant and algal genomes and transcriptomes reveals new insights into the evolution and function of the cellulose synthase superfamily. BMC Genomics 15:1–15
6. Aya K, Kobayashi M, Tanaka J, Ohyanagi H, Suzuki T, Yano K, Takano T, Matsuoka M (2014) De novo transcriptome assembly of a fern, lygodium japonicum, and a web resource database, ljtrans DB. Plant Cell Physiol 56: e5–e5
7. Hori K, Maruyama F, Fujisawa T, Togashi T, Yamamoto N, Seo M, Sato S, Yamada T, Mori H, Tajima N et al (2014) Klebsormidium flaccidum genome reveals primary factors for plant terrestrial adaptation. Nat Commun 5:3978
8. Richmond TA, Somerville CR (2000) The cellulose synthase superfamily. Plant Physiol 124:495–498
9. Yin Y, Huang J, Xu Y (2009) The cellulose synthase superfamily in fully sequenced plants and algae. BMC Plant Biol 9:99
10. Taujale R, Yin Y (2015) Glycosyltransferase family 43 is also found in early eukaryotes and has three subfamilies in Charophycean green algae. PLoS One 10:e0128409
11. Yin Y, Chen H, Hahn MG, Mohnen D, Xu Y (2010) Evolution and function of the plant cell wall synthesis-related glycosyltransferase family 8. Plant Physiol 153:1729–1746
12. Finn RD, Clements J, Eddy SR (2011) HMMER web server: interactive sequence similarity searching. Nucleic Acids Res 39: W29–W37
13. Katoh K, Standley DM (2013) MAFFT multiple sequence alignment software version 7: improvements in performance and usability. Mol Biol Evol 30:772–780
14. Price MN, Dehal PS, Arkin AP (2009) FastTree: computing large minimum evolution trees with profiles instead of a distance matrix. Mol Biol Evol 26:1641–1650
15. Mikkelsen MD, Harholt J, Ulvskov P, Johansen IE, Fangel JU, Doblin MS, Bacic A, Willats WG (2014) Evidence for land plant cell wall biosynthetic mechanisms in charophyte green algae. Ann Bot 114:1217–1236
16. Harholt J, Sorensen I, Fangel J, Roberts A, Willats WG, Scheller HV, Petersen BL, Banks JA, Ulvskov P (2012) The glycosyltransferase repertoire of the spikemoss Selaginella moellendorffii and a comparative study of its cell wall. PLoS One 7:e35846
17. Michel G, Tonon T, Scornet D, Cock JM, Kloareg B (2010) The cell wall polysaccharide metabolism of the brown alga Ectocarpus siliculosus. Insights into the evolution of

extracellular matrix polysaccharides in eukaryotes. New Phytol 188:82–97

18. Roberts E, Roberts AW (2009) A cellulose synthase (Cesa) gene from the red alga Porphyra Yezoensis (Rhodophyta). J Phycol 45:203–212
19. Altschul SF, Gish W, Miller W, Myers EW, Lipman DJ (1990) Basic local alignment search tool. J Mol Biol 215:403–410
20. Eddy SR (2011) Accelerated profile HMM searches. PLoS Comput Biol 7:e1002195
21. Letunic I, Bork P (2011) Interactive tree of life v2: online annotation and display of phylogenetic trees made easy. Nucleic Acids Res 39: W475–W478

Chapter 28

Automated Glycan Assembly of Plant Cell Wall Oligosaccharides

Fabian Pfrengle

Abstract

Synthetic cell wall oligosaccharides are promising molecular tools for investigating the structure and function of plant cell walls. Their well-defined structure and high purity prevents misinterpretations of experimental data, and the possibility to introduce chemical handles provides means for easier localization and detection. Automated glycan assembly as emerged has a powerful new method for the efficient preparation of oligosaccharide libraries. We recently made use of this technology to prepare a collection of plant cell wall glycans for cell wall research. In this chapter, detailed experimental procedures for the automated synthesis of oligosaccharides that are ready for use in biological assays are described.

Key words Plant cell wall oligosaccharides, Automated glycan assembly, Solid-phase synthesis, Arabinoxylan, Arabinogalactan, Xyloglucan

1 Introduction

Plant cell walls represent a cellular exoskeleton that provides mechanical support for growth and development [1]. Oligosaccharide fragments of cell wall polysaccharides are valuable molecular tools for investigating structure and function of this complex matrix of biopolymers. They can serve as acceptor molecules for glycosyltransferases [2], substrates for glycosyl hydrolases [3], or as epitopes for antibodies and glycan binding modules [4]. Despite the importance of such molecules, the availability of well-defined and pure oligosaccharide samples is very limited. Extraction of polysaccharides from plants requires the sequential use of a number of solvents followed by careful chromatographic separation. The obtained polysaccharides may be further treated with specific glycosidases, yielding oligosaccharides with purities up to 95% after extensive separation and purification steps. However, they are still mixtures of highly related structures, complicating the assignment of structure–activity relationships. Chemical synthesis is one possibility to approach this problem. It can provide well-defined and

Zoë A. Popper (ed.), *The Plant Cell Wall: Methods and Protocols*, Methods in Molecular Biology, vol. 2149,
https://doi.org/10.1007/978-1-0716-0621-6_28,

pure oligosaccharide samples in significant amounts. Although chemical synthesis of complex oligosaccharides is still not a routine task, the development of new reactions and techniques has resulted in impressive advancements in the past decades [5].

The synthesis of plant carbohydrates has received much less attention than the synthesis of mammalian and bacterial glycans. Reasons are the high structural complexity of plant glycans, which renders their synthesis very challenging, and low recognition of synthetic plant carbohydrates as important samples for biological research. Synthetic mammalian and bacterial glycans have become standard tools for binding studies on microarrays or serve as the antigenic component in a plethora of vaccine candidates [6]. Nevertheless, a number of syntheses have been published. Most researchers have aimed at the preparation of pectic oligosaccharides while hemicellulosic structures have received less attention [7, 8]. Despite these remarkable achievements, reports about the utilization of the prepared oligosaccharides as tools in plant biology have been very limited [9, 10]. Biological studies have often been hampered by the low availability of large sets of related structures in order to establish structure–activity relationships. Synthesizing a single carbohydrate let alone libraries is very challenging using conventional solution phase chemistry. A promising approach to tackle this problem is automated glycan assembly (AGA) using solid-phase synthesis.

Solid phase chemistry, when compared to solution-phase chemistry, has the distinct advantage in that it is possible to wash away reagents and unwanted side products by simple filtration. This advantage has led to the development of oligonucleotide and peptide synthesizers, which have become standard tools in biological research. Solid-phase synthesis has evolved over the past 45 years, becoming an efficient method for the preparation of oligosaccharides as well [11–14]. In an approach similar to peptide and oligonucleotide synthesis, a nascent oligosaccharide chain is attached to a solid support via a cleavable linker and elongated by the iterative addition of the building blocks. A large excess glycosylating agent ensures completion of all glycosylation reactions, ideally resulting in the formation of a single product after cleavage from the solid support.

Automation of the solid-phase approach holds great potential in assembling oligosaccharides in a short space of time. Oligosaccharide synthesizers (*see* Fig. 1) are able to carry out temperature-controlled reactions and washes in a fully automated fashion, saving the operator numerous manual operations [15]. These advances have allowed for a wide variety of large and complex glycans to be produced by automated glycan assembly [16, 17] since its initial invention in 2001 [18].

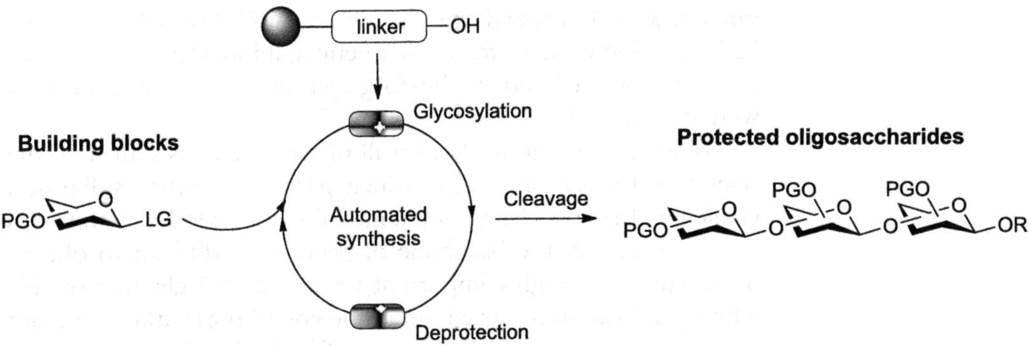

Fig. 1 Overview of the automated glycan assembly process

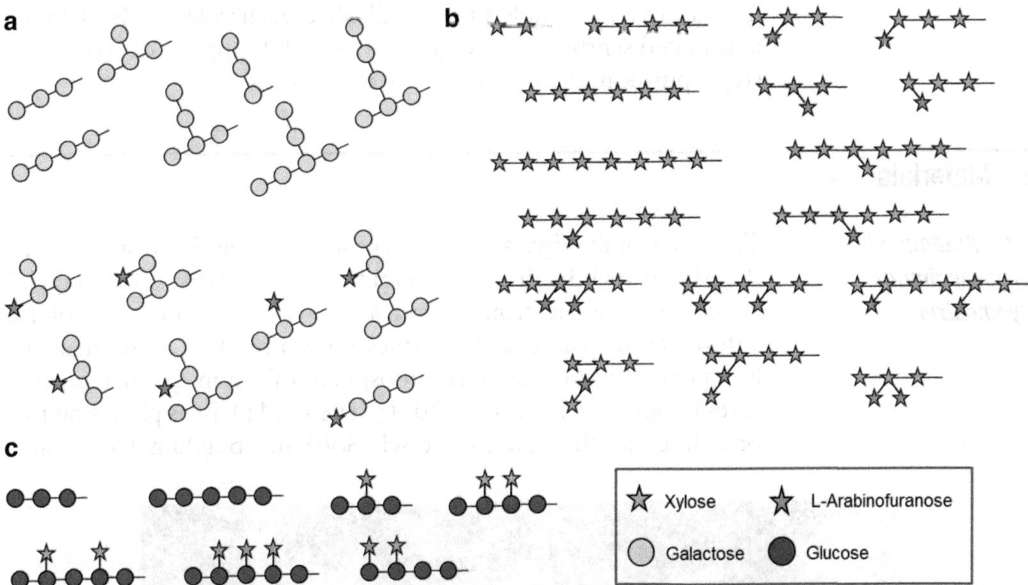

Fig. 2 Plant cell wall oligosaccharides obtained by automated glycan assembly: (**a**) type-II arabinogalactan-, (**b**) arabinoxylan-, and (**c**) xyloglucan oligosaccharides

Automated glycan assembly is particularly promising for the construction of carbohydrate libraries since practically any number of structurally related oligosaccharides can be obtained from common building blocks. We recently reported the automated glycan assembly of a collection of plant cell wall oligosaccharides (*see* Fig. 2). These include arabinoxylan-, type II arabinogalactan-, and xyloglucan-related structures [19–26]. The use of a standard photocleavable linker that allows the cleavage of the assembled products using light in a continuous-flow photoreactor provided milligram amounts of glycans after final deprotection steps and purification. The glycans are directly equipped with an aminopentyl linker at the reducing end for subsequent immobilization on glass slides or conjugation to proteins. All glycans were printed as microarray and probed with plant cell wall-directed antibodies [27]. Specific binding to the synthetic arabinoxylan oligosaccharides was observed and the binding epitopes for several antibodies were determined.

While the synthesis of cell wall oligosaccharides similar to the ones reported is meanwhile routinely performed, others still pose a challenge. In particular glycans containing multiple 1,2-*cis*-linkages such as found in the backbone of pectins are difficult to obtain. Stereocontrol is highly important for automated glycan assembly where purification is carried out at the end of the synthesis and not after each step. Improved protocols for the installation of "challenging" linkages are currently being introduced [28].

This chapter describes the necessary operations for the automated glycan assembly of cell wall oligosaccharides. This includes automated synthesis, cleavage from the solid support, deprotection steps, purification, and analysis of the glycans.

2 Materials

2.1 Automated Oligosaccharide Synthesizer

Three self-built oligosaccharide synthesizers (Fig. 3) are available at the Max Planck Institute of Colloids and Interfaces. A commercial system is available from *GlycoUniverse* [29]. Centerpiece of an automated oligosaccharide synthesizer is a reaction vessel that can be operated under an inert atmosphere of argon gas at temperatures ranging from −50 to +70 °C. The solid phase is placed on top of a filter in the reaction vessel. Solvents, building blocks and

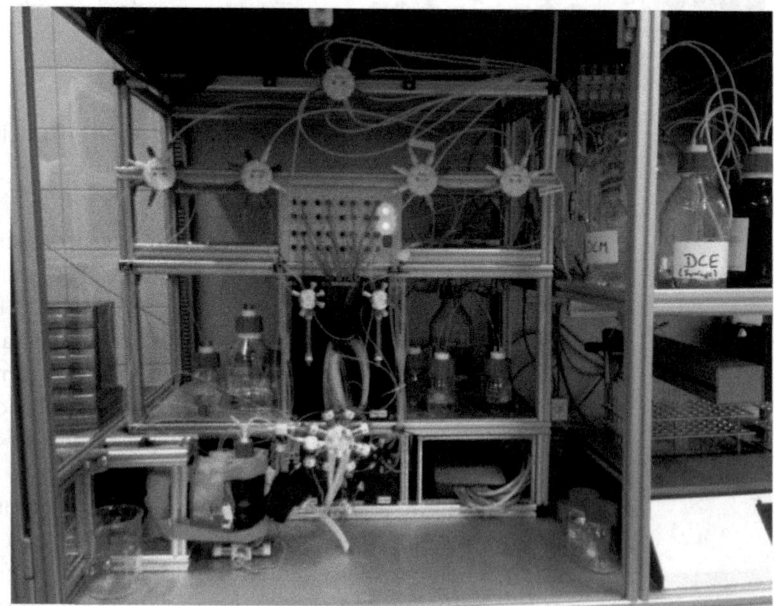

Fig. 3 Automated oligosaccharide synthesizer

reagents are added via syringe pumps or pressure driven and removed by an excess pressure of argon that pushes all liquids through the filter. Mixing of the reaction is accomplished by bubbling argon from the bottom through the reaction vessel.

2.2 Continuous-Flow Photoreactor

Cleavage of the oligosaccharides from the solid support is accomplished using a Vapourtec E-Series UV-150 photoreactor Flow Chemistry System equipped with a syringe pump (PHD2000, Harvard Apparatus) and a 10 mL reactor constructed of 1/8 in. o.d. FEP tubing. The medium pressure metal halide lamp is filtered using a commercially available red filter (Fig. 4).

2.3 Linker-Functionalized Resin

Merrifield resin equipped with a photolabile linker can be synthesized according to literature procedures [30] or is available from *GlycoUniverse* [29] (Fig. 5).

2.4 Building Blocks

Carbohydrate building blocks are synthesized according to literature procedures [19–22] or can be purchased from *GlycoUniverse* [24]. Suitably protected xylose, galactose, and arabinose building

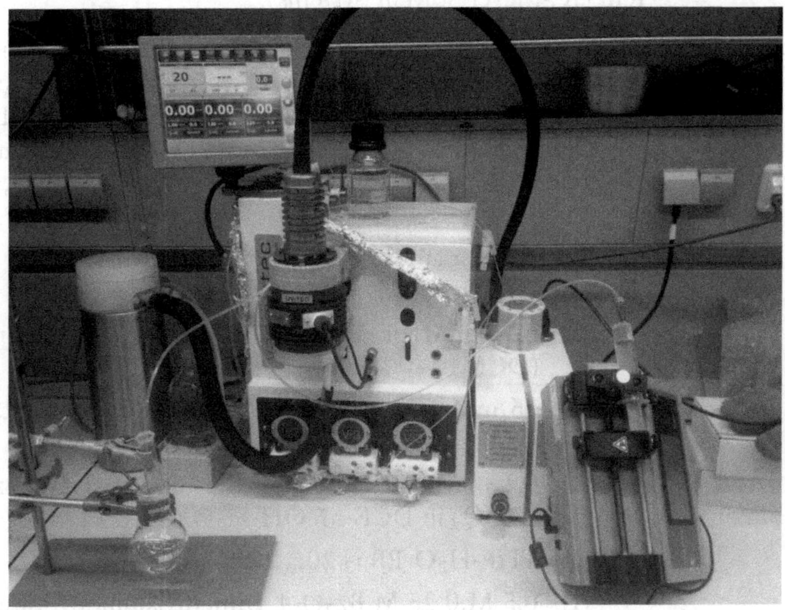

Fig. 4 Continuous-flow photoreactor

Fig. 5 Merrifield resin equipped with a photolabile linker

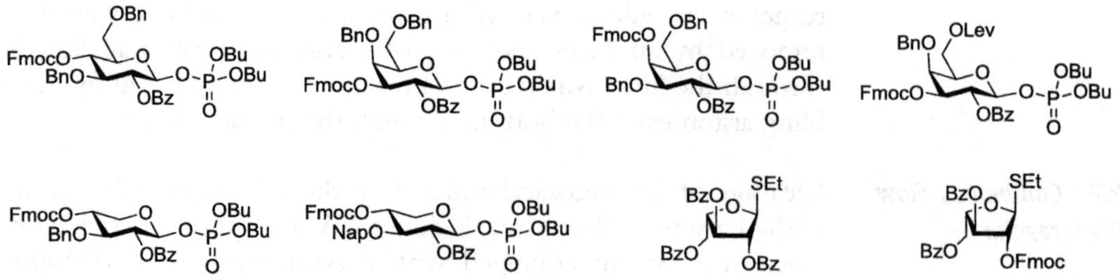

Fig. 6 Monosaccharide building blocks for the automated glycan assembly of plant cell wall oligosaccharides

blocks have been successfully used for the automated glycan assembly of plant cell wall oligosaccharides (Fig. 6). 9-Fluorenylmethoxycarbonyl (Fmoc) serves as the protecting group for chain elongation while levuloyl esters (Lev) and (2-naphtyl)methyl ethers (Nap) are used for the introduction of backbone substitutions.

2.5 Solvents and Reagents

Solvents and reagents are used as supplied without any further purification. Anhydrous solvents are taken from a dry solvent system (JC-Meyer Solvent Systems).

1. Solvents: 1,2-Dichloroethane (DCE) (anhydrous), Tetrahydrofuran (THF) (High pressure liquid chromatography (HPLC) grade), Dimethylformamide (DMF) (HPLC grade), Dioxane (anhydrous), EtOAc (HPLC grade), Hexane (HPLC grade), MeCN (HPLC grade), AcOH, Pyridine, Millipore water.
2. Reagents for automated glycan assembly:
   (a) 20% NEt_3 in DMF.
   (b) 62.5 mM or 125 mM TMSOTf in Dichloromethane (DCM).
   (c) 75 mM NIS/7.5 mM TfOH or 150 mM NIS/15 mM TfOH in DCM-dioxane 2:1.
   (d) 0.1 M 2,3-dichloro-5,6-dicyano-1,4-benzoquinone (DDQ) in DCE-MeOH-H_2O 64:16:1.
   (e) THF-H_2O-PBu_3 20:1:1.25.
   (f) 0.5 M:0.25 M Bz_2O-4-Dimethylaminopyridine (DMAP) in DCE.
   (g) Ac_2O.
3. Reagents for global deprotection:
   (a) NaOMe (0.5 M in MeOH).
   (b) Amberlite IR-120 resin.
   (c) Pd/C.
   (d) Cylinder with hydrogen gas.

2.6 HPLC

Analytical HPLC is performed using an YMC-Pack DIOL-300-NP column (150 × 4.6 mm), a Phenomenex Luna C5 column (250 × 4.6 mm), and a Thermo Scientific Hypercarb column (150 × 4.6 mm). Preparative HPLC is performed using a preparative YMC-Pack-Diol-300-NP (150 × 20 mm), a semipreparative Phenomenex Luna C5 column (250 × 10 mm), and a semipreparative Thermo Scientific Hypercarb column (150 × 10 mm).

2.7 Other Materials

1. Plastic pipette (3 mL, Roth).
2. Plastic syringe equipped with PTFE frit (10 mL, Biotage PP-Reactor).
3. Norm-Ject plastic syringe with cap (20 mL, Henke-Sass Wolf).
4. Syringe filter (RC 0.45 μm, Roth).

3 Methods

3.1 Automated Glycan Assembly of Protected Oligosaccharides

1. Fill the synthesizer with solvents (DCM, DMF, THF, 1,4-dioxane, DCE).
2. Prepare the required activator and deprotection solutions and attach them to the synthesizer.
3. Dissolve the building blocks at the appropriate concentration (2.5–10 eq.) in 1 mL anhydrous DCM per coupling cycle and attach the building block vessels to the synthesizer.
4. Transfer resin (12.5 or 25 μmol acceptor sites) into the reaction vessel and swell in 2 mL DCM for 30 min.
5. Program the synthesizer and start synthesis.

3.2 Cleavage of the Protected Oligosaccharide from the Resin Using a Continuous-Flow Photoreactor

1. Transfer the resin (suspended in the reaction vessel of the synthesizer in approx. 2 mL DCM) to a 20 mL plastic syringe sealed at the tip with a plastic cap using a 3 mL plastic transfer pipet.
2. Wash the reaction vessel with DCM and repeat the action until all resin is transferred. Fill up the syringe to 20 mL with DCM and insert the stamp carefully.
3. Turn the syringe upward, remove the cap, and place the syringe in a syringe pump. Connect the syringe to the reactor tubing of the photoreactor flow chemistry system.
4. Insert the outlet of the tubing into a filter syringe and place the filter syringe on top of a 100 mL flask.
5. Set the lamp power to 80%, start the cooling system to maintain the temperature in the reactor at 20 °C, ignite the UV lamp, and push the resin through the system with the syringe pump at a flow rate of 0.8 mL/min (*see* **Note 1**).

6. After the resin has passed the photoreactor wash the line with an additional 20 mL of DCM at a flow rate of 2 mL/min and subsequently push all remaining DCM through the system by using an air-filled syringe.
7. Remove the solvent using a rotary evaporator and analyze the product by MALDI mass spectrometry and analytical HPLC.

3.3 Deprotection and Purification of Oligosaccharides

1. Purify the oligosaccharides after cleavage from the solid support by normal phase HPLC using a preparative YMC Diol column and a linear gradient from 10% to 100% ethyl acetate in hexane (40 min, flow rate 1 mL/min).
2. Dissolve the purified compound in THF and add NaOMe (0.5 M in MeOH, 0.5 mL). Stir the resulting solution overnight and subsequently neutralize the reaction mixture by addition of prewashed IR-120 resin. Filter the resin off and wash the resin with MeOH.
3. Purify the resulting semiprotected oligosaccharides by normal phase HPLC using either a preparative YMC Diol column and linear gradients from 10% to 100% ethyl acetate in hexane (40 min, flow rate 1 mL/min), or a semipreparative C5 column and a linear gradient from 80% to 0% H_2O in MeCN (50 min, flow rate 1 mL/min) (*see* **Note 2**).
4. Dissolve the purified compound in a mixture of EtOAc–MeOH–AcOH–H_2O (4:2:2:1, 3 mL) and add the resulting solution to a round-bottom flask containing Pd/C (10% Pd, 8 mg). Flush the suspension with argon for 10 min, saturate it with hydrogen for 10 min, and stir it under a hydrogen atmosphere overnight. After filtration of the reaction mixture through a syringe filter remove the solvents using a rotary evaporator. Take up the residue in 1 m water and lyophilize the product to dryness.
5. Analyze the product by mass spectrometry, Nuclear magnetic resonance (NMR) spectroscopy and analytical HPLC using a Hypercarb column and a linear gradient from 97.5% to 30% water in MeCN (45 min, flow rate 0.7 mL/min) and from 30% to 0% H_2O in MeCN (10 min, flow rate 0.7 mL/min). A typical set of analytical data is shown in Fig. 7 (*see* **Note 3**).

4 Notes

1. Make sure that the resin quantitatively enters the photoreactor (it may be necessary to move and gently shake the syringe pump to accomplish this).

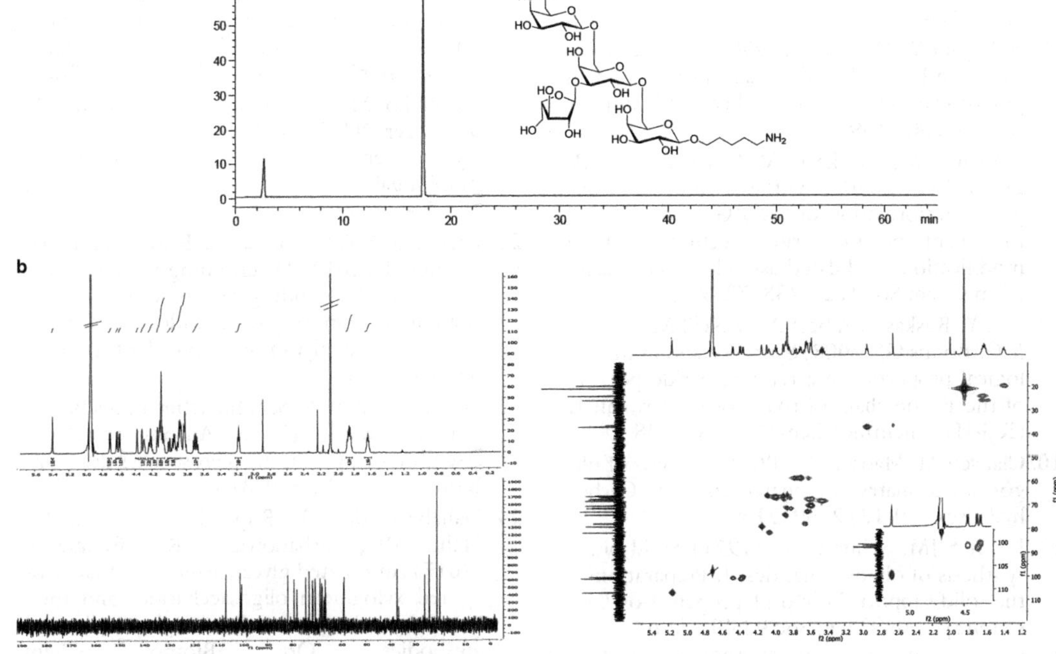

Fig. 7 Analysis of an exemplary synthetic cell wall oligosaccharide by (**a**) HPLC and (**b**) NMR spectroscopy. The sample does not contain significant detectable impurities

2. In many cases it will not be necessary to perform HPLC-purifications of both the fully protected and the semiprotected form of the oligosaccharide.
3. If any impurities are detected after final hydrogenolysis, purify the final product by reversed-phase HPLC using a semipreparative Hypercarb column and a linear gradient from 97.5% to 30% H_2O in MeCN (45 min, flow rate 0.7 mL/min) and from 30% to 0% H_2O in MeCN (10 min, flow rate 0.7 mL/min).

References

1. Albersheim P, Darvill A, Roberts K, Sederoff R, Staehelin A (2011) In: Masson S (ed) Plant cell walls. Taylor and Francis Publishing Group, LLC, Abington
2. Falk A, Price NJ, Raikhel NV, Keegstra K (2002) An *Arabidopsis* gene encoding an α-xylosyltransferase involved in xyloglucan biosynthesis. Proc Natl Acad Sci 99:7797–7802
3. Sampedro J, Pardo B, Gianzo C, Guitian E, Revilla G, Zarra I (2010) Lack of α-xylosidase activity in *Arabidopsis* alters xyloglucan composition and results in growth defects. Plant Physiol 154:1105–1115
4. Pedersen HL, Fangel JU, McCleary B, Ruzanski C, Rydahl MG, Ralet MC, Farkas V, von Schantz L, Marcus SE, Andersen MC, Field R, Ohlin M, Knox JP, Clausen MH, Willats WG (2012) Versatile high resolution oligosaccharide microarrays for plant glycobiology and cell wall research. J Biol Chem 287 (47):39429–39438
5. Boltje TJ, Buskas T, Boons GJ (2009) Opportunities and challenges in synthetic oligosaccharide and glycoconjugate research. Nat Chem 1(8):611–622
6. Kolarich D, Lepenies B, Seeberger PH (2012) Glycomics, glycoproteomics and the immune

system. Curr Opin Chem Biol 16 (1–2):214–220

7. Nakahara Y, Ogawa T (1990) Stereoselective total synthesis of dodecagalacturonic acid, a phytoalexin elicitor of soybean. Carbohydr Res 205:147–159
8. Scanlan EM, Mackeen MM, Wormald MR, Davis BG (2010) Synthesis and solution-phase conformation of the RG-1 fragment of the plant polysaccharide pectin reveals a modification-modulated assembly mechanism. J Am Chem Soc 132:7238–7239
9. Rao Y, Buskas T, Albert A, O'Neill MA, Hahn MG, Boons GJ (2008) Synthesis and immunological properties of a tetrasaccharide portion of the B side chain of rhamnogalacturonan II (RG-II). ChemBioChem 9(3):381–388
10. Clausen M, Madsen R (2004) Synthesis of oligogalacturonates conjugated to BSA. Carbohydr Res 339(13):2159–2169
11. Frechet JM, Schuerch C (1971) Solid-phase synthesis of oligosaccharides. I. Preparation of the solid support. Poly[p-(1-propen-3-ol-1-yl) styrene]. J Am Chem Soc 93:492–496
12. Seeberger PH, Haase W-C (2000) Solid-phase oligosaccharide synthesis and combinatorial carbohydrate libraries. Chem Rev 100:4349–4393
13. Seeberger PH (ed) (2001) Solid support oligosaccharide synthesis and combinatorial carbohydrate libraries. Wiley-VCH, New York, NY
14. Bennett CS (2014) Principles of modern solid-phase oligosaccharide synthesis. Org Biomol Chem 12:1686–1698
15. Kröck L, Esposito D, Castagner B, Wang C-C, Bindschädler P, Seeberger PH (2012) Streamlined access to conjugation-ready glycans by automated synthesis. Chem Sci 3:1617–1622
16. Seeberger PH (2008) Automated oligosaccharide synthesis. Chem Soc Rev 37:19–28
17. Seeberger PH (2015) The logic of automated glycan assembly. Acc Chem Res 48:1450–1463
18. Plante OJ, Palmacci ER, Seeberger PH (2001) Automated solid-phase synthesis of oligosaccharides. Science 291:1523–1527
19. Schmidt D, Schuhmacher F, Geissner A, Seeberger PH, Pfrengle F (2015) Automated synthesis of arabinoxylan-oligosaccharides enables characterization of antibodies that recognize plant cell wall glycans. Chem Eur J 21:5709–5713
20. Senf D, Ruprecht C, de Kruijff GHM, Simonetti SO, Schuhmacher F, Seeberger PH, Pfrengle F (2017) Active site-mapping of xylan-deconstructing enzymes with Arabinoxylan oligosaccharides produced by automated glycan assembly. Chem Eur J 23:3197–3205
21. Bartetzko MP, Schuhmacher F, Hahm HS, Seeberger PH, Pfrengle F (2015) Automated glycan assembly of oligosaccharides related to Arabinogalactan proteins. Org Lett 17:4344–4347
22. Bartetzko MP, Schuhmacher F, Seeberger PH, Pfrengle F (2017) Determining substrate specificities of β1,4-endo-galactanases using plant Arabinogalactan oligosaccharides synthesized by automated glycan assembly. J Org Chem 82:1842–1850
23. Dallabernardina P, Schuhmacher F, Seeberger PH, Pfrengle F (2016) Automated glycan assembly of xyloglucan oligosaccharides. Org Biomol Chem 14:309–313
24. Dallabernardina P, Ruprecht C, Smith PJ, Hahn MG, Urbanowicz BR, Pfrengle F (2017) Automated glycan assembly of galactosylated xyloglucan oligosaccharides and their recognition by plant cell wall glycan-directed antibodies. Org Biomol Chem 15:9996–10000
25. Dallabernardina P, Schuhmacher F, Seeberger PH, Pfrengle F (2017) Mixed-linkage glucan oligosaccharides produced by automated glycan assembly serve as tools to determine the substrate specificity of lichenase. Chem Eur J 23:3191–3196
26. Bartetzko MP, Pfrengle F (2019) Automated glycan assembly of plant oligosaccharides and their application in cell wall biology. ChemBioChem 20:877–885
27. Ruprecht C, Bartetzko MP, Senf D, Dallabernadina P, Boos I, Andersen MCF, Kotake T, Knox JP, Hahn MG, Clausen MH, Pfrengle F (2017) A synthetic glycan microarray enables epitope mapping of plant cell wall glycan-directed antibodies. Plant Physiol 175:1094–1104
28. Hahm HS, Schuhmacher F, Seeberger PH (2016) Automated assembly of oligosaccharides containing multiple cis-glycosidic linkages. Nat Chem 7:12482
29. www.glycouniverse.de
30. Eller S, Collot M, Yin J, Hahm HS, Seeberger PH (2013) Automated solid-phase synthesis of chondroitin sulfate glycosaminoglycans. Angew Chem Int Ed 52:5858–5861

Chapter 29

Computerized Molecular Modeling of Carbohydrates

Alfred D. French and Glenn P. Johnson

Abstract

Computerized molecular modeling continues to increase in capability and applicability to carbohydrates. This chapter covers nomenclature and conformational aspects of carbohydrates, perhaps of greater use to computational chemists who do not have a strong background in carbohydrates, and its comments on various methods and studies might be of more use to carbohydrate chemists who are inexperienced with computation. Work on the intrinsic variability of glucose, an overall theme, is described. Other areas of the authors' emphasis, including evaluation of hydrogen bonding by the atoms-in-molecules approach, and validation of modeling methods with crystallographic results are also presented.

Key words Carbohydrate, Disaccharide, Conformation, Puckering, Modeling, Quantum mechanics, Molecular mechanics

1 1. Introduction

The plant cell wall is composed of cellulose and often other polysaccharides that depend greatly on the particular plant source and its stage of growth. Other common cell wall polysaccharides include pectins, which contain α-D-galacturonic acid, and various hemicelluloses, for example, β-1,4-linked glucurono- and arabinoxylan and gluco- and galactomannan. Understanding the roles and interactions of these molecules strictly by experimentation has been difficult, and computerized modeling offers additional information to support or argue against various hypotheses. Thus, computer modeling is another tool that can be used to study plant cell walls.

To best take advantage of the capabilities of modeling, it is important to understand the primary hypothesis. Namely, the conformational structure with the lowest energy, compared to that of all other conformations, is the most likely structure. Also, the reaction pathway with the lowest barriers is the most likely one. Validation of experimental structures, such as those proposed from crystallographic studies, can be done by using the structure and modeling software to predict various spectra, such as IR or NMR

Zoë A. Popper (ed.), *The Plant Cell Wall: Methods and Protocols*, Methods in Molecular Biology, vol. 2149,
https://doi.org/10.1007/978-1-0716-0621-6_29,

[1, 2]. Beyond needing a model to visualize the structure at hand, it is useful to learn how to frame questions so that modeling can be applied to help with a problem.

Various computer modeling software systems provide a graphical user interface with a drawing function that lets the user start without any information other than a vision in his mind and a knowledge of the pattern of atom connectivity. Such software "knows" about atomic diameters, usual bond lengths and angles, and so on, so it will assist the user in creating a structure that may highly resemble the actual molecule as represented by balls and/or sticks. Different display options for the molecules provide a sense of artistic accomplishment, better convey information, and may even enhance the appeal of a particular research program. Still, the validity of such pictures of typical carbohydrate molecules is questionable because there are many alternative structures (conformations or shapes) that can be formed by simply changing the torsion angles. Figure 1 shows the three staggered conformations of *n*-butane and their torsion angles. Deciding which conformer best represents the "real" population of molecules requires a measure of

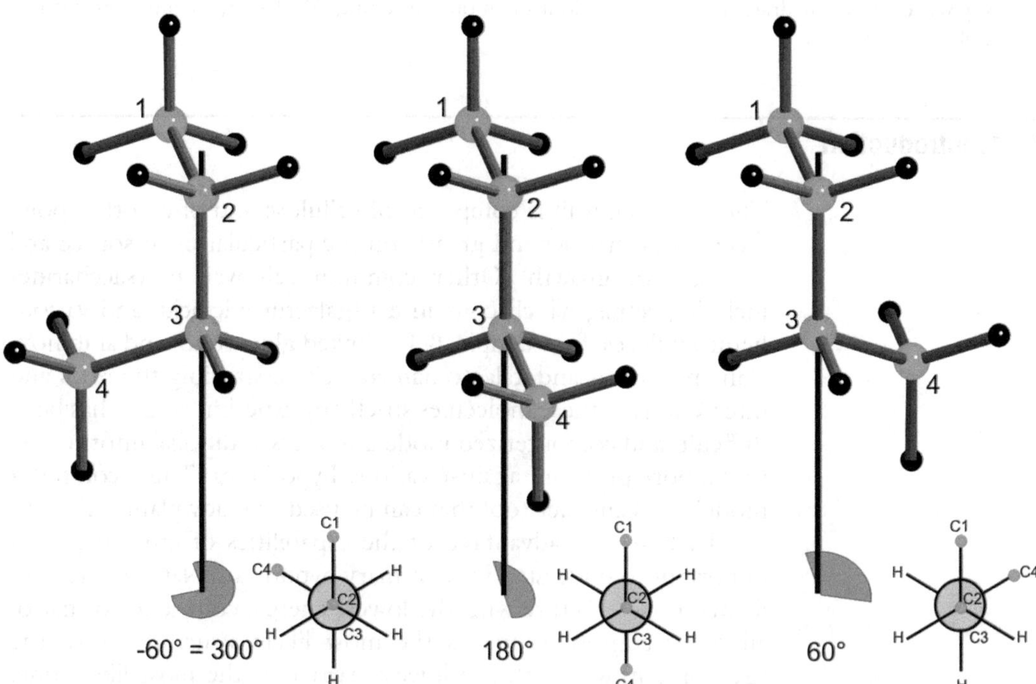

Fig. 1 Staggered conformations of butane, with their C1–C2–C3–C4 torsion angles indicated. Newman projections are also shown. The angles refer to the angle of the C1–C2 bond relative to the C3–C4 bond, when viewed down the C2–C3 bond. The C1–C2–C3–C4 torsion angle is 0° when all four atoms are in a plane and C1 and C4 are *cis*. The vertical rods indicate the axes about which the C4 groups rotate. All three conformations (−*gauche*, *trans* and +*gauche*) correspond to minima in the energy but the 180° conformer has the lowest energy, that is, it is the global minimum. The *gauche* forms are also called *syn*, and the *trans* form is also called *anti*

the relative free energy of each alternative. That may seem simple, but for a carbohydrate there are two complications. One is how to calculate the energy, and the other is to decide which alternative structures to consider and which to ignore. Both issues are typically approached with assumptions and approximations that are being reduced as computer power and software sophistication increase.

Regarding alternative structures, the pyranosyl ring of aldohexose sugars is typically assumed to be in a chair form, even though there are alternatives. In fact, idopyranose appears to have significant populations for both chairs and a skew form of the ring in solution [3]. Even glucose has conformational ambiguity, as some methylated or acylated cyclodextrin molecules have glucose residues that take alternative chair or skew forms in their experimentally determined crystal structures [4, 5].

Energy can be calculated in many ways, starting historically with a simple scan for short distances between atoms that are not bonded to each other. If such contacts exist, that kind of primitive analysis would assign a potential energy of infinity, reflecting complete improbability. One mainstream approach is molecular mechanics (MM). The energy terms of these empirical force-field models arise from our useful "cartoons" of molecular structure, with potential functions for bond stretching, angle bending and charge-charge interactions. Quantum mechanics (QM) calculations consider distributions and interactions among individual electrons. Thus, QM is also called electronic structure theory. QM calculations are so expensive in terms of computer time and memory that trisaccharides [6] are about the largest molecules to consider for reasonably complete studies. Some insights can, however, be obtained by applying QM to larger structures such as cycloamylose with 26 glucose residues [7] or even cellulose crystallites with 144 glucose residues [8]. Also, explicit treatments of neighbor molecules such as solvent or adjacent molecules in a crystal are seldom included in QM calculations, nor are most of the alternative structures, all because of the time required. Semiempirical quantum methods have not found much of a following for studies of carbohydrates [9].

The following material mentions some specific methods used in quantum mechanics studies. If the terminology is an obstacle, the book, "Exploring Chemistry with Electronic Structure Methods" [10], should be a great help. Sometimes errors in calculated energy cancel each other. When comparing two different molecular shapes, the same errors may affect both forms. Even if the absolute energy values are not correct, the relative energies could still be useful. Also, there is a useful but ultimately unreliable compensation for the lack of explicit treatment of electron correlation in the Hartree–Fock (HF) QM method. That error can often be offset by errors from using a small basis set such as 6-31G(d). Combining the HF method and the 6-31G(d) basis set gives HF/6-31G(d), a "magic" level of theory [11]. Since then, Density Functional Theory methods have rapidly advanced, with numerous possibilities such as B3LYP [12] or M062X [13].

Another approach lies in hybrid methods in which different calculations are used for different parts of the problem depending on how critical they are. The ONIOM approach is well-known [14]. Another fruitful approach is to use a small basis set for the less important carbon atoms and a larger basis set for the more determinative oxygen atoms [15]. Such calculations to optimize the geometry of the structure can be followed by a "single point" calculation with a more complete basis set to compute only the energy of the structure.

To some extent, the present chapter follows up on proceedings from a symposium on carbohydrates in 1989 (ACS Symposium Series 430). Besides many articles in chemical and carbohydrate journals, special issues have been at least partly dedicated to carbohydrate modeling (Molecular Simulation (vol. 4, issue 4); THEOCHEM (vol. 395–397), Carbohydrate Research (vol. 340, Issue 5), ACS Symposium Series 930). More recent reviews include Refs. (16–18). Some of the carbohydrate issues treated herein are covered in a more thorough review [19].

2 Structural Descriptors of Carbohydrates

Nomenclature. Because most carbohydrates have numerous asymmetric centers, they have kept their traditional nomenclature [20]. β-D-Glucopyranose should be easier to remember than (*2R, 3R, 4S, 5S, 6R*) 2,3,4,5 tetrahydroxy, 6 methoxy oxacyclohexane. Another point is that the first carbon atom in the parent acyclic sugar is number 1 instead of the heteroatom in the sugar ring as would be the case if standard organic chemistry nomenclature were used. Consider a disaccharide composed of two D-glucopyranose residues, linked at the 1′ and 4 positions. The particulars of the linkage between the glucose residues define the compound, that is, whether it is maltose or cellobiose. The configuration at C1, the reducing end's anomeric center, defines whether it is the α- or β-anomer. Thus, β-maltose and α-cellobiose both exist, despite the opposite configurations at the anomeric centers of their linkages. Rules for carbohydrates composed of more than two monosaccharides are covered by publications cited in Ref. 20.

Drawings. To aid recognizability of ring compounds, the anomeric center (the carbon bound to both the ring oxygen and a hydroxyl in the native sugar) is drawn on the far right and the bond between the ring oxygen and the nonanomeric carbon is parallel to the plane of the paper and indicated as being at the back of the ring. The ring oxygen of five-membered (furanosyl) rings is shown at the back.

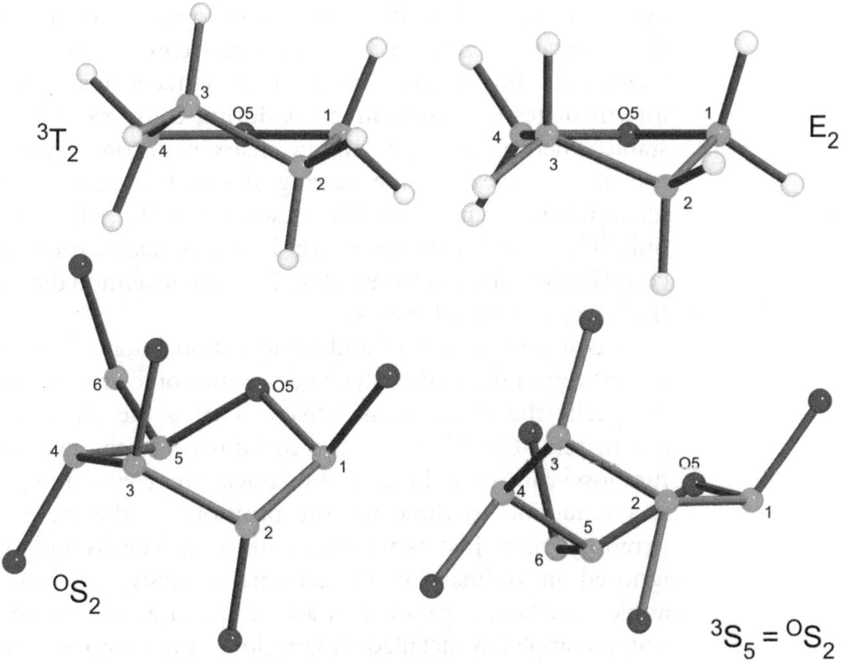

Fig. 2 Upper: Sample furanosyl rings with C1, O5, and C4 all coplanar. The $^3T_2$ drawing has C3 above the plane, and C2 below, where the E_2 drawing has only C2 out of plane. Lower: Both of the skew form β-D-glucopyranosyl rings (hydrogen atoms not shown) are exactly the same structures, correctly described as $^OS_2$ rings. The ring on the right has been rotated about a line between C1 and C4. Convention dictates that the ring is described as $^OS_2$ instead of $^3S_5$. There are two planes that contain four of the atoms. One contains C1, C3, C4, and C5, and the other contains C1, C2, C4, and O5

Ring Puckering. Carbohydrate rings are puckered, not flat. An extensive analysis of both furanose and pyranose ring shapes is presented in a study of the conformationally and configurationally ambiguous sugar, psicose [21]. Furanose rings can have envelope shapes, with four coplanar atoms and one atom out-of-plane. For example, a ring with its oxygen atom out of plane is a characteristic form, denoted ^OE if the oxygen is above the ring and E_O if below (when drawn in the above standard orientation). Otherwise, they are twists, with three coplanar atoms and one atom above the plane and one below (*see* Fig. 2 for $^3T_2$ and E_2 examples based on tetrahydrofuran). Each individual envelope (E) or twist (T) is a "characteristic form." Ten E forms of furanose rings exist, as well as 10 T forms, such as $^4T_3$, a favored form for β-D-fructofuranose. Facile transitions known as pseudorotation are permitted between the adjacent alternating E and T forms, such as ^4E, $^4T_3$, and E_3. The characteristic forms do not necessarily correspond to energy minima but are merely markers in ring-shape hyperspace for understanding the shape of a particular ring.

Characteristic pyranosyl forms and their designations include two chairs (C), six boats (B) and six skews (S, also called twist-boats), *see* Fig. 2. Also, there are 12 envelopes (E) and 12 half-chairs (H). Boeyens described 12 additional (screw-boat) characteristic forms [22]. For ordinary sugars, these latter forms correspond to intermediates or transition states during conversions and are not stable. There are only two unique chairs, with convention dictating use of the lowest-numbered ring atoms. For example, the same glucopyranosyl chair could be described as $^4C_1$, $^2C_5$, or $^OC_3$ but only $^4C_1$ is used. The other form is $^1C_4$. Similarly, each skew form could be described two ways (Fig. 2), but convention dictates use of the lower-numbered atoms.

Because experimental and computational sugar rings usually do not correspond exactly to a characteristic conformation, it is useful to specify the shape quantitatively. That is the job of puckering parameters. The 15 *x*, *y*, and *z* coordinates of the five atoms of a furanosyl ring are reduced to two puckering parameters, and there is a reduction to three for the positions of the six atoms of a pyranosyl ring. Just as variations in bond lengths and angles are ignored in ordinary conformational analysis based on torsion angles, puckering parameters are intended as overall descriptors, not a completely detailed description. Any computed puckering from an experimental or computed structure can be translated to a phrase such as "a conformation that is nearest in puckering space to $^OS_2$" (a skew form, with the ring oxygen above the plane and C2 below the plane that includes C1, C3, C4, and C5). Even when the shape is obvious, such as the frequently found $^4C_1$ (chair) form, assessment of puckering parameters allows the degree of distortion from the normal shape to be described to learn the effects of being packed in a crystal or complexed with an enzyme. In another example, the degree of puckering for glucose could be compared with that of tetrahydropyran to learn the effects of substituents on the ring shape.

Cremer–Pople puckering parameters [23] work well in most circumstances, but in cases where bond lengths are quite different, the translation from the computed parameters to a characteristic conformer may not meet expectations based on a visual assessment. Besides the Cremer-Pople puckering parameters, which apply to all sizes of rings, there are the Altoona–Sundaralingam puckering parameters for furanosyl rings [24]. Descriptors of ring puckering conformations have also been derived from endocyclic torsion angles [25, 26] or flap angles [27, 28].

Other measures of the ring shape are useful, especially when they are interrelated with the conformation of polysaccharides. For example, the distance between O1 and O4 of α-D-glucopyranose varies between 3.88 Å and 4.84 Å (see below) in crystals of mono- and oligosaccharides, despite a basic $^4C_1$ shape. If those residues are used to make models of the amylose polysaccharide, the resulting helix shape will vary widely. Recently, we observed a twist of the glucose ring, that is, the O1–C1. . .C4–O4 virtual torsion angle has a range of about 25° [29]. Again, this parameter governs the location of the adjacent residue in model polymers.

Exocyclic group orientation. The orientation of primary alcohol groups can be an important variable. Most often the three staggered rotamers are described as *gg (gauche-gauche)*, *gt (gauche-trans)*, and *tg (trans-gauche)*. These different conformers are shown in Fig. 4. The first of the two letters corresponds, in D-glucose, to O5–C5–C6–O6 torsion angles (ω) of –60° (-g), 60° (+*g*), or 180° (*t*) (*see* Fig. 1). The second letter corresponds to C4–C5–C6–O6 torsion angles of 60°, 180°, or -60°. Other authors prefer –*g*, +*g*, and *t* for the O5–C5–C6–O6 angle. Because C5 is sp^3 hybridized, the use of two letters could be considered to be redundant, but it avoids needing to remember the sign of the torsion angle. If considering L-glucopyranose, the signs of *gauche* forms must be reversed for the single-letter notation, but with the two-letter notation the mirror image of D-glucose with O6 *gt* is L-glucose with O6 *gt*. The *tg* conformation has been described as "forbidden" because it was not observed in early crystal structure studies of molecules having the *gluco* configuration at C4 (O4 equatorial), and some NMR-based analyses of sugars have yielded a sum of the *gt* and *gg* conformations slightly greater than 100%, implying negative amounts of *tg*. More recent experimental work has found examples of *tg* rotamers, especially in native cellulose I [30] and a derivative of cellobiose [31]. The *galacto* configuration, with O4 axial, has a low population of *gg* conformations [19]. IUPAC standards designate the reference atom in the ring to be the next-lower numbered ring atom instead of the oxygen atom. This reverses the definitions of *gt* and *tg*, so readers should be aware of both systems.

Secondary alcohol (hydroxyl) orientations are important to the calculated energy. They are often described in terms of clockwise and counterclockwise (reverse clockwise) systems of intraresidue hydrogen bonds that provide maximum stabilization for isolated molecules. Despite the significant lowering of the energy due to having all of the secondary hydroxyl groups appearing to participate in a continuous donor–acceptor–donor–acceptor network, the resulting long H...O distances and small O–H...O angles would yield weak attractions. According to Bader's atoms-in-molecules (AIM) theory [32] ("electron density gradient vector field analysis") these energy-lowering orientations do not result in the bond paths and bond critical points [33–35] needed for true hydrogen bonds [36]. The primary alcohol can hydrogen bond with O4, and two axial hydroxyl groups with 1,3 spacing (such as in $^1C_4$ glucose) can form better hydrogen bonds [36].

All staggered rotamers for glucose can be studied by fairly good QM. Putting each of the six rotatable groups into all three orientations, there are 3^6 (=729) combinations. The Jaguar program [37] was used with B3LYP/6-31+G(d,p) energy minimization on each

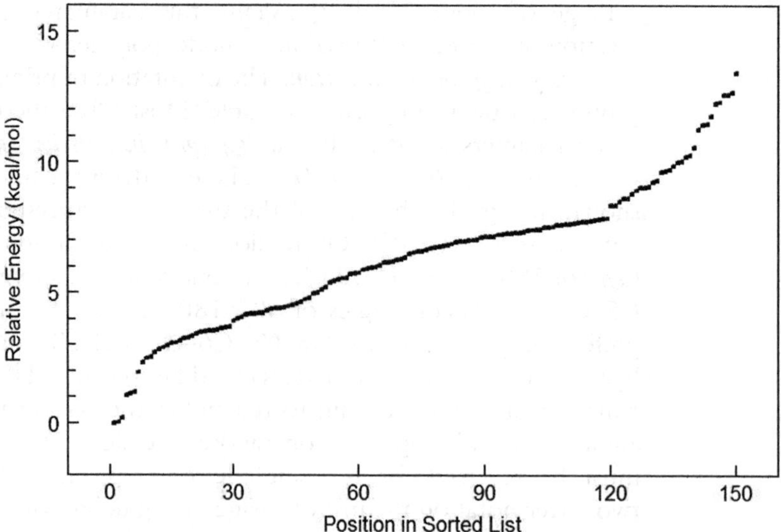

Fig. 3 Energy distribution for stationary shapes of glucose at the B3LYP/6-31+G(d,p) level. Twenty five of the energies above 4.3 kcal/mol correspond to "saddle-point" or "transition state" structures

of the conformers as isolated (gas phase) molecules. Most of the 729 combinations were unstable and one or more of their exocyclic groups rotated to a different staggered orientation. Still, there were 150 unique stationary structures, and their energy range (13.4 kcal/mol) is considerable. Figure 3 shows their nearly continuous distribution of energies. Figure 4 shows the six lowest-energy and the six highest-energy forms. The six lowest-energy structures included the *gt*, *gg*, and *tg* conformations of O6 and both reverse-clockwise and clockwise secondary hydroxyl groups. This would seem to support the notion of cooperative rings of hydrogen bonds, for which there is some experimental support [38], despite the above AIM studies. Four of the six highest-energy structures have the hydroxyl hydrogen on O1 located underneath the pyranose ring, a sterically disadvantageous orientation, and the rings of OH interactions are absent.

The exocyclic group orientations have a substantial effect on the molecular geometry, including the O1...O4 distance. Figure 5 shows answers to three different questions about this distance. The top curve shows the probability distribution predicted by $p_i = e^{-\Delta E/RT}$ for a glucose residue stretched and compressed with MM3 [39] calculations. The underlying energy curve obeys Hooke's law almost perfectly. The middle graph presents the O1...O4 distances in 2582 experimental examples of α-D-glucose and its derivatives having $^4C_1$ rings, obtained from a scan of the Cambridge Structural Database [40]. The lowest graph shows the O1...O4 distances within the 150 stable structures discussed above. The top and bottom theoretical analyses indicate, respectively, the elasticity of glucose and the intrinsic variability based on the exocyclic group orientations. The experimental data combine both these factors,

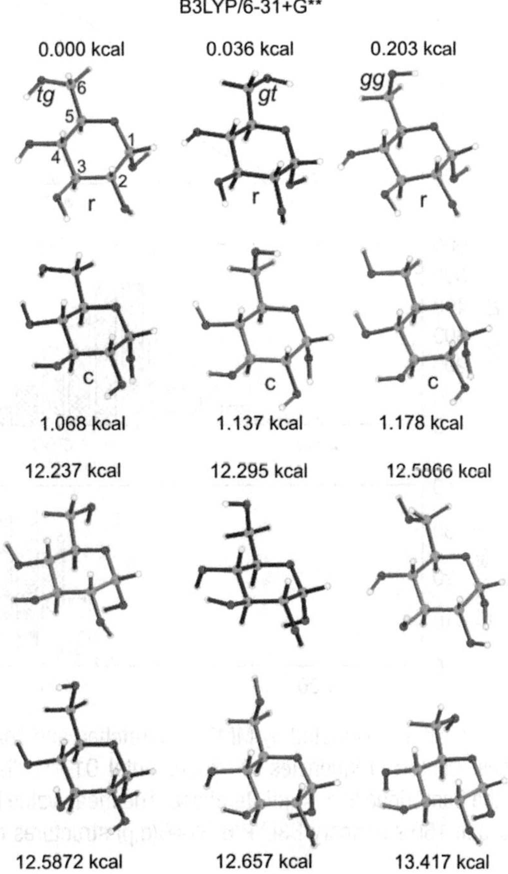

Fig. 4 Six lowest-energy and six highest-energy stationary forms of α-D-glucopyranose. Of these structures, only the lower-left is in a transition state according to frequency calculations

with the elasticity corresponding to deformations from random crystal packing and strain from being part of larger molecules, including macrocycles. The mean MM3 distance is longer than the mean experimental value, as is the shortest B3LYP/6-31+G (d,p) result.

Other methods for calculating the energy of the 729 conformers will give different numbers of final structures as well as energy values. All of these 150 structures have the $^4C_1$ shape; many more stationary points would be found for other ring shapes. Some of those alternative ring shapes with optimal hydroxyl orientations will give energies lower than the higher-energy $^4C_1$ shapes.

Anomeric centers. In aqueous solution, a single enantiomer such as D-glucose is five compounds: acyclic, and α- and β-pyranoses and furanoses. The populations of the furanose and acyclic forms are minimal for glucose, but must be considered for sugars such as the ketopyranose, psicose [21]. Opening and closing the ring allows the configuration of glucose to change at C1, and

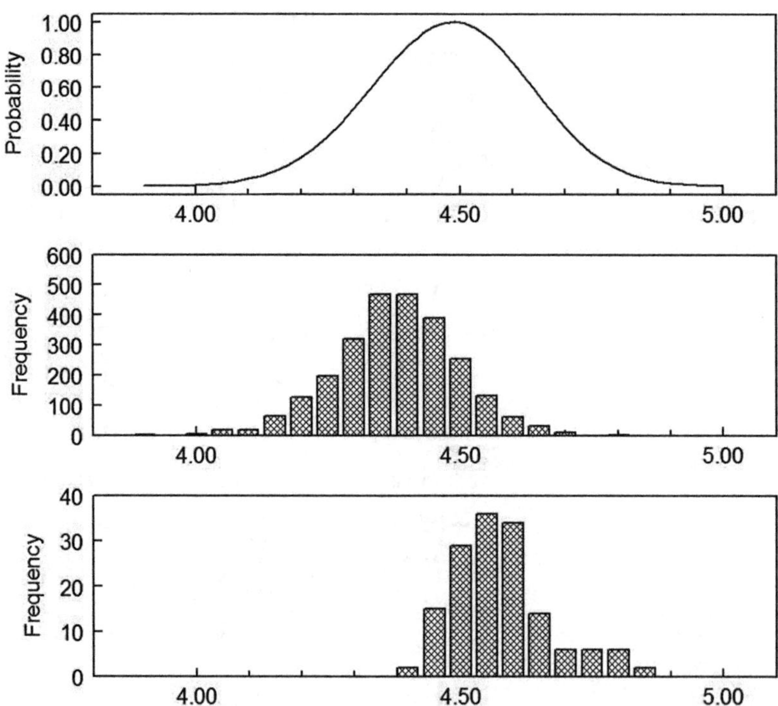

Fig. 5 Upper: Probabilities calculated by MM3 for stretched and compressed α-D-glucopyranose for different O1–O4 distances. Center: Frequencies of experimental O1–O4 distances in 2582 α-D-glucopyranose rings from a scan of the Cambridge Structural Database. The mean value is 4.356 Å. Lower: Frequencies of intrinsic O1–O4 distances in 150 stationary B3LYP/6-31+G(d,p) structures of α-D-glucopyranose

the resulting forms, α- and β-glucopyranose, are "anomers." Experimental data for reducing sugars are affected by this interconversion which is known as mutarotation. It can be avoided by substituting the hydroxyl hydrogen on the glycosidic oxygen with a methyl group.

Di-, oligo-, and polysaccharides. Atom numbers in the non-reducing residue of disaccharides are primed, while longer molecules have increasing values of Roman numerals for residues further from the reducing end. Linkages between monomeric units of larger saccharide molecules consist of either two or three bonds, typically with the oxygen atom attached to the anomeric carbon (the glycosidic oxygen) leaving during synthesis. Thus, in the formation of cellobiose, O4 remains. Disaccharide conformations are specified by the values of the torsion angles for the glycosidic (C1′-O*n*) (ϕ) and aglycon (O*n*-C*n*) (ψ) bonds ($n = 4$ for cellobiose). Three-bond linkages involve a primary alcohol group. Its conformation is described with letters, sometimes upper case, for example, *GG*, *GT*, and *TG*, or by the ω torsion angle. The central bond is specified by ψ, and ϕ is for rotation about the glycosidic bond. Polysaccharides have disaccharide linkages so the same descriptors apply. For polysaccharides composed of regular repeating units, helix nomenclature applies. Helices are described by the number of units per turn (***n***), and the rise, or advance, (***h***) along the helix axis.

Along with C1′, O4, and C4, the four-atom definition of the torsion angle ϕ in maltose or cellobiose could involve any of the three atoms H1′, O5′, or C2′. Many workers have used H1′, especially if they have a background in NMR and are thinking of using nuclear Overhauser effects to solve the structure. Others, mindful of the difficulties in accurately locating hydrogen atoms by X-ray crystallography, have opted for O5′. No examples of the use of C2′ in these definitions come to mind. The ψ torsion angle has been defined by all three possible atoms. For cellobiose, they would be H4, C3, or C5. The above reasons for favoring a hydrogen atom or a heavier atom also apply, but the above-cited nomenclature [20] uses the lower-numbered carbon atom. Thus, the standard definition of ψ for cellobiose is C1′–O4–C4–C3, along with the standard ϕ of O5′–C1′–O4–C4. Greek letter descriptors are italicized.

Standardization of the ends of 360° ranges of ϕ and ψ on plots of energies, experimental points and molecular dynamics (MD) trajectories would allow quick visual comparisons of different plots. We argue for plots that are equivalent to ϕ_H andψ_H (e.g., H1′–C1′–O4–C4 and C1′–O4–C4–H4) values from −180° to +180° for two reasons. Firstly, experimentally determined points for many reducing disaccharides are mostly in minima that have ϕ_H ndψ_H near 0° and therefore will fall in the center of the map. If the ranges were 0°–360°, the major populations could be separated into as many as four visual groups despite close structural similarity. Secondly, on such plots for maltose and cellobiose the diagonal line from the lower right to the upper left corresponds to helices with no chirality, separating right- and left-handed forms. For maltose, the diagonal line corresponds to polymers with ***h*** = 0. Long molecules having maltose-type linkages and ϕ,ψ values on the diagonal line would self-intersect, but molecules with 5–8 glucose residues could form cyclodextrins [31]. Cellulose polymers with ϕ,ψ values on the diagonal line are helices with ***n*** = 2. By adding or subtracting 120° as appropriate, the ranges of ϕ and ψ defined according to the nonhydrogen atoms can retain the central location of most experimental structures and the desirable diagonal line (*see* Fig. 6).

3 Special Problems of Carbohydrates

Because of their many hydroxyl groups and substantial flexibility, carbohydrates exaggerate many of the problems experienced in modeling other molecules. While some other molecules such as dimethoxymethane have sequences of atoms that correspond to anomeric centers, such centers are relentlessly prevalent in

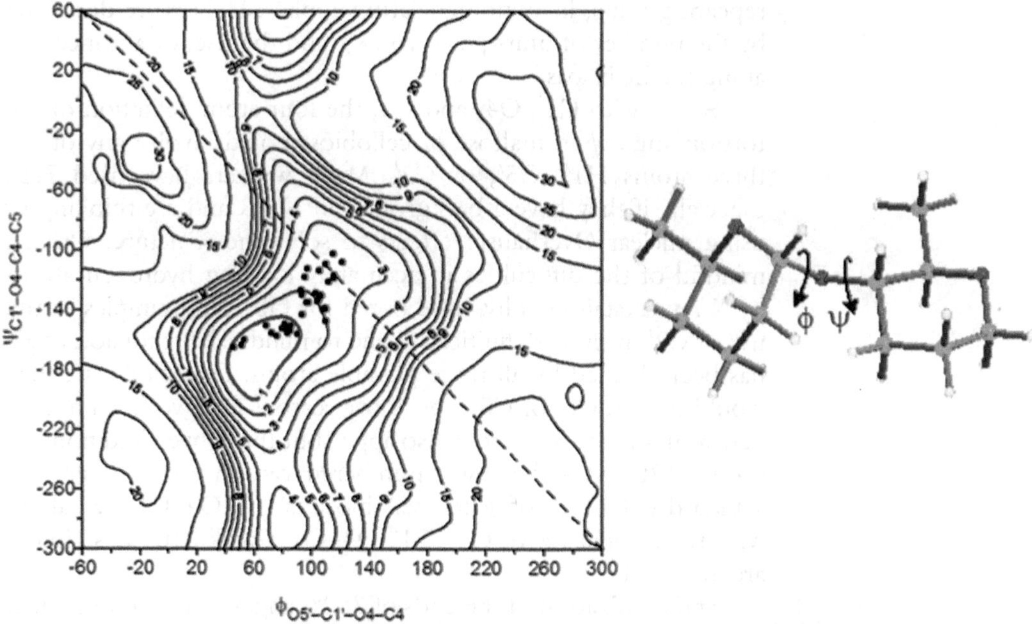

Fig. 6 HF/6-31G(d) energies from the depicted methylated maltose analog [29]. Observed conformations in experimental crystal structures of maltose and related structures are shown as points. Structures within the 1 kcal/mol contour do not possess the O2–O3′ hydrogen bond that is found for all other observed experimental structures. Axes correspond to ϕ_H and ψ_H (e.g., H1′–C1′–O4–C4 and C1′–O4–C4–H4) values from −180° to +180°. Note that a copy of this map can be placed on each edge to test for periodicity

carbohydrates. The following gives an overview of ways to cope with these problems.

Sampling. With all of the rotatable exocyclic substituents, it is necessary to ascertain that the modeling results are based on a sufficient exploration of the possible different states. How can one be certain that the final results depend only on the energy calculations of a particular modeling method and not on a failure to consider the best possible arrangements of these exocyclic groups? As seen above, there are 729 combinations of staggered exocyclic orientations for glucose, and cellobiose has 59,049. Again, most will not be stable, but many will and their stabilities are dependent on ϕ andψ. This leaves a large conformation space to sample. One approach is to use Simulated Annealing [41] or more modern techniques: Umbrella Sampling [42, 43], Replica Exchange Molecular Dynamics [44–46], Hyperdynamics [47], Metadynamics [48], Conformation Flooding [49], or accelerated molecular dynamics [50].

Two examples come from our recent work. In one [51], we developed ϕ,ψ energy maps to provide energies of distortion for conformations of substrates in complexes with hydrolyzing enzymes. The goal was to learn whether distortions of the ϕ,ψ torsion angles might be part of the catalytic function. To sufficiently sample conformation space with a "brute force" approach, we use different "starting structures," each with a particular combination of orientations of the exocyclic groups that will be likely to cover the lowest energy structures. The energy is computed for each of these starting structures at each ϕ,ψ location. The ϕ and ψ values were fixed at grid points in 20° intervals for a total of (18 × 18 =) 324 unique ϕ,ψ points. Initial studies, using Monte Carlo methods [52] and the OPLS-2005 [53] force field in MacroModel [39] identified 1863 stable starting structures for cellobiose when both rings were in the normal chair forms, 3485 when the reducing ring had the $^{2}S_{O}$ conformation, and 2871 when the reducing ring had the $^{3}S_{1}$ shape. Thiocellobiose (where sulfur replaces the interresidue oxygen) yielded 2277 different structures to test at each ϕ,ψ point.

In another project [54], methyl cellobioside, -tetraoside, and -hexaoside were investigated with replica exchange molecular dynamics studies in explicit TIP3P (Transferable Intermolecular Potential 3 Point Charge) water. (Reference 45 has a tutorial study on Replica Exchange MD of disaccharides in vacuum.) Depending on the size of the carbohydrate, 714–3741 molecules of water were included in the AMBER [55] calculations, using the GLYCAM-04 force field [56]. To assure complete sampling for the hexamer, replicas had temperatures of 297°–557° in 42 increments, with each simulation lasting more than 16 ns.

To make Ramachandran surfaces, the ϕ and ψ torsion angles of a disaccharide conformation are adjusted in increments of perhaps 10° or 20° and the energy calculated. Over the past 30 years, the energy has often been minimized for each ϕ,ψ conformation, holding ϕ and ψ constant but allowing all other parameters to find their nearest local minimum. Such studies are called "relaxed-residue" analyses. Their primary rationale is that they avoid collisions that would occur if the rings were kept rigid. (On one rigid-residue map, the crystallographic conformation of the sucrose moiety in raffinose corresponded to an unreasonable 100 kcal/mol [57].) Modeling software often provides for these calculations to step through ϕ,ψ space, using a tool called a "dihedral driver" or a "scan." Because the monosaccharide units of the disaccharide are flexible, there is the strong likelihood that an inelastic deformation of the molecule will occur during the scan. For example, a hydroxyl group may simply rotate to a new staggered form, or the ring might lose its chair form. If that deformed structure is used to start the subsequent minimization, the energy at –180° will be different from the one at +180°. This is a sampling issue because the points after the deformation occurs will not be tested with the intended starting structure. Our procedure for avoiding this problem is to use the same starting geometry at each ϕ,ψ point, with only rigid rotations to the point in question.

Our approach is not without problems. Not having energy-minimized structures at each preceding point can lead to an interpenetration of the two monosaccharide residues in regions of high energy. The interpenetration causes much higher calculated energies or failure of minimization altogether but can be partially avoided by increasing the glycosidic bond angle to 150° for each starting structure. A similar sampling problem is a concern during MD simulations because simulations may be too short. At least with MD, the deformed structures could recover if the simulation runs long enough.

Hydrogen bonding. Hydrogen bonding is especially important for most carbohydrates because of the high density of hydroxyl groups. If the stabilization from a typical moderate to weak hydrogen bond is 5.0 kcal/mol [58], model structures that have them would completely dominate those with otherwise similar structures. Further, the various acceptor oxygen atoms do not appear to accept hydrogen bonds with equal eagerness. The oxygen atom of the glycosidic linkage is a poor acceptor, and donations to the ring oxygen are less stabilizing than to hydroxyl groups. Further, there are cooperative effects. These effects are found in continuous donor–acceptor–donor–acceptor chains. Such sequences are more stable than an equal number of discontinuous hydrogen bonds and their H…O distances are shortened [59]. In modeling hydrogen bonds, some workers find that no further consideration is needed after the charges are assigned to the atoms in their empirical models. Other modelers have devised elaborate schemes to provide calculated energies whenever hydrogen bonding is present. With QM, most workers are finding that reliable calculation of hydrogen bonding geometries and energies requires fairly sophisticated techniques, such as post Hartree–Fock theory or Density Functional Theory and correction for Basis Set Superposition Error. Bader's AIM theory (see above) has been employed extensively in hydrogen bond studies [60].

Paradoxically, we have found that ϕ,ψ conformations in carbohydrate crystal structures can be predicted by isolated models with an elevated dielectric constant (e.g. 4.0–8.0) rather than the prescribed value of 1.0 for CHARMM and AMBER calculations or 1.5 for MM3 or MM4. Elevated dielectric constants reduce the strengths of interactions between charged atoms, including those in hydrogen bonds. This work-around appears to provide a potential of mean force similar to what might otherwise have been obtained by MD with explicit solvent. This approach is not appropriate for investigation of specific molecule-molecule interactions such as might be found in modeling an entire crystal, for example. It seems to work only for modeling condensed phase systems when the rest of the condensed phase is not explicitly present.

Less clear is the impact of C–H…O hydrogen bonds. Geometric criteria identified 14 such bonds in the crystal structure of dicyclohexyl cellobioside [33]. These fairly weak interactions have typically been emphasized less when developing empirical force fields.

Anomeric effects. The anomeric effect was the unexpected finding that the experimental α:β ratio for compounds such as D-glucopyranose favored the α-anomeric form more than would have been expected for an axial substituent on a cyclohexane ring [19]. The exoanomeric effect was named for the preference of the substituent in methyl glucopyranoside to take an orientation *gauche* to the ring oxygen. This increased stabilization affects ϕ_{O5} directly, favoring −60° and +60° angles but not the *trans*, 180° angle. An external anomeric torsional effect has also been proposed that affects ψ [61]. The term "general anomeric effect" covers the *gauche* preference for any R–X–C–Y atom sequence in organic chemistry where X denotes O, N, or S, and Y denotes any atom having lone pairs of electrons. Thus, dimethoxymethane (DMM), which has a C–O–C–O–C sequence, prefers the *g,g* conformation, while *n*-pentane prefers the all-*trans* form. The analogy between small molecules and the C5–O5–C1–O1–C_{Me} sequence in methyl glucosides was noted many years ago [62, 63].

Besides conformational preferences, anomeric effects cause differences in bond lengths and angles. In the very accurate multipole refinement of crystalline sucrose at 20K [64], the bonds from the anomeric carbons to the ring oxygens are 1.4192 and 1.4146 Å for the pyranosyl and furanosyl rings, respectively, whereas the distances between the ring oxygens and the other carbon atoms (C5 and C5′) are 1.4477 and 1.4543 Å, with standard deviations of 0.0005.

Often overlooked is the effect of an anomeric center, regardless of conformational details. It results in extra stability of the compound, as discussed by Tvaroška and Bleha [65]. A "bond and group enthalpy increment scheme" can be used to calculate heats of formation [66], either by QM or MM. The increment in MM3 for O–C–O is 6.62 kcal, much larger than the corrections for QM (0.505 kcal/mol for HF/6-31G(d) and −0.351 kcal/mol for B3LYP/6-31G(d)). MM3's steric energies do not consider anomeric centers so a large adjustment is needed. Fairly simple QM does calculate most of this enthalpy, but some correction is still needed.

Earlier, we used this method to calculate heats of formation for analogs of some disaccharides [67]. The analogs were based on the native sugars but all exocyclic groups were replaced by hydrogen. We also joined two tetrahydropyran molecules with an ether oxygen at the four positions (organic nomenclature), making isomeric pseudodisaccharide analogs with diaxial, axial-equatorial, and diequatorial linkages. Analogs of α,α-, α,β-, and β,β-trehalose had enthalpies of −149.9, −147.4, and −147.5 kcal/mol, respectively. The analogs of nigerose, laminarabiose, maltose, cellobiose, and galabiose gave −142.6, −141.7, −141.3, −140.6, and −140.4 kcal/mol and the diaxial, axial-equatorial, and diequatorial pseudosugar analogs had values of −135.1, −135.4, and −135.8 kcal/mol. These were the B3LYP/6-31G(d) values but quite similar results were computed by MM3 and HF/6-31G(d). The trehalose analogs had two anomeric centers and roughly 12 kcal of stabilization, and the analogs of the other disaccharides had one center and 6 kcal, relative to our pseudosugar analogs with no anomeric centers.

Anomeric effects are likely to have several different causes and are affected by different factors. Magnitudes vary in different solvents, suggesting that there is an electrostatic component, and considerable effort has been directed to analyses of the changes in electronic structure [19]. One group suggested that gauche conformers for anomeric sequences are stabilized by C–H...O hydrogen bonds and carried out Natural Bond Order calculations to confirm that result [68]. Our unpublished calculations with AIM did not find the proposed hydrogen bonds despite the relatively short H...O distances.

4 Quantum Mechanics Approaches

Empirical force fields (MM) summarize our knowledge of molecular structure and energy relationships for application to other molecules. On the other hand, electronic structure theory calculations (QM) are closer to experiment, with the possibility of increasing the resolution by using more computer time and memory. There are two other important concepts: balance between method and basis set, and density functional theory (DFT). Perdew [69] has introduced a "Jacob's Ladder" to rate the various DFT methods and Csonka et al. evaluated various DFT methods for applicability to carbohydrates [70]. The importance of choosing a QM method and basis set cannot be exaggerated. Consider the calculations of energy for four conformers of β-D-glucopyranose [11]. Some levels of the theory favored the seldom-observed chair ($^1C_4$) by as much as 17 kcal/mol, while others favored the dominant $^4C_1$ form by that same amount.

The analogs described above in the studies of the anomeric effect are much less demanding of a particular level of QM theory. We have made relaxed ϕ,ψ surfaces for the cellobiose analog at HF/6-31G(d), B3LYP/6-31G(d), MP2/6-31+G(d,p), and MP2/6-311+G(d,p) levels of theory, as well as B3LYP/6-311++G(d,p) calculations based on the B3LYP/6-31G(d) geometries. All of these maps are very similar, with the B3LYP maps being slightly flatter than the HF [71] and unpublished MP2 maps. Although it makes little difference for these analogs, the diffuse function (indicated by the +) is needed to avoid substantial over-estimation of hydrogen bond energies for the native saccharides with either the B3LYP, or MP2 methods and Pople basis sets.

Another lesson from those analogs was that their QM maps are fairly predictive of conformations of the native disaccharides in crystals. The addition of methyl groups at C5 to complete the carbon backbone improved predictions for cellobiose by reducing the size of the 1 kcal/mol region [72]. Figure 6 shows a map for the methylated maltose analog with the crystalline maltose linkages (except cyclodextrins). (This energy map was shown earlier with conformations from cyclodextrins [31].) Structures inside the 1.0 kcal/mol contour are in accord with the exoanomeric effect and do not have intramolecular hydrogen bonds in the crystals. Their hydroxyl groups are either acylated and cannot form hydrogen bonds, or they form intermolecular linkages. All other observed maltose-type structures have intramolecular, interresidue hydrogen bonds between O2′ and O3 that presumably compensate for the higher energy of the analog "backbone." The diagonal line corresponding to the helical parameter value of $h = 0$ is also shown. To its left are conformations that lead to left-handed helices, with right-handed ones on the right. The secondary minimum at $\phi = 80°$, $\psi = -280°$ is populated by some linkages in larger cycloamyloses, and is also compatible with the exoanomeric effect.

In at least one instance, the global minimum structure for the fully hydroxylated native disaccharide is not in the region of the crystal structures. That global minimum, for cellobiose, is stabilized by an exceptional hydrogen bonding network [73]. In an important validation of the computational finding with QM, the cellobiose structure and the global minimum for the closely related lactose molecule have been observed experimentally in the gas phase at very low temperature [72].

Quantum mechanics ϕ,ψ maps for cellobiose confirmed these earlier findings for the minimum energy conformation for the vacuum phase [74]. When the SMD continuum solvation model of Marenich et al. [75] was included in these B3LYP studies, however, the global minimum was found in the location favored by crystal structures. Another difference from the earlier work was that the various conformers were analyzed with Atoms-in-Molecules (AIM) theory. The vacuum global minimum is shown in Fig. 7. Analysis of the electron density with the AIMAll program [76] showed far fewer hydrogen bonds than had been reported earlier [74] for the various structures, as the adjacent hydroxyl groups did not meet the AIM criteria for intraresidue hydrogen bonds. Figure 7 also shows the AIM hydrogen bond lengths and the electron densities at the bond critical points. Although fewer O–H...O bonds were confirmed, a C–H...O hydrogen bond was found, despite the absence of such interactions in the earlier report. The electron density values for the O–H...O hydrogen bonds are about seven or eight percent of those for covalent O–H bonds. These analyses are useful for identifying van der Waals interactions as well [78]. Despite the absence of intraresidue clockwise or

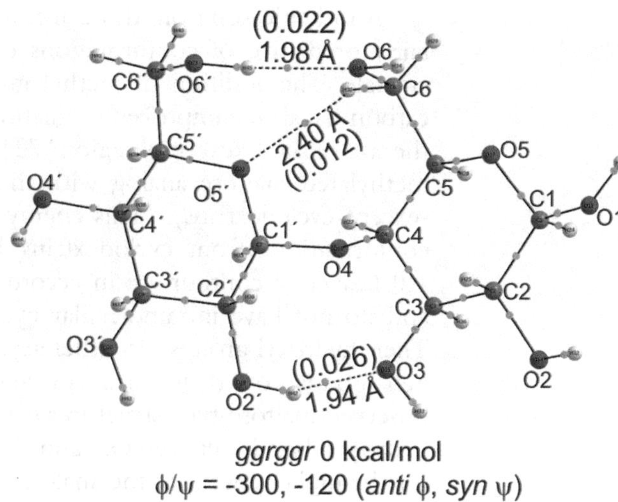

Fig. 7 Molecular graph for the structure giving the calculated global minimum energy for cellobiose in vacuum at very low temperature [74]. The representation is by the AIMAll program [75]; the gray, white, and red spheres represent carbon, hydrogen, and oxygen, and the small green spheres represent the Bond Critical Points that indicate the minimum in electron density along the bond path between any two atoms as well as a maximum in the plane perpendicular to the bond path. Solid lines indicate bond paths for covalent bonds, and dashed lines are for non–covalent-bonded interactions. The H. . .O distances are shown in angstroms, and the electron densities at the Bond Critical Points are in electrons per cubic Bohr, an atomic unit. The O6′–H. . .O6, O2′–H. . .O3, and C6–H. . .O5′ hydrogen bonds are shown, but no intraresidue hydrogen bonds are indicated by this analysis

counterclockwise hydrogen bonds according to AIM criteria, there does remain some lowering of energy when hydroxyl groups have these orientations, especially in vacuum calculations [77].

As might be expected because of the computer time required, there are only a few molecular dynamics simulations based on QM energies. The Car–Parinello method has been applied in a study of the distortion energies of glucopyranose [79]. The metadynamics approach was applied to force the ring into different puckering arrangements in a short time (44.4 ps). Our preliminary energy-minimization approach to mapping the energy of cellobiose against the linkage torsion angles was also computationally expensive, with 181 different starting geometries, and only a quarter of the entire ϕ,ψ space was covered [80].

5 Empirical Force Fields and Some Applications

Empirical, or MM, force fields can be used with either energy minimization methods or with MD. A minute's worth of MM time can be equivalent to a month or more of QM time. Therefore, many more issues can be evaluated with MM, making a more complete study possible. Even if it is desired to ultimately study the problem with QM, it can be worthwhile to study it first with MM.

Development of force fields that are "carbohydrate aware" continues. Besides the generally applicable MM3 and MM4 [81] programs, the GLYCAM-06 [82] force field primarily for AMBER is also intended for carbohydrates as well as proteins and lipids. Refinements of GLYCAM continue with the addition of explicit lone pairs on oxygen and nitrogen [83]. The CHARMM [84] program also has had carbohydrate force fields available (e.g., the Ha et al. parameterization [85] and the CSFF parameters [86]). New parameterizations for CHARMM that include carbohydrates have recently been released [87, 88] for pyranosyl rings, glycosidic linkages between pyranosyl rings, furanosyl rings, and linkages to furanosyl rings, respectively. The fact that so many parameterizations are needed for the CHARMM force field to model the various molecule types is a caution regarding the generality of the molecular mechanics approach, at least with a fairly simple potential energy function. Their force fields were developed with the aid of QM ϕ,ψ maps of the analogs. Their analog maps, typically at the MP2/cc-pVTZ level of theory, are quite similar to the HF/6-31G(d) maps published earlier [89, 90], indicating a robustness of the QM approach, especially for the analogs.

Some of the GROMOS parameter sets also consider carbohydrates [91]. Some of the force fields that incorporate these parameterizations were recently compared, as well as some older force fields [92]. There are still substantial variations among the various systems. One factor is that some force fields have been parameterized so that aqueous solution data can be reproduced with MD using explicit water molecules, especially TIP3P [93].

DeMarco and Woods [94] reviewed several state-of-the-art simulations and studies pertaining to carbohydrate–protein interactions, both conformational and energetic. That paper also advocates a new carbohydrate nomenclature that is more useful for glycomics and integration with the Protein Data Bank protocols and data formats. Another review of many force fields is provided by Foley et al. [95].

Krupička and Tvaroška have used a hybrid method for studying enzymatic action on carbohydrates, in which the substrate is systematically deformed in the protein. In that work, the active site atoms and ligand are represented by QM and the rest of the protein and water are represented with MM [96]. Schramm's group has a similar approach [97].

In our nonintegral hybrid method, QM maps for the analog furnish the conformational energy for the backbone of the disaccharide, while the hydrogen bonding and other steric considerations result from MM. The MM backbone map is subtracted from the MM disaccharide map and the QM map is added. This approach is useful when the torsional energies of a particular linkage are not well parameterized. Originally, the method was developed for studying sucrose [93], which has two adjacent anomeric centers. More recently, similar hybrid studies of acarviosinide and thiocellobiose compensated for incomplete parameters involving the nitrogen and sulfur atoms, respectively [53].

6 Evaluation of Energy Calculations with Crystal Structure Observations

Concerns regarding molecular distortion by crystal packing forces have discouraged many modelers from making detailed comparisons of their conformational modeling results with results from the extremely accurate and precise results from small molecule crystallography. Another countering argument is that the conformations in crystals are not relevant to the "real" molecular shapes, that is, those observed in aqueous solution or biological fluids. Finally, there is only one, or at best a very limited number, of crystal structures for a given molecule. Thus, a given crystal structure may not be representative of the lowest energy form. Experience with placing crystal structure conformations on energy surfaces such as ϕ,ψ maps suggests that a nuanced view is more appropriate. We have long argued that calculations must employ some sort of compensation for the condensed phase before making such comparisons. Earlier, we used elevated dielectric constants (not the extreme value of 78 for water, but 3.5 or 4.0 in MM3 calculations.) The above-mentioned results for QM studies of cellobiose confirmed the need to consider a condensed phase, such as by continuum solvation with a water model, before the conformations in crystal structures would be predicted from single molecules. The recent studies of a cellulose crystal [8] support this approach, finding that the van der Waals forces were more important in stabilizing the structure than hydrogen bonding forces.

We have also found that minor variations in molecular structure distant from the glycosidic linkage in disaccharides do not seem to have much influence on the linkage conformation. Consider that lactose and cellobiose crystals are isomorphous despite the axial disposition of the 4′ hydroxyl group in lactose vs. the equatorial orientation in cellobiose. Crystals of the methyl glycosides of these two molecules are also isomorphous with each other, although having a different location in ϕ,ψ space from the unsubstituted molecules. Generally speaking, a fair number of related molecules is available to be used in comparisons with the energy surface for a particular molecule [98]. There is overlap of conformations of the hydroxylated disaccharides with the peracetylated and other derivatives, for example, although the distributions do differ to some extent.

The fit of the experimentally observed conformations can be quantified by assuming that the distortions are random, in the same sense as the corresponding conformational energies of molecules in the gas phase will fit a Boltzmann distribution. The various crystal field forces in different crystals of related molecules will tug and push on the lowest energy form, there will be an exponentially decreasing number of conformations as the corresponding energies increase. The temperature of the distribution is a measure of fit, with our calculations often in the range of 400–500 K. Low values of the fitted temperature (e.g., near or below room temperature) indicate that the potential function may be too flat or that there are some external constraints on the molecular shape that are not included in the model. An example of an external constraint would be the fitting of conformations from cyclodextrins on an energy surface for maltose. The ϕ and ψ values for the cyclodextrins are constrained by the need to close the cycle of glucopyranosyl residues, not an issue for the disaccharide.

Figure 8 [7] shows the QM energy contours for the region of the crystal structures of cellobiose and its related molecules

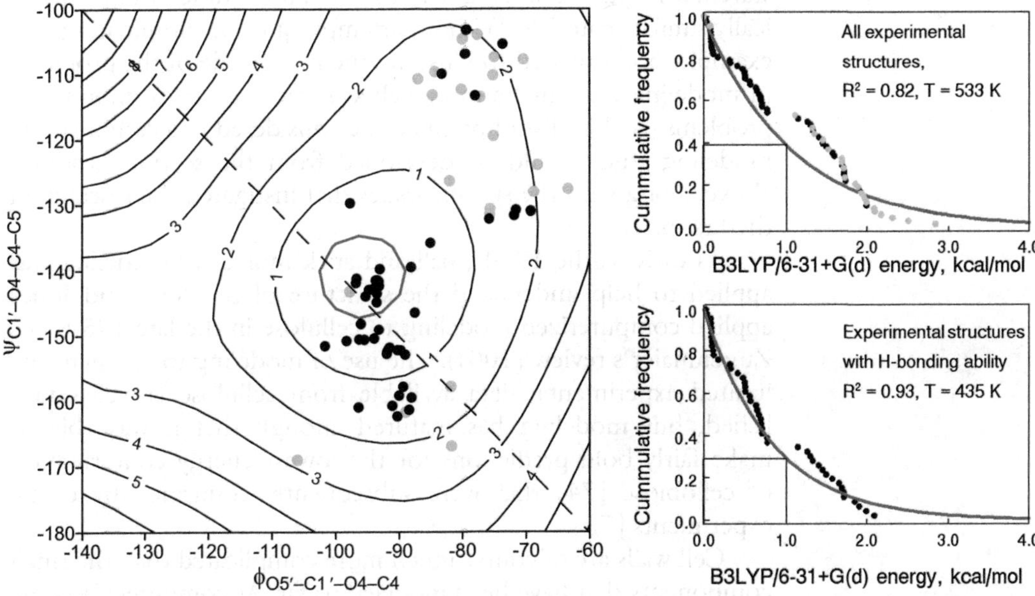

Fig. 8 Left, the B3LYP/6-31+G(d)/SMD conformational energies for β-cellobiose, adapted from Ref. 74. The inner contour (magenta) is for 0.25 kcal/mol. Black dots represent the ϕ,ψ conformations of observed crystal structures related to cellobiose that have the O2, O3, and O6 hydroxyl groups needed to form interresidue hydrogen bonds. The cyan dots are typically for substituted derivatives so that these same hydrogen bonds cannot form. On the right are cummulative frequency distribution plots for the energy data on the left (adapted from Ref. 7). Based on their corresponding energy values, each of the points is put in an ordered list. The value of the cummulative frequency is $(1 - i)/n$, where i is the position in the sorted list and n is the total number of ϕ,ψ points. An equation of the form $y = e^{-\beta x}$ is fitted to the plotted values and β corresponds to $1/(RT)$ where R is the universal gas constant, allowing for the derivation of the temperature of the distribution

[99]. Also shown are the exponential decay plots based on cumulative frequency listings of the energies. In the case where all experimental structures were included, the "temperature" was 533 K, and there were larger deviations from the fitted line. When only the structures able to form interresidue hydrogen bonds were included, the derived "temperature" was 435 K and some 70% of the structures had energies less than 1 kcal/mol above the minimum, based on the fitted line. Only 61% of all the structures had corresponding energies less than 1 kcal/mol based on the fit of all structures.

7 Conclusions

It is not possible to completely understand the plant cell wall without understanding the structures of the various carbohydrates that compose it. Cellulose is, of course, the main component of many cell walls, and other cell wall polysaccharides are closely related to it in regard to their backbone structures. In the present chapter, we have of course cited some of our papers on cellulose, but we have presented brief studies of the monomer and dimer of starch, α-D-glucose and maltose, as examples of different but chemically similar materials. Without attempting to be intimidating, the examples are intended to illustrate the magnitude of the problems in modeling carbohydrates as well as some ways to surmount the problems. A checklist of points to be considered in a carbohydrate modeling study could be developed from the section headings above, along with the specific issues that instigated the research in the first place.

As early as the 1920s, ball-and-stick molecular modeling was applied to help understand the structure of cellulose, and Jones applied computerized modeling to cellulose in the late 1950s (*see* Zugenmaier's review [100]). The use of modeling to augment the limited experimental data available from cellulose is well established, but modeling has matured enough that it was able to make fairly bold predictions for the lowest energy conformation of cellobiose [74] that were subsequently confirmed by exotic experiments [75].

Cell walls are of course much more complicated than the small components that have been modeled so far. As computers become even faster, larger assemblies of molecules can be studied, along with physical and chemical processes. For example, current studies include models of cellulose crystallites in water [101] and dynamic conversions of such crystallites [102] and models composed of up to a hundred thousand atoms are being studied. Larger structures may be treated with "coarse-grained" force fields [99, 103, 104] that have not otherwise been covered in this review. Thus, the future of carbohydrate modeling will include the study of previously undetermined structures, repeat studies of familiar structures but with ever-improving methods, and ever-larger and more complete representations of complex structures such as the plant cell wall.

Numerous resources are available online, and at least two should be mentioned. Interactive views of various mono- and oligosaccharides are also available. The GLYCAM (http://www.glycam.com) site at the Complex Carbohydrate Research Center at the University of Georgia in Athens provides several interactive tools for the carbohydrate modeler, including a builder for different structures. The www.glycoSCIENCES.DE website has numerous tools that take advantage of computed energy surfaces and conformational information in the Protein Data Bank. It is a good location to try out some modeling projects or to get an initial interaction with modeling.

References

1. Watts HD, Mohamed MNA, Kubicki JD (2014) A DFT study of vibrational frequencies and 13C NMR chemical shifts of model cellulosic fragments as a function of size. Cellulose **21**:53–70
2. Toukach FV, Ananikov VP (2013) Recent advances in computational predictions of NMR parameters for the structure elucidation of carbohydrates: methods and limitations. Chem Soc Rev **42**:8376–8415
3. Kurihara Y, Ueda K (2006) An investigation of the pyranose ring interconversion path of α-L-idose calculated using density functional theory. Carbohydr Res 341:2565–2574
4. Steiner T, Saenger W (1998) Closure of the cavity in permethylated cyclodextrins through glucose inversion, flipping, and kinking. Angew Chem Int Ed 37:3404–3407
5. Añibarro M, Gessler K, Usón I, Sheldrick GM, Harata K, Hirayama K, Abe Y, Saenger W (2001) Effect of peracylation of β-cyclodextrin on the molecular structure and on the formation of inclusion complexes: an X-ray study. J Am Chem Soc 123:11854–11862
6. Gould IR, Bettley HA-A, Bryce RA (2007) Correlated ab initio quantum chemical calculations of di- and trisaccharide conformations. J Comput Chem 28:1965–1973
7. French AD (2015) Computerized models of carbohydrates, in polysaccharides bioactivity and biotechnology. In: Mérillon JM, Ramawat KG (eds) Springer, pp 1397–1440
8. Devarajan A, Markutsya S, Lamm MH, Cheng X, Smith JC, Baluyut JY, Kholod Y, Gordon MS, Windus TL (2013) Ab initio study of molecular interactions in cellulose Iα. J Phys Chem B 117:10430–10443
9. Barnett CB, Naidoo KJ (2010) Ring puckering: a metric for evaluating the accuracy of AM1, PM3, PM3CARB-1, and SCC-DFTB carbohydrate QM/MM simulations. J Phys Chem B 114:17142–17154
10. Foresman J ., Frisch AE. Exploring chemistry with electronic structure methods, 3rd edn. http://www.gaussian.com/g_prod/explore3.htm
11. Barrows SE, Dulles FJ, Cramer CJ, French AD, Truhlar DG (1995) Relative stability of alternative chair forms and hydroxymethyl conformations of β-glucopyranose. Carbohydr Res 276:219–251
12. Becke, AD (1993) Density-functional thermochemistry. III. The role of exact exchange, J Chem Phys 98:5648–5652
13. Zhao Y, Truhlar DG (2008) The M06 suite of density functionals for main group thermochemistry, themochemical kinetics, noncovalent intereactions, excited states, and transition elements: two new functionals and systematic testing of four M06-class functionals and 12 other functionals Theor Chem Acc 120:215–241
14. Svensson M, Humbel S, Froese RDJ, Matsubara T, Sieber S, Morokuma K (1996) ONIOM: a multilayered integrated MO + MM method for geometry optimizations

and single point energy predictions. A test for Diels–Alder reactions and Pt(P(t-Bu)3)2 + H2 oxidative addition. J Phys Chem 100:19357–19363
15. Schnupf U, Momany FA (2011) Rapidly calculated DFT relaxed iso-potential ϕ/ψ maps: β-cellobiose. Cellulose 18:859–887
16. Mucs D, Bryce RA (2014) The application of quantum mechanics in structure-based drug design. Expert Opin Drug Discov 8:263–276
17. Ardèvol A, Rovira C (2015) Reaction mechanisms in carbohydrate-active enzymes: glycoside hydrolases and glycosyltransferases. Insights from ab initio quantum mechanics/molecular mechanics dynamic simulations. J Am Chem Soc 137:7528–7547
18. Frank M, Schloissnig S (2010) Bioinformatics and molecular modeling in glycobiology. Cell Mol Life Sci 67:2749–2772
19. Grindley TB (2008) In: Fraser-Reid BO, Tatsuta K, Thiem J (eds) Glycosciences: structure and conformation of carbohydrates. Springer, Berlin, pp 3–55
20. McNaught D (1996) Nomenclature of carbohydrates (IUPAC recommendations 1996). Pure Appl Chem 68:1919–2008. http://www.chem.qmul.ac.uk/iupac/2carb/00n01.html#00
21. French AD, Dowd MK (1994) Analysis of the ring-form tautomers of psicose with MM3 (92). J Comput Chem 15:561–570
22. Boeyens JCA (1978) The conformation of six-membered rings. J Cryst Mol Struct 8:317–320
23. Cremer D, Pople JA (1975) A general definition of ring puckering coordinates. J Am Chem Soc 97:1354–1358
24. Altona C, Sundaralingam M (1972) Conformational analysis of the sugar ring in nucleosides and nucleotides. A new description using the concept of pseudorotation. J Am Chem Soc 94:8205–8212
25. Haasnoot CAG (1992) The conformation of six-membered rings described by puckering coordinates derived from endocyclic torsion angles. J Am Chem Soc 114:882–887
26. Bérces A, Whitfield DM, Nukada T (2001) Quantitative description of six-membered ring conformations following the IUPAC conformational nomenclature. Tetrahedron 57:477–491
27. Joshi NV, Rao VSR (1979) Flexibility of the pyranose ring in α- and β-d-glucoses. Biopolymers 18:2993–3004
28. Hill D, Reilly PJ (2007) Puckering coordinates of monocyclic rings by triangular decomposition. J Chem Inf Model 47:1031–1035
29. French AD, Johnson GP (2007) Linkage and pyranosyl ring twisting in cyclodextrins. Carbohydr Res 342:1223–1237
30. Nishiyama Y, Langan P, Chanzy H (2002) Crystal structure and hydrogen-bonding system in cellulose Iβ from synchrotron X-ray and neutron fiber diffraction. J Am Chem Soc 124:9074–9082
31. Yoneda Y, Mereiter K, Jaeger C, Brecker L, Kosma P, Rosenau T, French A (2008) van der Waals versus hydrogen-bonding forces in a crystalline analog of cellotetraose: cyclohexyl 4′-O-cyclohexyl β-d-cellobioside cyclohexane solvate. J Am Chem Soc 130:16678–16690
32. Bader RFW (1990) Atoms in molecules — a quantum theory. Oxford University Press, Oxford
33. Csonka GI, Kolossváry I, Császár P, Éliás K, Csizmadia IG (1997) The conformational space of selected aldo-pyrano-hexoses. J Mol Struct Theochem 395-396:29–40
34. Klein RA (2002) Electron density topological analysis of hydrogen bonding in glucopyranose and hydrated glucopyranose. J Am Chem Soc 124:13931–19937
35. Klein RA (2006) Lack of intramolecular hydrogen bonding in glucopyranose: vicinal hydroxyl groups exhibit negative cooperativity. Chem Phys Lett 433:165–169
36. Koch U, Popelier P (1995) Characterization of C-H-O hydrogen bonds on the basis of the charge density. J Phys Chem 99:9747–9754
37. Schrodinger, Portland, OR. www.schrodinger.com
38. Çarçabal P, Jockusch RA, Hunig I, Snoek LC, Kroemer RT, Davis BG, Gamblin DP, Compagnon I, Oomens J, Simons JP (2005) Hydrogen bonding and cooperativity in isolated and hydrated sugars: mannose, galactose, glucose, and lactose. J Am Chem Soc 127:11414–11425
39. Allinger NL, Yuh YH, Lii J-H (1989) Molecular mechanics. The MM3 force field for hydrocarbons. J Am Chem Soc 111:8551–8567
40. Allen FH (2002) The Cambridge structural database: a quarter of a million crystal structures and rising. Acta Crystallogr B Struct Sci 58:380–388
41. Naidoo KJ, Brady JW (1997) The application of simulated annealing to the conformational analysis of disaccharides. Chem Phys 224:263–273

42. Schmidt RK, Teo B, Brady JW (1995) Use of umbrella sampling in the calculation of the potential of mean force for maltose in vacuum from molecular dynamics simulations. J Phys Chem 99:11339–11343
43. Kuttel MM, Naidoo KJ (2005) Free energy surfaces for the α(1→4)-glycosidic linkage: implications for polysaccharide solution structure and dynamics. J Phys Chem B 109:7468–7474
44. Sugita Y, Okamoto Y (1999) Replica-exchange molecular dynamics method for protein folding. Chem Phys Lett 314:141–151
45. Campen RK, Verde AV, Kubicki JD (2007) Influence of glycosidic linkage neighbors on disaccharide conformation in vacuum. J Phys Chem B 111:13775–13785
46. Shen T, Langan P, French AD, Johnson GP, Gnanakaran S (2009) Conformational flexibility of soluble cellulose oligomers: Chain length and temperature dependence. J Am Chem Soc 131:14786–14794
47. Voter AF (1997) Hyperdynamics: accelerated molecular dynamics of infrequent events. Phys Rev Lett 78:3908–3911
48. Laio A, Parrinello M (2002) Escaping free-energy minima. Proc Natl Acad Sci U S A 99:12562–12566
49. Grubmuller H (1995) Predicting slow structural transitions in macromolecular systems: conformational flooding. Phys Rev E 52:2893–2906
50. Hamelberg D, Mongan J, McCammon J (2004) Accelerated molecular dynamics: a promising and efficient simulation method for biomolecules A. J Chem Phys 120:11919–11929
51. Johnson GP, Petersen L, French AD, Reilly PJ (2009) Twisting of glycosidic bonds by hydrolases. Carbohydr Res 344:2157–2166
52. Mohamadi F, Richards NGJ, Guida WC, Liskamp R, Lipton M, Caufield C, Chang G, Hendrikson T, Still WC (1990) Macromodel – an integrated software system for modeling organic and bioorganic molecules using molecular mechanics. J Comput Chem 11:440–467
53. Kaminski GA, Friesner RA, Tirado-Rives J, Jorgensen WJ (2001) Evaluation and reparametrization of the OPLS-AA force field for proteins via comparison with accurate quantum chemical calculations on peptides. J Phys Chem B 105:6474–6487
54. Shen T, Langan P, French AD, Johnson GP, Gnanakaran S (2009) Conformational flexibility of soluble cellulose oligomers: chain length and temperature dependence. J Amer Chem Soc 131:14786–14794
55. Weiner SJ, Kollman PA, Case DA, Singh UC, Ghio C, Alagona G, Profeta S Jr, Weiner P (1984) A new force field for molecular mechanical simulation of nucleic acids and proteins. J Am Chem Soc 106:765–784
56. Woods RJ, Dwek RA, Edge CJ, Fraser-Reid B (1995) Molecular mechanical and molecular dynamic simulations of glycoproteins and oligosaccharides. 1. GLYCAM_93 parameter development. J Phys Chem 99:3832–3846
57. Ferretti V, Bertolasi V, Gilli G (1984) Structure of 6-kestose monohydrate, C18H31O16.H2O. Acta Crystallogr C 40:531–535
58. Jeffrey GA (1997) Introduction to hydrogen bonding. Oxford University Press, New York, NY, p 12
59. Parthasarathi R, Elango M, Subramanian V, Sathyamurthy N (2009) Structure and stability of water chains (H2O)n, n = 5–20. J Phys Chem A 113:3744–3749
60. Grabowski SJ (2006) Hydrogen bonding – new insights. Springer, Dordrecht, p 519
61. Lii J-H, Chen K-H, Johnson GP, French AD, Allinger NL (2005) The external-anomeric torsional effect. Carbohydr Res 340:853–862
62. Jeffrey GA, Pople JA, Radom L (1972) The application of ab initio molecular orbital theory to the anomeric effect. A comparison of theoretical predictions and experimental data on conformations and bond lengths in some pyranoses and methyl pyranosides. Carbohydr Res 25:117–131
63. Jeffrey GA, Pople JA, Radom L (1974) The application of ab initio molecular orbital theory to structural moieties of carbohydrates. Carbohydr Res 38:81–95
64. Jaradat DMM, Mebs S, Chęcińska L, Luger P (2007) Experimental charge density of sucrose at 20 K: bond topological, atomic, and intermolecular quantitative properties. Carbohydr Res 342:1480–1489
65. Tvaroška I, Bleha T (1979) Lone pair interactions in dimethoxymethane and anomeric effect. Can J Chem 57:424–435
66. Allinger NL, Schmitz LR, Motoc I, Bender C, Labanowski JK (1992) Heats of formation of organic molecules. 2. The basis for calculations using either ab initio or molecular mechanics methods. Alcohols and ethers. J Am Chem Soc 114:2880–2883
67. French AD, Kelterer A-M, Johnson GP, Dowd MK (2000) B3LYP/6-31G∗, RHF/6-31G∗ and MM3 heats of formation

of disaccharide analogs. J Mol Struct 556:303–313

68. Takahashia O, Yamasakia K, Kohnob Y, Uedab K, Suezawac H, Nishio M (2009) The origin of the generalized anomeric effect: possibility of CH/n and CH/π hydrogen bonds. Carbohydr Res 344:1225–1229
69. Perdew JP, Ruzsinszky A, Constantin LA, Sun J, Csonka GI (2009) Some fundamental issues in ground-state density functional theory: a guide for the perplexed. J Chem Theory Comput 5:902–908
70. Csonka GI, French AD, Johnson GP, Stortz CA (2009) Evaluation of density functionals and basis sets for carbohydrates. J Chem Theory Comput 5:679–692
71. French AD, Johnson GP (2004) Advanced conformational energy surfaces for cellobiose. Cellulose 11:449–462
72. Cocinero EJ, Gamblin DP, Davis BG, Simons JP (2009) The building blocks of cellulose: the intrinsic conformational structures of cellobiose, its epimer, lactose, and their singly hydrated complexes. J Am Chem Soc 131:11117–11123
73. Strati GL, Willett JL, Momany FA (2002) Ab initio computational study of β-cellobiose conformers using B3LYP/6-311++G∗∗. Carbohydr Res 337:1851–1859
74. French AD, Johnson GP, Cramer CJ, Csonka GI (2012) Conformational analysis of cellobiose by electronic structure theories. Carbohydr Res 350:68–76
75. Marenich AV, Cramer CJ, Truhlar DG (2009) Universal solvation model based on solute electron density and on a continuum model of the solvent defined by the bulk dielectric constant and atomic surface tensions. J Phys Chem B 113:6378–6396
76. Keith TA (2013) AIMAll (version 13.02.26). http://aim.tkgristmill.com
77. French AD, Csonka GI (2011) Hydroxyl orientations in cellobiose and other polyhydroxyl compounds: modeling versus experiment. Cellulose 18:897–909
78. French AD, Concha M, Dowd MK, Stevens ED (2014) Electron (charge) density studies of cellulose models. Cellulose 21:1051–1063
79. Biarnés X, Ardèvol A, Planas A, Rovira C, Laio A, Parrinello M (2007) The conformational free energy landscape of β-d-glucopyranose. Implications for substrate preactivation in β-glucoside hydrolases. J Am Chem Soc 129:10686–10693
80. French AD, Johnson GP (2006) Quantum mechanics studies of cellobiose conformations. Can J Chem 84:603–612
81. Lii J-H, Chen K-H, Allinger NL (2003) Alcohols, ethers, carbohydrates, and related compounds. IV. Carbohydrates. J Comput Chem 24:1504–1513
82. Kirschner KN, Yongye AB, Tschampel SM, González-Outeriño J, Daniels CR, Foley BL, Woods RJ (2008) J Comput Chem 29:622–655
83. Tschampel SM, Kennerty MR, Woods RJ (2007) TIP5P-consistent treatment of electrostatics for biomolecular simulations. J Chem Theory Comput 3:1721–1733
84. Brooks BR, Bruccoleri RE, Olafson BD, States DJ, Swaminathan S, Karplus M (1983) CHARMM: a program for macromolecular energy, minimization, and dynamics calculations. J Comput Chem 4:187–217
85. Ha SN, Giammona A, Field M, Brady JW (1988) A revised potential-energy surface for molecular mechanics studies of carbohydrates. Carbohydr Res 180:207–221
86. Kuttel M, Brady JW, Naidoo KJ (2002) Carbohydrate solution simulations: producing a force field with experimentally consistent primary alcohol rotational frequencies and populations. J Comput Chem 23:1236–1243
87. Guvench O, Greene SN, Kamath G, Brady JW, Venable RM, Pastor RW, Mackerell AD Jr (2008) Additive empirical force field for hexopyranose monosaccharides. J Comput Chem 29:2543–2564
88. Patel DS, He X, MacKerell AD Jr (2015) Polarizable empirical force field for hexopyranose monosaccharides based on the classical drude oscillator. J Phys Chem B 119:637–652
89. French AD, Kelterer A-M, Cramer CJ, Johnson GP, Dowd MK (2000) A QM/MM analysis of the conformations of crystalline sucrose moieties. Carbohydr Res 326:305–322
90. French AD, Kelterer A-M, Johnson GP, Dowd MK, Cramer CJ (2001) HF/6-31G∗ energy surfaces for disaccharide analogs. J Comput Chem 22:65–78
91. Oostenbrink C, Soares TA, van der Vegt NFA, van Gusteren WF (2005) Validation of the 53A6 GROMOS force field. Eur Biophys J 34:273–284
92. Stortz CA, Johnson GP, French AD, Csonka GI (2009) Comparison of different force fields for the study of disaccharides. Carbohydr Res 344:2217–2228
93. Jorgensen WL, Chandrasekhar J, Madura JD, Impey RW, Klein ML (1983) Comparison of simple potential functions for simulating liquid water. J Chem Phys 79:926–935

94. DeMarco ML, Woods RJ (2008) Structural glycobiology: a game of snakes and ladders. Glycobiology 18:426–440
95. Foley BL, Tessier MB, Woods RJ (2012) Carbohydrate force fields WIREs. Comput Mol Sci 2:652–697
96. Krupička M, Tvaroška I (2009) Hybrid quantum mechanical/molecular mechanical investigation of the β-1,4-galactosyltransferase-I mechanism. J Phys Chem B 113 (32):11314–11319
97. Zhang Y, Luo M, Schramm VL (2009) Transition states of plasmodium falciparum and human orotate phosphoribosyltransferases. J Am Chem Soc 131:4685–4694
98. French AD (2012) Combining computational chemistry and crystallography for a better understanding of the structure of cellulose. Adv Carbohydr Chem Biochem 67:19–93
99. Bu L, Beckham GT, Crowley MF, Chang CH, Matthews JF, Bomble YJ, Adney WA, Himmel ME, Nimlos MR (2009) The energy landscape for the interaction of the family 1 carbohydrate-binding module and the cellulose surface is altered by hydrolyzed glycosidic bonds. J Phys Chem B 113:10994–11002
100. Zugenmaier P (2008) Crystalline cellulose and derivatives. Characterization and structrures. Springer, Berlin, pp 8–38
101. Matthews JF, Skopec CE, Mason PE, Zuccato P, Torget RW, Sugiyama J, Himmel ME, Brady JW (2006) Computer simulation studies of microcrystalline cellulose Iβ. Carbohydr Res 341:138–152
102. Yui T, Hayashi S (2009) Structural stability of the solvated cellulose IIII crystal models: a molecular dynamics study. Cellulose 16:151–165
103. Shen T, Gnanakaran S (2001) The stability of cellulose: a statistical perspective from a coarse-grained model of hydrogen-bond networks. Biophys J 96:3032–3040
104. Wohlert J, Berglund LA (2011) A coarse-grained model for molecular dynamics simulations of native cellulose. J Chem Theory Comput 7:753–760

Index

A

B

C

Zoë A. Popper (ed.), *The Plant Cell Wall: Methods and Protocols*, Methods in Molecular Biology, vol. 2149, https://doi.org/10.1007/978-1-0716-0621-6, © Springer Science+Business Media, LLC, part of Springer Nature 2020

D

E

F

G

H

I

K

L

M

U

V

W

X

Y